Genetic Consequences of Nucleotide Pool Imbalance

BASIC LIFE SCIENCES

Alexander Hollaender, General Editor

Council for Research Planning in Biological Sciences, Inc., Washington, D.C.

Recent volumes in the series:

Volume 18 TRENDS IN THE BIOLOGY OF FERMENTATIONS FOR FUELS AND CHEMICALS
Edited by Alexander Hollaender, Robert Rabson, Palmer Rogers, Anthony San Pietro, Raymond Valentine, and Ralph Wolfe

Volume 19 GENETIC ENGINEERING OF MICROORGANISMS FOR CHEMICALS
Edited by Alexander Hollaender, Ralph D. DeMoss, Samuel Kaplan, Jordan Konisky, Dwayne Savage, and Ralph S. Wolfe

Volume 20 MOLECULAR AND CELLULAR MECHANISMS OF MUTAGENESIS
Edited by J. F. Lemontt and W. M. Generoso

Volume 21 GENETIC TOXICOLOGY: An Agricultural Perspective
Edited by Raymond A. Fleck and Alexander Hollaender

Volume 22 PLANT CELL CULTURE IN CROP IMPROVEMENT
Edited by S. K. Sen and Kenneth L. Giles

Volume 23 INDUCED MUTAGENESIS: Molecular Mechanisms and Their Implications for Environmental Protection
Edited by Christopher W. Lawrence

Volume 24 ORGAN AND SPECIES SPECIFICITY IN CHEMICAL CARCINOGENESIS
Edited by Robert Langenbach, Stephen Nesnow, and Jerry M. Rice

Volume 25 BASIC BIOLOGY OF NEW DEVELOPMENTS IN BIOTECHNOLOGY
Edited by Alexander Hollaender, Allen I. Laskin, and Palmer Rogers

Volume 26 GENETIC ENGINEERING OF PLANTS: An Agricultural Perspective
Edited by Tsune Kosuge, Carole P. Meredith, and Alexander Hollaender

Volume 27 NEUTRONS IN BIOLOGY
Edited by Benno P. Schoenborn

Volume 28 GENETIC CONTROL OF ENVIRONMENTAL POLLUTANTS
Edited by Gilbert S. Omenn and Alexander Hollaender

Volume 29 SISTER CHROMATID EXCHANGES: 25 Years of Experimental Research
Part A: The Nature of SCEs
Part B: Genetic Toxicology and Human Studies
Edited by Raymond R. Tice and Alexander Hollaender

Volume 30 PLASMIDS IN BACTERIA
Edited by Donald R. Helinski, Stanley N. Cohen, Don B. Clewell, David A. Jackson, and Alexander Hollaender

Volume 31 GENETIC CONSEQUENCES OF NUCLEOTIDE POOL IMBALANCE
Edited by Frederick J. de Serres

A Continuation Order Plan is available for this series. A continuation order will bring delivery of each new volume immediately upon publication. Volumes are billed only upon actual shipment. For further information please contact the publisher.

Genetic Consequences of Nucleotide Pool Imbalance

Edited by
Frederick J. de Serres
National Institute of Environmental Health Sciences
Research Triangle Park, North Carolina

Technical Editor
William Sheridan
National Institute of Environmental Health Sciences
Research Triangle Park, North Carolina

PLENUM PRESS • NEW YORK AND LONDON

Library of Congress Cataloging in Publication Data

Main entry under title:

Genetic consequences of nucleotide pool imbalance.

(Basic life sciences; v. 31)
''Proceedings of a conference on genetic consequences of nucleotide pool imbalance, held May 9–11, 1983, at the National Institute of Environmental Health Sciences, Research Triangle Park, North Carolina''—T.p. verso.
Includes bibliographical references and index.
1. Biochemical genetics—Congresses. 2. Nucleotide—Congresses. 3. DNA replication—Congresses. 4. Deoxyribonucleic acid repair—Congresses. I. De Serres, Frederick J. II. Title: Nucleotide pool imbalance. III. Series.
QH426.G468 1985 574.87′322 84-26430
ISBN 0-306-41902-5

Proceedings of a conference on Genetic Consequences of Nucleotide Pool Imbalance, held May 9–11, 1983, at the National Institute of Environmental Health Sciences, Research Triangle Park, North Carolina

A Division of Plenum Publishing Corporation
233 Spring Street, New York, N.Y. 10013

Printed in the United States of America

PREFACE

This book is concerned with the genetic consequences of nucleotide pool imbalance. There are a number of ways in which this state can be achieved and the majority of papers in this volume describe the effects of nucleotide pool imbalance on DNA replication and repair, and the induction of gene mutations in various organisms. This book grew out of a Conference on "Genetic Consequences of Nucleotide Pool Imbalance" which was held at the National Institute of Environmental Health Sciences on May 9-11, 1983. The Organizing Committee consisted of Drs. Diana Anderson (BIBRA), Robert Haynes (York University), Bernard Kunz (NIEHS), Barry Glickman (NIEHS), and Frederick J. de Serres (NIEHS), Chairman. The meeting was organized by the staff of the Office of the Associate Director for Genetics and was financed solely by the National Institute of Environmental Health Sciences.

Frederick J. de Serres, Ph.D.
Associate Director for Genetics

CONTENTS

SECTION I. INTRODUCTION

SECTION II. NUCLEOTIDE METABOLISM

SECTION III. DNA REPLICATION ERRORS CAUSED BY NUCELOTIDE IMBALANCE

SECTION VI. BIOLOGICAL IMPLICATIONS OF NUCLEOTIDE POOL IMBALANCES

MOLECULAR MECHANISMS IN GENETIC STABILITY AND CHANGE: THE ROLE OF DEOXYRIBONUCLEOTIDE POOL BALANCE

Robert H. Haynes

Department of Biology
York University
Toronto, Ontario
M3J 1P3, Canada

INTRODUCTION

For both theoretical and practical reasons, elucidation of the dialectical processes of genetic stability and change is one of the central problems of biology. The molecular basis of evolution resides in the intimately associated mechanisms of heredity and mutagenesis, which together have led gradually to the accumulation of a vast store of genetic variation within gene pools. Although the actual course of evolution is mediated by natural selection, in the absence of variation, selection can do nothing [7]. Heredity is a manifestation of the stability of genes from one generation to the next. It is necessary for the formation of normal progeny, and for the integrity and longevity of species. Genetic variation is a manifestation of the instability of genes, and of the 'fluidity' of the genome. It is responsible, ultimately, for the origin of species, but more immediately, for a substantial number of serious human diseases, including cancer.

Through the action of natural selection, unfavorable variations are lost, favorable variations proliferate, and species gradually become well-adapted to their environments. Thus, new mutations are far more likely to be deleterious than beneficial to the affected individuals and/or their progeny [36]. In most circumstances, strong selective pressures act against organisms with highly unstable genomes. However, selection has not driven rates of genetic change to zero. It therefore becomes necessary to identify, and assess quantitatively, the relative significance of those countervailing processes of genetic stability and change that give rise to the low, but non-zero, rates of mutation that normally are observed.

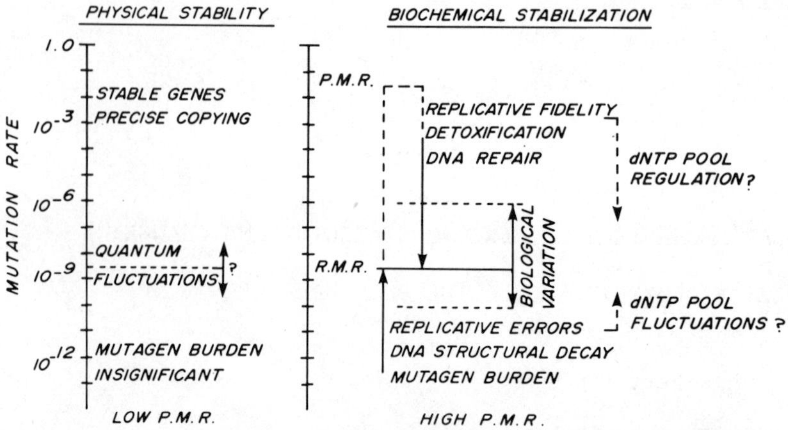

Fig. 1. Schematic diagram contrasting two extreme views on the molecular basis of genetic stability and change. Left hand panel: In 1935 Delbrück proposed a purely physical model based on the Polanyi-Wigner theory of molecular fluctuations [5]. On this model, the potential as well as the actual, mutation rate is very low because genes are assumed to be unusually stable molecules that are capable of extremely accurate replication. Mutations would arise from rare, quantum-statistical fluctuations between isomeric states of the genetic molecules. Right hand panel: The modern view attributes genetic stability primarily to biochemical mechanisms of fidelity, detoxification and repair. The potential mutation rate (P.M.R.) is very high, but these various biochemical devices suppress it to low residual mutation rates (R.M.R.). There is considerable biological variation in mutation rates, especially when mutator and anti-mutator genes are taken into account. The main question raised in this Symposium concerns the extent to which deoxyribonucleotide pool regulation contributes to genetic stability, and pool fluctuations to genetic change.

The remarkable stability of genes could be a reflection, in principle, of two different states of affairs, one characterized by very low, the other by high, potential rates of 'spontaneous' mutation (Fig. 1). The genetic material might have been composed of atypical molecules with unusually stable, unreactive structures, capable of accurate replication (Left-hand panel, Fig. 1). Under these circumstances, mutation rates would be controlled primarily by physical factors, such as rare, quantum-statistical fluctuations in molecular states, or exposure to high energy radiation, an agent whose intensity in terrestrial environments is very low [50]. Few, if any, chemicals would be expected to be mutagenic, and one would not expect to see the surprisingly large accumulation of standing variation that exists in contemporary populations. Alternatively, and as is now known to be the case, genes could be composed of rather ordinary molecular subunits, endowed with no peculiar physical stability. Various physicochemical factors would then have the potential to generate high error rates in DNA synthesis, and many chemical species, of both endogenous and exogenous origin, might be expected to be mutagenic. Under these circumstances, genetic stability would have to be maintained by other factors, such as mechanisms capable of promoting fidelity in DNA replication, of detoxifying mutagens, and of repairing induced damage in DNA (Right-hand panel, Fig. 1).

In recent years, several distinct routes to genetic change have been discovered, and much has been learned about the mechanisms of replicational fidelity and DNA repair that dynamically stablize the genome against the high mutation rates that otherwise would be expected to occur [11, 12]. However, much less thought has been given to the potential genetic consequences of disturbances in DNA metabolism associated with abnormal deoxyribonucleotide (dNTP) precursor pools. Biochemical experiments have shown that dNTP pool bias leads to reduced fidelity of DNA synthesis in vitro [33]. In vivo experiments suggest that unbalanced precursor pools can elicit the entire range of genetic effects commonly associated with physical and chemical mutagens: cell killing, nuclear and mitochondrial mutation, mitotic recombination, chromosome aberrations, aneuploidy, sister chromatid exchange, DNA strand breakage and degradation, prophage induction, sensitization to other mutagens, teratogenesis and oncogenic transformation [29]. In mammalian cells, and recently in yeast, it has been shown that mutants harboring defects in pyrimidine nucleotide biosynthesis exhibit mutator phenotypes [34, 35, 52]. It also has been reported that attack by certain mutagens produces specific dNTP pool expansions in both prokaryotes and eukaryotes [8, 41]. These latter data raise the possibility that dNTP pool imbalances may mediate even the genetic effects of standard mutagens in hitherto unexpected ways. Thus, three general questions are ripe for discussion in this Symposium and in future research: (1) Is deoxyribonucleotide pool regulation an important factor in the maintenance of genetic stability, and if so, how tightly are these pools controlled? (2) Do transient dNTP pool fluctuations occur during the cell cycle and could they be a source of spontaneous genetic change? and (3) Are pool effects large enought to be of significance in nature, and in genetic disease, or are they merely curiosities that arise from our ability to produce large, and highly artificial, pool disturbances in the laboratory?

THE CLASSIC PHYSICAL THEORY OF GENETIC STABILITY

Genetic stability is seen today more as a matter of biology than of physics. Unprogrammed genetic change emerges as a consequence of the malfunction, saturation or failure of various devices for mutation avoidance, and through the action of 'error-prone' processes of DNA synthesis that preserve cell viability at the expense of mutation. However, the purely physical theory of genetic stability has an honorable past and played an important role in the origin of molecular biology as a branch of genetical research. In 1909, Johannsen [25] showed that mutations are rare, sudden and discrete events that presumably cause genes to pass from one stable state to another. In 1919, Muller and Altenburg [38] concluded that the temperature coefficient (Q_{10}) for spontaneous mutation in *Drosophila* must be unusually high, a factor of 5 to 6, in contrast to values of 2 to 3 observed for other metabolic processes. Early searches for chemical mutagens gave negative or equivocal results, and in the early thirties, only the high energy quanta of X-rays or ultraviolet light were established mutagens. All of this led Delbrück, in 1935, to propose that both the remarkable stability of genes, and the high Q_{10} for mutagenesis, could be explained on the basis of the Polanyi-Wigner theory of molecular fluctuations [50]. It would be necessary to assume only that the molecules of which genes are made have unusually high energy thresholds for the appropriate isomeric transitions.* It is ironic, however, that in 1937, only two years after Delbrück's theory was announced, Demerec [9] discovered mutator genes in *Drosophila*. Soon other geneticists found that mutation rates depend on a variety of biological factors such as locus, species, assay procedure, stage of the life cycle and the nutritional status of the organisms. In addition, cold, as well as heat, was found to enhance mutation rates in certain cases, and finally, in war-time research, Auerbach and Robson showed unequivocally that a chemical, mustard gas, was a mutagen [2]. All of this led Muller, in 1952, to conclude [37] that mutation is the result of 'some biochemical disorganization in which processes *normally tending to hold mutation frequencies in check* are to some extent interfered with.' Discovery of the structure, replication and repair of DNA during the subsequent decade made it possible to identify many important biochemical mechanisms involved in genetic stability and change.

---

*This concept so impressed Erwin Schrödinger that in his famous book, "What is Life?", published in 1944 (Cambridge University Press), he stated, 'We may safely assert that there is no alternative to the molecular explanation of the hereditary substance. The physical aspect leaves no other possibility to account for its permanenance. If the Delbrück picture should fail, we would have to give up further attempts.' Even great physicists make poor prophets!

CURRENT VIEWS ON MOLECULAR MECHANISMS OF GENETIC STABILITY AND CHANGE

On thermodynamic and quantum mechanical grounds no real system for information processing can function with perfect accuracy: noise, alas, will be with us always, despite the many ingenious mechanisms that exist to combat it. Indeed, it has been persuasively argued that the existence of stochastic noise has given rise to the evolution of structural redundancy in the genetic machinery itself, and that 'the variety of life is a surface phenomenon, superimposed on mechanisms of genetic homeostasis by the very noise those mechanisms were designed to combat' [45].

Genetic noise arises from replicational errors and spontaneous physicochemical 'decay' of DNA structure. It occurs under all the varied conditions of rest or activity associated with DNA metabolism during the cell cycle. Under normal physiological conditions of temperature, pH and ionic strength, various chemical changes that affect normal DNA structure and replication can occur: for example, tautomeric shifts in nucleotide bases, spontaneous deamination of cytosine and purines, depurination and depyrimidization, phosphodiester bond hydrolysis, spontaneous methylation of adenine, base cleavage by free radicals and possibly various types of adventitous enzyme attack. Such events can have deleterious consequences, for chromosome structure, for the integrity of the encoded genetic information, and for replicational fidelity. In addition, cells are exposed naturally to many mutagenic agents of both endogenous and exogenous origin. A non-exhaustive list of these would include various free radicals, peroxides, singlet oxygen, reducing agents, alkyl nitrosamines, nitrites, photosensitizing and cross-linking agents, solar radiation, environmental radioactivity and heat. At present there is no straightforward way to estimate a cumulative effective 'dose' from all of these sources, and so we do not know whether the net mutagen burden on organisms is high, or actually very low, as is the case for environmental radioactivity. Finally, if recombinational processes associated with the transposition and rearrangement of DNA sequences are a normal feature of cell differentiation and gene expression, then one might expect mistakes to be made during the various processes of DNA strand alignment, breakage, synthesis and splicing that presumably are involved in such phenomena.

A variety of mechanisms have been discovered that serve to preserve the genetic integrity of DNA sequences. It is convenient to divide them into three categories, those that promote replicational fidelity, those that chemically protect DNA by detoxifying mutagens, and those that repair damaged DNA.

Fidelity Mechanisms Associated with DNA Replication

Fidelity mechanisms in polynucleotide synthesis have been essential for the evolution of the long polymers that constitute the genomes of even the simplest organisms [17]. Non-enzymatic copying of polynucleotides occurs with an error-rate of about 10^{-1} to 10^{-2} per base per replication, presumably determined by the difference in free energy (≈1-3 kcal/mol) between correct and incorrect base-pairings [43]. This rate would not allow chromosomes to expand stably much beyond 100 nucleotides in length. It is thought that a fundamental mechanism for enhancing fidelity beyond this physico-chemical level was the evolution of specific proteins for the enzymatic synthesis of RNA and DNA. Viral RNA replicases, the DNA polymerase I function, per se, of E. coli, and mammalian DNA polymerases, all reduce copy errors to frequencies of about 10^{-4}, though, in general, prokaryotic polymerases are the most accurate [24, 27]. It has been assumed that this increase in fidelity reflects an increase in the free energy difference between correct and incorrect base pairs in the presence of polymerase. Such an error rate would allow the development of DNA sequences several kilobases in length, but this clearly is inadequate for evolution of the much longer chromosomes that exist in contemporary organisms.

In addition to the enhanced fidelity obtained directly by enzymatic polymerization, there appear to be two error-correction devices closely associated with DNA synthesis in prokaryotes: 3'-exonucleolytic proofreading [6, 39] and methylation-instructed post-replicative mismatch correction [18]. In the case of E. coli, the former has the effect of reducing error rates to about 10^{-7}, and a further reduction of one to three orders of magnitude is achived by the latter [28, 40]. Thus, extraordinarily low error rates, on the order of 10^{-10} per nucleotide replicated, can be achieved in normal DNA replication in prokaryotes. It is interesting to contrast this frequency with the error-rate of one in five thousand for the most experienced and efficient human key-punch operators copying series of numbers or symbols into machines!

In eukaryotes, there is no clear evidence for the existence of proofreading or methylation-instructed mismatch correction. Indeed, on the basis of the genetic evidence that does exist for heteroduplex repair in yeast, it is not clear whether this process acts primarily to avoid or to fix mutations [16]. It is possible that eukaryotes reduce their mutation rates more by mitotic recombination and repair, than by fidelity mechanisms associated with replication.

Detoxification Mechanisms

It is clear that cells possess a variety of enzymes capable of protecting DNA from attack by free radicals, oxygen radicals in particular. These include superoxide dismutase, glutathione peroxidase,

DT-diaphorase, and the glutathione transferases. In addition, various small molecules act as 'chemical protectors' against the hazards of oxidative metabolism, and appear also to be antimutagens, for example, vitamin E, β-carotene, glutathione, ascorbic acid and uric acid, among others [1]. Such mechanisms would appear to be the 'first line of defense' against many mutagenic substances of both endogenous and exogenous origin. Once DNA actually suffers some kind of structural damage, additional mechanisms come into play to repair it.

DNA Repair Mechanisms

A variety of enzymatic mechanisms have evolved to repair or bypass damage to DNA caused by exposure to environmental mutagens, as well as by spontaneous physicochemical 'decay' [19, 23]. These mechanisms include photoreactivation of UV-damage, strand break ligation, base and nucleotide excision repair, replicative bypass of damaged bases, and post-replication recombinational repair. Operationally, these processes can be classified as being 'error-free' or 'error-prone'. On this basis the so-called 'DNA damage-repair' hypothesis has been developed to account for effects of high doses of mutagens that attack DNA, either directly, or after metabolic activation [22, 23]. Lesions handled by error-free repair have no genetically detectable consequences, whereas those channelled through error-prone processes can give rise to mutations [20]. Mutants defective in error-free repair are hypermutable, as well as hypersensitive to killing by the agent in question; those defective in error-prone repair also are hyper-sensitive, but hypomutable as well [23]. Thus, it would appear that repair can act as a causative, as as well as ameliorative factor in mutagenesis.

The classic and best studied error-prone process is SOS repair in *E. coli* [53], a mutagen-inducible response, that, among other things, seems to have the effect of allowing replication complexes to continue synthesis, though with reduced fidelity, past lesions in DNA that otherwise would block their progress and cause cell death [51]. Recently, both constitutive and inducible components of mutagenic repair have been demonstrated in yeast [15, 47]. The existence of mutagenic processes that normally are repressed suggests that cells are programmed to proliferate in rather noxious environments, even though this comes at the expense of elevated mutation rates. It has been argued that this capacity makes it possible for both cells and species to be buffered against substantial fluctuations in mutagen burden [44]. The enhanced resistance to killing associated with error-prone repair contributes to the survival of the individual. The concomitant increase in mutagenesis may enhance the fitness of the population in unfavorable circumstances, perhaps by stimulating further genetic events, such as gene amplification or polyploidy. In this way, resistant clones might emerge that can adapt to the added mutagen burden. What is most interesting about this pseudo-Lamarckian process, is the inducibility of error-prone repair. This has the effect of ensuring that lesions produced 'spontaneously', or by very small mutagen exposures, are acted upon primarily by error-free, rather than error-prone, processes; the inducible, error-prone process only come into play at moderate doses.

Redundancy and Recombination

In addition to the above-mentioned biochemical processes, the existence of structural redundancy at several levels in the design of the genetic machinery should be considered also as contributing to stability, especially in eukaryotes. For example, features such as codon redundancy, the ubiquitous utilization of double-stranded DNA with complementary base pairs as the genetic material of cells, gene reiteration, diploidy with its potential for recombinational repair, and the meiotic process itself, all appear to be adaptations for genetic stability [45]. The fact that eukaryotic DNA synthesis appears to be much less fastidious than prokaryotic replication, despite the greater lengths of eukaryotic genomes, suggests that diploidy, DNA repair and various recombinational mechanisms have been primarily responsible for genetic stability and change in higher organisms. Given the high replicational accuracy possible in prokaryotic anti-mutator strains, it is clear that prokaryotes could, in principle, develop genomes even longer than eukaryotes, but this has not occurred. On the other hand, the yeast *Saccharomyces cerevisiae*, presumably a very simple eukaryote, possesses a remarkably elaborate system of DNA repair and recombinational mechanisms, yet its haploid genome is only 3.4-fold longer than that of *E. coli* [23]. Thus, it is reasonable to imagine that relatively sloppy DNA synthesis, coupled with the genetic 'protection' and rapid variation inherent in diploidy, meiosis and recombination, has been responsible for the large accumulated variation upon which selection seems to have fed so effectively (from our point of view!) in eukaryotes [5].

Residual Mutation Rates

Spontaneous mutation rates per base pair per replication in prokaryotes and simple eukaryotes can vary over three orders of magnitude, though the rate per genome per replication is relatively constant ($\approx$0.003) [10]. This fact, together with the existence of error-prone repair processes and mutator genes, makes it clear that a rather wide range of error rates in DNA synthesis are consistent with the survival of individual organisms. Indeed, in *Drosophila*, mutation rates per diploid genome per sexual generation may approach unity [10]. Just what factors actually do control residual mutation rates are not known. The existence of anti-mutator alleles suggests that these rates are not as low as biochemically attainable. Thus, observed levels of genetic stability at the nucleotide level seem not to be determined by thermodynamic and quantum mechanical limitations on the efficiency of known fidelity and repair mechanisms.

From an evolutionary standpoint, it is not clear whether mutation rates are minimal or optimal. It has been argued that they may represent an optimal level determined by intragroup selection for a rate that is not so low as to compromise the evolutionary plasticity of the species in changing environments [26, 32]. Other controlling factors could be physiological limitations associated with the energetic cost-effectiveness of fidelity and repair mechanisms. Indeed, it has been suggested that the physiological and evolutionary limits may be similar in magnitude and that residual mutation rates may in fact be optimal from both points of view [26]. We will now summarize the evidence that suggests a role for the regulation of dNTP pools in genetic stability and change.

DEOXYRIBONUCLEOTIDE POOL DISTURBANCES

Biochemical Requirements for Accurate DNA Synthesis

The central process of DNA metabolism is the polymerization of deoxyribonucleotides. There are certain basic biochemical requirements for DNA replication, and it has become clear that if any of these conditions are not met, synthesis is disturbed in genetically significant ways [33]. Thus, genetic stability seems to depend not only on specific mechanisms for fidelity and repair, but on other biochemical factors as well.

The requirements for DNA synthesis *in vitro* are an appropriate polynucleotide template/primer, DNA polymerase, a suitable concentration of the divalent cation Mg^{2+}, and a supply of deoxyribonucleoside triphosphate precursors. Additional proteins are involved *in vivo*, including an enzyme system for dNTP biosynthesis. It is perhaps not surprising that infidelity of DNA synthesis *in vitro*, and genetic change *in vivo*, can arise from modification or damage to the template/primer, genetically altered enzymes, or even substitution of other divalent cations for magnesium. What is surprising is that deoxyribonucleotide pool bias or disturbance also reduces the fidelity of replication *in vitro* and produces essentially all known types of genetic change *in vivo*. The fact that nucleotide pool disturbances do have genetic consequences sheds new and unexpected light on the mechanisms involved in genetic stability and change. It suggests that all aspects of DNA metabolism, from precursor biosynthesis, through replication, repair, recombination and possibly degradation, may be regulated so as to reduce heritable genetic change, consistent with an over-riding priority for individual cell survival. Thus, virtually any disturbance to normal DNA metabolism should be regarded as a potential source of genetic change, and genetic stability should be viewed dynamically as a consequence of a variety of 'homeostatic' mechanisms that are an integral part of the reproductive processes of the cell.

The Nature of Deoxyribonucleotide Precursor Pools

It is important to realize that dNTP precursor 'pools' so far have not been well-defined in vivo, and the various means used to destabilize them may not produce the specific pool imbalances that we imagine. The relevant pools might be extremely small and compartmentalized near replication forks. Measurements indicate that gross intracellular concentrations of dNTP's in any case are small, only a few percent of the corresponding ribonucleotide pools. Such pools would be capable of maintaining DNA replication for only a few minutes at most [48]. Furthermore, it has been argued that the enzymes involved in dNTP synthesis might be part of a multi-enzyme replication complex, thus obviating the need for free dNTP pools at all [46]. On this basis, one might imagine the dNTP's being synthesized from pools of earlier precursors and then immediately utilized in replication. If the rates of DNA polymerization and dNTP synthesis should not be intimately and firmly coupled, then any exogenously produced slow-down or block in DNA synthesis would be expected to lead to dNTP pool expansion. However, more recently, evidence has been presented for the existence, at least, of a distinct dTTP pool [42].

What are called 'pool imbalances' can be produced experimentally in three distinct ways: (1) Inhibition of enzymes involved in deoxyribonucleotide synthesis by drugs such as FdUMP or antifolates [3, 30]; (2) Addition of excess dNTP precursors, or, in suitably permeable cells, the dNMP's themselves [4]; and (3) Blocking various steps in dNTP biosynthesis by genetic lesions [35, 52]. Genetic change can be produced in all of these ways and the most straightforward interpretation is that the effects arise from pool distubances. However, it should be borne in mind that it is impossible to rule out absolutely the possibility that these genetic phenomena arise from side effects of the drugs or added precursors, from unknown pleiotropies of the mutations, or from structural alterations in the enzymes and protein complexes involved in DNA replication and/or dNTP biosynthesis. Until such time as the relevant dNTP pools can be measured unambiguously, it would be wise not to overlook any of these possible sources of interpretive ambiguity. However, the simplest working hypothesis at present is that the pools are real, however small they may be, and that they are an important factor in genetic stability and change. Finally, of course, there can be little doubt regarding the authenticity of the dNTP pool bias effects on the fidelity of polynucleotide synthesis in vitro, even though their precise implications for the in vivo situation are not clearly established.

Genetic Effects of Deoxyribonucleotide Pool Disturbance

The biosynthetic pathways of the dNTP's are linked by nucleotide interconversions and controlled by various regulatory mecha-

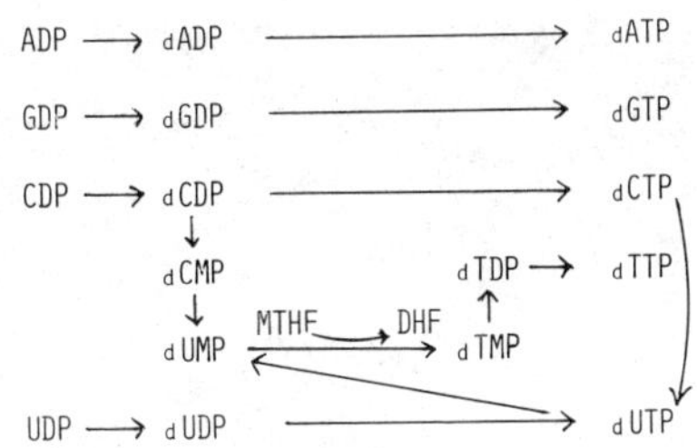

Fig. 2. A simplified scheme for DNA precursor synthesis. MTHF: N^5,N^{10}-methylene tetrahydrofolate. DHF: dihydrofolate.

nisms, one of the most important of which involves ribonucleotide reductase. This enzyme reduces ADP, GDP, CDP, and UDP to the corresponding deoxyribonucleoside diphosphates [49]. Figure 2 shows a simplified scheme of these relations, as they exist in yeast, which lacks the salvage pathways for dNTP synthesis. The purine deoxyribonucleoside diphosphates are converted directly to the corresponding triphosphates by nucleoside diphosphate kinase, whereas the pyrimidine interconversions, especially those involved in the synthesis of thymidylate are more complex. Note in particular the coupling of dTMP synthesis with folate metabolism: N^5,N^{10}-methylenetetrahydrofolate serves as the methyl donor in the conversion of dUMP to dTMP by thymidylate synthetase. Thus, pyrimidine nucleotide pools can be affected directly by genetic or drug-induced attack on pyrimidine nucleotide biosynthesis, and also by attack on folate metabolism. Spontaneous or induced alterations in any one dNTP pool can be expected to affect others, not only at the level of ribonucleotide reductase, but at other steps as well. Thus, the actual mechanism underlying the genetic effects produced by any given exogenously-induced pool distubance may be more complex than would appear at first sight. For example, dTTP and dCTP pools appear to be reciprocally related, and in general, elevation or depression of dTTP pools results in a corresponding elevation or depression of purine pools. Furthermore, thymidylate deprivation seems to result in expansion of the dUTP pool, even though this nucleotide normally is kept at very low levels through the action of dUTPase.

Kunz has published a detailed review [29] of what is presently known about the genetic effects of dNTP pool imbalance in all systems studied to date; thus there is no need to reiterate this material here. Although the effects of primary disturbances in pyrimidine nucleotide biosynthesis have been more extensively studied than those for purine nucleotides, a few generalizations emerge from his review. First, all possible pool stresses (pyrimidine or purine dNTP depletion or excess) have been found, in at least some organism, to lead to growth inhibition, cell killing, chromosome aberrations and sensitization to the effects of radiation and certain chemical

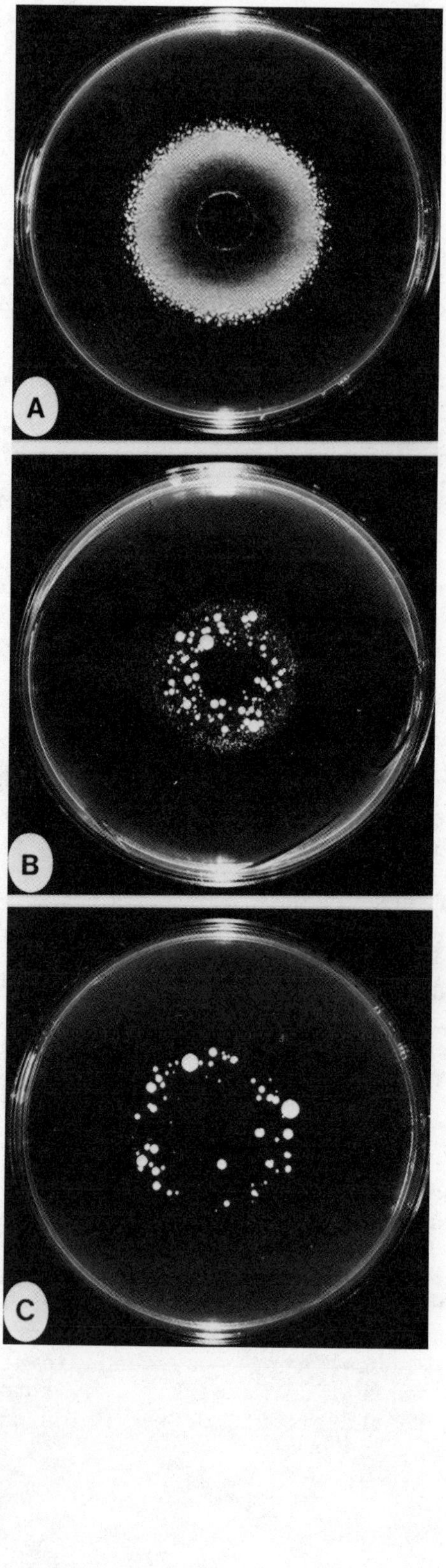
A
B
C

mutagens. Second, thymine nucleotide deprivation induces both mutation and recombination in prokaryotes. In yeast, thymidylate deprivation is a more efficient recombinagen, while thymidylate excess is a more efficiency mutagen (Fig. 3), though both mutation and recombination appear to be induced, at least to some extent, over the entire range of exogenous dTMP concentrations in dTMP auxotrophs [14]. Similar recombinagenic effects can be produced indirectly by inhibition of de novo folate biosynthesis (Fig. 4). In appropriate strains, the recombinagenic effects of thymidylate deprivation can be abolished specifically by provision of dTMP [3, 30]. Purine stress is clastogenic, but no data are available concerning any possible mutagenicity or recombinagenicity.

It seems likely that these various genetic effects of nucleotide pool disturbances in most cases are real, and are not the result of selection artifacts. This can be demonstrated most directly in yeast by measuring the increase in absolute yield of mitotic re-

Fig. 3. Simultaneous demonstration of mutation and mitotic recombination provoked by thymidylate stress in the yeast Saccharomyces cerevisiae, D7-T2 strain [3]. Plate A shows the growth of this strain on YPD medium [30] containing aminopterin plus sulfanilamide (ASA). The central well contains dTMP. In this way a gradient of thymidylate stress is set up across the plate growing from thymidylate excess near the well, to thymidylate starvation induced by ASA near the periphery of the plate. In the intermediate zone where cell growth is observed, the dTMP gradient counteracts the effect of the ASA. Note the zone of inhibition near the well caused by excess dTMP, and the absence of growth at the periphery caused by 'thymineless' death. Strain D7-T2 is homozygous for ilv1-92 and so reversion to prototrophy can be monitored at this locus. Plate B is a replica of A on isoleucine omission medium, and the induction of mutation by high thymidylate is clearly revealed by the ring of colonies growing close to the central well. The strain also is heterozygous for a recessive defect (cyh2) which confers resistance to cycloheximide. Both gene conversion and mitotic crossing-over can result in homozygosity for cyh2 which allows growth on cycloheximide-containing medium. Place C is a replicate of A on cycloheximide medium and the induction of mitotic recombination by low thymidylate concentrations is demonstrated by the ring of colonies more distantly located from the central well than the revertants in plate B.

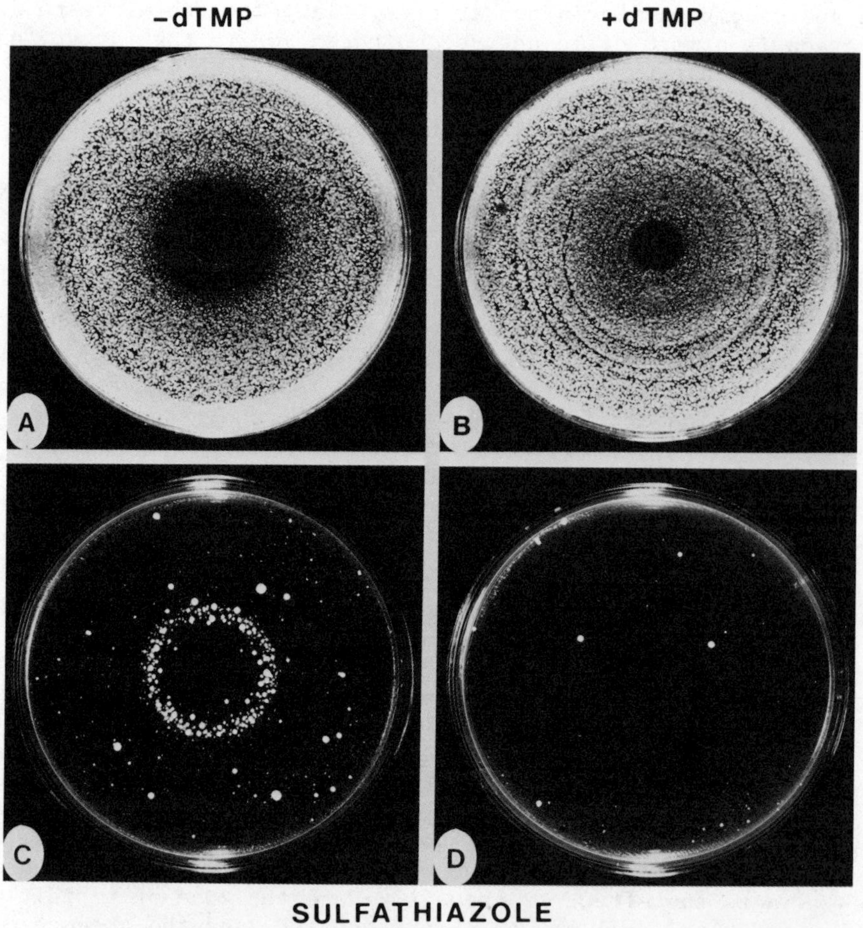

Fig. 4. Induction of mitotic recombination in yeast by sulfathiazole, an inhibitor of de novo folate biosynthesis and the abolition of this effect by provision of exogenous dTMP. The strain is D7-T2 (cf. Fig. 3) and it is permeable to dTMP. Plate A shows growth on YPD medium with the drug in the central well; note the zone of inhibition around the well. Plate B contains dTMP in the medium and the drug in the well; note the zone of inhibition around the well is greatly reduced. Place C is a replica of A on cycloheximide medium; the ring of colonies indicates the induction of mitotic recombination by the drug. The recombinagenic effect is abolished in B, as no such ring of recombinants appears when it is replica-plated onto cycloheximide medium (plate D).

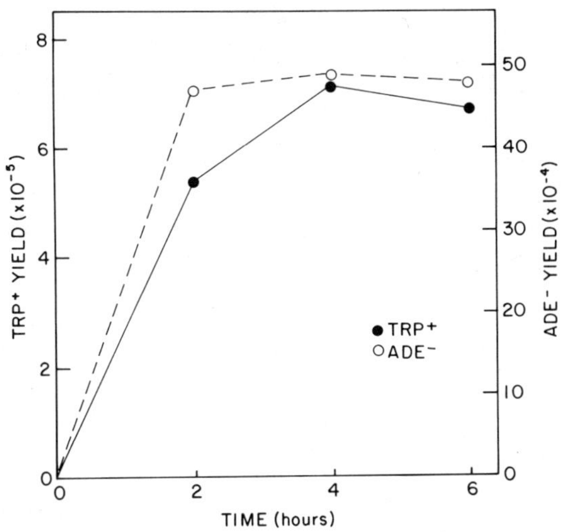

Fig. 5. Absolute yields (recombinants per cell exposed) of gene convertants (TRP^+) and mitotic cross-overs (ADE^-) in strain D7-T2 produced by incubation in medium containing sulfanilamide plus aminopterin. This plot shows that recombinants actually are induced by the treatment and the increases in recombination frequency are not selection artifacts.

combinants per cell exposed as a function of incubation time in FdUMP, a potent inhibitor of thymidylate synthetase (Fig. 5). Plots of yield, rather than frequency (recombinants per survivor), are not subject to misinterpretation based on selection, or differential lethal and genetic effects of the agent used [21]. It is possible also to compare the relative efficiencies of different agents by plotting mutant (or recombinant) yield versus lethal hits [13]. On this basis, it would appear that the relative efficiency of thymidylate deprivation, as compared with ultraviolet light or ionizing radiation, in provoking mitotic recombination, is about 0.1 for the data shown in Fig. 5; however, we have observed values approaching unity in other cases.

Finally, it should be noted that genes which control various modes of DNA repair also can affect cellular responses to deoxyribonucleotide pool imbalance. Examples of this are given in Fig. 6 where the effects of antifolate treatment (sulfanilamide plus aminopterin) on mitotic gene conversion and crossing-over are shown for RAD wild-type yeast (repair proficient), and also in strains defective in excision repair (rad3), error-prone repair (rad6) and recombinational repair (rad50) [31]. Unfortunately, we can only speculate as to how rad3 and rad6 deficiencies can protect against the recombinagenic effects of thymidylate deprivation, whereas rad50 sensitizes the cells. However, for the purposes of the present discussion, it is sufficient merely to realize that such effects occur.

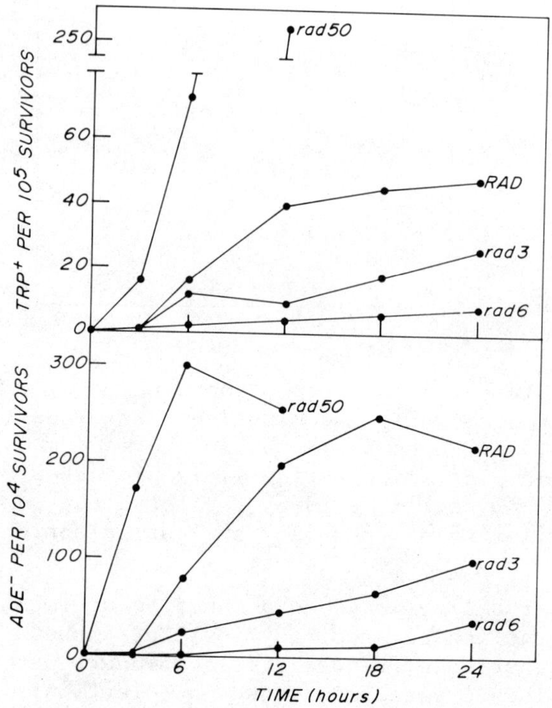

Fig. 6. Antifolate (sulfanilamide plus aminopterin) induced mitotic recombination in a repair-proficient wild-type strain (RAD) and three repair-defective mutants (rad3, 6, and 50). Upper panel: gene conversion frequencies. Lower panel: mitotic crossing-over frequencies. The abscissa gives the time of incubation in the presence of the antifolates. This demonstrates that genetic loci controlling various modes of DNA repair in yeast also affect cellular responses to thymidylate deprivation [31].

Thus, it would appear that regulation of dNTP pools may be involved in DNA repair, replication, and recombination in hitherto unexpected ways, and indeed, induced pool imbalance may mediate the genetic effects of standard mutagens as well.

The mechanisms for the lethal, mutagenic and recombinagenic effects of thymidylate stress in yeast have not been established. However, in the case of thymidylate depletion, DNA strand breaks are produced, and it is at least plausible to imagine that these breaks play an important role in stimulating the recombinagenic effect [3]. It is tempting also to suggest that such single strand breaks are produced through the action of AP-endonucleases acting subsequent to uracil-DNA-glycosylase at sites of uracil incorporated into newly replicated or repaired DNA. Such a mechanism is outlined schematically in Fig. 7, but it is important to realize that, at present, reiterative uracil insertion and excision has not been demonstrated in yeast.

CONCLUSIONS

There can be no doubt about the authenticity of the effects of deoxyribonucleotide pool bias in reducing the fidelity of DNA synthesis <u>in vitro</u>. Furthermore, essentially all known forms of genetic alterations, from point mutation to oncogenic transformation and teratogenesis, have been observed subsequent to induced dNTP pool disturbances in appropriate <u>in vivo</u> assay systems. While some of these effects could be selection artifacts, it is clear that not all are. The <u>in vivo</u> effects increase with exposure to the inducing agent or condition, and their magnitudes are not insignificant in comparison with those of other mutagens that attack DNA directly. Genetic loci known to control various modes of DNA repair also influence cellular responses to drug-induced dNTP pool imbalance. Certain mutator phenotypes in eukaryotes have been shown to arise from defects in enzymes involved in deoxyribonucleotide biosynthesis. The recombinagenicity of drugs that specifically inhibit certain enzymes involved in dNTP biosynthesis indicates that cells contain non-DNA primary targets for the induction of genetic change. Deoxyribonucleotide pool disturbances may even mediate the genetic effects of standard mutagens. Taken together, these various facts provide strong grounds for thinking that spontaneous or induced dNTP pool disturbances can be a source of genetic change, and that the regulation of such pools is an important mechanism in the maintenance of genetic stability. What we do not know is how stringently these pools are regulated, and whether pool imbalance is an important contributory factor in the accumulation of genetic variation in natural populations, or in the etiology of cancer and other diseases.

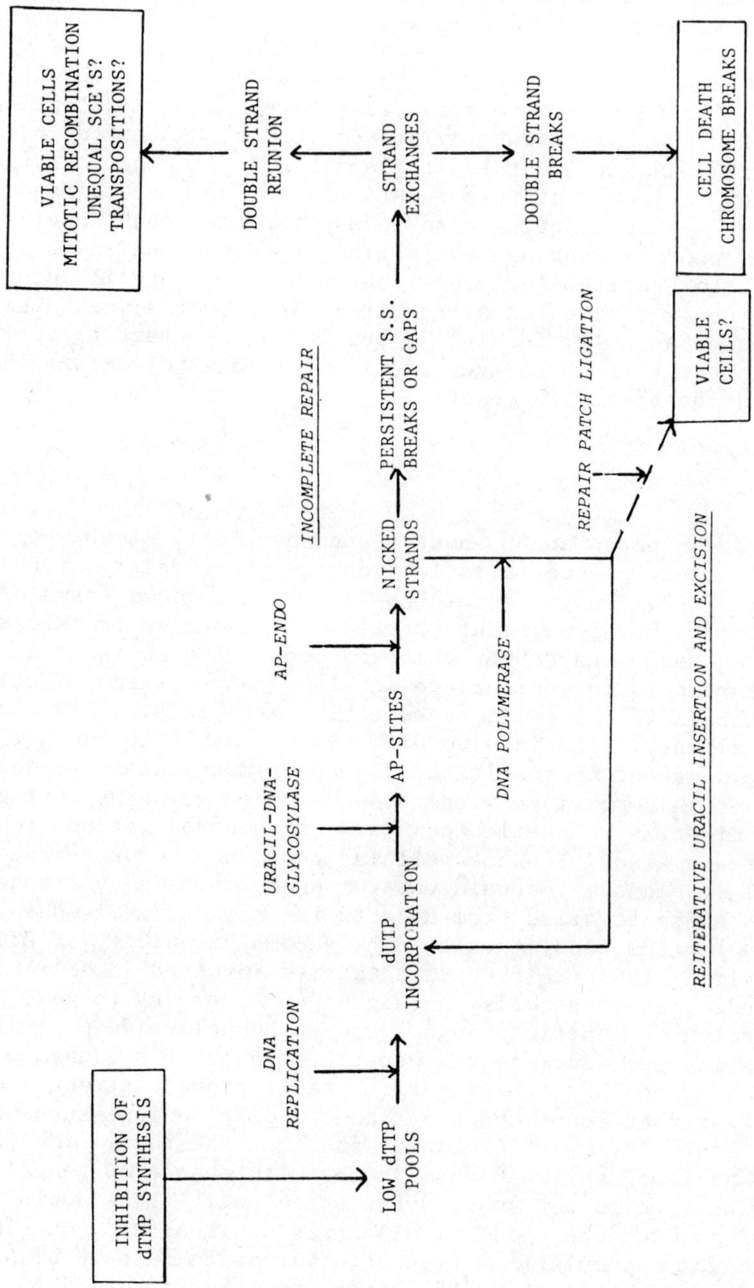
INHIBITION OF dTMP SYNTHESIS
LOW dTTP POOLS
DNA REPLICATION
dUTP INCORPCRATION
URACIL-DNA-GLYCOSYLASE
AP-SITES
AP-ENDO
NICKED STRANDS
INCOMPLETE REPAIR
PERSISTENT S.S. BREAKS OR GAPS
STRAND EXCHANGES
DOUBLE STRAND REUNION
VIABLE CELLS MITOTIC RECOMBINATION UNEQUAL SCE'S? TRANSPOSITIONS?
DOUBLE STRAND BREAKS
CELL DEATH CHROMOSOME BREAKS
DNA POLYMERASE
REPAIR PATCH LIGATION
VIABLE CELLS?
REITERATIVE URACIL INSERTION AND EXCISION

Fig. 7. Summary of the principal steps which are believed to mediate the various genetic consequences of thymidylate deprivation in yeast. The only endogenous source of thymidylate is via the methylation of dUMP to yield dTMP; tetrahydrofolate is the methyl donor for this reaction. Inhibition of dTMP biosynthesis will lower intracellular pools of dTTP. Under these conditions, dUTP pools will swell and uracil nucleotides can be incorporated into DNA in place of dTTP. Yeast contains a uracil-DNA glycosylase which can exicse these uracil moieties from DNA, thereby generating AP sites. Such sites normally are repaired by the concerted action of AP endonuclease, repair polymerase, and DNA ligase. Under conditions of dTTP deprivation, dUTP also can be inserted into repair patches. This would lead to a process of reiterative uracil insertion and excision with the result that few repair patches could be completed; those repaired would presumably allow cells to survive with elevated amounts of uracil in their DNA. Frequent incomplete episodes of repair would generate many single-strand (S.S.) breaks, or gaps, in daughter strand DNA. Such breaks could stimulate DNA strand exchanges. Attempted recombinational processes would generate additional breaks in parental strand DNA. At some places in the DNA, double-strand breaks would result; these would be manifested biologically as nonviable cells or, in appropriate systems, as chromosome aberrations. Completed strand exchange events could be detected biologically as mitotic recombination. It also is possible that some of these events would result in unequal sister-chromatid exchange (SCE's) or various kinds of illegitimate transposition of DNA fragments.

ACKNOWLEDGMENTS

I wish to thank F. Eckardt, B. A. Kunz, J. G. Little, and E. M. McIntosh for many helpful discussions on the subject of this paper. Work reported herein was supported by the Natural Sciences and Engineering Research Council of Canada.

REFERENCES

1. B. N. Ames, Dietary carcinogens and anticarcinogens, Science, 221:1256-64 (1983).
2. C. Auerbach and J. M. Robson, The production of mutations by chemical substances, Proc. Roy. Soc. Edinburgh, B62:271-283 (1947).
3. B. J. Barclay, B. A. Kunz, J. G. Little, and R. H. Haynes, Genetic and biochemical consequences of thymidylate stress, Can. J. Biochem., 60:172-194 (1982).
4. B. J. Barclay and J. G. Little, Mutation induction in yeast by deoxythymidine monophosphate: A model, Molec. Gen. Genet., 181:279-281 (1981).
5. J. F. Crow and M. Kimura, Evolution in sexual and asexual populations, Amer. Nat., 99:439-449 (1965).
6. D. Brutlag and A. Kornberg, Enzymatic synthesis of deoxyribonucleic acid. XXXVI. A proofreading function for the 3'→5'-exonuclease activity in deoxyribonucleic acid polymerases, J. Biol. Chem., 247:241-248 (1972).
7. C. R. Darwin, On the origin of Species by Means of Natural Selection, 1st Edition, John Murray, London (1859), p. 82.
8. S. K. Das, E. P. Benditt, and L. A. Loeb, Rapid changes in deoxynucleoside triphosphate pools in mammalian cells treated with mutagens, Biochem. Biophys. Res. Comm., 114:458-464 (1983).
9. M. Demerec, Frequency of spontaneous mutations in certain stocks of Drosophila melanogaster, Genetics, 22:469-478 (1937).
10. J. W. Drake, The role of mutation in microbial evolution, Symp. Soc. Gen. Microbiol., 24:41-58 (1974).
11. J. W. Drake, Fundamental mutagenic mechanisms and their significance for environmental mutagenesis, in: Progress in Genetic Toxicology (D. Scott, B. A. Bridges, and F. H. Sobels, eds.), Elsevier/North Holland, Amsterdam (1977), pp. 43-55.
12. J. W. Drake and R. H. Baltz, The biochemistry of mutagenesis, Ann. Rev. Biochem., 45:11-37 (1976).
13. F. Eckardt and R. H. Haynes, Quantitative measures of mutagenicity and mutability based on mutant yield data, Mutat. Res., 74:439-458 (1980).
14. F. Eckardt, B. A. Kunz, and R. H. Haynes, Variation of mutation and recombination frequencies over a range of thymidylate concentrations in a diploid thymidylate auxotroph, Curr. Genet., 7:399-402 (1983).

15. F. Eckardt, E. Moustacchi, and R. H. Haynes, On the inducibility of error-prone repair in yeast, in: DNA Repair Mechanisms (P. C. Hanawalt, E. C. Friedberg, and C. F. Fox, eds.), Academic Press, New York (1978), pp. 421-423.
16. F. Eckardt, S.-J. Teh, and R. H. Haynes, Heteroduplex repair as an intermediate step in UV mutagenesis in yeast, Genetics, 95:63-80 (1980).
17. M. Eigen and P. Schuster, The Hypercycle, Springer-Verlag, Berlin (1979).
18. B. W. Glickman and M. Radman, E. coli mutator mutants deficient in methylation instructed mismatch correction, Proc. Natl. Acad. Sci. U.S.A., 77:1063-1067 (1980).
19. P. C. Hanawalt, P. K. Cooper, A. K. Ganesan, and C. A. Smith, DNA repair in bacteria and mammalian cells, Ann. Rev. Biochem., 48:783-836 (1979).
20. P. J. Hastings, S.-K. Quah, and R. C. von Borstel, Spontaneous mutation by mutagenic repair of spontaneous lesions in DNA, Nature, 264:719-722 (1976).
21. R. H. Haynes and F. Eckardt, Analysis of dose-response patterns in mutation research, Can. J. Genet. Cytol., 21:277-302 (1979).
22. R. H. Haynes, F. Eckardt, and B. A. Kunz, The DNA damage-repair hypothesis in radiation biology: Comparison with classical hit theory, Brit. J. Cancer, 49 (Suppl. VI):81-90 (1984).
23. R. H. Haynes and B. A. Kunz, DNA repair and mutagenesis in yeast, in: Molecular Biology of the Yeast Saccharomyces: Life Cycle and Inheritance (J. N. Strathern, E. W. Jones, and J. R. Broach, eds.), Cold Spring Harbor Laboratory, Cold Spring Harbor, New York (1981), pp. 371-414.
24. J. Holland, K. Spindler, F. Horodyski, E. Graham, S. Nichol, and S. Vandepol, Rapid evolution of RNA genomes, Science, 215: 1577-1585 (1982).
25. W. Johannsen, Elemente der exakten Erblichkeitslehre, Carl Fischer, Jena (1909).
26. M. Kimura, On the evolutionary adjustment of spontaneous mutation rates, Genet. Res. Camb., 9:23-34 (1967).
27. T. A. Kunkel and L. A. Loeb, Fidelity of mammalian DNA polymerases, Science, 213:765-767 (1981).
28. T. A. Kunkel, R. R. Meyer, and L. A. Loeb, Single-strand binding protein enhances fidelity of DNA synthesis in vitro, Proc. Natl. Acad. Sci. U.S.A., 76:6331-6335 (1979).
29. B. A. Kunz, Genetic effects of deoxyribonucleotide pool imbalance, Environ. Mutagen., 4:695-725 (1982).
30. B. A. Kunz, B. J. Barclay, J. C. Game, J. G. Little, and R. H. Haynes, Induction of mitotic recombination in yeast by starvation for thymine nucleotides, Proc. Natl. Acad. Sci. U.S.A., 77:6057-6061 (1980).
31. B. A. Kunz and R. H. Haynes, DNA repair and the genetic effects of thymidylate stress in yeast, Mutat. Res., 93:353-375 (1982).
32. E. G. Leigh, The evolution of mutation rates, Genetics Supplement, 73s:1-18 (1973).

33. L. A. Loeb and T. A. Kunkel, Fidelity of DNA synthesis, Ann. Rev. Biochem., 52:429-457 (1982).
34. K. L. Maus, E. M. McIntosh, and R. H. Haynes, Defective dCMP deaminase confers a mutator phenotype on Saccharomyces cerevisiae, Environ. Mutagen., 6:415 (1984).
35. M. Meuth and N. L'Heureux-Huard, Characterization of a mutator gene in Chinese hamster ovary cells, Proc. Natl. Acad. Sci. U.S.A., 76:6505-6509 (1979).
36. H. J. Muller, Our load of mutations, Am. J. Hum. Genet., 2: 111-176 (1950).
37. H. J. Muller, The nature of the genetic effects produced by radiation, in: Radiation Biology, Vol. I, Part I (A. Hollaender, ed.), McGraw-Hill Book Co. Inc., New York (1954), p. 417.
38. H. J. Muller and E. Attenburg, The rate of change of hereditary factors in Drosophila, Proc. Soc. Exptl. Biol. Med., 17:10-14 (1919).
39. N. Muzyczka, R. L. Roland, and M. J. Bessman, Studies on the biochemical basis of spontaneous mutation, J. Biol. Chem., 247: 7116-7122 (1972).
40. P. Nevers and H.-C. Spatz, Escherichia coli mutants uvrD and uvrE deficient in gene conversion of λ-heteroduplexes, Molec. Gen. Genet., 139:237-243 (1975).
41. C. N. Newman and J. H. Miller, Mutagen-induced changes in cellular deoxycytidine triphosphate and thymidine triphosphate in Chinese hamster ovary cells, Biochem. Biophys. Res. Comm., 114: 34-40 (1983).
42. B. Nicander and P. Reichard, Dynamics of pyrimidine deoxynucleoside triphosphate pools in relationship to DNA synthesis in 3T6 mouse fibroblasts, Proc. Natl. Acad. Sci. U.S.A., 80: 1347-1351 (1983).
43. J. Ninio, Approaches Moléculaires de l'Evolution, Mason, Paris (1978).
44. M. Radman, Molecular mechanisms of mutagenesis: a summary review, in: Conference on Structural Pathology in DNA and the Biology of Ageing (Deutsche Forschungsgemeinschaft), H. Boldt Verlag, Boppard (1980), p. 21-26.
45. D. C. Reanney, D. G. MacPhee, and J. Pressing, Intrinsic noise and the design of the genetic machinery, Aust. J. Biol. Sci., 36:77-91 (1983).
46. G. P. V. Reddy and A. B. Pardee, Multienzyme complex for metabolic channeling in mammalian DNA replication, Proc. Natl. Acad. Sci. U.S.A., 77:3312-3316 (1980).
47. W. Seide, F. Eckardt, and M. Brendel, Analysis of mutagenic DNA repair in a thermoconditional repair mutant of Saccharomyces cerevisiae, Molec. Gen. Genet., 190:406-412 (1983).
48. L. Skoog and B. Nordenskjöld, Effects of hydroxyurea and 1-β-d arabinofuranosyl cytosine on deoxyribonucleotide pools in mouse embryo cells, Eur. J. Biochem., 19:81-89 (1971).
49. L. Thelander and P. Reichard, Reduction of ribonucleotides, Ann. Rev. Biochem., 48:133-58 (1979).

50. N. W. Timoféeff-Ressovsky, K. G. Zimmer, and M. Delbruck, Über die Natur den Genmutation und der Genstrucktur., Nachr. Ges. Wiss. Göttingen, 1:189-245 (1935).
51. G. S. Villani, S. Boiteux, and M. Radman, Mechanism of ultraviolet-induced mutagenesis. Extent and fidelity of _in vitro_ DNA synthesis on irradiated templates, Proc. Natl. Acad. Sci. U.S.A., 75:3037-3041 (1978).
52. G. Weinberg, B. Ullman, and D. W. Martin, Mutator phenotypes in mammalian cell mutants with distinct biochemical defects and abnormal deoxyribonucleoside triphosphate pools, Proc. Natl. Acad. Sci. U.S.A., 78:2447-2451 (1981).
53. E. M. Witkin, Ultraviolet mutagenesis and inducible DNA repair in _Escherichia coli_, Bacteriol. Rev., 40:869-907 (1976).

METABOLIC EFFECTS OF POLY(ADP-RIBOSE) INHIBITORS

Kathryn M. Milam and James E. Cleaver

Laboratory of Radiobiology and Environmental Health
University of California
San Francisco, CA. 94143

1. INTRODUCTION

3-Aminobenzamide (3-AB) and benzamide inhibit the synthesis of the biopolymer, poly(ADP-ribose), after exposure of cells to DNA-damaging agents [7]. Stimulation of poly(ADP-ribose) synthesis is believed to be necessary for normal DNA repair processes; its inhibition by 3-AB or benzamide delays ligation after excision of damaged DNA and repair synthesis at the site of excision [3, 5]. It has been proposed that 3-AB and benzamide alter poly(ADP-ribose) synthesis by specific inhibition of the enzyme poly(ADP-ribose) synthetase [6]. However, very little work has been done on the effects of 3-AB or benzamide on cellular metabolism; it has generally been assumed that they have no "profound metabolic effects of an unknown character in undamaged cells" [4].

In the present set of experiments, the effects of 3-AB or benzamide on cytotoxicity, glucose metabolism and DNA biosynthesis from labeled precursor molecules were examined in a lymphoid cell line. In addition, we studied the metabolic effects of m-aminobenzoic acid (m-ABOA) and benzoic acid (BOA), two analogues of benzamide that do not inhibit poly(ADP-ribose) biosynthesis [7].

2. METHODS

2.1. Lymphoid Cell Line (WIL-2)

WIL-2 cells were maintained in suspension culture in 75-mm flasks at 37°C in RPMI medium supplemented with 10% fetal calf serum plus penicillin (50 units/ml) and streptomycin (50 μg/ml). The medium was changed twice a week.

2.2. Cytotoxicity Assay

A spectrophotometric method [9] was used to measure cellular release of the cytoplasmic enzyme lactate dehydrogenase into the medium during a 24 hr incubation period with increasing concentrations of benzamide, 3-AB, m-ABOA or BOA (0-20 mM).

2.3. Glucose Metabolism

The rate of production of $^{14}CO_2$ from D-[^{14}C(U)]-glucose (348.2 mCi/mmol) was used to monitor the rate of glucose oxidation. Cells were incubated with 0.5 μCi/ml of [^{14}C]glucose for 24 hr. The reaction was stopped and $^{14}CO_2$ was released from the cells by the addition of 1 N H_2SO_4 to the medium. The $^{14}CO_2$ was trapped in hanging wells containing hyamine hydroxide, and radioactivity was determined by toluene-based liquid scintillation counting.

2.4. DNA Biosynthesis from Labeled Precursor Molecules

Cells were incubated for 24 hr with 0.1 μCi/ml [methyl-^{3}H]-thymidine (74.9 Ci/mmol) or [methyl-^{14}C]thymidine (53.4 mCi/mmol). This prelabeling of DNA allowed normalization of the data for DNA recovery. Another 24 hr incubation period immediately followed, with [^{14}C]glucose (0.5 μCi/ml), [^{3}H]thymidine (10 μCi/ml), or [methyl-^{3}H]methionine (10 μCi/ml, 80 Ci/mmol) in the presence of 0-5 mM 3-AB, benzamide, m-ABOA or BOA. DNA was then isolated from the cells by chloroform/isoamyl alcohol extraction and ethanol precipitation [1]. Radioactivity was determined by liquid scintillation counting.

2.5. Statistics

One-way analysis of variance was applied to data from each experiment. The Dunnett's test was used to determine which experimental groups differed from the control group [8].

3. RESULTS

At $\leqslant$20 mM in concentration, 3-AB, m-ABOA, and BOA had no effect on release of lactate dehydrogenase into the medium and were therefore not considered cytotoxic to these lymphoid cells (Fig. 1). However, at a concentration as low as 5 mM, benzamide caused a significant increase in release of lactate dehydrogenase, indicating that benzamide is cytotoxic to these cells.

Similarly, benzamide was the only chemical of the four to affect glucose oxidation to CO_2 (Fig. 2). The rate of metabolism of [^{14}C]-glucose to $^{14}CO_2$ was significantly decreased by the addition of as little as 3 mM benzamide and was reduced 55% by 7 mM benzamide.

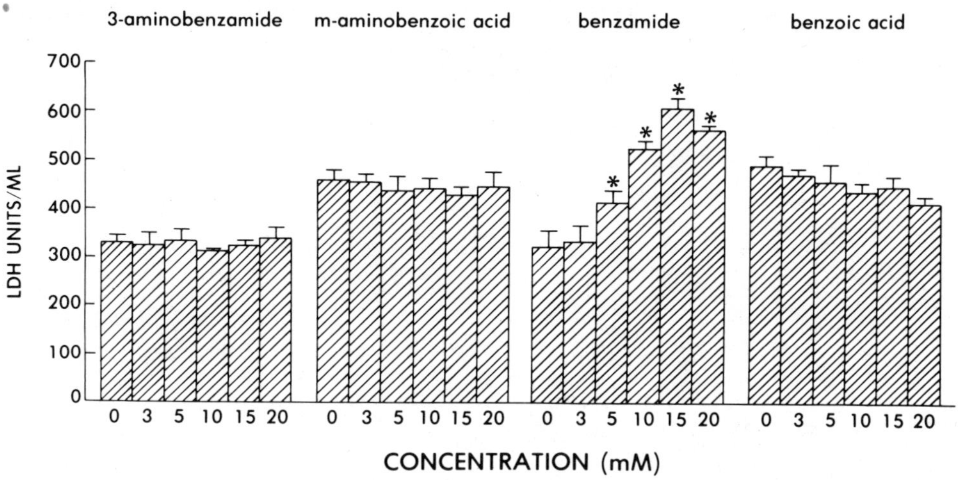

Fig. 1. Release of lactate dehydrogenase (LDH) from cells into the medium in the presence of benzamide, 3-AB, m-ABOA, or BOA (0-20 mM). Data are presented as the mean ±S.E.M. (bars); * indicates a significant difference from control, $p < 0.05$.

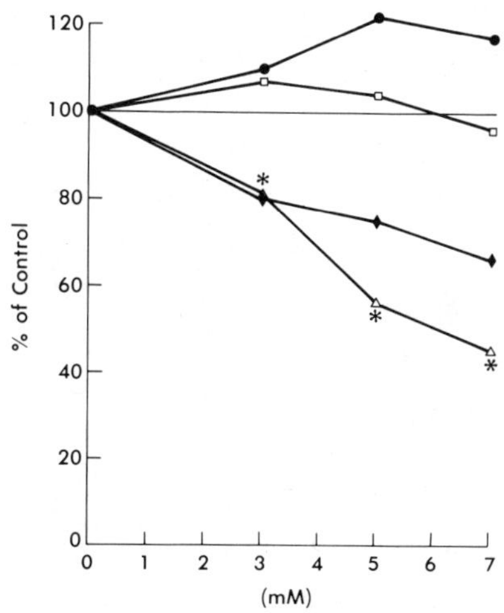

Fig. 2. The rate of glucose oxidation in the presence of benzamide (△), 3-AB (□), m-ABOA (♦), or BOA (●) (0-7 mM); * indicates a significant difference from control, $p < 0.05$.

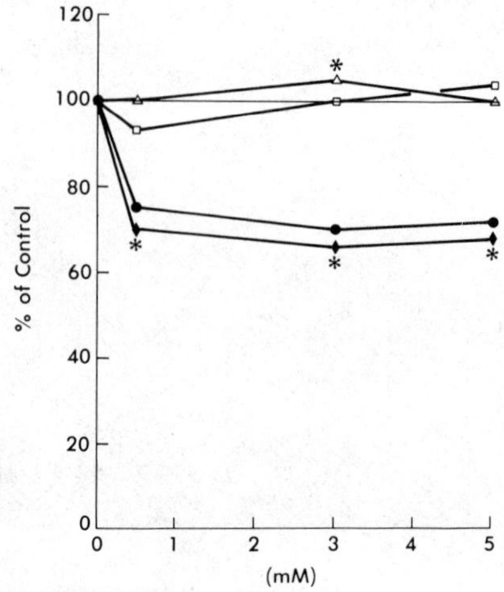

Fig. 3. Incorporation of [^{3}H]thymidine into DNA in the presence of benzamide (△), 3-AB (□), m-ABOA (♦), or BOA (●) (0-5 mM); * indicates a significant difference from control, $p < 0.05$.

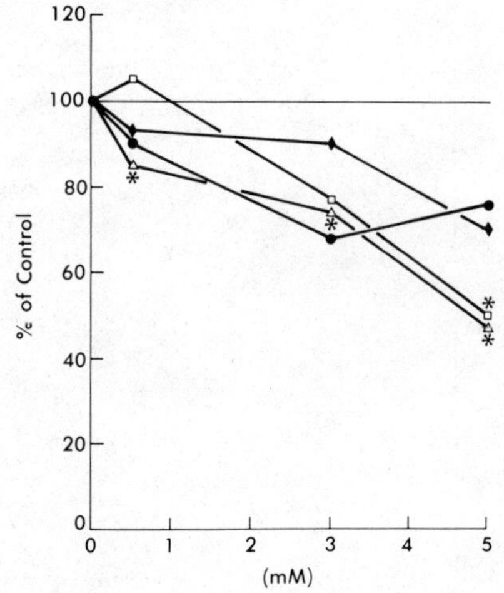

Fig. 4. Incorporation of [^{3}H]methionine into DNA in the presence of benzamide (△), 3-AB (□), m-ABOA (♦), or BOA (●) (0-5 mM); * indicates a significant difference from control, $p < 0.05$.

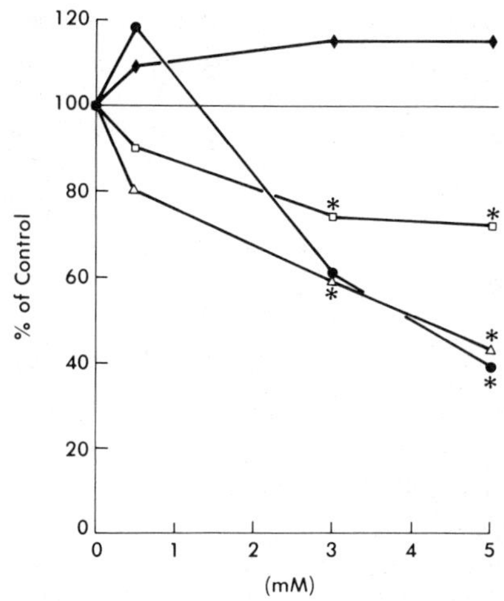

Fig. 5. Incorporation of [^{14}C]glucose into DNA in the presence of benzamide (△), 3-AB (□), m-ABOA (♦), or BOA (●) (0-5 mM); * indicates a significant difference from control, $p < 0.05$.

This inhibition of glucose oxidation is probably attributable to benzamide's cytotoxicity.

Neither inhibitor of poly(ADP-ribose) biosynthesis, i.e., 3-AB or benzamide, had a consistent effect on [^{3}H]thymidine incorporation into DNA (Fig. 3). In contrast, both m-ABOA and BOA reduced the amount of [^{3}H]thymidine incorporated into DNA; however, only the inhibitory effect of m-ABOA was significantly different from that of controls.

Both 3-AB and benzamide significantly reduced [^{3}H]methionine incorporation into DNA by roughly 50% at the 5 mM concentration, with inhibition by benzamide evident at a concentration as low as 0.5 mM (Fig. 4). m-ABOA and BOA had no significant effect on [^{3}H]-methionine metabolism at the doses used in this experiment.

In these experiments m-ABOA was the only benzamide analogue that did not significantly reduce the incorporation of [^{14}C]glucose into DNA (Fig. 5). 3-AB was moderately inhibitory at 3 and 5 mM, whereas both benzamide and BOA drastically reduced the incorporation of [^{14}C]glucose into DNA, with a 50-60% inhibition by these two chemicals at the 5 mM concentration.

4. CONCLUSIONS

Using doses well within the common experimental range (0-7 mM), we examined the metabolic effects of 3-AB, benzamide, m-ABOA, and BOA in a lymphoid cell line. Each of the chemicals exerted a significant inhibitory influence on some phase of metabolism.

Benzamide appeared to have the most profound effect on the metabolic parameters examined. This is probably due to its inhibition of metabolism per se, in addition to its cytotoxic impact on the cells. Thus, to combine 3-AB and benzamide under the simple rubric of specific poly(ADP-ribose) synthetase inhibitors is misleading, as they have very different effects on cellular metabolism. Furthermore, although 3-AB and benzamide profoundly inhibit the biosynthesis of poly(ADP-ribose), unlike the analogues m-ABOA and BOA, the specificity of the inhibition to poly(ADP-ribose) synthetase should be reconsidered in light of the other metabolic processes inhibited by these chemicals.

From the incorporation of [^{3}H]thymidine into DNA, it appears that the specific activity of the thymidine pool was not noticeably altered by the inhibitors of poly(ADP-ribose) synthetase. This suggests that the size of the thymidine pool itself was unchanged by 3-AB or benzamide. However, it has been reported that 3-AB does impair de novo synthesis of purines in HeLa cells [2]. Interpretation of the effect of m-ABOA and BOA on the thymidine pool is purely speculative. It is unlikely, but possible, that these two compounds stimulate endogenous thymidine synthesis, reducing the specific activity of [^{3}H]thymidine in the pool and consequwntly the amount of [^{3}H]thymidine incorporated into DNA. It is more likely that the thymidine pool size is unchanged and that m-ABOA and BOA exert an inhibitory influence on some enzyme(s) in the metabolic pathway of thymidine incorporation into DNA. It is possible that endogenous synthesis of pyrimidines and/or purines from several intermediary metabolites (e.g., glucose and methionine) is impaired by these chemicals, thus influencing nucleotide pool size. Such an effect may be detectable in the nucleotide pools only at higher concentrations or after more prolonged treatment with the benzamide analogues. The changes in DNA repair normally attributed to inhibition of poly-(ADP-ribose) biosynthesis may therefore be due in part to changes in the nucleotide pools. This cannot be verified, however, until the effect of these inhibitors on nucleotide pool sizes is measured directly.

REFERENCES

1. J. E. Cleaver, Methods for studying repair of DNA damaged by physical and chemical carcinogens, Methods Cancer Res., 11: 123-165 (1975).

2. J. E. Cleaver, W. J. Bodell, W. F. Morgan, and B. Zelle, Differences in the regulation by poly(ADP-ribose) of repair of DNA damage from alkylating agents and ultraviolet light according to cell type, J. Biol. Chem., 258:9054-9068 (1983).
3. D. Creissen and S. Shall, Regulation of DNA ligase activity of poly(ADP-ribose), Nature, 296:271-272 (1982).
4. B. W. Durkacz, O. Omidiji, D. A. Gray, and S. Shall, (ADP-ribose)$_n$ participates in DNA excision repair, Nature, 283:593-596 (1980).
5. W. F. Morgan and J. E. Cleaver, Effect of 3-aminobenzamide on the rate of ligation during repair of alkylated DNA, Cancer Res., 43:3104-3107 (1983).
6. M. R. Purnell and W. J. D. Whish, Novel inhibitors of poly-(ADP-ribose) synthetase, Biochem. J., 185:775-777 (1980).
7. J. L. Sims, G. W. Sikorski, D. M. Catino, S. J. Berger, and N. A. Berger, Poly(adenosinediphosphoribose) polymerase inhibitors stimulate unscheduled deoxyribonucleic acid synthesis in normal human lymphocytes, Biochemistry, 21:1813-1821 (1982).
8. R. G. D. Steele and J. H. Torrie, Principles and Procedures of Statistics, McGraw-Hill, New York (1960).
9. F. Wroblewski and J. S. LaDue, Lactic dehydrogenase activity in blood, Proc. Soc. Exp. Biol. Med., 90:210-213 (1955).

RIBONUCLEOTIDE REDUCTASE AND DEOXYRIBONUCLEOTIDE POOLS

Peter Reichard

Medical Nobel Institute
Department of Biochemistry I
Karolinska Institutet
Box 60400
S-104 01 Stockholm, Sweden

The replication of DNA requires a balanced supply of the four deoxyribonucleoside triphosphates (dNTPs) that is provided by reduction of the corresponding ribonucleotides through the action of the enzyme ribonucleotide reductase [24]. A single protein catalyzes the reduction of all four ribonucleotides. The evidence for this is two-fold: (i) Homogeneous proteins prepared from bacterial and mammalian sources catalyze the four reactions with about equal efficiency and (ii) single step mutations in the enzyme affect reductions of all ribonucleotides.

The balance between the four building blocks for DNA synthesis is regulated by an exquisite allosteric control of ribonucleotide reductases. In addition, a second enzyme deaminating a deoxycytidine nucleotides determines the ratio between dCTP and dTTP. In mammalian cells this enzyme acts on dCMP [20] while in bacteria dCTP [25] serves as substrate. I would like to point out that there is no allosteric control of ribonucleotide reductase distinguishing between the reduction of the two pyrimidine ribonucleotides and that the deaminase therefore occupies a key position for the control of the dCTP/dTTP ratio.

My main theme today is a discussion of the ways in which ribonucleotide reductase contributes to the supply of dNTPs. First I will give a brief description of our model for the allosteric regulation of the enzyme. This aspect is treated in more detail in earlier reviews [17, 18, 24]. I will then describe results obtained with intact mammalian cells that validate the model. Finally I will describe recent experiments that relate the formation of dNTPs to

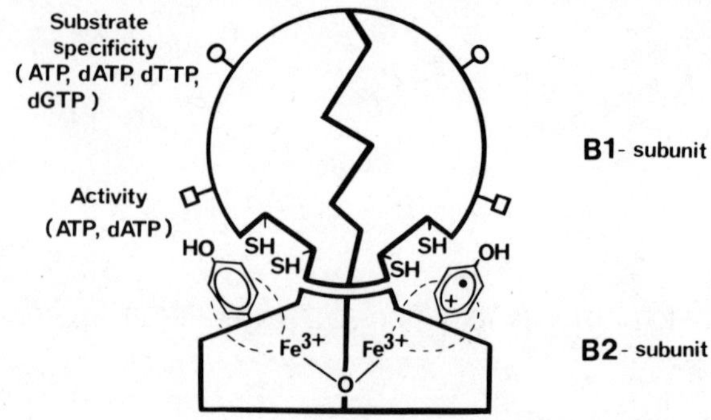

Fig. 1. Molecular model of ribonucleoside diphosphate reductase from E. coli.

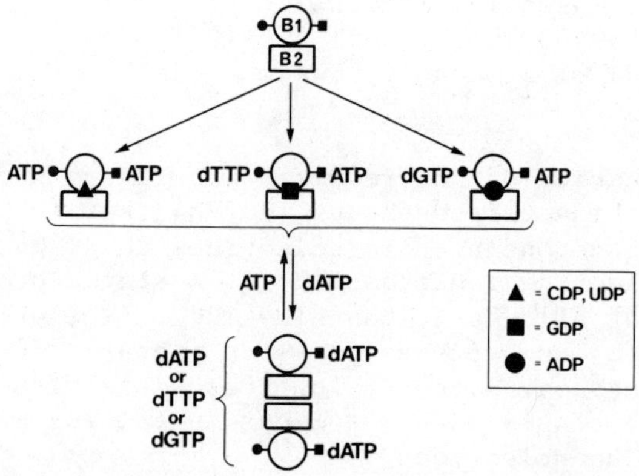

Fig. 2. Different conformational states of ribonucleotide reductase from E. coli. With ATP bound to the activity site an active enzyme ensues with its specificity determined by effector binding to the substrate specificity site: ATP binding gives an enzyme specific for CDP and UDP, dTTP binding for GDP, and dGTP binding for ADP. With dATP bound to the activity site the enzyme is unactive irrespective of binding to the substrate specificity site.

their utilization for DNA synthesis and study the interplay of the two processes.

A Model for the Allosteric Regulation of Ribonucleotide Reductases

Figure 1 describes our present understanding of ribonucleoside diphosphate reductase from E. coli. This enzyme consists of two nonidentical subunits named protein B1 and B2. Each subunit is a

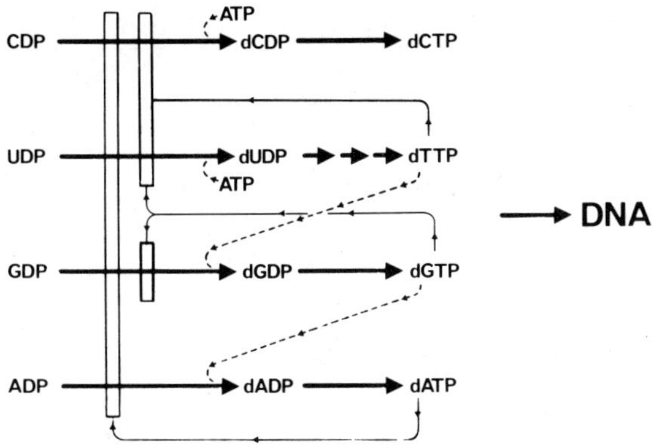

Fig. 3. Model for the metabolic activity of ribonucleotide reductases.

dimer consisting of two identical or nearly identical polypeptide chains. The catalytic site of the enzyme is constructed from parts of each subunit. Protein B1 mediates the allosteric control of the enzyme and is of main interest today. It does so through binding of effectors (nucleoside triphosphates) to two separate classes of binding sites. One class (activity site) binds only ATP or dATP. With ATP bound the enzyme is active, with dATP it is inactive. The second class of sites binds dTTP and dGTP in addition to ATP and dATP. In this case binding modulates the substrate specificity of the enzyme: binding of ATP or dATP results in an enzyme reducing pyrimidine ribonucleotides, binding of dTTP an enzyme reducing GDP, and binding of dGTP an enzyme reducing ADP. This second class has therefore been named specificity site. According to this model ribonucleotide reductase exists in many different conformational states, active or inactive, able to reduce the various ribonucleoside diphosphates depending on effector binding to the substrate specificity site. Some of the main conformational states are schematically represented in Fig. 2.

The model of the E. coli enzyme is based on binding studies [3] as well as kinetic experiments concerning the influence of different effectors on the reduction of the four ribonucleoside diphosphates [10, 11]. It leads to the scheme in Fig. 3 for the reduction of the four ribonucleotides in intact cells. There, a sequential process is started by the reduction of CDP and UDP when ATP binds to the substrate specificity site of the enzyme. This leads to the formation of dTTP which replaces ATP at the substrate specificity site, resulting in the reduction of GDP. The resulting dGTP replaces dTTP at the substrate specificity site and the enzyme reduces ADP. dATP, the final product of the sequence, turns off the whole process unless the nucleotide is consumed for the synthesis of DNA.

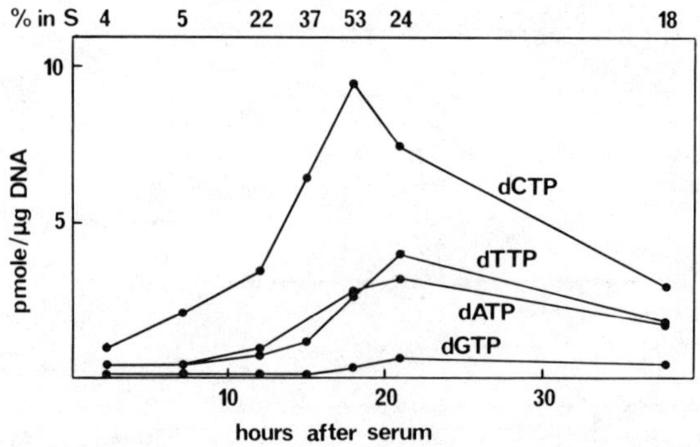

Fig. 4. dNTP pools in secondary mouse embryo cells after release from serum starvation (data from [21]).

This scheme was originally mainly based on experiments with the E. coli enzyme. However, it soon became apparent that also reductases of higher cells are regulated by a similar mechanism [14, 19]. Mammalian reductases also consist of two non-identical subunits, named M1 and M2 (in analogy to B1 and B2), with M1 being responsible for the allosteric control of the enzyme [6, 23]. Recent experiments with a highly purified mammalian reductase from calf thymus showed that the allosteric regulation of the activity and substrate specificity of this enzyme is in most respects identical to that of the E. coli enzyme. The scheme of Fig. 3 thus also applies to the mammalian enzyme. The molecular basis for the regulation of the mammalian enzyme has, however, not yet been worked out in detail.

Size of dNTP Pools

Changes in the size of the four dNTP pools in intact cells can as a first approximation serve as a measure of the activity and regulation of the reductase. During S-phase dNTPs are consumed in about equal amounts for DNA synthesis. Figure 4 shows an experiment in which quiescent secondary mouse fibroblasts started to enter S-phase 12 h after addition of serum [21]. In parallel, dNTP pools increased dramatically. Surprisingly large differences were found between the four pools, with dCTP having the largest pool and dGTP the smallest, in spite of the fact that equal amounts of each nucleotide are used up during DNA synthesis. The ratios between the four pools also varied during the cell cycle with the size of the dCTP pool being most closely related to the rate of DNA synthesis. The dCTP/dATP ratio is high in actively dividing cell populations or during S-phase of synchronously growing cells [22] while inhibition of DNA synthesis leads to a rapid drop [15]. Panel A of Fig. 5 shows a growth curve for 3T6 mouse fibroblasts with measurements of dNTP pools on 3 consecutive days after explantation [1]. This rapidly growing

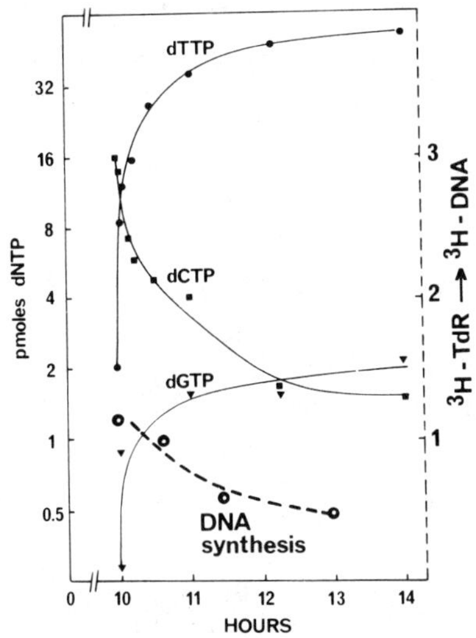

Fig. 5. Effect on dNTP pools and DNA synthesis of addition of 1 mM thymidine to CHO cells in S-phase (data from [2]).

culture with a generation time of 12-14 h shows dCTP/dATP ratios of between 3-4. Panel B of Fig. 5 shows a similar growth curve of a 3T6 cell line containing a fifteenfold higher activity of ribonucleotide reductase. This line was selected for resistance to hydroxyurea and the increase in reductase activity depended on an overproduction of M2 activity while the allosteric M1 subunit was unchanged [1]. This line thus had an inherent large capacity to overproduce dNTPs. A comparison of the dNTP analyses of two panels in Fig. 5 shows, however, that only the dATP pool was increased in the resistant cell line resulting in dCTP/dATP ratios of 0.6-0.7. These data suggest that the potential for an increased reduction of ribonucleotides in the resistant line is tempered by the binding of the excess of dATP to the allosteric M1 subunit (cf. Fig. 2).

Testing the Model in Intact Cells

Two main lines of evidence support our model for the regulation of dNTP pools in intact cells: (i) Perturbations of pools, either by administration of nucleosides [2, 8] or inhibitors [21] to cells in tissue culture, lead to predicted effects; and (ii) cells carrying regulatory mutations in ribonucleotide reductase show manifestations explainable by the model [4, 26].

Inhibition of cell growth by deoxyadenosine [9] and the discovery that this effect is related to an imbalance of dNTP pools depending on changes in the substrate specificity of ribonucleotide

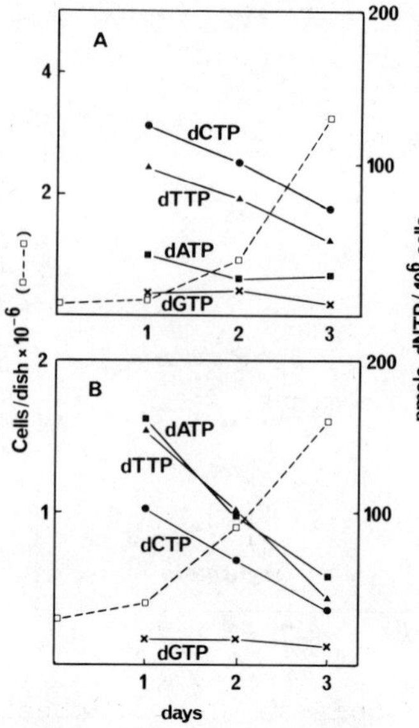

Fig. 6. dNTP pools during growth of 3T6 cells (Panel A) and a hydroxyurea resistant line of 3T6 cells with 15 fold increased ribonucleotide reductase activity (Panel B) [1].

reductase was the first example of this kind [19]. In general, addition of a deoxynucleoside leads to the expansion of the corresponding dNTP pool and locks the specificity of the reductase into one or a few of the available allosteric states, resulting in a deficiency of one or several dNTPs and inability to replicate DNA. This applies to three of the four dNTPs, while expansion of the fourth pool, dCTP, has no such effect since dCTP is not an allosteric effector of the reductase.

The effect of thymidine on dNTP pools and DNA synthesis in cultured CHO cells shown in Fig. 6 is a good example of such an effect [2]. Addition of the nucleoside leads to a more than twentyfold increase in the dTTP pool followed by a similar increase in dGTP and a decrease in dCTP. These results are in line with the model described in Fig. 2 where binding of dTTP to the regulatory subunit gives rise to a form of the reductase tuned for the reduction of DGP (but not CDP).

Many similar results obtained in experiments with cultured cells can be explained by the allosteric model. In addition, the manifestations of certain inherited immunodeficiency diseases [7, 13] derive from a disturbance of the normal allosteric regulation of the reductase [4, 26, 27]. In these cases the patients lack enzymes involved in the degradation of deoxyadenosine or deoxyguanosine which

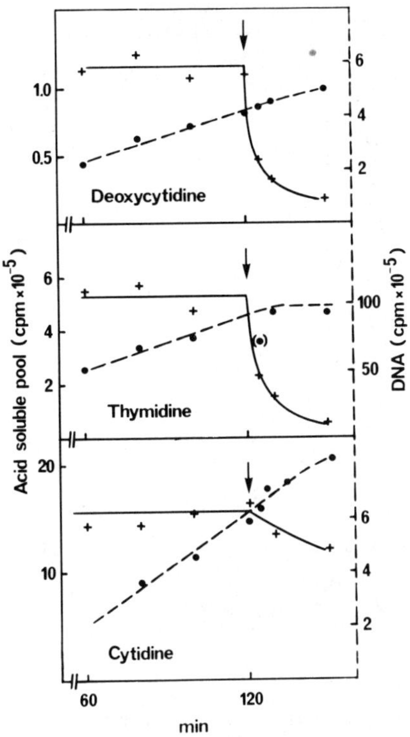

Fig. 7. Time curve of incorporation of labeled nucleosides into acid soluble nucleotides and DNA of growing 3T6 cells [16].

leads to an accumulation of dATP or dGTP in the immune system resulting in disturbances of dNTP pools and DNA synthesis.

The last example already exemplifies a contribution from genetics for the validation of the model. Additional convincing evidence along the same line comes from experiments with mutant cell lines of mouse T-lymphomas [4, 26]. Particularly impressive was the characterization of separate mutations for the activity site and the substrate specificity site of the M1 subunit of the reductase [5]. I presume that other speakers will address this point later in more detail.

Dynamics of dNTP Pools

Measurements of levels of dNTP pools give only a static picture of the reductase. We have started to analyze more dynamic aspects of the enzyme by measuring the turnover of dNTP pools and its coupling to DNA synthesis [16]. In these experiments we supply tracer amounts of [^{3}H]-labeled cytidine, deoxycytidine or thymidine to the

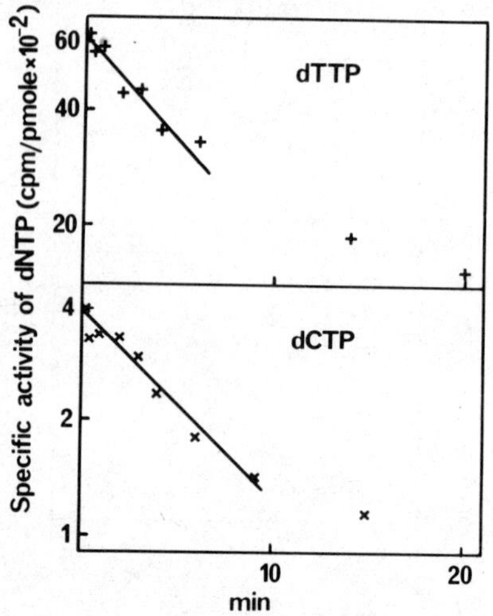

Fig. 8. Decay of radioactivity from dTTP and dCTP pools labeled from thymidine and deoxycytidine, respectively, after change to cold medium [16].

medium of rapidly growing cultured 3T6 cells. Thirty to 60 minutes thereafter the dNTP pools have reached a constant specific activity due to a steady state equilibrium between incorporation of nucleoside, de novo synthesis and linear incorporation of dNTP into DNA (Fig. 7). At this stage the rate of DNA synthesis can be calculated from the specific activity of the dNTP pool and isotope incorporations into DNA. During the second part of the experiment, starting at the arrow in Fig. 7, the radioactive medium was replaced by cold medium in order to measure the decay of the dNTP pool. Except in the cytidine experiment, this resulted in a rapid loss of isotope from the dNTP pool due to incorporation into DNA without replenishment of isotope from the medium. In the case of cytidine, isotope continued to flow into the dCTP pool from the large labeled intracellular CTP pool.

A more detailed analysis of the decay of the dTTP and dCTP pools is shown in Fig. 8 where the logarithms of the specific activities of the two dNTP pools are plotted against time. The linearity of the initial part of the curves demonstrates that the decay followed first order kinetics. From these curves and from the known pool sizes the turnover of the two dNTP pools can be calculated.

Results from the experiments depicted in Figs. 7 and 8 are summarized in Table 1. Rates of DNA synthesis calculated from the incorporation of cytidine or thymidine were almost identical, while the deoxycytidine experiment gave a lower value. In this connection

TABLE 1. Rate of DNA Synthesis and Turnover of dNTP Pools in Rapidly Growing 3T6 Cells Measured by Incorporation of Labeled Nucleosides [16]

	dNTP pool[1]		DNA synthesis[1]
	Spec. activ. cpm/pmole	decay pmole/min	(pmole labeled dNMP/min
[5-^{3}H]Deoxycytidine	580	10	5.5
[5-^{3}H]Cytidine	375	2	16
[methyl-^{3}H]Thymidine	3900	14	14

[1]All values are given for 10^6 cells.

TABLE 2. Rate of DNA Synthesis in Growing 3T6 Cells Determined from the Incorporation of [^{3}H]-thymidine in Normal or HAT Medium [16]

	dTTP pool[1]		DNA synthesis[1]
Medium	pmole	cpm/pmole	pmole dTMP/min
Normal	90	5900	13
HAT[2]	150	280	14

[1]Values are given for 10^6 cells.

[2]HAT medium (12) was used from the start of the experiment.

it should, however, be recalled that mouse DNA contains about 1.3 times more thymine than cytosine. Crucial to the calculations is the assumption that the measured specific activity of a dNTP pool is identical with the specific activity of the immediate precursor for DNA synthesis. However, a given dNTP may be compartmentalized in the cell, be present in more than one pool and have a different specific activity in each. It is also conceivable that one such pool, derived, e.g., by de novo synthesis via reduction of ribonucleotides, is more directly used for DNA synthesis and that the specific activity of this particular pool differs from the measured specific activity obtained by labeling from thymidine or deoxycytidine.

This problem can be circumvented by forcing cells to label a given dNTP pool via only one specific pathway. The de novo synthesis of dTTP can be blocked with amethopterin by cultivating cells in HAT medium [12]. When [^{3}H]-thymidine is added to such cells all dTTP is synthesized from the labeled nucleoside. In the experiment illustrated in Table 2 two parallel cultures in normal or HAT medium were labeled from [^{3}H]-thymidine and the specific activities of the dTTP pools and incorporation into DNA were measured. Identical rates of DNA synthesis were obtained for both cultures in spite of the fact that the specific activities as well as the sizes of the dTTP pools showed large differences. From this experiment we concluded that only one kinetic dTTP pool exists in 3T6 cells and that the true rate of DNA synthesis can be calculated form the incorporation of a trace amount of labeled thymidine into dTTP and DNA.

Let us return to the experiment described in Table 1. In view of the conclusion just arrived the data can now be interpreted as follows: at least two separate pools of dCTP exist, one preferentially labeled from deoxycytidine, the other from cytidine. The measured specific activity is an average for the two pools which appear to be in slow equilibrium with each other. DNA is synthesized preferentially from the pool labeled from cytidine, i.e., via the reductase.

Table 1 also gives the values calculated for the decay of the dNTP pools. With dTTP the decay equals the incorporation of the nucleotide into DNA which suggests that all dTTP is exclusively used for DNA synthesis. The results for the dCTP pool are complicated by the fact that this nucleotide is incorporated into DNA not only as such but also after deamination and transformation to dTTP. However, during this transformation the isotope is lost since deoxycytidine was labeled with tritium exclusively at C5 of the pyrimidine ring. This effect may explain faster decay of dCTP compared to its incorporation into DNA.

CONCLUSIONS AND GENERAL SUMMARY

Ribonucleotide reductase plays a crucical role in the maintenance of dNTP pools. A molecular model for the allosteric regulation of the enzyme is substantiated by the effects of nucleosides or inhibitors on pool sizes in intact cells as well as by manifestations of mutations of the reductase and other enzymes involved in deoxynucleoside metabolism. Studies of the dynamics of dNTP pools in 3T6 cells suggest compartmentation of dCTP pools, with preferential DNA synthesis from a pool labeled from cytidine. In contrast, no kinetic evidence for more than one dTTP pool was obtained. Our experiments also suggest that all dTTP formed in growing 3T6 cells is used up for DNA synthesis.

ACKNOWLEDGMENTS

This investigation was supported by grants from the Swedish Medical Research Council and Knut and Alice Wallenberg's Foundation.

REFERENCES

1. L. Åkerblom, A. Ehrenberg, A. Gräslund, H. Lankinen, P. Reichard, and L. Thelander, Overproduction of the free radical of ribonucleotide reductase in hydroxyurea-resistant mouse fibroblast 3T6 cells, Proc. Natl. Acad. Sci. USA, 78:2159-2163 (1981).
2. G. Bjursell and P. Reichard, Effects of thymidine on deoxyribonucleoside triphosphate pools and deoxyribonucleic acid synthesis in Chinese hamster ovary cells, J. Biol. Chem., 248: 3904-3909 (1973).
3. N. C. Brown and P. Reichard, Role of effector binding in allosteric control of ribonucleoside diphosphate reductase, J. Mol. Biol., 46:39-55 (1969).
4. S. Eriksson, L. J. Gudas, B. Ullman, S. M. Clift, and D. W. Martin, Jr., DeoxyATP resistant ribonucleotide reductase of mutant mouse lymphoma cells; evidence for heterozygosity for the protein M1 subunits, J. Biol. Chem., 256:10184-10188 (1981).
5. S. Eriksson, L. J. Gudas, S. M. Clift, J. W. Caras, B. Ullman, and D. W. Martin, Jr., Evidence for genetically independent allosteric regulatory domains of the protein M1 subunit of mouse ribonucleotide reductase, J. Biol. Chem., 256:10193-10197 (1981).
6. S. Eriksson, L. Thelander, and M. Åkerman, The allosteric regulation of calf thymus ribonucleoside diphosphate reductase, Biochemistry, 18:2948-2952 (1979).
7. E. R. Giblett, J. E. Anderson, F. Cohen, B. Pollara, and H. J. Meuwissen, Adenosine-deaminase deficiency in two patients with severely impaired cellular immunity, Lancet, 2:1067-1069 (1972).
8. M. T. Hakala and E. Taylor, The ability of purine and thymine derivatives and of glycine to support the growth of mammalian cells in culture, J. Biol. Chem., 234:126-128 (1959).
9. H. Klenow, On the effect of some adenine derivatives on the incorporation in vitro of isotopically labeled compounds into the nucleic acids of Ehrlich ascites tumor cells, Biochem. Biophys. Acta, 35:412-421 (1959).
10. A. Larsson and P. Reichard, Enzymatic synthesis of deoxyribonucleotides. IX. Allosteric effects in the reduction of

pyrimidine ribonucleotides by the ribonucleoside diphosphate reductase system of Escherichia coli, J. Biol. Chem., 241: 2533-2539 (1966).

11. A. Larsson and P. Reichard, Enzymatic synthesis of deoxyribonucleotides. X. Reduction of purine ribonucleotides; allosteric behavior and substrate specificity of the enzyme system from Escherichia coli B, J. Biol. Chem., 241:2540-2549 (1966).

12. J. W. Littlefield, Selection of hybrids from matings of fibroblasts in vitro and their presumed recombinants, Science, 145: 709-710 (1964).

13. D. W. Martin, Jr., and E. W. Gelfand, Biochemistry of diseases of immunodevelopment, Ann. Rev. Biochem., 50:845-877 (1981).

14. E. C. Moore and R. B. Hurlbert, Regulation of mammalian deoxyribonucleotide biosynthesis by nucleotides as activators and inhibitors, J. Biol. Chem., 241:4802-4809 (1966).

15. B. Nicander and P. Reichard, Aphidicolin sensitivity of variant 3T6 cells selected for changes in ribonucleotide reductase, Biochem. Biophys. Res. Commun., 103:148-155 (1981).

16. B. Nicander and P. Reichard, Dynamics of pyrimidine deoxynucleoside triphosphate pools in relationship to DNA synthesis in 3T6 mouse fibroblasts, Proc. Natl. Acad. Sci. USA, 80:1347-1351 (1983).

17. P. Reichard, Control of deoxyribonucleotide synthesis in vitro and in vivo, in: Advances in Enzyme Regulation (G. Weber, ed.), Vol. 10, pp. 3-16, Pergamon Press, Oxford and New York (1972).

18. P. Reichard, From deoxynucleotides to DNA synthesis, Fed. Proc., 37:9-14 (1978).

19. P. Reichard, Z. N. Canellakis, and E. S. Canellakis, Regulatory mechanisms in the synthesis of deoxyribonucleic acid in vitro, Biochim. Biophys. Acta, 41:558-559 (1960).

20. E. Scarano, The enzymatic deamination of 6-aminopyrimidine deoxyribonucleotides. I. The enzymatic deamination of deoxycytidine 5'-phosphate and of 5'-methyldeoxycytidine 5'-phosphate, J. Biol. Chem., 235:706-713 (1960).

21. L. Skoog and B. Nordenskjöld, Effects of hydroxyurea and 1-β-Arabinofuranosyl cytosine on deoxyribonucleotide pools in mouse embryo cells, Eur. J. Biochem., 19:81-89 (1971).

22. L. Skoog, B. Nordenskjöld, and G. Bjursell, Deoxyribonucleoside-triphosphate pools and DNA synthesis in synchronized hamster cells, Eur. J. Biochem., 33:428-432 (1973).

23. L. Thelander, S. Eriksson, and M. Åkerman, Ribonucleotide reductase from calf thymus. Separation of the enzyme into two nonidentical subunits, protein M1 and M2, J. Biol. Chem., 255: 7426-7432 (1980).

24. L. Thelander and P. Reichard, Reduction of ribonucleotides, Ann. Rev. Biochem., 48:133-158 (1979).

25. F. Tomita and I. Takahashi, A novel enzyme, dCTP deaminase, found in bacillus subtilis infected with phage PBS 1, Biochim. Biophys. Acta, 179:18-27 (1969).

26. B. Ullman, L. J. Gudas, S. M. Clift, and D. W. Martin, Jr., Isolation and characterization of purine-nucleoside phosphorylase-deficient T-lymphoma cells and secondary mutants with altered ribonucleotide reductase: Genetic model for immunodeficiency disease, Proc. Natl. Acad. Sci. USA, 76:1074-1078 (1979).
27. B. Ullman, L. J. Gudas, J. W. Caras, S. Eriksson, G. Weinberg, M. A. Wormsted, and D. W. Martin, Jr., Demonstration of normal and mutant protein M1 subunits of deoxyGTP resistant ribonucleotide reductase from mutant mouse lymphoma cells, J. Biol. Chem., 256:10189-10192 (1981).

ENZYMATIC CHANNELING OF DNA PRECURSORS

Christopher K. Mathews

Department of Biochemistry and Biophysics
Oregon State University
Corvallis, Oregon 97331, USA

1. INTRODUCTION: FUNCTIONAL COMPARTMENTATION VIA THE ACTION OF MULTIENZYME COMPLEXES

The idea that metabolic intermediates can be channeled through the action of organized multienzyme complexes is well established in cellular biochemistry. However, application of these ideas to reaction pathways in nucleic acid biosynthesis represents a quite recent development. In part this is because most of the enzymes of nucleic acid metabolism are readily isolated in soluble form, whereas the well-understood complexes, such as pyruvate dehydrogenase or the mitochondrial electron transport chain, consist either of tightly associated enzymes or proteins firmly anchored in a membrane. It is noteworthy that a conference volume published in 1978 [49], "Microenvironments and Metabolic Compartmentation," made no mention of DNA precursor biosynthesis. Yet the collected papers of that volume include the major concepts to be developed in this chapter - kinetic advantages resulting from complexation, maintenance of metabolite concentration gradients, distinct regulatory mechanisms affecting channeled pathways, and so forth. Although the enzymes of DNA precursor biosynthesis are not tightly associated with one another or with the replication apparatus, the interactions which do exist evidently allow for metabolite channeling similar to that which has been so elegantly demonstrated in other systems.

---

*The following abbrevations are used: dNTP, deoxyribonucleoside 5'-triphosphate; rNDP, ribonucleoside 5'-diphosphate; hm-dCMP and hm-DCT, 5-hydroxymethyldeoxycytidine 5'-mono- and triphosphate, respectively; ts, temperature-sensitive, dNMP, deoxyribonucleoside 5'-5'-monophosphate.

This discussion will focus primarily upon chromosomal DNA replication. This process occurs at a limited number of intracellular sites, using specialized precursors, the dNTP's*, whose metabolic roles are limited to DNA synthesis, a process which occurs only within a defined portion of the cell cycle. Since DNA chains are elongated very rapidly during replication, for example, about 800 nucleotides per second in *E. coli* at 37° [26], one could be tempted *a priori* to postulate the existence of machinery for generating dNTP's *in situ*, in some process coordinated with DNA replication itself. Considerable evidence has accumulated, mostly since 1974, in support of such a hypothesis. Basically the model postulates the existence of (1) a small, replication-active pool of dNTP's generated at replication sites through the action of multienzyme complexes and rapidly turned over as these are used for replicative chain elongation; and (2) a much larger and more slowly replenished pool of dNTP's, which may be dispersed through much of the cell's volume and which is used for other DNA transactions, such as DNA repair or replication of extrachromosomal DNA. Whether these latter processes involve the action of multienzyme complexes or functional compartmentation has not yet been investigated. However, since (1) these processes are not highly regulated temporally, (2) they involve just a small fraction of the dNTP's generated within a cell, and (3) DNA chain growth rates are relatively low, the arguments applied to chromosomal DNA replication in this discussion need not apply.

The model as stated above is greatly oversimplified. For example, it ignores the fact that in eukaryotic cells membrane-limited organelles can superimpose physical compartmentation upon the postulated kinetic compartmentation; thus, there may exist more than two pools of DNA precursors. In addition the model says nothing about the extent to which the several pools can mix with one another. Also, as stated, the model ignores substantial differences between replicative processes in prokaryotes, where most of the early evidence was developed, and eukaryotes, where much confirmatory evidence is now accumulating. In particular, since DNA chain growth rates in eukaryotic replication are some one to two orders of magnitude lower than those observed in prokaryotes [26], simple diffusion may suffice to supply growing DNA chains at adequate rates. Nevertheless, evidence is accumulating, mostly from animal cell studies, for some type of DNA precursor channeling. Much of this evidence is indirect, because of the impossibility of isolating the posulated replication-active pools. In the aggregate, however, it is persuasive.

Before discussing this evidence, it is worth listing several other well-studied examples of multifunctional proteins or multienzyme complexes in nucleotide metabolism. In animal cells it is now well-established that the first three activities of pyrimidine nucleotide biosynthesis are carried on one protein [12], while the last two activities are carried on a separate protein [52]. Also in animal cells three activities of folate coenzyme interconversion are carried on one protein [39]. This trifunctional protein forms a noncovalently bonded complex with serine transhydroxymethylase and with the two transformylase enzymes involved in purine nucleotide biosynthesis [48]. Thus, both pyrimidine and purine nucleotide production involve at least some metabolic channeling. Finally, the deoxypyrimidine kinase encoded by the herpes simplex virus genome is tightly complexed with the enzyme catalyzing the next reaction, namely, dTMP kinase [8]. In all of these cases kinetic experiments make it clear that juxtaposition of catalytic sites of enzymes mediating sequential reactions allows for rapid flux through a pathway with very low accumulation of intermediates. In the following sections I will discuss the extent to which the same principles might apply to the much more complex processes of deoxyribonucleotide production and DNA replication.

2. EVIDENCE FOR dNTP CHANNELING IN DNA REPLICATION

2.1. DNA Precursor Labeling Kinetics

The widespread use of thymidine as a labeled DNA precursor has revealed several peculiarities in its quantitative utilization for DNA synthesis. In some cases exogenous thymidine is poorly incorporated into DNA, and its rate of incorporation does not reflect rates of DNA synthesis [13, 24]. Yet in many other cases exogenous thymidine can bypass substantial endogenous thymidylate pools and be incorporated into DNA at full specific activity [27, 30, 33]. This strongly suggests the existence of a channeling mechanism for preferential utilization of salvage-derived nucleotides over that of nucleotides formed *de novo*. So far it has not been possible to predict circumstances under which this channeling will occur, or to explain its mechanism.

One of the earliest indications of dNTP compartmentation came from Werner's studies of DNA labeling kinetics by thymine or thymidine in *E. coli* [55]. Label reached its maximal rate of incorporation into DNA long before labeling of the dTTP pool had reached its highest value. Similar results were obtained by Pato [38]. One way to rationalize these data is to postulate that the small, replication-active pool of dTTP fills almost immediately after administration of the isotope, and that the prolonged incorporation of label into dTTP represents filling of the much larger, less rapidly replenished pool. Data suggestive of similar compartmentation in human lymphoblasts were published by Fridland [21].

Scott and Forsdyke [45] have carried out an extensive precursor labeling study, using the isotope dilution method, which is consistent with the existence of at least two pools of dNTP's in rat thymus cells. In the isotope dilution method a constant quantity of a radiolabeled DNA precursor is added to cells along with varying quantities of the corresponding unlabeled precursor. One plots the total concentration of precursor as a function of the reciprocal of

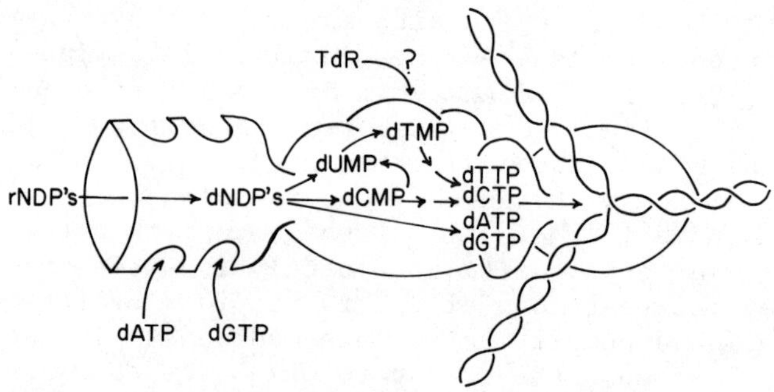

Fig. 1. Model of functional compartmentation of DNA precursors. As discussed by Scott and Forsdyke [45], dATP and dGTP derived from salvage of exogenous nucleosides can bind to allosteric sites controlling the activity of rNDP reductase. However, they cannot readily mix with the pools generated through the action of rNDP reductase and associated enzymes. The variable status of the thymidine salvage pathway is also indicated.

the radioactivity incorporated into DNA. The slope of the resultant straight line gives a measure of the rate-limiting step (in this case, the rate of DNA synthesis), while the intercept at the ordinate gives a measure of the pool of compounds in the cell which compete with the radioactive precursor before the rate-limiting step.

Using this method, Scott and Forsdyke asked whether deoxyadenosine or deoxyguanosine could reverse the inhibition of DNA synthesis by hydroxyurea. Since the major effect of hydroxyurea is to deplete purine dNTP pools [47], one would expect purine deoxyribonucleosides to replete these pools and to restore DNA synthesis. Using labeled pyrimidine nucleosides, Scott and Forsdyke found that purines did not restore DNA synthesis and that intracellular pools of pyrimidine precursors decreased. They interpret these data as schematized in Fig. 1. dATP and dGTP formed by salvage of the added deoxyribonucleosides cannot readily associate with replication sites because of the presumed existence of a complex of dNTP-synthesizing and replication activities. However, they can bind to control sites affecting ribonucleotide reductase, where they can inhibit UDP and CDP reduction, and hence, decrease the replication-active pools of dTTP and dCTP.

Figure 1 also illustrates the uncertain relationship between thymidine kinase and this complex, as suggested by the variable results of thymidine-labeling experiments alluded to earlier.

2.2. Enzyme Activities Estimated in vivo

An experimental approach developed in Greenberg's laboratory [51] has been particularly fruitful in developing the concept of multienzyme complexes acting to coordinate dNTP synthesis with DNA replication. The pyrimidine bases in T4 phage DNA are thymine and 5-hydroxymethylcytosine. Both bases are synthesized at the nucleotide level, by reactions which transfer a single-carbon unit to position 5 on the pyrimidine ring, with concomitant displacement of hydrogen at that position. By administering 5[^{3}H]uridine to infected cells, Tomich et al. [51] could measure tritium release in vivo, and hence, determine the intracellular activities of the two enzymes involved - thymidylate synthase and deoxycytidylate hydroxymethylase. Either enzyme could be assayed independently of the other by use of appropriate T4 mutants. Tomich et al. found that enzyme protein is synthesized very early in infection, but that the enzymes do not begin to function until several minutes later, when DNA replication begins. Protein synthesis is essential for this metabolic activation to occur, suggesting that additional elements of a multienzyme complex must be formed in order for dNTP synthesis to begin in vivo. Using this assay and other methods Flanegan and Greenberg and their colleagues [11, 17, 18] were able to determine rates of both dTMP and hm-dCMP biosynthesis in vivo. Remarkably, the rates were maintained at a ratio of 2.1:1 (dTMP/hm-dCMP), precisely the ratio at which these nucleotides are found in T4 DNA. This ratio was maintained over a wide range of conditions, suggesting that activities of the complex were highly regulated. Two exceptions were noted: (1) infection with a dCMP deaminase-negative mutant [11], suggesting a critical role for this enzyme in regulation, and (2) infection in the presence of DNA-reactive substances such as proflavine [17], suggesting a link between DNA or the replication machinery and the enzymes synthesizing DNA precursors.

Various investigators have applied this approach to mammalian cells, where, of course, one can monitor only thymidylate synthase activity. The most detailed investigation was that of Rode et al. [44], who found that activity as measured in extracts was relatively constant throughout a cycle of synchronized growth, but that the enzyme was active in vivo only during S phase. Product inhibition by dTMP was proposed as a mechanism which, along with the existence of a multienzyme complex, helped coordinate flux through dTMP synthase with DNA replication.

The other investigations to be mentioned in this section pertain to the intracellular activity of ribonucleoside diphosphate (rNDP) reductase. Early studies suggested that the activity of the enzyme in cell-free extracts was far too low to account for the observed rate of DNA synthesis in vitro. However, in ether-permeabilized E. coli cells [53] or in very gently lysed bacteria [15, 28] it was found that activity was much higher, suggesting that main-

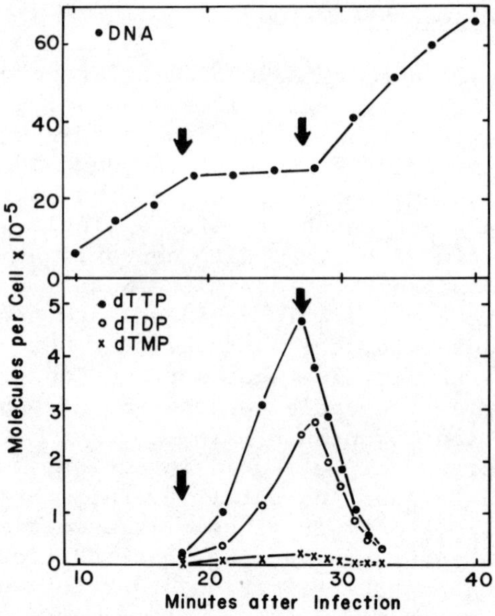

Fig. 2. Effects of reversible inhibition of DNA synthesis on thymine nucleotide pools (lower panel) and DNA synthesis (upper panel). Data are from reference 31. E. coli was infected at 30° with T4, under conditions where all thymine nucleotides are derived from exogenous thymidine. The phage bore a ts mutation in gene 45, essential for DNA replication. At 18 min DNA replication was arrested by shifting the culture from 30° to 42°. The culture was sampled for thymine nucleotide pool determinations before and after restoration of DNA synthesis by a shift back to 30°.

tenance of intracellular associations was essential if the enzyme was to function at the calculated *in vivo* rate.

2.3. dNTP Pool Dynamics

Several different analyses of dNTP pool kinetics have contributed to the concept of compartmentalized DNA precursor supplies and metabolic channeling. Many studies of cells treated with antimetabolites affecting dNTP synthesis, such as fluorodeoxyuridine or methotrexate, show a failure to completely deplete dNTP's as expected from the known enzymatic targets for these agents. Yet other studies point to DNA replication as a metabolic process critically affected by these agents. Representative of these studies, and perhaps the most fully developed, are investigations of megaloblastic anemia by Wickremasinghe, Hoffbrand, and colleages [50, 56, 57]. Whether caused by deficiency of folate or of vitamin B_{12}, anemias of this type are thought to result from a deficit of thymine nucleotides. Paradoxically, measured dTTP levels are normal in peripheral lymphocytes of patients, yet DNA fork movement rates are slowed, as expected if there were DNA precursor deficiencies. Again, these observations have been rationalized in terms of the idea that most of the dNTP's in a cell are not immediately available to support DNA replication, and that the deficiency in folate coenzymes affects primarily the much smaller replication-active dNTP pools. In support of this model, Wickremasinghe et al. [57] have described an enzyme aggregate in human lymphoblastoid cells which has some of the properties expected of a dNTP-DNA synthesis complex.

Several years ago I followed thymine nucleotide pool kinetics in T4 phage-infected bacteria as a function of time after reversible inhibition of DNA replication in infection by a T4 mutant with a temperature-sensitive mutation affecting DNA replication [31]. The results, summarized in Fig. 2, showed that dTDP accumulated almost as rapidly as did dTTP. This was surprising, because in extracts the activity of nucleoside diphosphokinase, which converts dTDP to dTTP, was many-fold higher than that of dTMP kinase. Nucleoside diphosphate kinase is encoded by the host genome, and I suggested that only a small fraction of this enzyme is available, via complex formation with virus-coded enzymes, to participate in dNTP synthesis. Subsequent experiments with cell-free enzyme aggregates [1, 40, 43] supported this idea.

One of the most provocative experiments in this area was described by Manwaring and Fuchs [29], who studied an *E. coli* mutant with a temperature-sensitive *nrdA* mutation, affecting one of the subunits of rNDP reductase. *A priori* one would expect DNA replication to continue after a temperature upshift of a culture, until the supply of the critical dNTP was exhausted. In fact, DNA replication was inhibited almost immediately, even though measured dNTP pools did not decrease significantly for at least one hour. Again, these observations were interpreted in terms of relative inaccessibility to replication sites of most of the dNTP's present within a cell. Inactivation of rNDP reductase would affect primarily the replenishment of the replication-active pool, since this is the one presumably generated at replication sites.

2.4. DNA Replication in vitro

In vitro replication systems, both permeabilized cells and cell-free preparations, have been used in a variety of ways to study DNA substrate channeling. One of the first observations was made both in Greenberg's laboratory and in mine [37, 60], in studies on sucrose-plasmolyzed T4-infected bacteria. When infection was carried out with a gene 42 mutant, defective in dCMP hydroxymethylase production, the expectation was that this would not affect DNA synthesis in the permeabilized cells, since the metabolic block would be

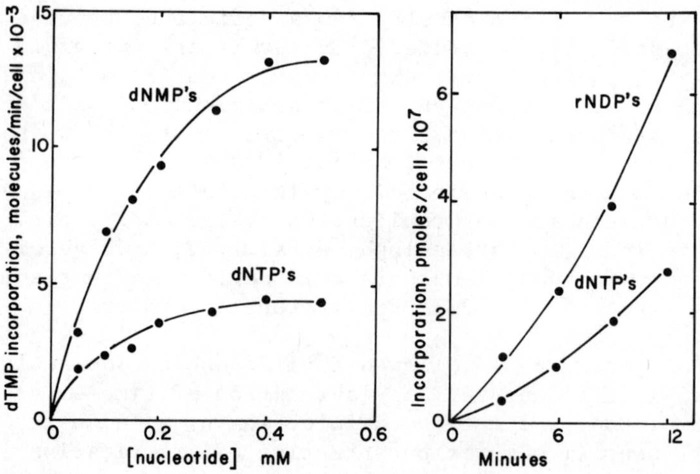

Fig. 3. Preferential utilization of distal precursors for DNA replication in T4 phage-infected E. coli rendered permeable by sucrose plasmolysis. Data are from reference 40.

circumvented by provision of the affected precursor, hm-dCTP. However, DNA replication was defective under these conditions in vitro, even when the initial infection had been carried out at permissive temperature with a ts gene 42 mutant [37]. Our interpretation, subsequently supported by kinetic evidence with a cell-free enzyme aggregate [40], was that inactivation of this enzyme functionally inactivated the complex, and that only those precursors generated in situ were readily available to support DNA replication.

Both Greenberg's laboratory and ours have exploited the useful sucrose plasmolysis system in different ways. Wovcha et al. [59] obtained evidence for two different channeled pathways to DNA in these permeabilized infected cells. One such pathway was readily labeled by nucleosides and seemed to represent monitoring of ongoing replication from de novo-synthesized nucleotides in this system. The other pathway used added dNTP's, occurred at a slower rate, and was assumed to represent the pathway which utilizes host cell DNA breakdown products. Both pathways represented semiconservative replication and were dependent upon T4 replication gene products. Since T4 is unusual in having these two major routes to DNA precursor synthesis, this two-pathway phenomenon is unlikely to be seen in other systems.

In our laboratory Prem Reddy [40] found that plasmolyzed T4-infected cells use distal DNA precursors, either dNMP's or rNDP's, more efficiently for DNA synthesis than do they use dNTP's (Fig. 3).

In fact, dNTP's rapidly break down in this system and must be rephosphorylated before they can be incorporated into DNA. When dNMP's were used as DNA precursors, incorporation occurred at substantial rates, even though no detectable dTTP was formed in these cells. Since intracellular protein-protein associations are presumed to be maintained in these permeabilized cells, we interpret these observations in terms of (1) preferential utilization of distal DNA precursors, which are not compartmentalized, because they are accessible to catalytic sites on a multienzyme complex; (2) obligatory breakdown of added dNTP's, probably to dNMP's, before they can be utilized by the complex and shuttled to replication sites; and (3) very small size of the replication-active dNTP pools, such that incorporation _in situ_ can occur even when total dNTP levels are too low to measure.

Although the replication-active pools may be small in terms of molecules per cell, there is reason to believe that effective concentrations are high. The T4 purified protein DNA synthesis system of Alberts and colleagues requires at least 220 μM dTTP in order to elongate DNA chains at maximal rates [33]. Yet DNA replication rates _in vivo_ reach their maximal rates when the average intracellular dTTP concentration is only 65 μM [33]. Note from Fig. 2 that the rate of DNA replication was not greatly increased following a huge accumulation of thymine nucleotides, suggesting that replication forks are saturated, or nearly saturated, with dNTP's _in vivo_. Thus, it appears that proteins in the vicinity of replication forks must be capable of maintaining concentration gradients of DNA precursors.

In Pardee's laboratory, Reddy has now carried out comparable experiments with permeabilized mammalian cells [42]. This system incorporates rNDP's into DNA up to 1.5-fold more efficiently than dNTP's. That this represents channeled synthesis is suggested first by the fact that deoxycytidine nucleotides, which must be intermediates, accumulate only to low levels during this process, and secondly by the failure of 0.2 mM dCTP to significantly inhibit incorporation of label into DNA from radiolabeled CDP. This latter suggests that the labeled dCTP pool derived from CDP does not readily mix with the much larger amount of dCTP added as the nonradioactive material. However, dTTP added to this system does inhibit DNA labeling, presumably via allosteric inhibition of the reduction CDP by rNDP reductase. Recently Noguchi, Reddy, and Pardee [36] described rapid rNDP incorporation by a cell-free preparation, but they have not yet shown that this system uses rNDP's more readily than dNTP's.

2.5. Genetic Interactions

Two types of genetic investigations have contributed to our understanding of the relations between DNA precursor synthesis and

TABLE 1. Antimutator Phenotype of ts Gene 1 Mutation

	Number of Revertants/10^8 cells	
	gene 1 allele	
rII mutation	wild type	temperature-sensitive (JW5)
UV183	221	90
UV163	44	1.1
HB74	387	1.9
x655	1497	10.4

Double mutants, containing the ts gene 1 mutation and an rII mutation, were prepared, and reversion frequencies determined, as described by Williams and Drake (58).

DNA replication: first, observations from the T4 system which point to intracellular interactions closer than those observed for other precursor-macromolecular synthetic systems; and (2) the much wider arena, covered in the rest of this volume, which points to replication errors occurring as a result of dNTP pool imbalances.

Both Greenberg's and Drake's laboratories [10, 48] have described changes in spontaneous mutation frequency which result from ts mutations in gene 42. Since this gene codes for dCMP hydroxymethylase, these findings point directly to a link between DNA precursor synthesis and the replication machinery. A comparable conclusion results from the observation of Chao et al. [7] that a particular UGA mutation in gene 42 could be suppressed by certain gene 43 mutations. None of these observations can yet be explained mechanistically. Some gene 42 mutations are mutagenic, while others are antimutagenic [58]. Both transitions and transversions are seen, suggesting that pool alterations at replication sites cannot be solely responsible for the observed changes in spontaneous mutation frequencies. Recently our laboratory obtained from John Wiberg a T4 strain with a ts mutation in gene 1, which codes for another enzyme of dNTP synthesis, a deoxyribonucleoside monophosphate kinase. As shown in Table 1, this mutation exerts a strong antimutagenic effect against reversion of certain rII mutations. Although this does not yet increase our mechanistic insight, it adds to the idea that the enzymes of DNA precursor biosynthesis interact closely with the replication machinery.

As detailed elsewhere in this volume, mutations in animal cells which alter gross dNTP pools also increase spontaneous mutation frequencies at several loci. Of particular interest are the cell lines developed by Meuth [34], where pool imbalances may result from alterations in regulation of CTP synthetase; and those described by Weinberg et al. [54], where imbalances result from altered regulation of either dCMP deaminase or ribonucleotide reductase. Relevant to this discussion is the fact that spontaneous mutation frequencies in these cell lines are often elevated more than are alterations in total dNTP pools, and hence, more than expected if mass action effects during replication represented the sole link between dNTP pools and mutation. The situation is obviously too complex to analyze until one can identify specific molecular changes accompanying mutagenesis. However, the discrepancies between pool alterations and mutation frequencies might relate to replication-active dNTP pool compositions which are quite different from those determined in total pool measurements.

2.6. Isolated Multienzyme Aggregates

Of course the ultimate test of the functional compartmentation model must be isolation of a complex containing dNTP-synthesizing activities physically and kinetically linked both to one another and to enzymes and proteins of DNA replication. Such a complex must display the attributes expected from the model as adduced from experiments with intact or permeabilized cells: reduced transient time for catalysis of a multistep pathway, low steady-state levels of intermediates, resistance to dilution by added intermediates, and regulation comparable to that observed *in vivo*. Parts of these goals have been achieved, but much remains to be done.

In 1973 Baril et al. [4] described a membrane fraction from animal cells which contained DNA polymerase, rNDP reductase, and thymidine kinase, and they suggested the existence of a precursor-replication complex. The first kinetic evidence came from the T4 system, when we found that gently lysed infected cells yield an aggregate containing most of the enzymes of dNTP biosynthesis [43]. These activities were kinetically coupled, as shown by catalysis of multi-step reaction sequences *in vitro* with little or no detectable lag between addition of initial substrate and formation of product at its maximal rate, and accumulation of intermediates to levels far below those expected if the enzymes were not coupled. Analysis of one such sequence, dUMP → dTMP → dTDP → dTTP, suggested that intermediates were maintained at active sites some one to two orders of magnitude more concentrated than their average levels in the reaction mixture. The longest sequence for which coupling has been demonstrated is the six-step pathway UDP → dUDP → dUTP → dUMP → dTMP → dTDP → dTTP [2]. Figure 3 shows that this activity sedimented in a sucrose gradient with the aggregate, at some 15-20S (representing a molecular weight of about one million). The slowly sedimenting fraction of ribonucleotide reductase, which probably represents the host cell enzyme, displays negligible kinetic linkage to the other reactions. In more recent experiments [1] we have found that this aggregate demonstrates at least some normal regulatory behavior. When dCMP was presented as substrate, dTMP and hm-dCMP were synthesized in vitro at a ratio of 0.5:1, quite different from their relative rates of synthesis in vivo. However, when dCTP, an allosteric activator of dCMP deaminase, was added, the ratio changed to 2:1, very close to the in vivo rate as determined by Chiu et al. [11].

In our hands the enzymes of dNTP biosynthesis are relatively closely associated and remain so through several fractionation steps: differential centrifugation, isoelectric focusing, gel filtration, and dye-ligand affinity chromatography. However, we have not yet been able to clearly demonstrate either physical or kinetic association to DNA polymerase or to other proteins of DNA replication. Recently Chiu et al. [9] described the isolation of an aggregate, also from T4-infected bacteria, which also exhibits channeling behavior in vitro. Their preparation seems to be somewhat larger than ours and also contains DNA polymerase, T4 topoisomerase, and gene 32 protein. However, these proteins have not yet been shown to be bound to dNTP-synthesizing enzymes in the preparation.

As stated earlier, two laboratories have isolated dNTP-synthesizing enzyme aggregates from animal cells [36, 41, 57], and available data suggest that they represent biologically significant complexes. The Noguchi-Reddy-Pardee preparation [36, 41] evidently becomes assembled in nuclei just at the onset of S phase in synchronized cells. As isolated the preparation contains both nascent and template DNA, as well as rNDP reductase, thymidylate synthase, DNA polymerase, and several other enzymes. Although not highly purified, the preparation shows spherical particles in the electron microscope whose dimensions suggest an aggregate molecular weight of about five million.

A fascinating extension of the work with mammalian cell complexes has recently been described by Ayusawa et al. [3]. These workers transformed, using human DNA, a thymidylate synthase-negative cell line of FM3A mouse cells. The thymidine prototrophs were found to have acquired human thymidylate synthase. However, this enzyme was not found in a dNTP-synthesizing aggregate when gentle lysates were analyzed by sedimentation, whereas the corresponging mouse enzyme in wild-type cells was found in this aggregate. This suggests both that specific protein-protein interactions are involved in assembly of this complex, and also that complete integrity of the complex is not essential for viability of a cell.

What factors are involved in the genesis of kinetically coupled enzyme aggregates? Mosbach's laboratory has approached this question by chemically cross-linking two enzymes catalytizing sequential reactions - malate dehydrogenase and citrate synthase [25]. Presumably the enzymes were linked in random orientation to each other. A kinetic advantage from this coupling was not seen when these artificial aggregates were in solution, but was seen when the aggregates

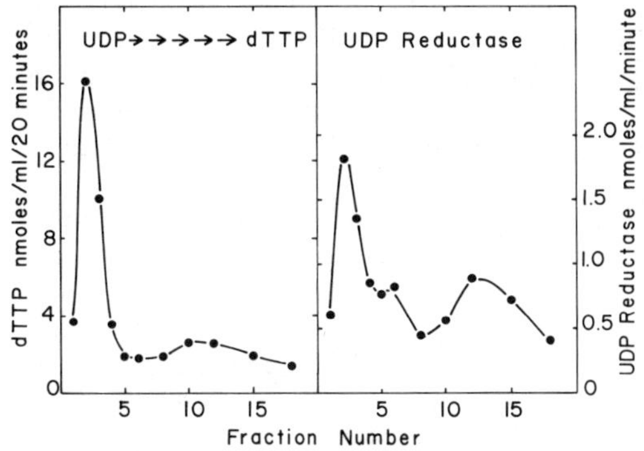

Fig. 4. Sedimentation through a 5-20% sucrose gradient of a kinetically coupled enzyme aggregate. Sedimentation is from right to left. Data are from reference 2. The left-hand panel depicts analysis of the multi-step conversion of UDP to dTTP, while the right-hand panel shows sedimentation behavior of ribonucleotide reductase alone, assayed with UDP as substrate.

were immobilized on Sephadex beads. This suggests that both physical proximity of catalytic sites and a favorable microenvironment are essential features contributing to what is probably a widespread biological phenomenon. However, much more must be involved in generating the ability to catalyze coupled long, complex, and regulated pathways. Fulton [22] has suggested that the ability of specific proteins to interact with each other is highly selected evolutionarily and strongly favored within the intracellular environment.

3. DISCUSSION; RELATIONS TO REPLICATION FIDELITY

The accumulated evidence indicates that DNA precursor pools are compartmentalized, even in prokaryotic systems where membrane-limited compartments probably do not exist. Whether that compartmentation is of the type described here - minute, highly concentrated, rapidly-turning-over pools at replication sites, and much larger, more dilute, and slowly replenished pools elsewhere in the cell - remains to be established. Indirect experimental approaches will continue to be important because of the virtual impossibility of physically isolating the putative replication-active pool.

The extent to which the presumed two pools can mix with each other is an important subject for more detailed study. Some mixing must occur, because I do not know of any case where DNA replication is completely inhibited in the presence of measurable intracellular pools. Also, note from Fig. 2 that the expanded dTDP and dTTP pools formed during inhibition of DNA synthesis return to normal within six minutes after restoration of DNA synthesis, presumably by being used for DNA replication. Thus, it appears that the pools can slowly mix with one another under some conditions, but that they cannot equilibrate during normal DNA replication.

The question of the effective concentrations of dNTP's at replication sites is another one which will probably be impossible to answer by direct methods. As Fulton [22] has pointed out, cells consist of two phases of water - a bulk phase and an ordered phase, which represents water of hydration around proteins. Since intracellular protein concentrations are extremely high - 17 to 26% by weight - the ordered phase represents a large fraction of the water, and the kinetic behavior of enzymes in dilute solution may provide little or no help in understanding how metabolic intermediates behave in an environment consisting mostly of ordered water. Nevertheless, it does appear from evidence now available that effective dNTP levels are increased at replication sites, whatever the concentrations in chemical terms. How might this phenomenon affect the accuracy of biological DNA replication? It is certainly well known, from studies of DNA replication in vitro, that dNTP concentration imbalances can force base substitution errors in replication. As shown by Fersht [16], among others, there are at least two dNTP concentration-dependent steps affecting accuracy - (1) competition between correct and incorrect nucleotide for insertion, and (2) competition between editing 3' exonuclease activity, which removes a mispaired base, and DNA polymerase activity, which would insert the next nucleotide. If the effect of dNTP functional compartmentation is simply to increase the steady state levels of all four dNTP's near replication sites, then there may not be a general effect on replication fidelity in the global sense. However, it clearly behooves us to learn more about the nature of this functional compartmentation. The gene interaction effects seen with T4 and described in Section 2.5 indicate that the linkage between dNTP metabolism and replication fidelity is more complex than simple mass action relationships.

Finally, we should consider some similarities and differences between the two biological systems used for most of these investigations - T4 phage-infected bacteria and cultured mammalian cells. The functional compartmentation model was developed for T4 partly because it seemed that DNA precursors could not be supplied by simple diffusion to replication sites fast enough to account for observed rates of replicative chain elongation. Whereas that may be true for prokaryotes, it is probably not true in mammalian cells. First, as noted earlier, eukaryotic DNA chains elongate far more slowly, and at many more intracellular sites, than do prokaryotic DNA chains. Second, it appears from studies on DNA replication in vitro that mammalian replication complexes have much higher affinity for dNTP's than do those prepared from T4 phage-infected bacteria. Table 2

TABLE 2. Affinity of *In vitro* Replication Systems for dNTP's

Reference	System	[dNTP] µmoles/liter half-saturating	[dNTP] µmoles/liter minimal saturating
Mathews & Sinha, 1982[33]	Purified T4 replication proteins	35	220
Imae & Okazaki, 1976[23]	Cellophane disc system; T4-infected bacteria	35	200
Miller *et al*, 1973[35]	Toluene-treated T4-infected bacteria	∿200	>1000
Reddy & Mathews, 1978[40]	Sucrose-plasmolyzed T4-infected bacteria	80	400
Berger & Johnson, 1976[5]	Permeabilized L cells	12.5	n.r.
Francke, 1977[19]	HSV-infected BHK cell lysate	n.r.	10(dATP,dTTP) 50(dCTP,dGTP)
Fraser & Huberman, 1977[20]	HeLa cell lysate	n.r.	50
Eliasson & Reichard, 1979[14]	Polyoma-infected 3T6 cell nuclei	0.1(dATP) 0.2(dGTP) 0.3(dTTP) 0.4(dCTP)	5.0
Reddy & Pardee, 1982[42]	Permeabilized Chinese hamster embryo fibroblasts	10	80

n.r., not reported

shows some of these data. Finally, there is the potential for physical compartmentation of DNA precursors in nuclei, mitochondria, and chloroplasts of eukaryotic cells. Indeed, we have recently shown that HeLa cells contain mitochondrial dNTP pools which are distinct from the bulk cellular pools in several respects [6]. The more important question of the extent to which nuclei can physically compartmentalize dNTP's is harder to answer with current methods because of the relative porosity of the nuclear membrane. One report [46], from 1974, indicates that dNTP's do become concentrated in nuclei during S phase. Those results require confirmation and extension by other methods. Even if confirmed, however, the question will remain whether this concentration in the nucleus results primarily from selective permeability of the nuclear membrane or from the action of, presumably, thousands of dNTP-synthesizing, DNA replication complexes within each nucleus. Further investigation of this apparent physical compartmentation might yield insight into cell cycle control, as well as into determinants of the accuracy of DNA replication.

ACKNOWLEDGMENTS

Research in our laboratory is supported by NIH research grants AI-08230 and CA-24323. I thank many students and associates, who are identified as coauthors on our cited papers, for their contributions to our work.

REFERENCES

1. J. R. Allen, G. W. Lasser, D. A. Goldman, J. W. Booth, and C. K. Mathews, T4 Phage deoxyribonucleotide-synthesizing enzyme complex, J. Biol. Chem., 258:5746-5753 (1983).
2. J. R. Allen, G. P. V. Reddy, G. W. Lasser, and C. K. Mathews, T4 Ribonucleotide reductase. Physical and kinetic linkage to other enzymes of deoxyribonucleotide biosynthesis, J. Biol. Chem., 255:7583-7588 (1980).
3. D. Ayusawa, K. Shimizu, H. Koyama, K. Takeishi, and T. Seno, Unusual aspects of human thymidylate synthase in mouse cells introduced by DNA-mediated gene transfer, J. Biol. Chem., 258: 48-53 (1983).
4. E. Baril, B. Baril, H. Elford, and R. B. Luftig, Aggregated mammalian enzymes in deoxyribonucleotide and DNA replication, in: "Mechanism and Regulation of DNA Replication" (A. R. Kolber and M. Kohiyama, eds.), pp. 275-291, Plenum Press, New York (1973).
5. N. A. Berger and E. S. Johnson, DNA synthesis in permeabilized mouse L cells, Biochim. Biophys. Acta, 425:1-17 (1976).
6. R. K. Bestwick, G. L. Moffett, and C. K. Mathews, Selective expansion of mitochondrial nucleotide pools in antimetabolite-treated HeLa cells, J. Biol. Chem., 257:9300-9304 (1982).

7. J. Chao, M. Leach, and J. Karam, In vivo functional interaction between DNA polymerase and dCMP hydroxymethylase of bacteriophage T4, J. Virol., 24:557-563 (1977).
8. M. S. Chen, J. Walker, and W. H. Prusoff, Kinetic studies of herpes simplex virus type 1-encoded thymidine and thymidylate kinase, a multifunctional enzyme, J. Biol. Chem., 254:10747-10753 (1979).
9. C.-S. Chiu, K. S. Cook, and G. R. Greenberg, Characteristics of a bacteriophage T4-induced complex synthesizing deoxyribonucleotides, J. Biol. Chem., 257:15087-15097 (1982).
10. C.-S. Chiu and G. R. Greenberg, Mutagenic effects of temperature-sensitive mutants of gene 42 of bacteriophage T4, J. Virol., 12:199-201 (1973).
11. C.-S. Chiu, T. Ruettinger, J. B. Flanegan, and G. Greenberg, Role of deoxycytidylate deaminase in deoxyribonucleotide synthesis in bacteriophage T4 DNA replication, J. Biol. Chem., 252:8603-8608 (1977).
12. R. I. Christopherson and M. E. Jones, The overall synthesis of L-5,β-dihydroorotate by multienzymatic protein pyrol-3 from hamster cells, J. Biol. Chem., 255:11381-11395 (1980).
13. J. C. Drach, M. A. Thomas, J. W. Barnett, S. H. Smith, and C. Shipman, Jr., Tritiated thymidine incorporation does not measure DNA synthesis in ribavarin-treated human cells, Science, 212:549-551 (1981).
14. R. Eliasson and P. Reichard, Replication of polyoma DNA in isolated nuclei, J. Mol. Biol., 129:393-409 (1979).
15. S. Eriksson, Ribonucleotide reductase from E. coli: Demonstration of a highly active form of the enzyme, Eur. J. Biochem., 56:289-294 (1975).
16. A. R. Fersht, Fidelity of replication of phage ϕX174 DNA by DNA polymerase III holoenzyme: spontaneous mutation by misincorporation, Proc. Natl. Acad. Sci. USA, 76:4946-4950 (1979).
17. J. B. Flanegan, C.-S. Chiu, and G. R. Greenberg, Inhibitory effect of agents altering the structure of DNA on the synthesis of pyrimidine deoxyribonucleotides in bacteriophage T4 DNA replication, J. Biol. Chem., 252:6031-6037 (1977).
18. J. B. Flanegan and G. R. Greenberg, Regulation of deoxyribonucleotide biosynthesis during in vivo bacteriophage T4 DNA replication, J. Biol. Chem., 252:3019-3027 (1977).
19. B. Francke, Cell-free synthesis of herpes simplex virus DNA: Conditions for optimal synthesis, Biochemistry, 16:5655-5664 (1977).
20. J. M. K. Fraser and J. A. Huberman, In vitro HeLa cell DNA synthesis similarity to in vivo replication, J. Mol. Biol., 117: 249-272 (1977).
21. A. Fridland, DNA precursors in eukaryotic cells, Nature, 243: 105-107 (1973).
22. A. B. Fulton, How crowded is the cytoplasm? Cell, 30:345-347 (1982).
23. Y. Imae and R. Okazaki, Replication of bacteriophage T4 DNA in vitro, J. Virol., 19:435-445 (1976).

24. D. E. Kizer and B. A. Howell, Evidence that incorporation of inorganic [^{32}P]phosphate into DNA of regenerating rat liver is stimulated 6 h earlier than similar incorporation of labeled thymidine, Biochim. Biophys. Acta, 561:276-293 (1979).
25. A. Koch-Schmidt, B. Mattiasson, and K. Mosbach, Aspects on microenvironmental compartmentation, Eur. J. Biochem., 81:71-78 (1977).
26. A. Kornberg, DNA Replication, W. H. Freeman and Co., San Francisco (1980).
27. D. Kuebbing and R. Werner, A model for compartmentation of de novo and salvage thymidine nucleotide pools in mammalian cells, Proc. Natl. Acad. Sci. USA, 72:3333-3336 (1975).
28. C. A. Lunn and V. Pigiet, Characterization of a high activity form of ribonucleoside diphosphate reductase from E. coli, J. Biol. Chem., 254:5008-5014 (1979).
29. J. D. Manwaring and J. A. Fuchs, Relationship between deoxyribonucleoside triphosphate pools and DNA synthesis in an nrdA mutant of E. coli, J. Bacteriol., 138:245-248 (1979).
30. C. K. Mathews, Giant pools of DNA precursors in sea urchin eggs, Exptl. Cell Res., 92:47-56 (1975).
31. C. K. Mathews, Biochemistry of DNA-defective mutants of bacteriophage T4. Thymine nucleotide pool dynamics, Arch. Biochem. Biophys., 172:178-187 (1976).
32. C. K. Mathews, T. W. North, and G. P. V. Reddy, Multienzyme complexes in DNA precursor biosynthesis, Adv. Enz. Regul., 17:133-156 (1979).
33. C. K. Mathews and N. K. Sinha, Are DNA precursors concentrated at replication sites?, Proc. Natl. Acad. Sci. USA, 79:302-306 (1982).
34. M. Meuth, Sensitivity of a mutator gene in Chinese hamster ovary cells to deoxyribonucleoside triphosphate pool alterations, Mol. Cell. Biol., 1:652-660 (1981).
35. R. C. Miller, D. M. Taylor, K. MacKay, and H. W. Smith, Replication of T4 DNA in E. coli treated with toluene, J. Virol., 12:1195-1203 (1973).
36. H. Noguchi, G. P. V. Reddy, and A. B. Pardee, Rapid incorporation of label from ribonucleoside diphosphates into DNA by a cell-free high molecular weight fraction from animal cell nuclei, Cell, 32:443-451 (1983).
37. T. W. North, M. E. Stafford, and C. K. Mathews, Biochemistry of DNA-defective mutants of bacteriophage T4. VI. Biological functions of gene 42, J. Virol., 17:973-982 (1976).
38. M. L. Pato, Alterations of deoxyribonucleoside triphosphate pools in E. coli: Effects on DNA replication and evidence for compartmentation, J. Bacteriol., 140:518-524 (1979).
39. J. L. Paukert, L. D. Straus, and J. C. Rabinowitz, Formyl-methenyl-methylenetetrahydrofolate synthetase-(combined), J. Biol. Chem., 251:5104-5111 (1976).
40. G. P. V. Reddy and C. K. Mathews, Functional compartmentation of DNA precursors in T4 phage-infected bacteria, J. Biol. Chem., 253:3461-3467 (1978).

41. G. P. V. Reddy and A. B. Pardee, Multienzyme complex for metabolic channeling in mammalian DNA replication, Proc. Natl. Acad. Sci. USA, 77:3312-3316 (1980).
42. G. P. V. Reddy and A. B. Pardee, Coupled ribonucleoside diphosphate reduction, channeling, and incorporation into DNA of mammalian cells, J. Biol. Chem., 257:12526-12531 (1982).
43. G. P. V. Reddy, A. Singh, M. E. Stafford, and C. K. Mathews, Enzyme associations in T4 phage DNA precursor synthesis, Proc. Natl. Acad. Sci. USA, 74:3152-3156 (1977).
44. W. Rode, K. J. Scanlon, B. A. Moroson, and J. R. Bertino, Regulation of thymidylate synthetase in mouse leukemia cells (L1210), J. Biol. Chem., 255:1305-1311 (1980).
45. F. E. Scott and D. R. Forsdyke, Isotope-dilution analysis of the effects of deoxyguanosine and deoxyadenosine on the incorporation of thymidine and deoxycytidine by hydroxyurea-treated thymus cells, Biochem. J., 190:721-730 (1980).
46. L. Skoog and G. Bjursell, Nuclear and cytoplasmic pools of deoxyribonucleoside triphosphates in Chinese hamster ovary cells, J. Biol. Chem., 249:6434-6438 (1974).
47. L. Skoog and B. Nordenskjöld, Effects of hydroxyurea and 1-β-D-arabinofuranosylcytosine on deoxyribonucleotide pools in mouse embryo cells, Eur. J. Biochem., 19:81-89 (1971).
48. G. K. Smith, W. T. Mueller, G. F. Wasserman, W. D. Taylor, and S. J. Benkovic, Characterization of the enzyme complex involving the folate-requiring enzymes of de novo purine biosynthesis, Biochemistry, 19:4313-4321 (1980).
49. P. S. Srere and R. W. Estabrook (eds.), Microenvironments and Metabolic Compartmentation, Academic Press, New York (1978).
50. M. R. Taheri, R. G. Wickremasinghe, and A. V. Hoffbrand, Alternative metabolic fates of thymine nucleotides in human cells, Biochem. J., 194:451-461 (1981).
51. P. K. Tomich, C.-S. Chiu, M. G. Wovcha, and G. R. Greenberg, Evidence for a complex regulating the in vivo activities of early enzymes induced by bacteriophage T4, J. Biol. Chem., 249:7613-7622 (1974).
52. T. W. Traut and M. E. Jones, Inhibitors of orotate phosphoribosyltransferase and orotidine 5'-phosphate decarboxylase from mouse Ehrlich ascites cells, Biochem. Pharm., 26:2291-2296 (1977).
53. H. R. Warner, Properties of ribonucleoside diphosphate reductase in nucleotide-permeable cells, J. Bacteriol., 115:18-22 (1973).
54. G. Weinberg, B. Ullman, and D. W. Martin, Jr., Mutator phenotypes in mammalian cell mutants with distinct biochemical defects and abnormal deoxyribonucleoside triphosphate pools, Proc. Natl. Acad. Sci. USA, 78:2447-2451 (1981).
55. R. Werner, Nature of DNA precursors, Nature, 233:99-103 (1971).
56. R. G. Wickremasinghe and A. V. Hoffbrand, Reduced rate of DNA replication fork movement in megaloblastic anemia, J. Clin. Invest., 65:26-36 (1980).

57. R. G. Wickremasinghe, J. C. Yaxley, and A. V. Hoffbrand, Solubilization and partial characterization of a multienzyme complex of DNA synthesis from human lymphoblastoid cells, Eur. J. Biochem., 126:589-596 (1982).
58. W. E. Williams and J. W. Drake, Mutator mutations in bacteriophage T4 gene 42, Genetics, 86:501-511 (1977).
59. M. G. Wovcha, C.-S. Chiu, P. K. Tomich, and G. R. Greenberg, Replicative bacteriophage DNA synthesis in plasmolyzed T4-infected cells, J. Virol., 20:142-156 (1976).
60. M. G. Wovcha, P. K. Tomich, C.-S. Chiu, and G. R. Greenberg, Direct participation of dCMP hydroxymethylase in synthesis of bacteriophage T4 DNA, Proc. Natl. Acad. Sci. USA, 70:2196-2200 (1973).

A POSSIBLE SECONDARY ROLE FOR THYMINE-CONTAINING DNA PRECURSORS

Robert J. Melamede and Susan S. Wallace

Department of Microbiology
New York Medical College
Valhalla, New York 10595

1. INTRODUCTION

Multienzyme complexes are involved in DNA replication and have been described in both uninfected [50] and T4-infected [59] Escherichia coli. These complexes comprise a variety of proteins including DNA binding, unwinding, priming, and polymerizing activities. The existence of multienzyme complexes for generating DNA precursors and/or channeling such DNA precursors has also been demonstrated in several systems [41, 52, 63, 65, 80, 96, 97] including T4-infected cells [4, 5, 20, 21, 38, 66, 71, 73, 74, 94, 95, 103, 106, 115, 116]. The presence of a DNA precursor complex greatly reduces the diffusion of sequentially synthesized reactants and results in an efficient funneling of DNA precursors to the replication apparatus [94]. The nature of the coupling between DNA precursor production and utilization during DNA synthesis has not yet been established.

Chiu and Greenberg [19] first suggested that there might be a tight coupling between precursor production and DNA replication. While characterizing dCMP hydroxymethylase mutants (gene 42) of T4, they observed that some mutants could not synthesize DNA even though cell extracts from these mutants contained dCMP hydroxymethylase activity. A more formal link between precursor production and DNA replication in T4-infected cells was demonstrated by the isolation of gene 42 suppressor mutants that mapped in the T4 DNA polymerase gene [18]. This link was further strengthened when Manoil et al. [64] isolated DNA/protein complexes from T4-infected cells that contained dCMP hydroxymethylase in addition to many T4 replication proteins. Recently, DNA precursor complexes have been isolated from T4-infected cells and were partially characterized in both Mathews' [4, 5, 66, 94, 95] and Greenberg's [19] laboratories. Both groups found the complexes to contain ribonucleotide reductase, dCMP hydroxymethylase, thymidylate synthetase and the required kinase activities for the synthesis of pyrimidine deoxynucleoside triphosphates.

Precursor complexes do not appear to be unique to T4-infected cells. In E. coli [63], a "high activity form" of ribonucleotide reductase has been isolated after gentle lysis of spheroplasts. The "high activity form" of ribonucleotide reductase was found associated with thymidylate synthetase and DNA polymerase activities in a rapidly sedimenting fraction. Further, temperature sensitive mutants in the ribonucleotide reductase gene exhibited an immediate cessation of DNA synthesis when the temperature was raised to nonpermissive levels [39]. Manwaring and Fuchs [65] recently demonstrated in a ribonucleotide reductase mutant that, under nonpermissive conditions, DNA synthesis stops without a concomitant decrease in the nucleotide triphosphate pools. These observations indicate that DNA synthesis is coupled to precursor production not simply precursor utilization. Indirect evidence suggests that, in E. coli, the functioning of the sequential precursor producing enzymes may be coupled. For example, when ribonucleotide reductase activity was measured in crude extracts, the levels found were below those theoretically required for DNA synthesis [65]. Yet, when immological methods were used, 10-fold more enzyme (crossreacting material) was found than could be enzymatically detected [35]. An interesting but unexplained phenomenon was observed in an early study by Biswas et al. [13] who failed to find detectable levels of ribonucleotide reductase activity in a strain that lacked thymidylate synthetase when UDP was used as a substrate, although, reductase activity was observed with ADP, CDP, and GDP. Perhaps the reduction of UDP is coupled with the utilization of dUMP by thymidylate synthetase.

Precursor complexes have also been reported in mammalian systems [80, 96, 97]. Most recently Noguchi et al. [80] have isolated a sedimentable nuclear fraction from Chinese hamster embryo fibroblasts that catalyzes the incorporation of CDP into DNA.

While attempting to characterize the T4 recombination repair defective mutants, uvsX, uvsY, and uvsW in vivo [69-74], we observed a number of anomalies with respect to thymidine incorporation. We felt that the explanation for these results might lie in a relationship between precursor complexes and recombination specific DNA synthesis [71]. Further studies on incorporation of thymine-containing DNA precursors in plasmolyzed T4-infected cells [73, 74] as well as in uninfected E. coli, have convinced us that there is a high affinity, low capacity pathway by which thymidine has unique and preferential access to the replication fork and that this pathway is related to the DNA precursor complex.

A number of additional fundamental questions may now be asked. How are precursors complexes linked to leading and lagging strand DNA synthesis? Are there separate complexes for leading and lagging strands or might only one strand, presumably the leading strand, be more tightly coupled to precursor production? What advantage might a cell gain from being able to feed differentially leading or lagging strands? What is the significance of the widespread occurrence of thymidine kinase? We shall attempt to answer some of these questions and link our findings on the incorporation of various thymine-containing DNA precursors, both in vivo and in situ, to a general model that describes a secondary role for pyrimidine DNA precursors.

The model we present is based on variable precursor availability and precursor "push" to leading or lagging strands. A chemical network is described that, in the case of E. coli, may serve as the detector of genetic damage and as the effector of homeostatic responses such as SOS induction. In addition, we feel that the network we describe may represent a dissipative structure [79, 89]. The synthesis of DNA is a condition far from equilibrium. Precursors are produced and utilized under nonequilibrium conditions where the manifestation of self organizational principles would be expected [79, 89]. We shall discuss some of the implications of nonequilibrium thermodynamic principles for DNA synthesis and DNA precursor production. We recognize that other points of view, in particular those involving precursor transport and/or catabolism, can be used to explain some of the observations considered in this paper. Because we are attempting to integrate a number of experimental results, within a new conceptual framework we have chosen, for the sake of clarity, not to include these alternatives.

2. MATERIALS AND METHODS

2.1. Phage Strains

T4+ (T4D) was supplied by A. H. Doermann, T4 uvsX by W. Harm, and T4 uvsW by N. Hamlett D. Hall provided the T4 thymidylate synthetase mutant (T4 Do td8) and the T4 thymidine kinase mutant (T4 Do tk1). The T4 DNA binding protein amber mutant in gene 32 (am A853), the T4 helicase amber mutant in gene 41 (am N81) and the T4 primase amber mutant in gene 58-61 (am E21g) were supplied by G. Mosig. The T4 td uvsX double mutant was constructed in our laboratory by J. DeVries.

2.2. Bacterial Strains

The strain of E. coli B used in our laboratory was originally supplied by A. H. Doermann. The E. coli uracil DNA N-glycosylase mutant, ung (BW 271) and its parent strain (BW 273) supplied by H. Warner. The E. coli TAMT strain ($thy^- arg^- met^- trp^- drm^-$) was provided by D. Denhardt.

2.3. Chemicals

Methyl-^{3}H-thymidine was purchased from International Chemical and Nuclear. Methyl-^{3}H-thymine and ethyl-^{3}H-thymidine monophosphate and triphosphate were purchased from Amersham and Searle. ^{32}P labeled thymidine monophosphate was generously given to us by Amersham and Searle. All unlabeled bases, deoxynucleosides, deoxynucleoside monophosphates and triphosphates were purchased from P and L Biochemicals.

2.4. Plasmolysis

Both T4-infected and uninfected cells were plasmolyzed by a modification [73, 74] of the procedure of Stafford et al. [103]. Cultures were pelleted at a concentration of 2 × 10^8/ml and resuspended in fresh M9 medium at a concentration of 4 × 10^8/ml. T4 phage-infected cells were pelleted after 15 min of infection at 37°C. Pellets from 3 ml of infected or uninfected cells were resuspended in 20 μl plasmolysis solution (2 M sucrose, 40 mM Tris pH 8 and 10 mM EGTA). After vigorous vortexing, the plasmolyzed cells were left on ice for 45-50 min before use; shorter times gave inconsistent plasmolysis.

2.5. Incorporation of Labeled DNA Precursors

Reactions were initiated by mixing 5 μl plasmolyzed cells or cells resuspended in M9 medium (for in vivo measurements) with 25 μl of reaction mix. The reaction mix was the same as Stafford et al. [103] consisting of 32 mM Tris-HCl pH 8.4, 1.6 mM magnesium acetate, 400 mM sucrose, 64 mM KCl. The concentrations of thymine, thymidine, dTMP or dTTP (labeled and/or unlabeled) used are stated in the figure legends. Reactions were carried out at 37°C. Samples (5 μl) were spotted onto Whatman GF/A filter papers and immersed in 10% TCA. When necessary, reaction sizes and sample sizes were scaled up. The filters were batch-washed with 5% TCA followed by ethanol. Samples were counted in Liquiscint scintillation fluid.

2.6. Alkaline Sucrose Sedimentation Analysis of "Okazaki" Pieces

T4-infected cells were harvested 15 min after infection at 37°C. T4-infected or uninfected cells (3 ml) were centrifuged for 10 min in a Sorvall RC5B centrifuge using a S24 head. The pellet was resuspended in 20 μl of plasmolysis solution by vigorous vortexing. From combined pellet resuspensions, 10 μl of cells were added to 50 μl of reaction mix as indicated above. After incubation at 37°C for the times indicated in the figure legends, the reactions were terminated by icing and simultaneously adding 4 volumes of cold 0.3 N NaOH, 10 mM EDTA, 0.5% sodium lauryl sarcosinate. The mixture was left on ice for approximately 15 min and then layered on an alkaline

TABLE 1. Thymidine Incorporation in T4 Mutant-Infected Cells Relative to Wild Type

Concentration of Labeled Substrate	$T4^+$	T4*uvs*X	T4*uvs*W	T4*td*	T4*td* *uvs*X
1 μM ^{3}HTdR	1.0(1.0)*	0.32(0.5)	2.5(1.5)	(1.2)	(1.0)
1 μM ^{3}HTdR + Hydroxyurea	0.6	0.95	0.89	----	----
20 μM ^{3}HTdR	1.0	0.72	0.78	----	----
20 μM ^{3}HTdR + Hydroxyurea	0.22	0.22	0.22	----	----

*Data in ( ) refer to relative TdR incorporation in T4-infected plasmolyzed cells.

sucrose gradient (5–20% sucrose, 0.3 M NaoH, 10 mM EDTA, 0.5 M NaCl, 0.3% sodium lauryl sarcosinate) and centrifuged in polyallomer tubes in a Beckman SW41 rotor at 27,500 rpm for 15.5 h. Approximately 30 fractions were collected, precipitated with 10% TCA, and counted in Liquiscent scintillation fluid.

3. RESULTS AND DISCUSSION

3.1. Thymidine Incorporation in Cells Infected with the T4 Recombination Defective Mutants, uvsX and uvsW

Table 1 summarizes the thymidine incorporation in $T4^+$, T4 uvsX and T4 uvsW mutant-infected cells [70-74]. When high concentrations of thymidine (20 μM) were used to measure DNA synthesis, incorporation was reduced approximately 25% in the recombination defective mutants compared to wild type. When DNA synthesis was measured with high concentrations of thymidine in the presence of hydroxyurea, both wild type and the recombination-deficient mutant phages showed the same low level of incorporation [71]. Table 1 also shows the incorporation of 1 μM thymidine in wild-type and the recombination mutant-infected cells in the presence or absence of hydroxyurea. With low levels of exogenous thymidine, the incorproation patterns in cells infected with the recombination-deficient mutants are strikingly different from those infected with wild type; the incorporation of 1 μM thymidine in uvsW mutant-infected cells is much greater than with wild type whereas incorporation of 1 μM thymidine in uvsX mutant-infected cells is much less than with wild type. Furthermore, when 1 μM thymidine incorporation was measured in the presence of hydroxyurea, uvsW directed incorporation is decreased, wild type incorporation is only slightly decreased, but uvsX directed incorporation is increased. It can also be seen in Table 1 that the presence of a thymidylate synthetase mutation in the uvsX background, like hydroxyurea, suppresses the low thymidine incorporation phenotype of the uvsX mutant.

These data led to the postulate that DNA specific activity differences that result from the ready access of thymidine to the thymine-containing precursor flow to the replication fork account for the observed differences in incorporation of low concentrations of thymidine. Thus, the degree of incorporation at high concentrations of thymidine more likely reflects the true synthetic differences between the recombination-deficient mutants and wild type T4. [We have purposely avoided the use of the terms "pools" in order to underscore the dynamic properties associated with coupled chemical reactions.]

Within the framework of these considerations, the data of Table 1 can be reduced to the following question: How do the recombination repair defects of the T4 uvsX and uvsW mutants result in a 25% decrease in DNA synthesis as measured by incorporation of high concentrations of thymidine while, when incorporation is measured with low concentrations of thymidine, uvsX directed incorporation is decreased while not with uvsW directed incorporation is increased? We proposed [74] that both the uvsX and uvsW directed rates of incorporation at high levels of thymidine result from their decreased ability to undergo genetic recombination and its associated DNA synthesis. In the case of uvsW, the state of DNA synthesis appears somewhat comparable to that observed in cells infected with the T4 DNA delay mutants [68] where there are fewer replicating forks that move at a slightly faster rate. This increased rate of fork progression may be the result of fewer intramolecular recombinational events. Furthermore, Wirak and Greenberg [115] have shown that the DNA delay gene products must function for the DNA precursor complex to assemble. The effect would be a net reduction of effective endogenous precursors relative to fork movement. Hence, low concentrations of exogenous thymidine would be more readily incorporated in T4 uvsW mutant-infected cells than in wild type, and would account for the observed increased incorporation with 1 μM thymidine. In contrast, we proposed that in T4 uvsX mutant-infected cells, recombination is initiated but not completed resulting in a reduced rate of fork progression. This reduced rate of fork movement would create a surplus of endogenous thymine DNA precursors relative to fork movement. In this case, the incorporation of low levels of thymidine in uvsX mutant-infected cells would be diluted by the surplus endogenous supply and would result in the observed reduction of incorporation.

These explanations are critically dependent on the sequential functioning of the $uvsW^+$ and $uvsX^+$ gene products. Double mutants in genes 46 and uvsW do not exhibit the DNA arrest phenotype that characterizes mutants of gene 46. The recombination deficient phenotype however, remains [24]. Double mutants of gene 46 and uvsX express the DNA arrest phenotype [8]. The following sequence of events offers a possible explanation. The $uvsW^+$ gene product initiates a recombinational event that is dependent for its completion on the subsequent functioning of both the $uvsX^+$ and the wild type gene 46 products. In uvsW gene 46 double mutants, a component of wild type gene 46 recombination function is eliminated by the lack of uvsW initiation into the recombination pathway causing the gene 46 mutant DNA arrest phenotype to be suppressed. The uvsX mutation in contrast does not suppress the DNA arrest phenotype of gene 46 mutants because the pathway requiring the wild type gene 46 function has already been initiated. Thus, the gene product $uvsW^+$ appears to act ahead of the gene products of $uvsX^+$ and 46.

TABLE 2. Percent Incorporation of ³HTdR or d³HTMP in T4-Infected Plasmolyzed Cells

Concentration of Labeled Substrate	Concentration [μM] of Competing Unlabeled Precursor 50	100	200	400
	Competing dTMP			
T4⁺ 1 μM ³HTdR	96	108	126	----
T4⁺ 10 μM ³HTdR	89	107	93	93
T4⁺ 100 μM ³HTdR	91	138	115	89
	Competing TdR			
T4⁺ 1 μM d³HTMP	40	21	17	----
T4tk 1 μM d³HTMP (in tk⁺host)	21	21	21	----

The results observed with thymidine incorporation in the presence of hydroxyurea are similarly compatible with precursor feed effects. The proposed increased rate of fork movement relative to precursor production in T4 uvsW mutant-infected cells would limit precursors more severely than in either wild type- or T4 uvsX mutant-infected cells. Hydroxyurea, because of its effect on ribonucleotide reductase [11], would critically amplify this imbalance. Hence, the uvsW mutant-infected cells would be expected to show reduced rate of DNA synthesis in the presence of hydroxyurea (Table 1). T4 uvsW mutants are in fact sensitive to hydroxyurea when viability is measured [42].

In contrast, cells infected by uvsX mutants would have recombination blocks to fork movement and thus a relative abundance of precursors per replicating fork. The inhibition of precursor production by the addition of hydroxyurea or by the absence of thymidylate synthetase would decrease the production of precursors somewhat but not critically. Thus, the T4 uvsX mutants would not be expected to be sensitive to hydroxyurea. In fact, there is an increase in the incorporation of low concentrations of thymidine in T4 uvsX mutant-infected cells in the presence of hydroxyurea compared to incorporation in its absence (Table 1). This observation is in keeping with a reduction of the ratio of precursor production to fork movement which led to the proposed specific activity effects in untreated uvsX mutant-infected cells.

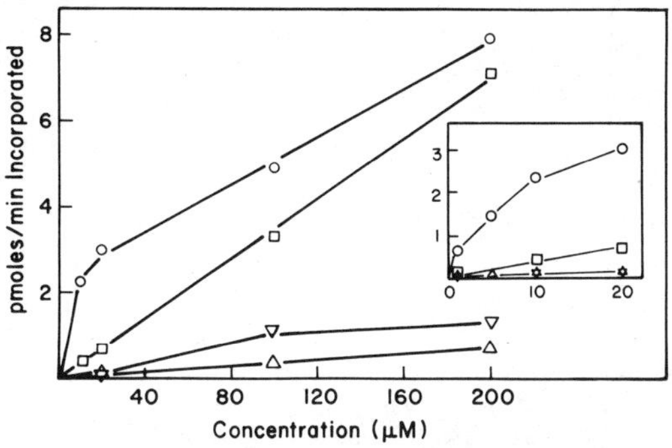

Fig. 1. The concentration dependence of the rate of incorporation of thymine-containing DNA precursors in wild type T4-infected plasmolyzed cells. Cells were grown, plasmolyzed and labeled as described in Materials and Methods. Incorporation rates were calculated using the linear portions of curves generated by the incorporation of increasing concentrations of tritiated: (▽) thymine; (○) thymidine; (□) thymidine monophosphate; (△) thymidine triphosphate. (Modified from Melamede and Wallace [73].)

3.2. Incorporation of Thymine-Containing DNA Precursors in Plasmolyzed T4-Infected Cells

The incorporation of various thymine-containing DNA precursors in plasmolyzed T4-infected cells is shown in Fig. 1. These results indicate that at low concentrations, thymidine is the preferred precursor. However, as precursor concentrations increase, dTMP incorporation approaches that of thymidine. Table 2 shows the results of competition experiments where either dTMP was used to compete with the incorporation of thymidine or thymidine was used to compete with the incorporation of dTMP. As can be seen, even 400 times the concentration of dTMP does not reduce thymidine incorporation. In contrast, thymidine is able to reduce the incorporation of dTMP. When dTMP incorporation was examined in T4tk mutant-infected hosts lacking thymidine kinase (Table 2), the levels of d^{3}HTMP incorporation were only 20% those observed with wild type phage-infected wild type hosts. Thus 80% of the incorporation of dTMP appears to reflect incorporation as thymidine while about 20% actually represents incorporation as dTMP.

Since thymidine appeared to be the preferred thymine-containing DNA precursor in T4-infected cells, not only in vivo but also in situ in plasmolyzed cells, we decided to examine whether thymidine was incorporated into continuous or discontinuous nascent DNA in plasmolyzed infected cells. Figure 2 shows, as has been demonstrated in vivo [82, 84], that in T4-infected plasmolyzed cells,

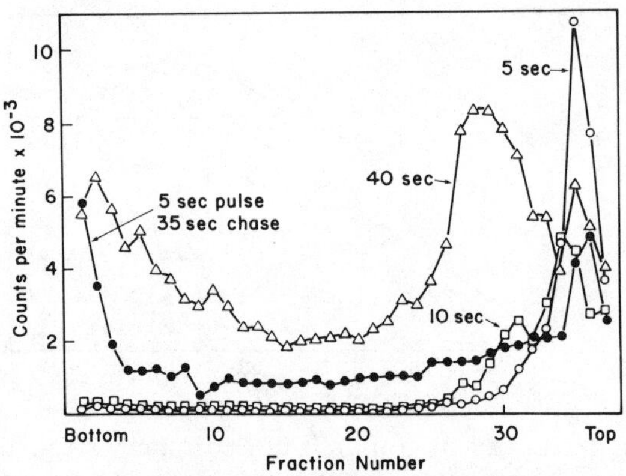

Fig. 2. Preferential incorporation of thymidine into short DNA pieces. T4-infected cells were grown, infected and plasmolyzed as described in Materials and Methods. The plasmolyzed cell suspension (10 μl) was diluted six-fold into 1 μM ^{3}H thymidine reaction mix and incubated for the indicated times at 37°C. The reactions were stopped by adding four volumes (240 μl) of cold 1 N NaOH and placing in ice. The entire 300 μl of lysed cells were layered on a 5-20% alkaline sucrose gradient (0.3 N NaOH, 0.01 M EDTA, 0.5 M NaCl) and centrifuged in a Beckman SW41 rotor at 27,500 rpm for 15.5 h. Symbols: (○) 5 sec, (□) 10 sec, (△) 40 sec, (●) 5 sec labeling followed by 35 sec incubation in cold 200 μM thymidine.

short pulses of thymidine are exclusively incorporated into Okazaki pieces and subsequently can be chased into larger ones. Thus, either DNA synthesis is discontinuous for both T4 strands or exogenously added thymidine does not have access to the leading strand. Leading strand synthesis can therefore not be readily measured.

3.3. Incorporation of Thymine-Containing DNA Precursors in Uninfected Plasmolyzed *E. coli*

Table 3 shows that when incorporation of 1 μM thymidine is compared to incorporation of 1 μM dTMP in the tritiated form, thymidine is incorporated at about three times the rate of dTMP. When incorporation of 1 μM thymidine is compared to dTM^{32}P incorporation, thymidine is incorporated at about 1000 fold greater rate than dTMP. In Table 3 it can also be seen that 100 μM thymidine can effectively block the incorporation of 1 or 100 μM d^{3}HTMP. In contrast, in the presence of 100 μM dTMP, significant incorporation of 1 μM thymidine occurs. When the data obtained with plasmolyzed T4-infected cells (Table 1) are compared with those from uninfected cells (Table 3), dTMP appears to reduce thymidine incorporation in uninfected cells

TABLE 3. Percent Incorporation of Labeled dTMP or TdR in Uninfected Plasmolyzed E. coli

Concentration of Labeled Substrate	Concentration [μM] Competing Unlabeled Precursor 0	10	100
	Competing TdR		
1 μM d^3HTMP	100 (0.0095)*	50	5
1 μM $dTM^{32}P$	(0.000036)		
10 μM d^3HTMP	100 (0.095)	60	5
100 μM d^3HTMP	100 (0.42)	61	12
	Competing dTMP		
1 μM ^3HTdR	100 (0.026)	67	33

*Data in ( ) refer to the rate of incorporation of labeled substrate in pmoles/min/sample.

to a greater extent than in infected cells. In addition, incorporation of 1 μM thymidine is about 14-fold less in uninfected cells than in T4-infected cells, whereas, incorporation of 1 μM $dTM^{32}P$ is about 50-fold reduced in uninfected cells compared with T4-infected cells. We offer the following interpretation. Exogenous thymidine is the preferred precursor in both T4-infected and uninfected cells. However, because of the high rate of DNA synthesis in T4-infected cells, replication is more precursor limited during T4 development [106, 115] than it is during normal E. coli growth. The effective concentrations of endogenous precursors at an E. coli replication fork would be functionally higher than at a T4 replication fork. Exogenous precursors therefore, would be less preferred in uninfected E. coli than in T4-infected cells. A relative abundance of precursors to fork movement in uninfected E. coli would minimize dTMP incorporation and maximize the ability of dTMP to compete with thymidine.

There are numerous examples in the literature of what might be pool related anomalies associated with thymine and thymidine incorporation, especially in thymine requirers [7, 12, 16, 17, 31, 32, 43, 69-74, 99, 110-112]. We therefore examined the effect of thymine on the incorporation of 1 μM thymidine in both wild type E. coli B and in the $thyA^-$ TAMT strain of E. coli K12. Preliminary data, de-

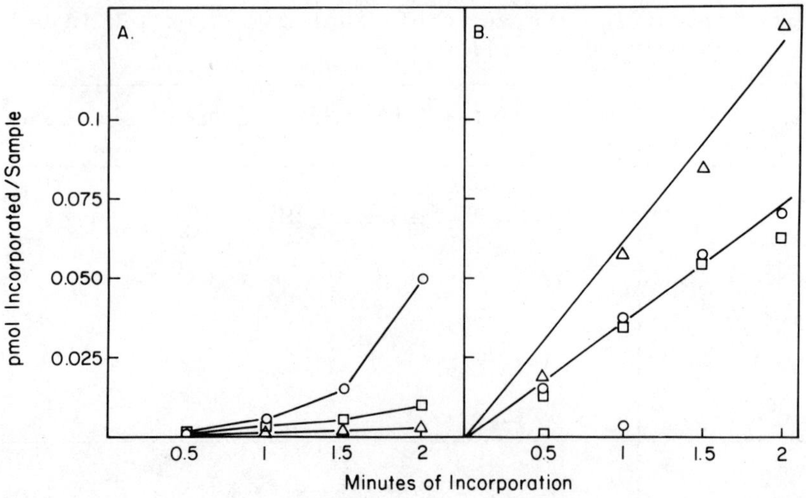

Fig. 3. The effects of thymine on the kinetics of 1 μM ^{3}H thymidine incorporation in E. coli B and E. coli TAMT in vivo. Cells were grown and labeled as described in Materials and Methods except that the TAMT strain was grown in M9 medium supplemented with 40 μg/ml each of tryptophan, arginine and methionine and 2 μg/ml of thymine. A. E. coli TAMT B. E. coli B. Symbols: (○) no added thymine; (□) plus 10 μM cold thymine; (△) plus 100 μM cold thymine.

picted in Fig. 3B, show that 1 μM thymine stimulates the incorporation of 1 μM thymidine in E. coli B. In contrast, in the TAMT strain (Fig. 3), 10 μM thymine almost completely inhibits 1 μM thymidine incorporation. In E. coli B, we assume the enchancing effect of thymine on thymidine incorporation to be due to the thymine-induced inhibition of thymidine phosphorylase [93] as well as to mass action effects. Thymine incorporation is known to be very poor in wild type E. coli cells [57]. In the TAMT strain, which lacks a functional thymidylate synthetase and thus is grown in the presence of thymine, it appears that exogenous thymine is incorporated in a manner analogous to thymidine. It is not clear whether in the TAMT strain both thymine and thymidine are taken up by a preferred pathway, as thymidine is in wild type E. coli, or if this pathway no longer exists in strains lacking thymidylate synthetase. It may be that either of these possibilities are correct. Jackson and Lark [45] found that thymidine appeared to be preferentially incorporated over thymine in E. coli TAMT since cold thymidine reduced the incorporation of a ^{3}H thymine pulse. These results appear to be in direct contradiction to our data (Fig. 3B) that show thymidine not to be preferred over thymine in the TAMT strain. However, Jacobson and Lark pulsed with thymine in the presence of the thymine

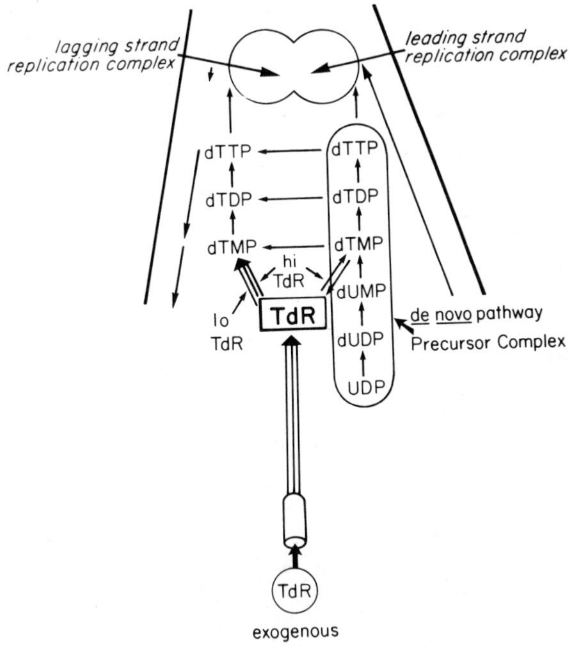

Fig. 4. A model describing a proposed preferential feed to the leading strand of thymine-containing DNA precursors, synthesized de novo by a multienzyme precursor complex. Over extension of the leading strand relative to the lagging strand, or leading strand replicative blocks, would shift the precursors to the lagging strand until the source of the precursor imbalance is eliminated. In thymine prototrophs low concentrations of exogenously added thymidine would preferentially supply the lagging strand. High concentrations of thymidine could inhibit de novo thymine precursor production with the result that thymidine would supply both the leading and lagging strands. In thymine auxotrophs, low concentrations of thymidine would preferentially supply the lagging strand only during the continued presence of exogenous thymine. If thymine auxotrophs are starved, exogenously supplied thymine would supply both the leading and lagging strands.

used for growth. We, however, resuspended our cells in the absence of cold thymine. The continued presence of thymine may be required for differential thymidine/thymine utilization in thymine auxotrophs. The difference between the kinetics of thymidine incorporation in E. coli B and the TAMT strain is probably due to the existence of thymine-containing precursor pools in the mutant TAMT whereas the precursors are dynamically provided by a "precursor complex" in wild type *E. coli* B. This interpretation agrees with the data of Diaz and Werner [31] which show similar incorporation kinetics to ours.

3.4. Possible Secondary Role for Thymidine

Figure 4 presents a model that is consistent with our incorporation studies in T4-infected and uninfected plasmolyzed cells as well as for T4-infected cells *in vivo*. We specifically hypothesize that there are differential precursor feeds for leading and lagging strands. The principal points of this model can be described as follows:

1. The pyrimidine precursor complex is more tightly coupled to the leading than to the lagging strand. As a result, low concentrations of exogenous thymidine would have preferential access to the lagging strand.

2. When the leading strand extends beyond a critical distance from the concomitantly synthesized lagging strand, or if there is a leading strand block, pyrimidine precursors would backup and feed the lagging strand.

3. Internally generated thymidine would help to modulate lagging strand synthesis by providing an additional thymine containing precursor reservoir that can expand and accentuate lagging strand synthesis.

4. Externally added thymidine has preferred access to the replication fork. At low concentrations, thymidine preferentially feeds the lagging strand. At higher concentrations, thymidine feeds both strands and the *de novo* production of thymine-containing DNA precursors is proportionately reduced. We believe that the preferred access pathway by which low concentrations of thymidine are incorporated occurs by the direct phosphorylation of thymidine by thymidine kinase [69-74]. When high concentrations of thymidine are used to measure incorporation over longer periods of time, thymidine is first broken down to thymine.

5. When a leading strand block is present, the lagging strand could be pushed by precursor supply beyond its normal distance from the leading strand and perhaps beyond the leading strand block. This could result in reinitiation on the leading strand beyond the damage, a template switch that could bypass the damage, recombinational events that could lead to damage bypass or damage bypass by an altered polymerase activity.

6. Abortive recombinational events that result from inactive T4 uvsX gene product could serve as leading strand blocks. (The uvsX+ gene product has a number of strand pairing activities analogous to those of the RecA protein [10].) The resulting enhanced precursor feed to the lagging strand would reduce the utilization of exogenously added thymidine. In T4 uvsW mutant-infected cells, where a class of recombinational events would not be initiated, the recombination-induced leading strand blocks would not occur and incorporation of low concentrations of thymidine is enhanced.

7. In uninfected *E. coli*, a leading strand block, with its associated pool backup, could lead to high dATP levels [88, 98] and increased production of ribonucleotide reductase [13, 37]. In turn, the elevated dATP levels could act in concert with the single stranded regions generated by the leading/lagging strand imbalance to induce the SOS response.

3.5. Attempts to Test the Hypothesis

Because of our observations that thymidine appears to be a preferred precursor *in vivo* and *in situ* in plasmolyzed cells and because thymidine seems to be exclusively incorporated into small pieces in T4-infected cells, we attempted to test the above hypothesis by labeling with a precursor that might perhaps have access to the leading strand through the precursor complex. We chose to label with dTM^{32}P for a number of reasons. First, the ^{32}P label would eliminate any ambiguities caused by the incorporation of thymidine from the breakdown of d^{3}HTMP. Secondly, it provides a very high specific activity label which enabled us to use short pulse times. Third, studies from Mathews' group [66, 95, 103] suggested that dTMP might have access to the precursor complex and our own studies, using d^{3}HTMP, showed that dTMP was the second most preferred thymine-containing DNA precursor in plasmolyzed cells. The idea was to set up conditions whereby we might simultaneously label the lagging strand with tritiated thymidine and the leading strand with dTM^{32}P.

In Fig. 5 it can be seen that dTM^{32}P is linearly incorporated into the acid precipitable material in T4-infected plasmolyzed cells and that this incorporation is greatly reduced in a gene 32 mutant that cannot synthesize DNA. Incorporation is also appropriately lower in uninfected *E. coli*. Although the absolute incorporation was low, in view of the apparently normal relative rates of dTM^{32}P incorporation, it was surprising to find that virtually all of this label was incorporated into small nonchasable pieces in T4-infected plasmolyzed cells (Fig. 6). In order to eliminate the possibility that these small pieces may have resulted from strand break induction by ^{32}P beta decay, double labeling experiments were performed in which tritiated thymidine and dTM^{32}P were incorporated and centrifuged in

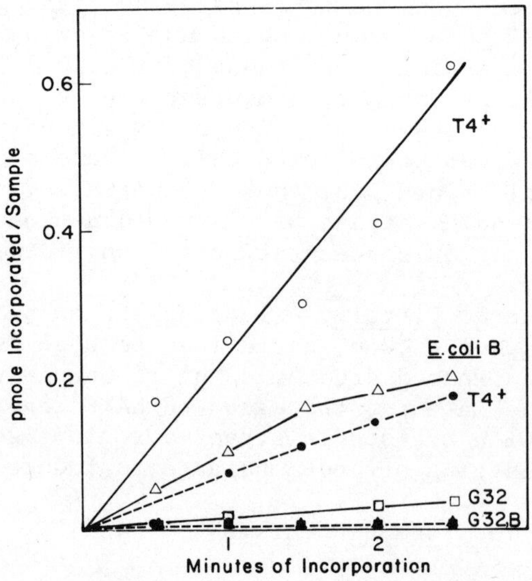

Fig. 5. Incorporation of 1 μM ^{3}H TdR and 1 μM dTM^{32}P in plasmolyzed E. coli B, uninfected or infected for 15 min with T4$^+$ or T4 am A853 (amber gene 32). Labeling and plasmolysis were as described in Materials and Methods. Symbols: open symbols depict ^{3}H thymidine incorporation, solid symbols depict dTM^{32}P incorporation. (○●) T4$^+$; (□■) T4 am A853; (△▲) uninfected E. coli.

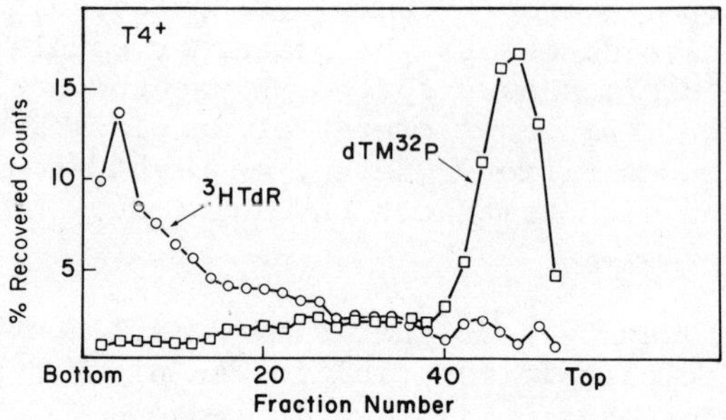

Fig. 6. Preferential incorporation of dTM^{32}P into small nonchasable DNA pieces. Alkaline sucrose gradient sedimentation analysis of DNA from T4-infected cells double labeled for 5 min at 37°C with 1 μM dTM^{32}P. Labeling and sedimentation analysis were as described in the legend of Fig. 2. Symbols: (○) ^{3}H thymidine (□) dTM^{32}P.

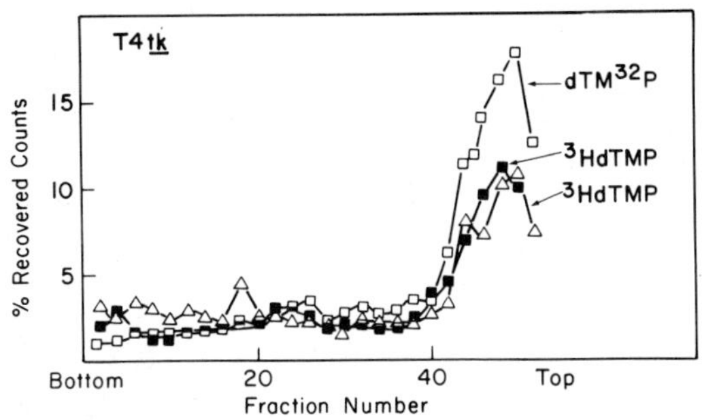

Fig. 7. Preferential incorporation of d^{3}HTMP into small nonchasable DNA pieces in plasmolyzed cells infected with a T4 thymidine kinase mutant. Alkaline sucrose sedimentation analysis of DNA from T4 do tk1 infected cells labeled with d^{3}HTMP or dTM^{32}P. Plasmolyzed infected cells were either double labeled for 5 min with 0.5 μM d^{3}HTMP plus 0.5 μM dTM^{32}P or labeled with 1 μM d^{3}HTMP. Labeling and sedimentation analysis were described in the legend of Fig. 2. Symbols: Plasmolyzed infected cells double labeled with (△) d^{3}HTMP and (□) dTM^{32}P; (■) or d^{3}HTMP alone.

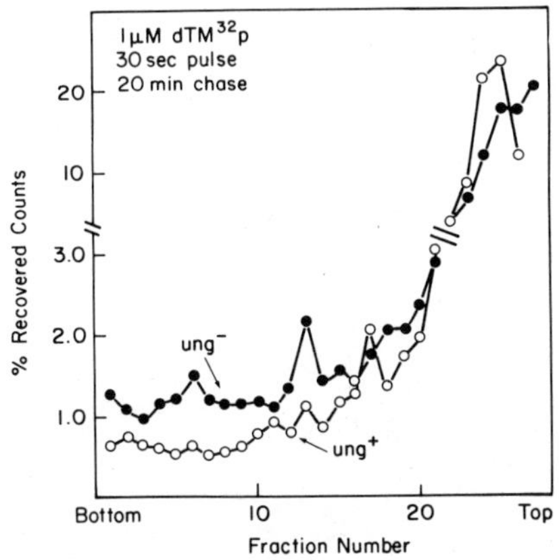

Fig. 8. Alkaline sucrose sedimentation analysis of DNA from a T4-infected *E. coli* *ung*$^-$ mutant that lacks uracil DNA N-glycosylase. *E. coli* (●) BW 271 *ung*$^-$ or its parent (○) BW 273 *ung*$^+$ were pulse labeled for 30 sec with 1 μM dTM^{32}P and then chased for 20 min. Labeling and sedimentation analysis were described in the legend of Fig. 2.

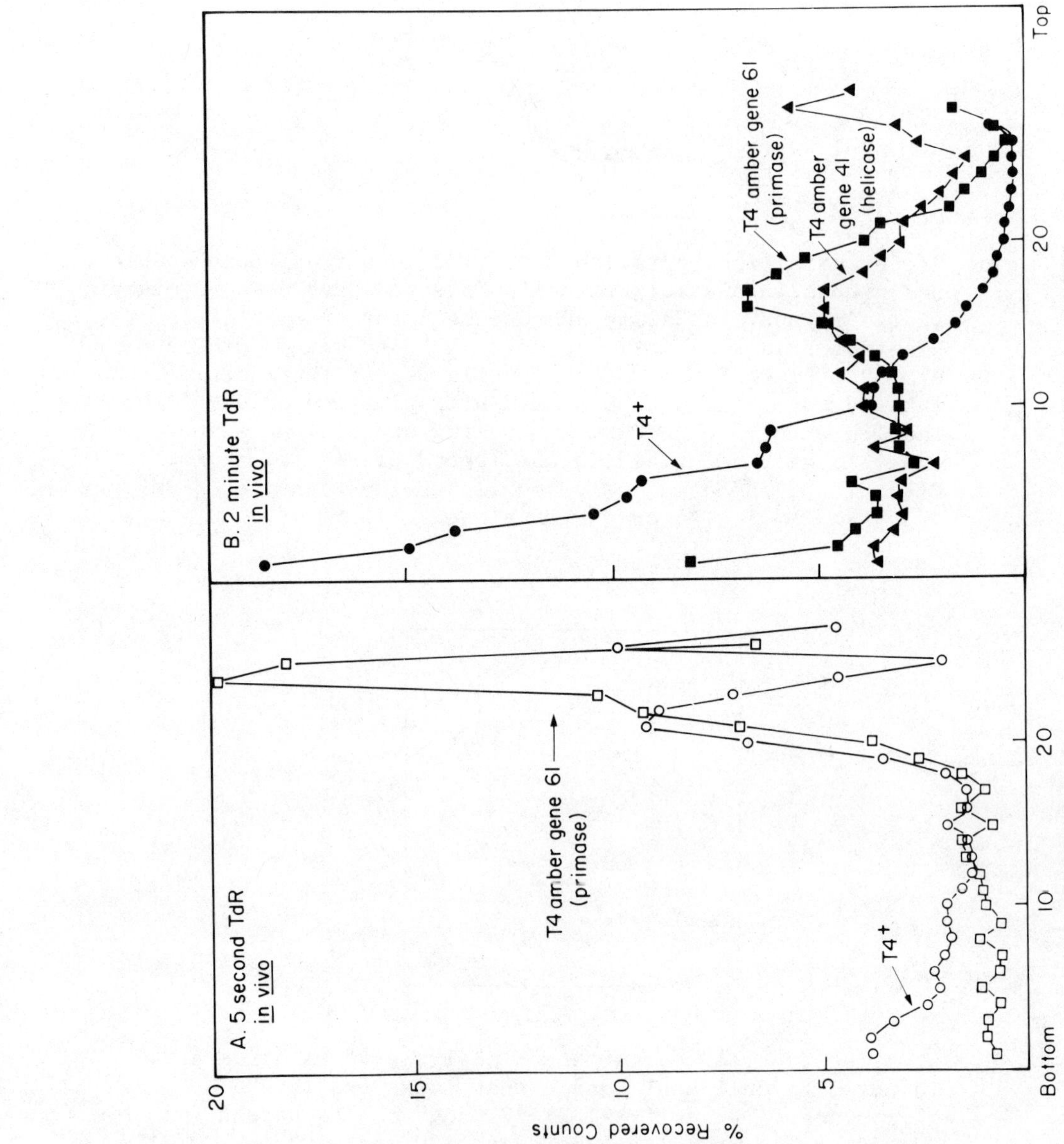
A. 5 second TdR
in vivo
T4+
T4 amber gene 61
(primase)
B. 2 minute TdR
in vivo
T4+
T4 amber gene 61
(primase)
T4 amber
gene 41
(helicase)
% Recovered Counts
20
15
10
5
Bottom
10
20
Top
Fraction Number

Fig. 9. Alkaline sucrose sedimentation analysis of ^3H thymidine labeled DNA from cells infected with: wild T4, a T4 helicase mutant or a T4 primase mutant. Labeling and sedimentation analysis were described in the legend of Fig. 2. A) Cells were pulse labeled *in vivo* for 5 sec with 1 μM ^3H thymidine. B) Cells were pulse labeled *in vivo* for 2 min with 1 μM ^3H thymidine. Symbols: (○●) T4$^+$; (□■) T4 am E21g (amber in gene 61, primase) (▲) T4 am N81) (amber in gene 41, helicase).

the same or in separate tubes (data not shown). The results of this experiment are the same as those shown in Fig. 6. After 5 min of labeling, tritiated thymidine is found in rapidly sedimenting pieces while $dTM^{32}P$ is found in small pieces. The finding that dTMP is only incorporated into small nonchasable pieces was confirmed by examining the sedimentation properties of d^3HTMP labeled DNA from T4tk mutant-infected cells (Fig. 7). In this experiment we biased for the incorporation of dTMP since the tk mutant phage does not have a functioning thymidine kinase. As was observed with the incorporation of $dTM^{32}P$, all of the d^3HTMP label is found in small nonchasable pieces. In contrast, when $T4^+$-infected cells were labeled with d^3HTMP for 5 min, the label was found in larger DNA pieces (data not shown) as if thymidine had been used as label. These results clearly demonstrate that dTMP and thymidine are incorporated into different DNA molecules. However, dTMP does not selectively label continuous leading strands. The possibility remained that small dTMP labeled pieces might be due to dTMP incorporation into alkali-labile DNA generated by uracil DNA N-glycosylase activity. This explanation seems unlikely since most of the label is still found in small nonchasable pieces in $dTM^{32}P$ labeled T4-infected ung^- mutants (Fig. 8) although there is a slight increase in rapidly sedimenting $dTM^{32}P$ labeled DNA. Further, when $dTM^{32}P$ labeled DNA from T4-infected plasmolyzed cells was examined on neutral sucrose gradients in the presence of 7 M urea, the label was again found only in small nonchasable pieces (data not shown).

In another attempt to bias measurements for leading strand synthesis in T4-infected cells, we examined the sedimentation characteristics of DNA labeled with ^3H thymidine (Fig. 9) or $dTM^{32}P$ (data not shown) in cells infected with either a T4 helicase or a T4 primase mutant. Once again dTMP went into small nonchasable pieces. However, we made two interesting observations using these mutants. Both primase and helicase mutants incorporate short pulses of thymidine exclusively into small pieces (Fig. 9A). It should be noted that the source of primer for the small thymidine labeled chasable pieces in a primase mutant is unknown. Luder and Mosig [62] have shown that initiation at primary T4 origins seems to be dependent on host RNA polymerase. Presumably the small chasable pieces that we see in helicase and primase mutants may also be dependent on host RNA polymerase for priming. Figure 9B shows that in both helicase and primase mutant-infected cells labeled with thymidine, an intermediate size DNA accumulates that is not observed in wild type-infected cells. This may reflect the involvement of genes 41 and 61 in increasing the number of replicating forks [68] and the reported reduced replicational repair activities of the gene 41 mutant [25]. In addition, the helicase mutant appears to have a reduced ability to chase the small pieces into the intermediate size pieces mentioned above, thus suggesting that helicase activity is required for optimal processing of Okazaki pieces.

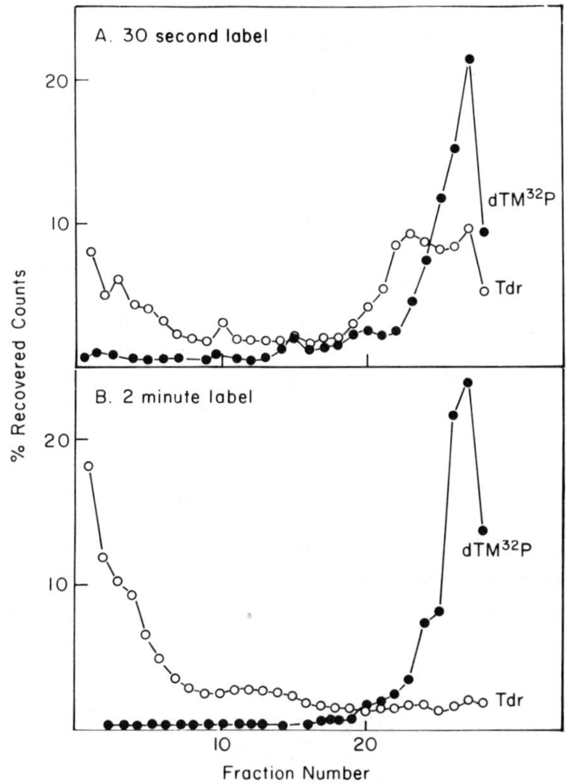

Fig. 10. Alkaline sucrose sedimentation analysis of DNA from plasmolyzed uninfected E. coli labeled for 2 min at 37°C with 1 μM ^{3}H thymidine and 1 μM dTM^{32}P. Labeling and sedimentation analysis were described in the legend of Fig. 2. A) Cells labeled for 30 sec; B) cells labeled for 2 min. Symbols: (○) ^{3}H thymidine; (●) dTM^{32}P.

As was found in T4-infected cells, dTM^{32}P is also incorporated almost entirely into small DNA pieces in uninfected plasmolyzed E. coli (Fig. 10A and B). Figure 10B shows that even after 2 min of labeling, these pieces remain small. When plasmolyzed E. coli were labeled with 1 μM ^{3}H thymidine for 30 sec (Fig. 10A) the label is found in both long and short pieces and after 2 min (Fig. 10B) the label is found almost entirely in large pieces.

The physiological role, if any, of the dTM^{32}P labeled small pieces remains unknown. Small nonchasable DNA pieces have been identified in prokaryotic [6, 16, 18, 29, 31, 32, 45, 54, 76, 92, 110-112] and eukaryotic [29, 99] organisms. A variety of techniques have been used to identify small pieces of DNA in an attempt to characterize them. In general, small DNA pieces may be divided into two classes: those that constitute nascent intermediates of DNA replication, Okazaki pieces, and those that do not [29]. How the nascent/nonnascent classification relates to size and origin has not been clearly defined. Using endlabeling, Miyamoto and Denhardt [76] found a large number of small 2s molecules in various strains of E. coli. E. coli TAMT, unlike other strains they tested (mutants in genes polA, polB, rep, recA, recC, dnaA, dnaB, dnaC, dnaE, and dnaG), seemed to have a reduced number of the small 2s molecules. The biological significance of these DNA pieces remains unknown as does any relationship to the small $dTM^{32}P$ labeled pieces we have found. However, since the small nonchasable pieces examined by Denhardt's group were endlabeled, we assume that our $dTM^{32}P$ labeled small pieces must constitute at least a subset of theirs. Although we have not examined $dTM^{32}P$ incorporation in uninfected ung^- strains, Kwoh et al. [54] have not observed any significant effect of the ung mutation on the small pieces they examined in thymine prototrophs.

Small nonchasing DNA pieces have not been extensively investigated in T4-infected cells. Mosig (personal communication) has by ^{32}P labeling, found a class of small DNA pieces that specifically hybridizes to a T4 replication origin. We intend to determine if our $dTM^{32}P$ pieces from both T4-infected and uninfected cells are origin specific.

Clearly our attempts to demonstrate differential leading and lagging strand precursor feeds have thus far been unsuccessful although dTMP and thymidine do not appear to be incorporated into the same DNA molecules. However, the ability of the proposed model to resolve many inconsistencies that exist in the literature warrants its further investigation.

4. SUMMARY AND PERSPECTIVES

4.1. Summary

Historically the investigation of DNA synthesis has preceded and outweighed the investigation of DNA precursor channeling and precursor complexes. A formal link between these two areas now seems possible. Recently, there have been a number of observations showing that nucleotide pool imbalances can have genetic consequences in organisms ranging from T4 bacteriophage-infected cells to mammalian cells [53]. The isolation and characterization of DNA precursor synthesizing complexes and in vitro studies on the fidelity of DNA polymerases [60] have provided insights into possible mechanisms whereby nucleotide pool imbalances might influence an organism's genetic integrity. We have developed a model, based on our studies on incorporation pathways of thymine-containing DNA precursors in T4-infected and uninfected E. coli, that attempts to provide a framework for understanding relationships between pool imbalances and their effects. We feel that the potential general usefulness of the concepts introduced in this paper warrants our highly speculative presentation.

We have examined the incorporation and sedimentation characteristics of DNA labeled with thymine, thymidine, dTMP, and dTTP in both T4-infected and uninfected plasmolyzed cells and found that as in vivo, thymidine has unique access to the replication fork. In sucrose plasmolyzed cells, thymidine is preferentially incorporated into DNA over other thymine-containing precursors even when they are present in large excess (Table 2). However, thymidine seems to be selectively incorporated into Okazaki pieces (Fig. 2). In light of these and related data, we believe that the possible existence of separate precursor feeds to leading and lagging strands has not been adequately explored.

4.2. DNA Precursor Feeds and Discontinuous DNA Synthesis

There has been little experimental demonstration of continuous strand DNA synthesis in cells in vivo [3, 27, 67, 81, 104, 110]. In polA$^+$ strains of *E. coli*, Lourn and Bird [61] found that after a short pulse of labeled thymidine the labeled short DNA pieces preferentially hybridized to the lagging strand template, whereas the longer DNA pieces hybridized to the leading strand template. The longer pieces were still shorter than what would be expected from a fully continuously synthesized strand. In a similar experiment with a polA$^-$ strain, all of the labeled DNA was found in short pieces that hybridized with both templates [61, 84] Ogawa and Okazaki [81] interpreted the above results as demonstrating discontinuous synthesis on both strands with the lagging strand having a slower rate of piece joining. This interpretation is complicated by the presence of uracil DNA N-glycosylase activity [86, 105, 107]. In strains that lack thymidylate synthetase, dUTP pools expand and, as shown in dut mutants that lack dUTPase, the repair of incorporated dUMP can create small pieces post replicatively [33, 86, 105, 107]. The production of small pieces by this mechanism does not seem to be important in thymine prototrophs [54, 107].

An alternative, but not exclusive, interpretation to a double strand discontinuous mode of DNA synthesis is implied by the model presented here. In the above experiments Lourn and Bird [61] used thymine auxotrophs. DNA synthesis in these strains may tend to be discontinuous on both strands because of the lack of thymidylate synthetase in the precursor complex and the resulting reduced funneling of thymine precursors to the leading strand. Further, Lourn and Bird noted that they consistently recovered more thymidine labeled lagging strand DNA, a phenomenon they called "overreplication." Their protocol consisted of pulsing with thymidine in the presence of cold thymine. In contrast Raggenbass and Caro [92] using a similar experimental design, could not detect "overreplication" of the lagging strand, nor did they have cold thymine present during the thymidine pulse. The differential strand feed that we propose can be used to resolve this discrepancy. When cold thymine was present during the thymidine pulse, labeled thymidine could be preferentially used for lagging strand synthesis. Raggenbass and Caro depleted thymine-containing precursor pools by synchronizing their cells with thymine starvation. Since they only provided thymidine during their pulse labeling, the thymidine had to be utilized for synthesis of both strands.

Additional support for the above interpretation based on differential leading and lagging strand pyrimidine feeds may be found in early data from Werner [111, 112]. He showed that when E. coli 15-TAMT (a thy⁻ strain) was labeled with thymine in the presence of cold thymidine, no label was found in Okazaki pieces. Presumably the cold thymidine supplied the thymine-containing precursors for the lagging strand.

The mechanism of thymine utilization in thymine auxotrophs appears to be more complex than might be expected. The competitive effect that thymine exerts on thymidine incorporation in the TAMT strain (Fig. 3) [111], that was grown in the presence of thymine, implies that the metabolic machinery has flexibility. Not only can thymine be an effective DNA precursor but thymidine can lose its preferred access to the replication fork. This possibility will be elaborated on later in the text (4.4).

4.3. DNA Precursor Feeds and SOS Induction

Aside from resolving some apparently conflicting data, the model of differential precursor feeds for leading and lagging strands describes what might be a fundamental process that is easily and uniquely adaptable to the genetic needs of a given organism. For example, T4 DNA replication may exhibit an exaggerated "Okazaki" mode of synthesis that is well suited to the highly developed recombination process intrinsic to T4 development. The work of Wirak and Greenberg [115] supports this idea. They demonstrated that the exponential phase of T4 DNA synthesis coincides with the assembly of the T4-directed DNA precursor complexes and the functioning of wild type DNA delay gene products. These genes code for the additional topoisomerase and primase functions that are required as the number of replication forks increase, or for the number of replication forks to increase. In contrast, E. coli may utilize a balanced mode of leading and lagging strand synthesis. Imbalances could serve to trigger basic molecular homeostatic responses such as SOS induction [91, 113]. This imbalance would in turn be a response to a specific metabolic state characterized by the interaction between the ability to produce DNA precursors and the ability to use them limited only by the physical condition of the DNA. Thus, SOS induction may be a response not to the DNA damage itself but rather to the damage-induced precursor imbalance and the single stranded regions of DNA that result from the imbalance. This possibility is in keeping with the observation that many different kinds of lesions induce the SOS response [114]. Recently, many of the molecular events that characterize the SOS response have been elucidated [58, 88, 98]. The precise nature, however, of the SOS inducing signal is not known. Since single stranded DNA is apparently required for induction, it has been suggested that discontinuous strand DNA synthesis may be involved in the process [98]. Our model proposes that precursor "push" coupled to replicative blocks would create both the single stranded DNA and the dATP cofactors for RecA protease activation.

4.4. DNA Precursor Feeds and Thymineless Death

In the previous section we proposed an obligatory link between thymine precursor "push" that would exaggerate lagging strand synthesis and cause SOS induction. Here we shall consider how this concept can include thymineless death and SOS induction that result from thymine starvation. When *E. coli* [22, 23] or *Bacillus subtilis* [36] cells, deficient in thymidylate synthetase, are grown in liquid culture lacking thymine or thymidine they undergo thymineless death. The mechanism of this phenomenon is unknown. Nakayama and Hanawalt [77] have shown a strain-dependent production of single strand breaks in thymine starved cells that is enhanced when DNA polymerase I is missing. In fact, Valenzuela and Inman [109] found gaps in both newly synthesized strands after thymine starvation. A fundamental role for DNA polymerase I in minimizing the effects of thymine starvation is strongly indicated by recent findings of Ahmad et al. [2] who isolated an ultraviolet light resistant mutant of *E. coli* K12 that overproduces DNA polymerase I (35-fold), and endonuclease I (12-fold) and is resistant to thymineless death. DNA polymerase I is assumed to have a role in the joining Okazaki pieces [61, 84]. We interpret these characteristics of thymineless death in the following manner. As mentioned earlier, there are indications that in thy^- cells, DNA synthesis is discontinuous on both strands. The degree of double strand discontinuous synthesis may be an inverse function of thymine concentration and not simply a result of uracil N-glycosylase activity. We propose that below certain critical concentrations of thymine, as the ratio of elongation to initiation of DNA strands decreases, cells would acquire an additional DNA polymerase I requirement. In other words, the leading strand would have a length of continuous synthesis that would be somewhat proportional to thymine availability. During starvation the length of continuous synthesis would decrease and oscillations between leading and lagging strand synthesis would increase per unit length of synthesis as the thymine precursor pools are depleted, that is, leading and lagging strands would be shorter. The mechanism whereby this proposed sequence of events could lead to death is not clear. In view of the proposed secondary role for pyrimidine precursors, the pool perturbations that occur when thymine or thymidine is added back to starved cultures may be as important as the starvation itself in causing thymineless death. A stalled replication fork could, by virtue of gaps and perhaps misincorporated bases, be analogous to a damage-induced replicative block. The readdition of a thymine precursor could then provide the pool responses that we described in the previous section to account for SOS induction. Since thymine starvation also releases repressor control of ribonucleotide reductase [13, 37], dATP levels would be high thus potentially amplifying induction. A recent paper by Valenzuela and Inman [109] supports this precursor imbalance hypothesis. They examined multiply branched DNA that is induced by amino acid and thymine starvation in both uninfected and bacteriophage lambda-infected E. coli cells, and proposed that these structures result either from an activated secondary origin caused by an accumulation of initiation proteins during the starvation period or from DNA snap-back due to starvation-induced single stranded regions. Some of their data however suggest an explanation based on the precursor model we have presented. They found that lambda could replicate in the absence of exogenously added thymidine presumably because enough was present in the medium used to grow them. Under these conditions multiply branched structures were minimal (1%). If thymidine was added to the medium and incubation continued for an additional 5 min, the proportion of multiply branched structures increased to 11%. We therefore suggest that starvation creates replication blocks and that at least some of the multiple branches are the results of reinitiation past the blocks. The reinitiations would be induced by the pool perturbations created when thymidine is added back to the medium. Excessive initiations have been associated with thymineless death [90] and stable replication [47-49].

The proposed control of replicative reinitiations by thymine-containing precursor mediated lagging strand "push" is a generalization of a demonstrated phenomenon. In both uninfected [34] and bacteriophage lambda-infected [108] E. coli, auxotrophic for thymine, whether DNA synthesis is uni- or bidirectional is determined by the thymine concentration. At low concentrations of thymine, replication is undirectional. At high thymine concentrations replication becomes bidirectional presumably as a result of the first lagging strand becoming a leading strand in the opposite direction. Another example of precursor "push" may be found in the experiments of Olivera and Bonhoeffer [85] who showed, in a cellophane disc system, that the size of Okazaki pieces was reduced when the concentration of deoxynucleoside triphosphates was lowered.

4.5. DNA Precursor Feeds and Thymidine Mutagenesis

Thymidine-induced mutagenesis has been observed in a number of cellular systems [1, 14, 26, 28, 46, 55, 87] including T4-infected E. coli [9, 30]. MutD mutants of E. coli are extremely sensitive to thymidine mutageneis [28]. Measurable yields can be obtained at a concentration of 0.04 μM. The extreme sensitivity of these mutants to thymidine mutagenesis might be related to the preferred access of thymidine to the replication fork. If, as proposed, thymidine preferentially feeds the lagging strand we predict that, in mutD thymine prototrophs, mutation induced by low concentrations of thymidine would occur with a lagging strand bias. Furthermore in mutD thymine auxotrophs the bias should be a function of growth conditions. If thymine and thymidine are present, mutagenesis should be lagging strand specific whereas if only thymidine is present the strand specificity should be lost. These general predictions would be additionally dependent on the relative concentrations of thymine and/or thymidine used.

4.6. DNA Precursor Feeds and Agents that Affect Them

Agents that affect DNA precursor biosynthesis are used extensively for both research and therapuetic purposes. Many of them are mutagenic. Some novel possibilities that could result from assymetric strand feeds are worth considering. The concepts developed in this paper emphasize that there is no *a priori* reason to assume that mutagenic events, or the correction of potentially mutagenic events, should occur with equal probability on leading and lagging strands. One possibility is that precursor analogs may be differentially incorporated into, or may differentially inhibit, leading or lagging strands much the same as thymine and thymidine seem to incorporate differentially under appropriate circumstances. Some analogs may feed through a precursor complex and others may not. Of those that do go through a complex, some may be directed to one strand or the other perhaps with significantly different effects. The proposed "spillover" feed for the lagging strand further implies that when precursor pools are examined an additional degree of complexity may be required for interpreting any data.

4.7. DNA Precursor Feeds and Dissipative Structures

We have attempted to develop the notion that DNA precursor production involves a complex chemical network consisting of spatially localized enzymes and substrates all interacting in functional harmony. The ability of this network to detect and effect change at the genetic level may constitute a fundamental homeostatic mechanism. Historically biology and the Second Law of Thermodynamics seemed at odds. Simple probabilistic considerations of DNA or protein sequences were incompatible with the emergence of life [79, 89]. However, the development of nonequilibrium thermodynamics and statistical mechanics cast living systems in a new light. The formation and maintenance of highly ordered structures is fully compatible with the Second Law by virtue of the additional entropy that is generated in the universe by the presence of ordered open systems.

Spontaneous order formation in an open system is dependent upon an appropriate entropic gradient. In other words, there must be a negentropic source and an entropic sink in order to create a flow dependent structure (dissipative structure) [79, 89]. An interesting and relevant characteristic of a dissipative structure is that non-physically limiting parameters of an open system define the borders of a flow dependent structure as opposed to simply having a physical constraint such as a wall or a membrane. We propose that the DNA precursor enzyme/substrate network constitutes a dissipative structure. The "shape" of the structure in turn affects and is affected by its negentropic source, metabolism, and its entropic sink, the polymerization of monomers into the ordered DNA polymer.

The concept of dissipative structures serves not only to characterize the physical nature of precursor production and DNA synthesis but also provides an understanding of how formal precursor complexes may have evolved. We suggest that when enzymes in a sequential pathway can become part of a dissipative structure, natural selection can be accelerated. Once proteins of related function become localized in time and space, genetic changes that alter the proteins can be more efficiently tested for fitness. Direct physical association of DNA precursor enzymes would provide an evolutionary advantage because of the potential for control that could be selected for. It should be emphasized that we are not suggesting that the formal structuring of an enzyme association, as occurs in a protein complex, precludes the complex from participating in a flow dependent structure. For example, because dTTP is synthesized from UDP by a unique biochemical pathway it is possible that the dTTP producting precursor complex (an equilibrium structure) could control a flow dependent structure that supplies the other DNA precursors to the replication fork.

4.8. Conclusions

In conclusion, our understanding of the relationship between DNA precursor production and its potential secondary effects is rudimentary. The precursor model that we have presented provides a new way that DNA precursor production can be integrated with the basic genetic processes of replication, recombination repair, and mutagenesis. The control of gene expression may also be a potential area of applicability. The putative relationship between active transcription on leading or lagging strand templates and DNA synthesis is largely unknown but nevertheless open for speculation [102]. If a very actively transcribed gene, in particular on a leading strand template, could serve as a replicative block, it could result in directed mutagenesis or directed chromosomal rearrangement by the mechanism we have described in this paper. In prokaryotes, this phenomenon, for example, could bear on the evolution of plasmids carrying resistance factors. In eukaryotes, transscriptionally induced replicative blocks could perhaps lead to directed gene amplification and/or functional chromosomal rearrangements.

The model that we have presented is an example of a molecular mechanism for genetic homeostasis. Homeostasis is a misnomer when applied to genetics because of the requirement for change by the evolutionary process. The concept of fitness homeostasis seems relevant since it encompasses the concepts of the dynamics we are promoting. In other words the ability to adapt has been selected for and we feel that the model we have presented may be an example of the response to this selection.

As the needs and complexity of organisms evolve so should the specifics of any basic molecular process. The recent demonstration that thymidine kinase is required for the interferon-induced antiviral response [56] and that ribonucleotide reductase may be a herpes virus oncogene [44] points to the possibility that a DNA precursor/enzyme network may have been adapted to many different functions.

ACKNOWLEDGMENTS

We are grateful to Brian Baker and Amersham-Searle for generously supplying the $dTM^{32}P$ free of charge and to Felix Wassermann for critically reading the manuscript. This work was supported by Grant CA33657 awarded by the National Cancer Institute, DHHS.

REFERENCES

1. S. I. Ahmad and A. Eisenstark, Thymidine sensitivity of certain strains of Escherichia coli K12, Molec. Gen. Genet., 172:229-237 (1977).
2. S. I. Ahmad, A. Atkinson, and A. Eisenstark, Isolation and characterization of a mutant of Escherichia coli K12 synthesizing DNA polymerase I and endonuclease I constitutively, J. Gen. Micro., 117:419-422 (1980).
3. B. Alberts and R. Sternglanz, Recent excitement in the DNA replication problem, Nature, 269:655-660 (1977).
4. J. R. Allen, G. P. V. Reddy, G. W. Lasser, and C. K. Mathews, T4 ribonucleotide reductase, Physical and kinetic linkage to other enzymes of deoxyribonucleotide biosynthesis, J. Biol. Chem., 255:7583-7588 (1980).
5. J. R. Allen, G. W. Lasser, D. A. Goldman, G. W. Booth, and C. K. Mathews, T4 Phage deoxyribonucleotide synthesizing enzyme complex, J. Biol. Chem., 258:5749-5753 (1983).
6. M. L. M. Anderson, Number and size distribution of the DNA chains in Escherichia coli, J. Mol. Biol., 118:227-240 (1978).

7. I. R. Beacham, K. Beacham, A. Zaritsky, and R. H. Prichard, Intracellular thymidine triphosphate concentrations in wild-type and in thymine requiring mutants of Escherichia coli 15 K12, J. Mol. Biol., 60:75-86 (1971).
8. M. T. Behme, G. D. Zilley, and K. Ebisuzaki, Post infection control by bacteriophage T4 of Escherichia coli rec BC nuclease activity, J. Virol., 18:20-25 (1976).
9. C. Bernstein, H. Bernstein, S. Mufti, and B. Strom, Stimulation of mutation in phage T4 by lesions in gene 32 and thymidine imbalance, Mutat. Res., 16:113-119 (1972).
10. C. Bernstein and S. S. Wallace, DNA repair, in: "Bacteriophage T4." C. K. Mathews, E. Kutter, G. Mosig, and P. B. Berget, eds.), ASM Press, Washington (1983).
11. O. Berglund and B. M. Sjoberg, Effect of hydroxyurea on T4 ribonucleotide reductase, J. Biol. Chem., 245:253-254 (1979).
12. R. E. Bird, J. Lourn, J. Mastuscelli, and L. Caro, Origin and sequence of chromosome replication in Escherichia coli, J. Mol. Biol., 70:549-566 (1972).
13. C. Biswas, J. Hardy, and W. S. Beck, Release of repressor control of ribonucleotide reductase by thymine starvation, J. Biol. Chem., 240:3631-3640 (1965).
14. M. O. Bradly and W. A. Sharkey, Mutageneicity of thymidine to cultured Chinese hamster cells, Nature, 274:607-608 (1978).
15. S. Bresler, M. Mosevitsky, and L. Vyacheslavov, Complete mutagenesis in a bacterial population induced by thymine starvation on solid media, Nature, 225:746-766 (1970).
16. N. Brewin, Origin and fate of the small DNA chains synthesized in bacteria after thymine deprivation, J. Mol. Biol., 111:343-352 (1977).
17. N. Brewin and J. Cairns, State of the DNA replication fork during thymine deprivation of Escherichia coli, as observed by pulse-labeling with ^{3}H thymine, J. Mol. Biol., 111:353-363 (1977).
18. J. Chao, M. Leach, and J. Karam, In vivo functional interaction between DNA polymerase and dCMP hydroxymethylase of bacteriophage T4 V1. Biological functions of gene 42, J. Virol., 24: 557-563 (1977).
19. C.-S. Chiu and R. G. Greenberg, Evidence for a possible direct role of dCMP hydroxymethylase in T4 phage DNA synthesis, Cold Spring Harbor Symp. Quant. Biol., 33:351-359 (1968).
20. C.-S. Chiu, T. Reuttinger, J. B. Flanegan, and R. G. Greenberg, Role of deoxycytidylate deaminase in deoxyribonucleotide synthesis in bacteriophage T4 DNA replication, J. Biol. Chem., 252:8603-8608 (1977).
21. C.-S. Chiu, K. S. Cook, and R. G. Greenberg, Characteristics of a bacteriophage T4-induced complex synthesizing deoxyribonucleotides, J. Biol. Chem., 257:15087-15097 (1982).
22. S. S. Cohen and H. P. Barner, Enzymatic adaptation in thymine requiring strain of Escherichia coli, J. Bact., 69:59-66 (1955).

23. L. V. Crawford, Thymine metabolism in strains of Escherichia coli, Biochem. et Biophys. Act., 30:428-429 (1958).
24. R. P. Cunningham and H. Berger, Mutations affecting genetic recombination in bacteriophage T4. II. Pathway Analysis, Virol., 80:67-82 (1977).
25. M. Cupido, S. Grimbergen, and B. De Groot, Participation of bacteriophage T4 gene 41 in replication repair, Mutat. Res., 70:131-138 (1980).
26. R. L. Davidson and E. R. Kaufman, Bromodeoxyuridine mutagenesis in mammalian cells is stimulated by thymidine and suppressed by deoxycytidine, Nature, 276:722-723 (1978).
27. M. L. De Pamphilis, and P. M. Wassarman, Replication of eukaryotic chromosomes: A close-up of the replication fork, Ann. Rev. Biochem., 49:627-666 (1980).
28. G. E. Degnen and E. Cox, Conditional mutator gene in Escherichia coli: Isolation, mapping and effector studies, J. Bact., 117: 477-487 (1974).
29. D. T. Denhardt and C. Miyamoto, Characteristics of the nascent and non-nascent small DNA molecules found in Escherichia coli, J. Mol. Biol., 165:419-442 (1983).
30. J. K. DeVries and S. S. Wallace, Reversion of bacteriophage T4 rII mutants by high levels of pyrimidine deoxyribonucleosides, Mol. Gen. Genet., 186:101-105 (1982).
31. A. T. Diaz and R. Werner, Mechanism of DNA chain growth, J. Mol. Biol., 95:63-70 (1975).
32. A. T. Diaz, D. Wiener, and R. Werner, Synthesis of small polynucleotide chains in thymine depleted bacteria, J. Mol. Biol., 95:45-61 (1975).
33. B. K. Duncan, P. A. Rockstroh, and H. R. Warner, Escherichia coli K12 mutants deficient in uracil-DNA glycosylase, J. Bacteriol., 134:1039-1045 (1978).
34. T. Edlund, P. Gustafsson, and H. Wolf-Katz, Effect of thymine concentration on the mode of chromosomal replication in Escherichia coli K12, J. Mol. Biol., 108:295-303 (1976).
35. S. Eriksson, B. Sjoberg, S. Hahne, and O. Karlstrom, Ribonucleoside diphosphate reductase from Escherichia coli. An immunological assay and a novel purification from an overproducing strain lysogenic for phage λ dnrd, J. Biol. Chem., 252: 6132-6138 (1977).
36. J. L. Farmer and F. J. Rothman, Transformable thymine requiring mutant of Bacillus subtilis, J. Bact., 89:262-263 (1965).
37. D. Filpula and J. A. Fuchs, Regulation of ribonucleoside diphosphate reductase synthesis in Escherichia coli: Increased enzyme enzyme synthesis as a result of inhibition of deoxyribonucleic acid synthesis, J. Bact., 130:107-113 (1977).
38. J. B. Flanegan and G. R. Greenberg, Regulation of deoxyribonucleotide biosynthesis during in vivo bacteriophage T4 DNA replication. Intrinsic control of synthesis of thymine and 5-hydroxymethylcytosine deoxyribonucleotides at precise ratio found in DNA, J. Biol. Chem., 252:3019-3027 (1977).

39. J. A. Fuchs, O. H. Karlstrom, R. H. Warner, and P. Reichard, Defective gene product in dnaF mutant of Escherichia coli, Nature New Biol., 238:69-71 (1972).
40. A. R. Grivell and J. F. Jackson, Thymidine kinase: Evidence for its absence from Neurospora crassa and some other microorganisms, and the relevance of this to specific labeling of deoxyribonucleic acid, J. Gen. Microbiol., 54:307-317 (1968).
41. M. Greene and W. Firschein, Role of deoxyribonucleic acid ligase in a deoxyribonucleic acid membrane fraction extracted from pneumonocci, J. Bact., 126:777-784 (1976).
42. M. V. Hamlett and H. Berger, Mutations altering genetic recombination and repair of DNA in bacteriophage T4, Virology, 63:539-562 (1975).
43. W. J. Harris, The occurrence of two types of synthesis of deoxyribonucleic acid during normal growth of Bacillus subtilis, Biochem. J., 135:315-325 (1973).
44. D. Huszar and S. Bacchetti, Is ribonucleotide reductase the transforming function of herpes simplex virus 2, Nature, 302: 76-79 (1983).
45. M. K. Jacobson and K. G. Lark, DNA replication in Escherichia coli: evidence for two classes of small deoxyribonucleotide chains, J. Mol. Biol., 73:371-376 (1973).
46. J. H. Kim, I. H. Kim, and M. L. Eidinoff, Cell viability and nucleic acid metabolism after exposure of HeLa cells to excess thymidine and deoxyadenosine, Biochem. Pharmacol., 14:1821-1829 (1975).
47. T. Kogoma and K. G. Lark, DNA replication in Escherichia coli: replication in the absence of protein synthesis after replication inhibition, J. Mol. Biol., 52:143-164 (1970).
48. T. Kogoma, A. novel Escherichia coli mutant capable of DNA replication in the absence of protein synthesis, J. Mol. Biol., 121:55-69 (1978).
49. T. Kogoma, T. A. Torrey, and M. J. Connaughton, Induction of UV-resistant DNA replication in Escherichia coli: Induced stable DNA replication as an SOS function, Mol. Gen. Genet., 176:1-9 (1979).
50. A. Kornberg, DNA Replication, W. H. Freeman and Co., San Francisco (1980).
51. Y. Kurosawa and R. Okazaki, Mechanism of DNA chain growth XIII. Evidence for discontinuous replication of both strands of P2 phage DNA, J. Mol. Biol., 94, 229-241 (1975).
52. D. Kuebbing and R. Werner, A. model for compartmentation of de novo and salvage thymidine nucleotide pools in mammalian cells, Proc. Natl. Acad. Sci., 72:3333-3336 (1975).
53. B. A. Kunz, Genetic effects of deoxyribonucleotide pool imbalances, Environ. Mut., 4:695-725 (1982).
54. T. J. Kwoh, P. T. Chan, and M. H. Patrick, Examination of newly synthesized DNA in Escherichia coli, Mol. Gen. Genet., 173:85-93 (1979).

55. S. S. Lee, B. C. Giovanella, and J. S. Stehlin, Selective lethal effect of thymidine on human and mouse tumor cells, J. Cell Physiol., 92:401-405 (1977).
56. L. A. Lewis, E. Mengheri, and M. Estaban, Induction of an antiviral response by interferon requires thymidine kinase, Proc. Natl. Acad. Sci., 80:26-30 (1983).
57. J. Lichtenstein, H. D. Barner, S. S. Cohen, The metabolism of exogenously supplied nucleotides by Escherichia coli, J. Biol. Chem., 235:457-465 (1960).
58. J. Little, Control of the SOS regulatory system by the level of recA protease, Biochemie, 64:585-589 (1982).
59. C.-C. Liu, R. L. Burke, U. Hibner, J. Barry, and B. M. Alberts, Probing DNA replication mechanisms with the T4 bacteriophage in vitro system, Cold Spring Harbor Sym. Quant. Biol., 43: 469-487 (1978).
60. L. A. Loeb and T. A. Kunkel, Fidelity of DNA synthesis, Ann. Rev. Biochem., 52:429-457 (1982).
61. J. M. Lourn and R. E. Bird, Size distribution and molecular polarity of newly replicated DNA in Escherichia coli, Proc. Natl. Acad. Sci., 71:329-333 (1974).
62. A. Luder and G. Mosig, Two alternative mechanisms for initiation of DNA replication forks in bacteriophage T4: Priming by RNA polymerase and by recombination, Proc. Natl. Acad. Sci., 79:1101-1105 (1982).
63. C. A. Lunn and V. Pigiet, Characterization of a high activity form of ribonucleotide diphosphate reductase from Escherichia coli, J. Biol. Chem., 254:5008-5014 (1979).
64. C. Manoil, N. Sinha, and B. Alberts, Intracellular DNA-protein complexes from bacteriophage T4-infected cells isolated by a rapid two-step procedure, J. Biol. Chem., 252:2734-2741 (1977).
65. J. D. Manwaring and J. A. Fuchs, Relationship between deoxyribonucleoside triphosphate pools and deoxyribonucleic acid synthesis in nrdA mutant of Escherichia coli, J. Bact., 138: 245-248 (1979).
66. C. K. Mathews, T. W. North, and G. P. V. Reddy, Multienzyme complexes in DNA precursor biosynthesis, Adv. Enzyme Regul., 17:133-156 (1979).
67. M. Matthes, D. T. Denhardt, The mechanism of replication of ϕX174 DNA XVI. Evidence that the ϕX174 viral strand is synthesized discontinuously, J. Mol. Biol., 136:45-63 (1980).
68. D. McCarthy, C. Minner, H. Bernstein, and C. Bernstein, DNA elongation rates and growing point distributions of wild-type phage T4 and a DNA-delay amber mutant, J. Mol. Biol., 106: 963-981 (1976).
69. R. J. Melamede and S. S. Wallace, Properties of the nonlethal recombinational repair x and y mutants of bacteriophage T4 II. DNA synthesis, J. Virol., 24:28-40 (1977).
70. R. J. Melamede and S. S. Wallace, The effect of exogenous deoxyribonucleosides on thymidine incorporation in T4-infected cells, FEBS Let., 87:12-15 (1978).

71. R. J. Melamede and S. S. Wallace, Properties of the nonlethal recombinational repair deficient mutants of bacteriophage T4 III. DNA replicative intermediates and T4w, Mol. Gen. Genet., 177:501-509 (1980).
72. R. J. Melamede and S. S. Wallace, Phenotypic differences among the alleles of the T4 recombination defective mutants, Mol. Gen. Genet., 179:327-330 (1980).
73. R. J. Melamede and S. S. Wallace, Incorporation of thymine-containing DNA precursors in wild-type and mutant T4-infected plasmolyzed cells, Mol. Gen. Genet., 191:382-388 (1983).
74. R. J. Melamede and S. S. Wallace, Incorporation of thymine-containing DNA precursors in plasmolyzed cells infected by the T4 non-lethal recombination defective mutants, Mol. Gen. Genet., 191:389-392 (1983).
75. T. Minagawa, A. Murakami, Y. Ryo, and W. Yamagishi, Structural features of very fast sedimenting DNA formed by gene 49 defective T4, J. Mol. Biol., 126:183-193 (1983).

ribonucleotides at the 5' termini of some DNA molecules isolated from Escherichia coli polA ex2, J. Mol. Biol., 116:681-707 (1977).
77. Nakayama and P. Hanawalt, Sedimentation analysis of deoxyribonucleic acid from thymine starved Escherichia coli, J. Bact., 121:537-547 (1975).
78. J. Neuhard and E. Thomasen, Turnover of deoxyribonucleoside triphosphates in Escherichia coli 15T during thymine starvation, Eur. J. Biochem., 20:36-43 (1971).
79. G. Nicolis and J. Prigogine, Self-organization in nonequilibrium systems. From Dissipative Structures to Order through Fluctuations, John Wiley and Sons, New York (1977).
80. H. Noguchi, G. P. V. Reddy, and A. B. Pardee, Rapid incorporation of label from ribonucleotide diphosphates into DNA by a cell-free high molecular weight fraction from animal cell nuclei, Cell, 32:443-451 (1983).
81. T. Ogawa and T. Okazaki, Discontinuous DNA replication, Ann. Rev. Biochem., 49:421-457 (1980).
82. R. Okazaki, T. Okazaki, K. Sakabe, K. Sugimoto, R. Kainuma, A. Sugino, and N. Iwatsuki, In vivo mechanisms of DNA chain growth, Cold Spring Harbor Symp. Quant. Biol., 33:129-143 (1968).
83. R. Okazaki, K. Sugimoto, T. Okazaki, Y. Imae, and A. Sugino, Mechanism of DNA chain growth III. Equal annealing of T4 nascent short DNA chains with the separated complementary strands of the phage DNA, Nature, 228:223-226 (1970).
84. R. Okazaki, M. Arisnwa, and A. Sugino, Slow joining of newly replicated DNA chains in DNA polymerase I deficient Escherichia coli mutants, Proc. Natl. Acad. Sci., 68:2954-2957 (1971).
85. B. M. Olivera and F. Bonhoeffer, Discontinuous DNA replication in vitro: I two distinct size classes of intermediates, Nature New Biol., 240:233-235 (1972).

86. B. M. Olivera, DNA intermediates at the Escherichia coli replication fork: Effect of dUTP, Proc. Natl. Acad. Sci. USA, 75:238-242 (1978).
87. A. R. Peterson, J. R. Landolph, H. Peterson, and C. Heidelberger, Mutagenesis of Chinese hamster cells is facilitated by thymidine and deoxycytidine, Nature, 276:508-510 (1978).
88. E. M. Phizicky and J. W. Roberts, Induction of SOS functions: Regulation of proteolytic activity of E. coli recA protein by interaction with DNA and nucleoside triphosphate, Cell, 25: 259-267 (1981).
89. I. Prigogine, From Being to Becoming, Time and Complexity in the Physical Sciences, W. H. Freeman and Co., San Franscisco (1980).
90. K. H. Pritchard and K. G. Lark, Induction of replication by thymine starvation at the chromosome origin in Escherichia coli, J. Mol. Biol., 9:288-307 (1964).
91. M. Radman, Phenomenology of an inducible mutagenic DNA repair pathway in Escherichia coli: SOS repair hypothesis, in: "Molecular and Environmental Aspects of Mutagenesis" (L. Prakash, F. Sherman, M. Miller, C. Lawrence, and H. W. Tabor, eds.), Charles C. Thomas, Publisher, Springfield, Illinois (1974).
92. M. Raggenbass and L. Caro, Intermediates of chromosomal DNA replication in Escherichia coli, J. Mol. Biol., 159:273-301 (1982).
93. W. E. Razzel and P. Casshyap, Substrate specificity and induction of thymidine phosphorylase in Escherichia coli, J. Biol. Chem., 239:1789-1793 (1964).
94. G. P. V. Reddy, A. Singh, M. E. Stafford, C. K. Mathews, Enzyme associations in T4 phage DNA precursor synthesis, Proc. Natl. Acad. Sci. USA, 74:3152-3156 (1977).
95. G. P. V. Reddy and C. K. Mathews, Functional compartmentation of DNA precursors in T4 phage-infected bacteria, J. Biol. Chem., 253:3461-3467 (1978).
96. G. P. V. Reddy and A. B. Pardee, Multienzyme complex for metabolic channeling in mammalian DNA replication, Proc. Natl. Acad. Sci., 77:3312-3316 (1980).
97. G. P. V. Reddy and A. B. Pardee, Coupled ribonucleoside diphosphate reduction, channeling and incorporation into DNA of mammalian cells, J. Biol. Chem., 257:12526-12531 (1982).
98. C. W. Roberts and P. L. Moreau, A brief consideration of the SOS inducing signal, Biochemie, 64:805-807 (1982).
99. E. Schandl, Oligodeoxyribonucleotides from pulse labeled bacterial cells, Biochem. Biophys. Acta, 262:420-430 (1972).
100. D. W. Siegmann and R. Werner, Origin and characterization of short DNA chains in Escherichia coli, Eur. J. Biochem., 120: 497-509 (1981).

101. D. W. Siegmann and R. Werner, Novel structure at 5'-ends of nascent DNA chains, Proc. Natl. Acad. Sci., 73:3438-3442 (1976).
102. O. Smithies, The control of globin and other eukaryotic genes, J. Cell. Physiol. Suppl. 1:137-143 (1982).
103. M. F. Stafford, G. P. V. Reddy, C. K. Mathews, Further studies on bacteriophage T4 DNA synthesis in sucrose-plasmolyzed cells, J. Virol., 23:53-60 (1977).
104. R. Sternglanz, H. F. Wang, and J. D. Donegan, Evidence that both growing DNA chains at a replication fork are synthesized discontinuously, Biochem., 15:1838-1843 (1976).
105. F. Tamanoi and T. Okazaki, Uracil incorporation into nascent DNA of thymine-requiring mutant of *Bacillus subtilis* 168, Proc. Natl. Acad. Sci. USA, 75:2195-2199 (1978).
106. P. K. Tomich, C. S. Chiu, M. G. Wovcha, and G. R. Greenberg, Evidence for a complex regulating the *in vivo* activities of early enzymes induced by bacteriophage T4, J. Biol. Chem., 249:7613-7632 (1974).
107. G.-K. Tye, P.-O. Nyman, I. R. Lehman, S. Hochhauser, and B. Weiss, Transient accumulation of Okazaki fragments as a result of uracil incorporation into nascent DNA, Proc. Natl. Acad. Sci. USA, 74:154-157 (1977).
108. M. S. Valenzuela and R. B. Inman, Direction of bacteriophage DNA replication in a thymine requiring *Escherichia coli* K-12 strain. Effect of thymidine concentration, Nucleic Acids Res., 9:6975-6984 (1981).
109. M. S. Valenzuela and R. B. Inman, Multiply branched replicative intermediates in *E. coli* and bacteriophage λ, Mol. Gen. Genet., 184:450-456 (1981).
110. H. F. Wang and R. Sternglanz, Thymine-labeled deoxyoligonucleotide involved in DNA chain growth in *Bacillus subtilis*, Nature, 248 147-150 (1974).
111. R. Werner, Mechanism of DNA replication, Nature, 230:570-572 (1971).
112. R. Werner, Nature of DNA precursors, Nature New Biol., 233: 99-103 (1971).
113. E. M. Witkin, Ultraviolet mutagenesis and inducible DNA repair in *Escherichia coli*, Bacteriol. Rev., 40:869-907 (1976).
114. E. M. Witkin, From Gainsville to Toulouse: the evolution of a model, Biochemie, 64:549-555 (1982).
115. O. Wirak and G. R. Greenberg, Role of bacteriophage T4 DNA-delay gene products in deoxyribonucleotide synthesis, J. Biol. Chem., 255:1896-1904 (1980).
116. M. G. Wovcha, C.-S. Chiu, P. K. Tomich, and G. R. Greenberg, Replicative bacteriophage DNA synthesis in plasmolyzed T4-infected cells: evidence for two independent pathways to DNA, J. Virol., 20:142-156 (1976).

MULTIPLE MECHANISMS OF ADENOSINE TOXICITY IN AN ADENOSINE SENSITIVE MUTANT OF BABY HAMSTER KIDNEY (BHK) CELLS

Voon-Loong Chan and H. Jeok Ho

Department of Microbiology
Fitzgerald Building
The University of Toronto
Toronto, Ontario
Canada M5S 1A8

ABSTRACT

A class of arabinosyladenine-resistant baby hamster kidney (BHK) cell mutants, isolated in our laboratory, shows cross-resistance to deoxyadenosine, alteration of adenosine kinase, elevation of spontaneous mutation rate, and extreme sensitivity to adenosine. One of these adenosine sensitive mutants, ara-s10d, was isolated spontaneously and studies with Ado^r revertants suggest the involvement of a single pleiotropic mutation. The enhanced adenosine toxicity in ara-s10d cells can be attributed to pyrimidine nucleotide starvation and to at least one other mechanism, which is associated with a 200-fold elevation of IMP, 3-5 fold elevation of ATP, GTP, S-adenosylmethionine (AdoMet) and methylthioadenosine (MeSAdo).

INTRODUCTION

Adenosine (Ado), a naturally occurring purine nucleoside can be extremely cytotoxic [6, 9]. The recent increased interest in the mechanisms of adenosine toxicity can be attributed to the discovery of a severe combined immunodeficiency disease caused by a deficiency in adenosine deaminase [7], a purine salvage enzyme. At least three mechanisms have been proposed for adenosine toxicity. These include alteration of adenylate cyclase activity, inhibition of pyrimidine nucleotide synthesis and inhibition of methylation via the formation of S-adenosylhomocysteine (AdoHcy) [6, 9].

Recently we [1] reported the isolation of a class of arabinosyladenine resistant ($araA^r$) baby hamster kidney (BHK) cell mutants that shows cross-resistance to deoxyadenosine (dAdo) and extreme adenosine sensitivity. One of these Ado^s mutants, ara-s10d, was isolated without mutagenic treatment. Studies carried out with Ado^r revertants of ara-s10d cells showed that the Ado^s/araA^r/dAdo^r phenotype and the adenosine kinase alteration are probably due to a single mutational event [3]. The nature of the pleiotropic mutation is intriguing but is presently unknown.

In this communication, evidence of the involvement of multiple mechanisms of adenosine toxicity in ara-s10d cells are presented. Toxicity can be attributed to pyrimidine nucleotide starvation and to at least one other novel mechanism, which is associated with an elevation of methylthioadenosine, S-adenosylmethionine and purine nucleotides.

MATERIALS AND METHODS

Chemicals: Adenosine (Ado), AMP, ADP, and ATP were from Calbiochem; GDP, GTP, UTP, CTP, uridine (Urd), deoxycytidine (dCyd), IMP, AdoMet, AdoHcy, and MeSAdo were purchased from Sigma Chemicals, erythro-9-(2-hydroxy-3-nonyl) adenine (EHNA) from Burrough Welcome Co.

Cell Lines, Media, and Culture Conditions

BHK wild-type and mutant cells were routinely cultivated in plastic tissue culture flasks at 34°C in α-minimal essential medium (MEM) lacking ribonucleosides and deoxyribonucleosides [22] supplemented with 5% fetal calf serum. In all plating experiments to determine Ado toxicity, α-MEM was supplemented with 10% horse serum plus 5 μM EHNA. Plating experiments to assess Ado toxicity by colony formation were carried out in 6 multiwell culture dish or 60 mm petri culture dish with 50-5000 or 100-50,000 cells respectively.

Preparation of Acid-Soluble Cell Extracts

BHK wild-type and Ado^s mutant cells were subjected to treatment with 50 μM Ado plus 5 μM EHNA for a period of up to 8 h. About 6×10^6 cells were plated in 80 ml incubation medium in 150 cm^2 plastic tissue culture flasks. At the end of the treatment period, the medium was removed and the cells washed three times with 80 ml ice-cold 0.9% sodium chloride solution containing 20 μM EHNA. The cells were next scraped with a rubber policeman and collected into a centrifuge tube. Centrifugation was carried out at 710 g for 5 min to separate the pellet from the supernatant. The pellets were kept for protein determination [16] and the supernatant neutralized with 2 N $KHCO_3$. The supernatants were again centrifuged at 710 g for 5 min to remove the potassium perchlorate precipitate, freeze-dried and made up to 0.2 to 0.3 ml with distilled water. Cell extracts were stored at -70°C.

High Performance Liquid Chromatography (HPLC)

Analysis was performed using the Waters Associates solvent delivery system (model 6000A and M-45), a universal liquid chromatography injector (model U6K) and an absorbance detector (model 440). Chromatographic runs were recorded on Waters 730 Data Module. Quantitation of intracellular pool size levels of AdoHcy and MeSAdo were carried out with Waters Associates μBondapak C18 column (0.4 × 30 cm) at a flow rate of 1 ml/min using a 30 min linear gradient from a low buffer at 100 mM phosphate, pH 6.0 to a high buffer of 100 mM phosphate, pH 6.0 in 25% methanol. The final condition was maintained for a further 15 min. The retention time for AdoHcy and MeSAdo was 21 min and 38.5 min respectively.

Intracellular AdoMet pools were analyzed on a paired-ion coupling exchange μBondapak C18 column eluting isocratically at a flow rate of 2 ml/min with 6 mM heptane sulfonate in 20% methanol; titrated to pH 3.3 with acetic acid. AdoMet was eluted at 8.5 min using this system.

AdoHcy, AdoMet, and MeSAdo were identified by chromatographing with standards as well as determining the incorporation of [^{3}H] and [^{35}S]-label in the respective peaks when cells were treated in the presence of [^{3}H]-Ado and [^{35}S]-methionine.

Purine and pyrimidine nucleotides were separated on a Partisil-SAX anionic exchange column (0.46 × 25 cm), eluting isocratically for 10 min at a flow rate of 2 ml/min with 10 mM $NH_4H_2PO_4$ pH 4.0 followed by a 30 min linear gradient to 250 mM $NH_4H_2PO_4$ pH 5.2 containing 10% acetonitrile. The final conditions was maintained for a further 15 min. All buffers used were filtered through 0.45 μ filter and degassed under reduced pressure before use. The retention time for the various nucleotides are as follows: GTP - 54 min, ATP - 43 min, CTP - 29 min, UTP - 27 min, GDP - 25 min, ADP - 23 min, and IMP - 11 min.

RESULTS

An arabinosyladenine resistant mutant, ara-s10d, of BHK cells was shown to be about 200-fold more sensitive to adenosine [1]. To determine if the enhanced adenosine sensitivity can be attributed to pyrimidine starvation, we plated ara-s10d cells in various concentrations of adenosine in the presence of 5 μM uridine (Urd) and deoxycytidine (dCyd). As shown in Fig. 1, the addition of pyrimidine nucleosides resulted in a partial prevention of Ado toxicity in ara-s10d cells. Addition of Urd and dCyd to a concentration of 20 μM did not further reduce the toxicity of adenosine (data not shown). Inability of pyrimidine nucleosides to abolish the enhanced adenosine toxicity in ara-s10d cells suggest that the increased Ado toxicity is not solely due to pyrimidine starvation. The toxicity of Ado for

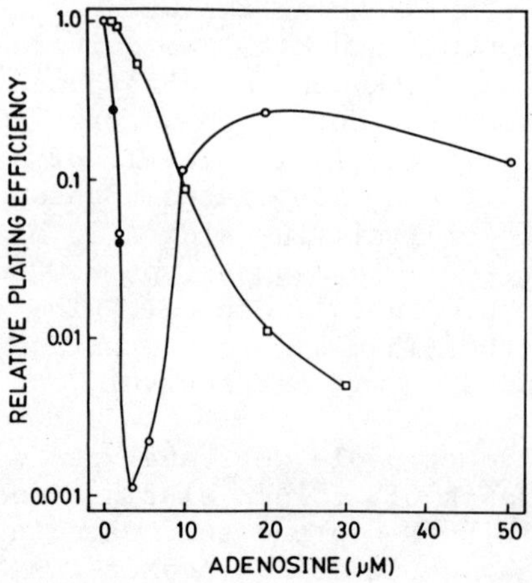

Fig. 1. Effects of Ado on the relative plating efficiency of BHK wild-type cells (○---○), ara-s10d cells (●---●), and ara-s10d cells in the presence of 5 μM Urd and dCyd (□---□). Figure shown is a representative of 3 separate experiments in which identical results were obtained.

TABLE 1. Effect of Exogenous Ado on Intracellular Purine and Pyrimidine Nucleotides. Treatment was Carried out for 8 h in the Presence of 5 μM EHNA

Cell Type	Addition (μM Ado)	ATP	GTP	IMP	UTP + CTP
		nmole/mg protein			
BHK	0	17.0	4.0	0.07	9.7
BHK	50	23.1	4.3	0.06	7.7
ara-s10d	0	15.5	3.9	0.07	10.5
ara-s10d	50	52.5	19.5	15.3	0.3
c8a100	0	20.0	4.5	0.1	10.4
c8a100	50	27.0	5.3	0.1	8.0
ara-s10d501A	0	16.3	4.3	0.1	9.4
ara-s10d501A	50	55.2	17.2	4.3	4.9

wild-type BHK cells for concentration below 5 μM can be completely prevented by a concurrent addition of 0.5 μM uridine (Archer, Ho, and Chan submitted for publication).

The effects of Ado on purine and pyrimidine nucleotides intracellular levels can be followed more directly by quantitating the nucleotides using high performance liquid chromatography (HPLC). The results for both the wild type BHK and ara-s10d cells are summarized in Table 1. As expected, the levels of UTP and CTP were drastically reduced in ara-s10d cells. The combined level of CTP and UTP in control cells was 10.5 nmole/mg protein and upon treatment with 50 μM Ado plus 5 μM EHNA for 8 h the CTP and UTP level was reduced to 0.3 nmole/mg protein. In contrast, similar treatment of BHK wild-type cells only resulted in a slight reduction in UTP and CTP levels.

Ado treatment also drastically altered purine metabolism in ara-s10d cells, resulting in an elevation of ATP and GTP by 3.4 and 5 fold respectively, and a dramatic 200-fold increase in IMP. Similarly ADP and GDP were also elevated 4.3 and 2.9-fold respectively (data not shown). In contrast, similar treatment of wild-type BHK cells caused only a slight increase in adenine and guanine nucleotides and the level of IMP did not change significantly from that of control cells.

To determine if the elevation of adenine and guanine nucleotide and the dramatic accumulation of IMP in ara-s10d cells are associated with the enhanced sensitivity to Ado, we performed similar treatment and pool sizes analyses with an adenosine resistant revertant, C8a100, of ara-s10d. C8a100, is one of the 7 independently derived single-site- revertants of ara-s10d cells and in all cases the reversal mutation completely abolished the AK activity [3]. Treatment of C8a100 cells with 50 μM Ado plus 5 μM EHNA for 8 h resulted in only a slight elevation of ATP and GTP pools amounting to 1.35 and 1.17 fold respectively when compared to untreated C8a100 cells (Table 1). IMP levels in the C8a100 treated cells like wild-type BHK cells did not vary from untreated control cells. The reversal mutation also abolished the pyrimidine starvation effect of Ado (Table 1).

A nucleotide independent mode of Ado toxicity that is associated with an elevation of intracellular AdoHcy, has been proposed [11-13]. AdoHcy is presumed to inhibit one or more essential AdoMet-dependent methylation reactions. We thus analyzed the levels of AdoHcy, AdoMet, and MeSAdo in BHK wild-type, ara-s10d, and C8a100 cells treated with 50 μM Ado plus 5 μM EHNA. As shown in Table 2 all three show an elevation of AdoHcy after adenosine treatment as has been observed in other mammalian cells [11-13]. It is apparent that the AdoHcy levels of BHK wild-type cells treated with Ado for 1 h (300 picomole/mg protein) and 8 h (337 picomole/mg protein) are significantly higher than that of ara-s10d cells for both 1 h (199

TABLE 2. Effects of Ado on Intracellular AdoHcy, AdoMet, and MeSAdo Pool Size Levels

Cell Type	Additions	AdoHcy (picomole/mg protein)		AdoMet (picomole/mg protein)		MeSAdo (picomole/mg protein)	
		1 hr	8 hr	1 hr	8 hr	1 hr	8 hr
BHK	0 μM Ado, 5 μM EHNA	9.5	11.6	1350	1360	111	126
BHK	50 μM Ado, 5 μM EHNA	300.0	337.0	1340	1370	108	130
ara-s10d	0 μM Ado, 5 μM EHNA	10.0	11.5	1340	1370	116	121
ara-s10d	50 μM Ado, 5 μM EHNA	199.0	189.0	1550	3720	176	325
ara-16c	0 μM Ado, 5 μM EHNA	ND	9.8	ND	1420	ND	70
ara-16c	50 μM Ado, 5 μM EHNA	ND	105.0	ND	4960	ND	404
c8a100	0 μM Ado, 5 μM EHNA	ND	12.7	ND	1300	ND	125
c8a100	50 μM Ado, 5 μM EHNA	ND	284.0	ND	1358	ND	118

ND - not done

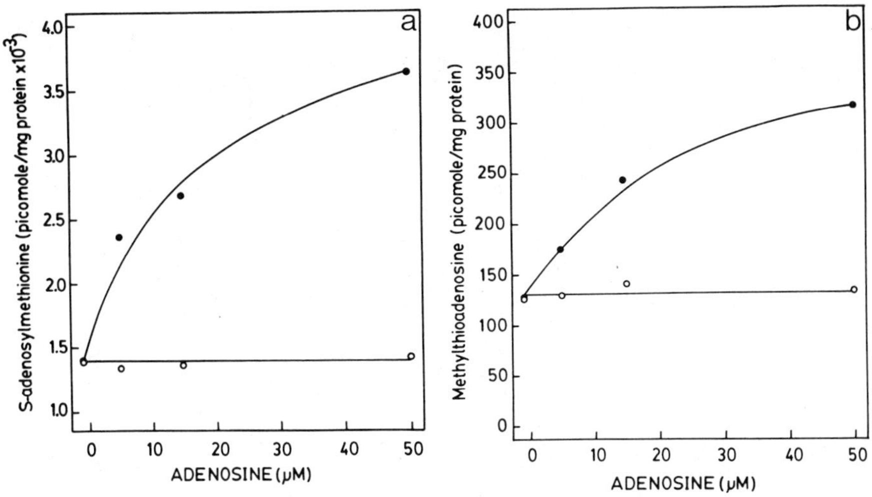

Fig. 2. Effects of Ado on intracellular (a) AdoMet and (b) MeSAdo levels in BHK (○---○) and ara-s10d (●---●) cells. Both cell types were treated for 8 h.

picomole/mg protein) and 8 h (189 picomole/mg protein). Thus the enhanced Ado toxicity in ara-s10d cells is not due to an overproduction of AdoHcy. To determine if the decreased level of AdoHcy in ara-s10d cells treated with Ado is associated with the Ado^s mutation we did the same treatment and analyses with another Ado^s mutant, ara-16c [1] and an Ado^r revertant of ara-s10d, C8a100. Like ara-s10d, ara-16c cells also showed a lesser increase in AdoHcy after treatment with Ado. Furthermore, AdoHcy levels of Ado-treated C8a100 cells are similar to that of wild-type BHK cells, consistent with a notion that the Ado^s mutation has a significant effect on the metabolism of AdoHcy and perhaps also AdoMet.

Measurement of intracellular AdoMet and a catabolite of AdoMet, MeSAdo showed that the levels of these two compounds were not altered in BHK wild-type cells when treated with 50 μM Ado plus 5 μM EHNA for 8 h (Table 1).

Ado-treated ara-s10d and ara-16c cells, on the other hand, showed a substantial elevation in AdoMet and MeSAdo pools and the accumulation can be observed as early as 1 h after the addition of Ado (in ara-s10d cells). The level of elevation of AdoMet and MeSAdo correlates with increasing concentration of Ado added as shown in Fig. 2a and b, respectively. Recently, MeSAdo has been shown to be toxic to lymphocytes culture [23], SV-3T3 fibroblast cells [18], and BHK cells [20].

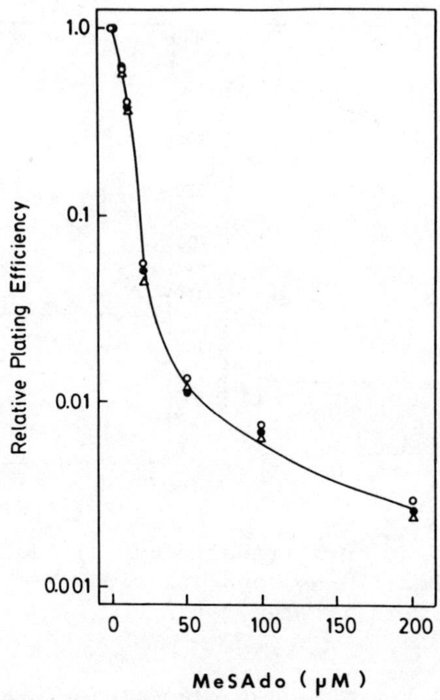

Fig. 3. Relative plating efficiencies of BHK wild-type (○---○), ara-s10d (●---●) and ara-16c (△---△) cells in various MeSAdo concentrations in the presence of 5 μM EHNA.

Levels of AdoHcy, AdoMet, and MeSAdo were determined in the Ado^r revertant of ara-s10d cells, C8a100, following treatment with 50 μM Ado plus 5 μM EHNA for a preriod of 8 h as seen in Table 2. Ado treatment did not increase the AdoMet or MeSAdo level in C8a100 cells. Thus a reversal of Ado^s to Ado^r phenotype is accompanied by a loss in the elevation of AdoMet as well as MeSAdo suggesting an association between Ado^s phenotype and the accumulation of AdoMet and MeSAdo.

MeSAdo is generated during polyamine synthesis or by direct cleavage of AdoMet. Together these two pathways account for the major proportion of MeSAdo productions in mammalian cells [5].

The toxicity of MeSAdo in BHK wild-type and Ado^s mutant cells was assessed by determining the relative plating efficiency of the cells in increasing MeSAdo concentrations. As shown in Fig. 3, both wild-type and mutant cells are relatively sensitive to MeSAdo with a D_{10} value of approximately 17 μM (D_{10} being the concentration of drug needed to achive 90% killing of the cells).

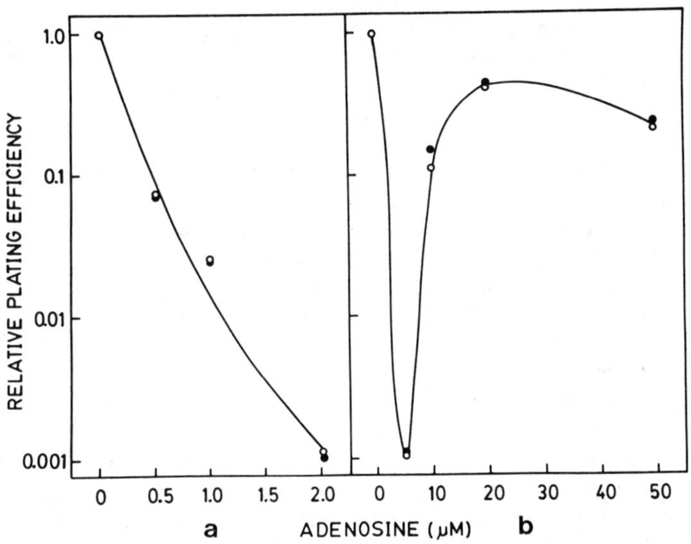

Fig. 4. Ado toxicity in BHK wild-type cells in the presence of 0 μM (○---○) and 1.0 μM (●---●) MeSAdo.

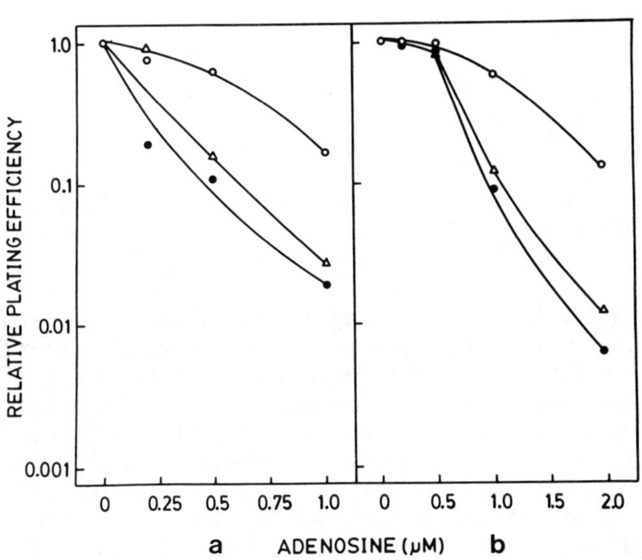

Fig. 5. Ado toxicity in (a) ara-s10d and (b) ara-16c cells in the presence of 0 M (○---○), 0.5 μM (△---△) and 1.0 μM (●---●) MeSAdo.

The high degree of sensitivity exhibited by BHK wild-type and Ado^s mutant cells toward MeSAdo suggest that MeSAdo may be involved in Ado toxicity in ara-s10d and ara-16c cells. Plating of wild-type and Ado^s mutant cells were carried out in the presence of non-toxic MeSAdo concentrations and various concentrations of Ado. MeSAdo at 0.5 and 1.0 μM was previously shown to be non-toxic to the cells.

Relative plating efficiency of wild-type cells in Ado was not altered in the presence of 1.0 μM MeSAdo at any of the Ado concentrations tested (Fig. 4a and b). Toxicity encountered at low Ado concentrations has previously been shown to be due to depletion of pyrimidine nucleotides (S. M. Archer, H. J. Ho, and V. L. Chan, manuscript submitted). Toxicity at higher Ado concentrations is presumably due to accumulation of AdoHcy and as shown in Fig. 4b, MeSAdo at 1.0 μM again did not enhance AdoHcy-mediated Ado toxicity in BHK wild-type cells.

Effects of 0.5 μM and 1.0 μM MeSAdo on Ado toxicity in ara-s10d and ara-16c cells are shown in Fig. 5a and b, respectively. Non-toxic MeSAdo concentrations could enhance Ado toxicity in these mutant cells in contrast to wild-type cells. The degree of enhancement was dependant upon the concentrations of MeSAdo added to the Ado-treated ara-s10d and ara-16c cells, being more pronounced at 1.0 μM than at 0.5 μM.

DISCUSSION

The $araA^r$ BHK mutants that show at 100 to 200-fold increase in sensitivity to Ado, increased resistance to dAdo, altered AK activity [1, 3] and an increased spontaneous mutation rate [2], represent a novel and an interesting group of mammalian cell mutants. One of these Ado^s mutants, ara-s10d, was isolated without mutagenic treatment and studies with 7 independently derived Ado^r revertants suggest the involvement of a single pleiotropic mutation [3].

The increase in Ado toxicity in ara-s10d cells clearly involves pyrimidine nucleotide starvation as demonstrated directly with pool sizes analyses after treatment with a toxic concentration of Ado. However, pyrimidine starvation cannot be the sole basis for enhanced Ado toxicity because uridine and deoxycytidine at a concentration of 5 and 20 μM each can only partially reduce the toxicity of Ado. In contrast toxicity of Ado in wild type BHK cells at concentration below 5 μM can be completely abolished by 0.5 μM uridine (S. M. Archer, H. J. Ho, and V. L. Chan, manuscript submitted). Similarly in other mammalian cells, cellular toxicity of Ado attributed to pyrimidine starvation can be prevented by low concentration of uridine or uridine plus deoxycytidine [8, 10]. Addition of 5 μM Urd and dCyd together with 50 μM Ado plus 5 μM EHNA to ara-s10d cells can prevent a reduction in the levels of UPT and CTP (data not shown) but cannot totally abolish the toxicity of Ado represents the strongest suggestion for the involvement of pyrimidine nucleotide starvation and the presence of at least one other mechanism of Ado toxicity in causing the 200-fold increase in Ado sensitivity of ara-s10d cells.

Adenosine induced pyrimidine starvation in ara-s10d cells is presumably caused by a reduction in the level of phosphoribosylpyrophosphate (PRPP). PRPP synthetase is known to be inhibited by various purine nucleotide including AMP, ADP, GDP, GTP, and IMP all of which are elevated in Ado-treated ara-s10d cells. Consistent with the notion that a reduction in PRPP is induced in Ado-treated cells is the observation that cells of a thioguanine resistant ($HGPRT^-$) mutant of ara-s10d, ara-s10d 501A, are significantly more resistant to Ado compared to the parental ara-s10d cells (unpublished observation). Furthermore, the drastic reduction of pyrimidine nucleotide levels observed in Ado-treated ara-s10d cells was not observed with ara-s10d 501A cells; UTP and CTP being only reduced by about 50% (see Table 1). If there is a significant reduction of PRPP level in Ado-treated ara-s10d cells the *de novo* rate of synthesis of purine nucleotides, which is comprised of 10 consecutive steps starting with the synthesis of 5-phosphoribosyl-1-amine from PRPP and glutamine, should be substantially reduced. The rate of intracellular purine *de novo* synthesis was determined by measuring the incorporation of $[^{14}C]$-formate into the cellular nucleotide pools. The amount of label incorporated into ATP and GTP were indeed inhibited by about 95% in Ado-treated ara-s10d cells while similar Ado treatment of wild-type BHK cells reduced the synthesis of ATP and GTP to 85% and 72% levels respectively of untreated cells (unpublished observations).

In an earlier publication [1] we postulated that the Ado sensitivity of the Ado^s/ara^r mutants may be due to a mutation which increases the level of S-adenosylhomocysteine hydrolase, thereby causing a higher production of AdoHcy, a potent inhibitor of many methyltransferases. The increased resistance to araA was presumed to be attributed to the sequestering effect of SAH hydrolyase. This hypothesis is not supported by the observation reported here; the level of AdoHcy in ara-s10d cells treated with Ado increased to a level lower than that of wild-type BHK cells (Table 2). Furthermore, the levels of SAH hydrolase in BHK and ara-s10d cells are quite comparable in both Ado-treated and untreated cells (unpublished data of S. M. Archer). Thus elevation of SAH hydrolase with concurrent increase in SAH is not the basis of enhanced adenosine toxicity in ara-s10d cells.

The significance of the 200-, 3.4, and 5-fold elevation of IMP, ATP, and GTP respectively in contributing to the enhanced Ado toxcity in ara-s10d cells remains to be established. However, studies made with revertants of ara-s10d showed that a reversal of Ado^s to Ado^r phenotype is accompanied by a loss in the sharp elevation of purine nucleotides upon treatment with Ado suggesting an association between enhanced Ado sensitivity and the elevation of purine nucleotides.

The dramatic accumulation of purine nucleotides in ara-s10d cells treated with Ado is partly attributed to a higher level of phosphorylation of exogenous Ado coupled with increased degradation of RNA or inhibition of RNA synthesis (Ho and Chan submitted for publication).

AdoMet is a methyl donor of transmethylation reactions which have numerous roles in cellular metabolism including gene regulation [4, 21]. The 5-fold elevation of AdoMet may be of significance in determining the enhanced toxicity of Ado in ara-s10d and ara-16c cells. It is likely that the increase in AdoMet levels in the Ado^s mutant cells is a result of increased synthesis (due to a higher level of ATP, a limiting substrate). Blockage of methylations in Ado-treated cells would lead to an accumulation of AdoMet and a lowered increase in AdoHcy [12] as observed for ara-s10d and ara-16c cells treated with Ado.

The observation of a 3-4-fold accumulation of MeSAdo in the Ado^s mutants, ara-s10d, and ara-16c treated with Ado while similar treatment do not cause any accumulation of MeSAdo in wild-type BHK and an Ado^r revertant of ara-s10d, C8a100 cells, suggesting an association of Ado^s phenotype with an accumulation of MeSAdo. Both wild-type and ara-s10d cells are sensitive to the toxic effect of MeSAdo with a D_{10} value of approximately 17 μM. Based on a cell volume of 4.5 $\mu l/10^6$ BHK cells [17], MeSAdo concentrations were elevated from less than 3.4 μM in control cells to 10.0 and 12.5 μM in ara-s10d and ara-16c cells respectively upon treatment with 50 μM Ado plus 5 μM EHNA for 8 h. From the above discussion and the observation that a nontoxic concentration of MeSAdo can enhance the toxicity of Ado in Ado^s mutant cells, ara-s10d, and ara-16c, but not that of wild-type BHK cells suggest a distinct possibility that part of the basis of the enhanced Ado toxicity in the Ado^s mutants is caused by an elevated level of MeSAdo.

The molecular basis of MeSAdo toxicity in mammalian cells is not completely understood but it has been shown to inhibit mRNA [15], r-RNA [14] synthesis, inhibit activity of spermine and spermidine synthetase [18] and SAH hydrolase [19].

The results presented here indicate the involvement of multiple mechanisms in determining the 100-200-fold increase in sensitivity to adenosine by ara-s10d and ara-16c cells. The mutation is pleiotropic [13], causing increased resistance to araA, dAdo, increased sensitivity to Ado, increased spontaneous mutation rate [2] and an alteration in the AK enzyme [3]. The mutation alters Km and pH optimum of the AK enzyme as demonstrated with crude [3] and purified enzume (unpublished observation of P. Juranka). The level of AK activity does not correlate with the degree of sensitivity to Ado, since all 6 Ado^s mutants have very similar level of sensitivity but their AK values range from 59% to non detectable at pH 5.5 [1] or from 300% to less than 0.1% at pH 7.4 [1, 3]. It is possible that the AK enzyme may have an unidentified regulatory function, which is altered in the Ado^s mutants. Alternatively all the Ado^s mutants have a common structural mutation, insertion or translocation, which has a polar or pleiotropic effect.

ACKNOWLEDGMENTS

This work was supported by the University of Toronto Connaught Fund, the National Cancer Institute and Medical Research Council of Canada. We are grateful to Dr. W. H. Lewis and Mr. P. Juranka for valuable discussions, Mrs. S. Guttman for excellent technical assistance and Ms. Trudy Carrol for excellent typing.

REFERENCES

1. V. L. Chan and P. Juranka, Isolation and preliminary characterization of 9-β-D-arainofuranosyladenine-resistant mutants of baby hamster cells, Somat. Cell Genet., 7:147-160 (1981).
2. V. L. Chan, S. Guttman, and P. Juranka, Mutator genes of baby hamster kidney cells, Mol. Cell Biol., 1:568-571 (1981).
3. V. L. Chan, F. Meffe, S. Guttman, P. Juranka, and S. M. Archer, in: "Manipulation and Expression of Genes in Eucaryotes" (P. Nagley, A. W. Linnane, W. J. Peacock, and J. A. Pateman, eds.), pp. 55-58, Acad. Press, Sydney (1983).
4. M. Ehrlich and R. H. Wang, 5-methylcytosine in eukaryotic DNA, Science, 212:1350-1357 (1981).
5. A. J. Fero, in: "Transmethylation" (E. Usdin, R. T. Borchardt, and C. R. Crevelling, eds.), pp. 117-126, Elsevier North Holland, Inc., New York (1979).
6. I. H. Fox and W. N. Kelley, The role of adenosine and 2'-deoxyadenosine in mammalian cells, Ann. Rev. Biochem., 47:655-686 (1978).
7. E. R. Giblett, J. E. Anderson, F. Cohen, B. Pollara, H. J. Meuwissen, Adenosine-deaminase deficiency in two patients with severely impaired cellular immunity, Lancet, 2:1067-1069 (1972).
8. L. J. Gudas, A. Cohen, B. Ullman, and D. W. Martin, Jr., Analysis of adenosine-mediated pyrimidine starvation using cultured wild-type and mutant mouse T lymphoma cells, Somat. Cell. Genet., 4:201-219 (1978).
9. J. F. Henderson, Effects of nucleoside analoges on purine metabolism, Pharmacol. Ther., 2:751-769 (1978).
10. K. Ishii and H. Green, Lethality of adenosine for cultured mammalian cells by interference with pyrimidine biosynthesis, J. Cell Sci., 13:429-439 (1973).

11. J. M. Johnston and N. M. Kredich, Inhibition on methylation by adenosine in adenosine deaminase inhibited, phytohemagglutinin-stimulated human lymphocytes, J. Immunol., 123:97-103 (1979).
12. N. M. Kredich and M. S. Hershfield, Pertubation in S-adenosylhomocysteine and S-adenosylmethionine metabolism: effects on transmethylation, Adv. Enzy. Regul., 18:181-191 (1980).
13. N. M. Kredich and D. W. Martin, Jr., Role of S-adenosylhomocysteine in adenosine-mediated toxicity in cultured mouse T lymphoma cells, Cell, 12:931-938 (1977).
14. R. E. Law and L. Miller, Effect of 5'-methylthioadenosine on nuclear morphology and RNA metabolism in cultured Xenopus laevis cells, Exp. Cell Res., 135:435-438 (1981).
15. R. E. Law, R. M. Sinibaldi, M. R. Cummings, and A. J. Ferro, Inhibition of RNA synthesis in salivary glands of Drosophila melanogaster by 5'-methylthioadenosine, Biochem. Biophys. Res. Commun., 73:600-606 (1976).
16. O. H. Lowry, N. F. Rosenburg, A. L. Farr, and R. J. Randall, Protein measurement with folin phenol reagent, J. Biol. Chem., 193:265-275 (1951).
17. T. McDonald, H. G. Sachs, C. W. Orr, and J. D. Ebert, External potassium and baby hamster kidney cells: intracellular ions, ATP, growth, DNA synthesis and membrane potential, Develop. Biol., 28:290-303 (1972).
18. A. E. Pegg, R. T. Borchardt, and J. K. Coward, Effects of inhibitors of spermidine and spermine synthesis on polyamine concentrations on growth of transformed mouse fibroblasts, Biochem. J., 194:79-89 (1981).
19. F. D. Ragione and A. E. Pegg, Effect of analogues of 5'-methylthioadenosine on cellular metabolism, Biochem. J., 210:429-435 (1983).
20. A. Raina, K. Tuomi, and R. L. Pajula, Inhibition of the synthesis of polyamines and macromolecules by 5'-methylthioadenosine and 5'-alkylthiotubercidine in BHK 21 cells, Biochem. J., 204:697-703 (1982).
21. A. Razin and A. D. Riggs, DNA methylation and gene function, Science, 210:604-610 (1980).
22. C. P. Stanners, G. I. Eliceri, and H. Green, Two types of ribosomes in mouse-hamster hybrid cells, Nature New Biol., 230:52-54 (1971).
23. A. A. Vandenbark, A. J. Ferro, and G. L. Barney, Inhibition of lymphocyte transformation by a naturally occurring metabolite 5'-methylthioadenosine, Cell. Immunol., 49:26-33 (1980).

EFFECTS OF ALTERED NUCLEOTIDE CONCENTRATIONS

ON THE FIDELITY OF DNA REPLICATION

Shishir K. Das, Thomas A. Kunkel*,
and Lawrence A. Loeb

Joseph Gottstein Memorial Cancer Research Laboratory
Department of Pathology, SM-30
University of Washington
Seattle, Washington 98195

*NIEHS
P. O. Box 12233
Research Triangle Park, North Carolina 27709

1. INTRODUCTION

Until recently, studies on the molecular basis of mutageneis have focused primarily on changes caused by mutagenesis that modify DNA. In addition, limited efforts have been directed at asking whether modification of the DNA-replicating proteins affects the accuracy of DNA replication. Recently there has been a growing realization that deoxynucleoside triphosphate pools themselves also might influence the accuracy of DNA replication. One of the most direct approaches to the problem is to study the effects of variations in the concentrations of deoxynucleoside triphosphates on the frequency of in vitro misincorporation by purified DNA polymerases. The concept is that errors made at the point of DNA replication by these enzymes are rate-limiting with respect to mutagenesis, and that misincorporated nucleotides are not excised with complete efficiency. In this manuscript we will review studies on the fidelity of DNA synthesis in vitro as a function of the relative concentration of nucleotides. We will also consider recent studies suggesting that DNA-damaging agents alter nucleotide metabolism in cells.

2. MISINCORPORATION WITH POLYNUCLEOTIDE TEMPLATES

Early studies on the effects of variations in deoxynucleotide concentration on the fidelity of DNA replication were carried out with polynucleotide templates. These templates contain only one or two types of deoxynucleotides; thus, correct and incorrect incorporation can be identified simply as those nucleotides that are complementary and noncomplementary to nucleotides on the template. With the use of radioactive substrates containing different isotopes, it has been possible to simultaneously measure both correct and incorrect incorporation, and thus calculate error rates [1]. In fact, in the initial studies by Hall and Lehman (1968) on the fidelity of T4 DNA polymerase, it was shown that error rate was influenced by the ratio of correct and incorrect nucleotides [10].

2.1. Studies with Eucaryotic DNA Polymerases

A direct approach to analyzing this relationship was provided by studies on the error-prone DNA polymerase from avian myeloblastosis virus [1]. This enzyme lacks any 3' → 5' exonucleolytic activity and is thus unable to excise noncomplementary nucleotides immediately after incorporation. Moreover, the enzyme is error-prone in copying both synthetic polynucleotides and natural DNA templates [8, 18]. The error rates with this enzyme are 1/300 to 1/5000, and approach within an order of magnitude of the value predicted on the basis of the ΔG for discrimination between correct and incorrect nucleotides in the absence of the polymerase. This value is estimated to be about 1-2 kcal, and would account for a discrimination frequency of approximately 1 in 10 to 1 in 100 [18]. In Table 1, it is shown that the ratio of incorporation of dT to dC using poly(A) as a template is proportional to the concentration of the deoxynucleoside triphosphates in the reaction mixture. Furthermore, the product of the ratio of correct to incorrect incorporation times the relative substrate concentration varies less than 20% under conditions in which the ratio of substrate varies greater than 7.5-fold. This dependency of error rate on the relative concentrations of complementary to noncomplementary nucleotides is compatible with models of base selection in the absence of proofreading in which the enzyme encounters nucleotide substrates based on mass action.

2.2. Studies with Procaryotic DNA Polymerases

Studies on the effects of substrate concentration on the fidelity with procaryotic DNA polymerases are more complex, since these enzymes possess a 3' → 5' exonucleolytic activity. The contribution of this proofreading activity to accuracy has been limited in studies with polynucleotides, since such assays are relatively insensitive [18]. Nevertheless, measurements of nucleotide turnover, monophosphate generation, and base analog incorporation have yielded important information about the mechanism of proofreading [6, 7].

TABLE 1. Effect of Relative Deoxynucleoside Triphosphate Concentrations on Copying Errors with Poly(rA) Oligo(dT) by Avian Myeloblastosis Virus DNA Polymerase

Ratio of correct (TTP) to incorrect (dCTP) Substrates	Incorporation of T (pmol)	Incorporation of C (pmol)	Discrimination $\frac{\text{T inc}}{\text{C inc.}} : \frac{[\text{dTTP}]}{[\text{dCTP}]}$
3.33	65.5	0.020	983
2.00	59.0	0.036	820
1.25	63.0	0.046	1096
0.833	63.0	0.092	834
0.602	65.0	0.107	1007
0.407	64.0	0.154	1023

Adapted from the data of Battula and Loeb [1], as calculated by Bernardi and Ninio [3].

TABLE 2. Effect of Deoxynucleoside Triphosphate Concentration on DNA Polymerase-β Copying Errors Using φX174 am3 DNA

Ratio of correct (dTTP) to total incorrect substrates	Reversion Frequency ($\times 10^6$)	Discrimination: Rev. Freq. $\times \frac{(\text{dTTP})}{(\text{dCTP,dATP,dGTP})} \times 10^6$
30.0	10.8	324
15.0	27.1	406
3.0	126.0	378
0.6	742.0	445
0.3	1350.0	405

Collected from the data in Kunkel et al.(16)

3. MISINCORPORATION WITH NATURAL DNA TEMPLATES

The ϕX174 fidelity assay was developed to quantitate single-base substitution errors made during in vitro DNA synthesis [26]. The assay utilizes a mutant ϕX174 single-stranded DNA containing an amber mutation. To this template, a single fragment, produced by restriction endonuclease treatment of ϕX174 RFI DNA is hybridized to serve as a primer for initiation of in vitro DNA synthesis. The frequency of misincorporation in the process of copying past the amber mutation is quantitated by transfection of E. coli spheroplasts with the copied DNA and plating on Su^+ and Su^- hosts. With a template containing an am3 site, substitution of all three incorrect substitutions (dATP, dGTP, or dCTP) can occur at position 587 in place of the normally correct dTTP [13].

3.1. Studies with Eucaryotic DNA Polymerases

An analysis of the fidelity of copying natural DNA by DNA polymerase-β, the putative repair enzyme in eucaryotes, is shown in Table 2. The reversion frequency is proportional to the ratio of incorrect to correct substrates over a 100-fold range. The discrimination factor was calculated in a manner similar to that described by Bernardi and Ninio, in the case of polynucleotide templates [3]. This consistency of reversion frequencies as a function of the ratio of correct to incorrect substrates is also observed with DNA polymerase-α, the putative DNA-replicating enzyme is eucaryotes [13].

3.2. Studies with Procaryotic DNA Polymerases

An analysis of the ratio of error rate to nucleotide concentration with procaryotic DNA polymerase must take into account the presence of 3' → 5' proofreading activity. Studies with natural DNA templates using procaryotic DNA polymerases have permitted an assessment of the relative contributions of base selection and proofreading to accuracy. Of particular interest with respect to nucleotide pools and mutagenesis are studies on the "next-nucleotide effect." Considering that proofreading involves selective hydrolysis of the terminally incorrect base during the course of DNA synthesis, one can intuitively reason that more rapid synthesis would result in decreased fidelity. This concept is embodied in several mathematical models of proofreading involving competition between excision and polymerization [11, 21, 22]. The first experimental demonstration of this with natural DNA was performed by Fersht [5]. He observed that during in vitro DNA synthesis in a multiprotein system catalyzed by E. coli DNA polymerase III holoenzyme, the frequency of production of a specific wild-type revertant of the amber3 mutation in ϕX174 DNA increased as a square of the concentration of the incorrect nucleotide when this same nucleotide was complementary to the next template nucleotide. Thus when the enzyme is forced to increase the rate of polymerization due to an absolute increase in substrate concentration, larger numbers of mistakes are retained, presumably due to decreased time available for excision by proofreading. This concept was expanded using the ϕX174 system in which one can measure the contributions to fidelity by DNA polymerase alone. For example, with am3 ϕX174 DNA as a template, increasing the concentration of dATP, the next correct nucleotide, under conditions in which C is misincorporated opposite the A at position 587, results in 50-fold increase in errors [15]. Sequencing revertants produced from the product of the reaction reveals that indeed dCMP is incorporated as a result of increasing the concentration of the next nucleotide, dATP [15]. This dependency on the absolute concentration of the next nucleotide suggests that effects of changes in pool sizes with procaryotic enzymes will be complex and not simply the result of the ratio of correct to incorrect nucleotides. It may account in part for preferential mutagenesis at particular sites. At the moment there is inadequate evidence for enhanced fidelity by proofreading in eucaryotic cells, and thus such a mechanism may not apply. The next-nucleotide effect provides a logical approach for determining the presence of putative proofreading activities in eucaryotes.

Another approach to quantitating the contribution of proofreading to fidelity is provided by using deoxynucleoside thiotriphosphates. These analogs contain a sulfur atom in place of oxygen on the α-phosphorus [4]. They are incorporated by eucaryotic DNA polymerases, with a similar fidelity to that obtained with unsubstituted nucleotides [14]. In contrast, fidelity of procaryotic DNA polymerases is markedly diminished using phosphorothiolated dCTP as an incorrect substrate. The diminution in accuracy appears to correspond to that which results from turning off the 3' → 5' exonuclease activity by increasing the concentration of the next-nucleotide [14, 15]. The supposition is that the diminution in accuracy is due to a lack of excision of incorporated phosphorothiolate analogs. This was suggested by several experiments [2, 4], including one demonstrating a clear difference in hydrolysis of normal vis. αS-containing DNA for both E. coli DNA polymerase I (Pol I) and T4 DNA polymerase [14]. In light of the recent evidence of Gupta et al. [9], demonstrating turnover of αS substrates by T4 DNA polymerase, we have reexamined the ability of these enzymes to digest DNA containing three of the four phosphorothiodiester bond linkages. As previously demonstrated and shown in Figs. 1A and B, respectively, neither exonuclease III nor Pol I hydrolyzes the αS-containing DNA [14]. However, the exonuclease associated with T4 DNA is able to do so at a rate similar to that for unsubstituted DNA. The reason(s) for the previous failure of this enzyme to hydrolyze αS-containing DNA are not known [14], but may represent some differences in the enzyme preparations used in these experiments. The results in Fig. 1C have been repeated with three different preparations of enzyme. It has been known for some time

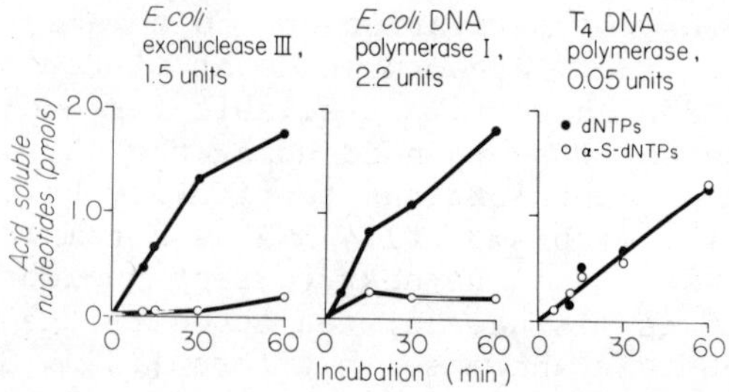

Fig. 1. Hydrolysis of control and phosphorothiolated DNA by 3' → 5' exonucleases. The DNA substrate was synthesized in a reaction mixture (0.2 ml) containing the following: ϕX174 single-stranded DNA annealed to a Hae III Z-10 restriction fragment at a 1 to 5 template to primer ratio, 3.5 μg; Tris-HCl, pH 7.8, 40 mM; $MgCl_2$, 10 mM, dithiothreitol, 2 mM; 50 μM of α-S-dGTP, α-S-dCTP, and α-S-dATP; 50 μM [α-^{32}P]dTTP (Sp. Act., 10,000 dpm/pmol); and E. coli DNA polymerase I, 3 μg. In control reaction, all four nucleotides were unsubstituted and each was present at 50 μM. Incubation was for 30 min at 37°C, and the reaction was stopped by the addition of 0.02 ml of 0.1 M EDTA. Synthesis based on incorporation corresponded to 30% of the ϕX174 template. After phenol extraction, the DNA was precipitated with ethanol and isolated by chromatography on 50 × 0.5 cm^2 columns containing Sephadex G-100, which was equilibrated and eluted with 50 mM Tris-HCl, pH 7.8. The hydrolysis reactions in a volume of 0.05 ml contained: ϕX174 resynthesized DNA (50 ng); 30 mM Tris-HCl (pH 7.8); 2 mM dithiothreitol; 6 mM $MgCl_2$; and the indicated amounts of enzyme. Reactions were incubated at 37°C, and release of acid-soluble radioactivity was determined as previously described [14].

that the exonuclease-to-polymerase ratio is much higher for T4 DNA polymerase than Pol I or any other DNA polymerase [21]. Thus these data may, in fact, indicate a fundamental differences in the mechanism of hydrolysis by the exonucleolytic activities associated with Pol I and T4 DNA polymerase. Alternatively, the S at this position might inhibit the 3' → 5' exonuclease associated with E. coli Pol I and not T4 DNA polymerase.

Finally, the ability of T4 DNA polymerase-associated exonuclease to hydrolyze αS-substituted DNA must be interpreted in light of the highly mutagenic effect of substituting αS-dCTP for normal dCTP during in vitro DNA synthesis by T4 DNA polymerase [14]. It is possible that phosphorothiodiester bonds involving αS-Cyt are more resistant to hydrolysis, the rate difference being enough to account for the observed mutagenicity.

TABLE 3. Changes in Deoxynucleoside Triphosphates in a Subclone of V79 Cells After MNNG Treatment

	pmol/10^6 cells			
Treatment	dATP	dTTP	dGTP	dCTP
None	7.0	2.8	<0.2	448.0
MNNG*	55.0	48.0	<0.2	616.0

Deoxynucleoside triphosphates were measured by the method of Skoog et al.,(24) using E. coli DNA polymerase I.

*Nucleotide pools were measured 4 hr after a 1-hr treatment with 5.0 μM MNNG at 37°C.

4. CHANGES IN NUCLEOTIDE POOLS DUE TO DNA DAMAGE

We have recently investigated the relationship between treatment of eucaryotic cells in culture by DNA-damaging agents and changes in nucleotide pools. Treatment of V79 cells with 5.0 μM N-methyl-N'nitro-N-nitrosoguanidine (MNNG) results in a rapid increase in dATP and dTTP pools. In fact, the increase in pool sizes can be detected as early as 2 h after the beginning of drug treatment. After MNNG treatment, the dATP pool increases for 4 h, remains at the elevated level for as long as 8 h, and then returns to the level of untreated cells by 16 h. In three separate experiments, the increase in dTTP pool follows different kinetics compared to dATP; maximum levels are attained 2 h after addition of 5.0 μM MNNG. In five different experiments, the maximum increase in dATP and dTTP pools ranged from 3-fold to 8.5-fold, and 1.6-fold to 4.5-fold, respectively. During the same time interval, dCTP and dGTP pools changed by less than 50%.

The magnitude of pool expansion by MNNG is similar to that reported during the cell cycle [24], but the specificity appears to be different. The concentrations of MNNG used in these studies killed approximately 10% of the V79 cells, as determined by clonogenic assay, and resulted in a mutation frequency of 50×10^{-5} at the HGPRT locus (thioguanine resistance), compared to 1.6×10^{-6} spontaneous frequency [12, 23]. Thus the effect on dNTP pools occurs at concentrations of MNNG which allow relatively high survivals and are highly mutagenic.

A different class of mutagen, ultraviolet irradiation (UV), which is known to terminate DNA chain elongation [17, 20], also increases dATP and dTTP pools. In UV-treated cells, the levels of dGTP and dCTP did not increase significantly. The doses of ultraviolet light used in this study have also been shown by Stone-Wolff and Rossman [25] to cause substantial mutagenesis at the ouabain locus in V79 cells.

We have extended these studies to a subclone of V79 cells which has large dCTP pools. This subclone exhibits dNTP pool alterations without mutagen treatment which are reminiscent of the thy$^-$ mutator phenotype described by Meuth et al. [19] in Chinese hamster ovary cells. For example, the dTTP pool in these cells is smaller compared to the original V79 cells, and dGTP pools are barely detectable. When these cells were treated with 5.0 μM MNNG, a time-dependent expansion of dATP and dTTP pools is also observed. As shown in Table 1, the maximum increases in dATP and dTTP pools were 7.5-fold and 16-fold, respectively; dCTP pool changed only by 50%. It is interesting that UV treatment of E. coli also causes a rapid increase in the concentrations of dATP and dTTP. Twenty minutes after treatment with 105 $J.m^{-2}$, there are 7- and 9-fold changes in the concentrations of dATP and dTTP, without significant changes in dCTP and dGTP. The increase in dATP is of particular interest, since dATP activates the recA protease in vitro, and may signal the induction of SOS repair functions.

V. DISCUSSION

The results presented indicate that both the ratio and concentration of deoxynucleoside triphosphates affect the fidelity of catalysis using purified DNA polymerases. In fact, the frequency of misincorporations with eucaryotic DNA polymerases is directly proportional to the ratio of incorrect to correct substrates. Even in the presence of proofreading by procaryotic DNA polymerases or a hypothesized associated nuclease in eucaryotes, this proportionality should pertain to overall error rates. Differences in neighboring template nucleotides could yield exponential effects at particular sites [5], and thus might be a contributing factor for mutagenic hot spots. The supposition in these in vitro studies is that misincorporation by DNA polymerase is a rate-determining step in mutagenesis, i.e., all errors are not excised by nucleases that hydrolyze mismatched base pairs.

On the other hand, changes in cellular nucleotide pools have been associated with enhanced mutagenesis. In this chapter, we have presented evidence that exposure of cells to known mutagens alters nucleotide pools. An analysis of changes in nucleotide pools and mutation rates at different loci should yield information on the contribution of nucleotide pools to mutagenesis. Alterations in nucleotide pools may provide an explanation for nontargeted mutageneis that is nucleotide change not in immediate proximity to damaged nucleotides on DNA. The increased dATP pool that we report may be particularly relevant to the induction of SOS repair, and to the selective incorporation opposite depurinated sites on DNA [18].

ACKNOWLEDGEMENTS

This investigation was supported by grants from the National Institutes of Health (CA24845), the National Science Foundation (PCM-7680439), the Environmental Protection Agency (R809-623-010), and the Department of Energy (76RL02225).

REFERENCES

1. N. Battula and L. A. Loeb, The infidelity of avian myeloblastosis virus deoxyribonucleic acid polymerase in polynucleotide replication, J. Biol. Chem., 249:4086-4093 (1974).
2. S. J. Benkovic and K. J. Schray, in: "Transition States of Biochemical Processes" (R. D. Gandour and R. L. Schowen, eds.), pp. 493-526, Plenum Press, New York (1978).
3. F. Bernardi and J. Ninio, The accuracy of DNA replication, Biochimie, 60:762-778 (1978).
4. P. M. J. Burgers and F. Eckstein, A study of the mechanism of DNA Polymerase I from Escherichia coli with diastereomeric phosphorothiolate analogs of deoxyadenosine triphosphate, J. Biol. Chem., 254:6889-6893 (1979).
5. A. R. Fersht, Fidelity of replication of phage ϕX174 DNA by DNA polymerase III holoenzyme: Spontaneous mutation by misincorporation, Proc. Natl. Acad. Sci. USA, 76:4946-4950 (1979).
6. D. J. Galas and E. W. Branscomb, The enzymatic determinants of DNA polymerase accuracy: Theory of T4 polymerase mechanisms, J. Mol. Biol., 124:653-687 (1978).
7. M. F. Goodman, W. C. Grove, N. Muzyczka, and M. J. Bessman, Studies on the biochemical basis of spontaneous mutation. III. Rate model for DNA polymerase-effected nucleotide misincorporation, J. Mol. Biol., 88:423-435 (1974).
8. K. P. Gopinathan, L. A. Weymouth, T. A. Kunkel, and L. A. Loeb, Mutagenesis in vitro by DNA polymerase from an RNA tumor virus, Nature, 278:857-859 (1979).
9. A. P. Gupta, P. A. Benkovic, and S. J. Benkovic, A comparison of the exonuclease activities of T4 DNA polymerase and Pol I towards thiophosphoryl analogs, Fed. Proc., 42:2147 (Abstr. No. 2275) (1983).
10. Z. W. Hall and I. R. Lehman, An in vitro transversion by a mutationally altered T4-induced DNA polymerase, J. Mol. Biol., 36:321-333 (1968).
11. J. J. Hopfield, Kinetic proofreading: A new mechanism for reducing errors in biosynthetic processes requiring high specificity, Proc. Natl. Acad. Sci. USA, 71:4135-4139 (1974).

12. A. W. Hsie, P. A. Brimer, T. J. Mitchell, and D. G. Goslee, The dose-response relationship for ethyl methanesulfonate-induced mutations at the hypoxanthine-guanine phosphoribosyl transferase locus in Chinese hamster ovary cells, Somatic Cell Genetics, 1:247-261 (1975).
13. T. A. Kunkel and L. A. Loeb, On the fidelity of DNA replication X. Effect of divalent metal ion activators and deoxyribonucleoside triphosphate pools on in vitro mutagenesis, J. Biol. Chem., 254:5718-5725 (1979).
14. T. A. Kunkel, F. Eckstein, A. S. Mildvan, R. M. Koplitz, and L. A. Loeb, Deoxynucleoside[1-thio]triphosphates prvent proofreading during in vitro DNA synthesis, Proc. Natl. Acad. Sci. USA, 78:6734-6738 (1981).
15. T. A. Kunkel, R. M. Schaaper, R. A. Beckman, and L. A. Loeb, On the fidelity of DNA replication: Effect of the next nucleotide on proofreading, J. Biol. Chem., 256:9883-9889 (1981).
16. T. A. Kunkel, J. R. Silber, and L. A. Loeb, The mutagenic effect of deoxynucleotide substrate imbalances during DNA synthesis with mammalian DNA polymerases, Mut. Res., 94:413-419 (1982).
17. A. R. Lehmann, Postreplication repair of DNA in mammalian cells, Life Sci., 15:2005-2016 (1974).
18. L. A. Loeb and T. A. Kunkel, Fidelity of DNA synthesis, Ann. Rev. Biochem., 52:429-457 (1982).
19. M. Meuth, M. Trudel, and L. Siminovitch, Selection of Chinese hamster cells auxotrophic for thymidine by 1-β-D-Arabino furanosyl cytosine, Somatic Cell Genetics, 5;303-318 (1979).
20. P. Moore and B. S. Strauss, Sites of inhibition of in vitro DNA synthesis in carcinogen- and UV-treated ϕX174 DNA, Nature, 278:664-666 (1979).
21. N. Muzyczka, R. L. Poland, and M. J. Bessman, Studies on the biochemical basis of spontaneous mutation. I. A comparison of the deoxyribonucleic acid polymerases of mutator, antimutator, and wild type strains of bacteriophage T4, J. Biol. Chem., 247:7116-7122 (1972).
22. J. Ninio, Kinetic amplification of enzyme discrimination, Biochimie, 57:587-595 (1975).
23. E. I. Shaw and A. W. Hsie, Conditions necessary for quantifying ethyl methanesulfonate-induced mutations to purine-analogue resistance in Chinese hamster V79 cells, Mutat. Res., 51:237-254 (1978).
24. K. L. Skoog, B. A. Nordenskjöld, and K. G. Bjursell, Deoxyribonucleoside triphosphate pools and DNA synthesis in synchronized hamster cells, Eur. J. Biochem., 33:428-432 (1973).
25. D. S. Stone-Wolff and T. G. Rossman, Effects of inhibitors of de novo protein synthesis on UV-mutagenesis in Chinese hamster cells. Evidence against mutagenesis via inducible, error-prone DNA repair, Mutat. Res., 82:147-157 (1981).
26. L. A. Weymouth and L. A. Loeb, Mutagenesis during in vitro DNA synthesis, Proc. Natl. Acad. Sci. USA, 75:1924-1928 (1978).

RESPONSE OF CHO CELL DNA POLYMERASE ALPHA TO dCTP AND dTTP POOL IMBALANCE: RELATION TO DNA SYNTHESIS INHIBITION, SURVIVAL AND MUTATION*

C. N. Newman and J. H. Miller

Pacific Northwest Laboratory
Richland, Washington 99352

INTRODUCTION

The roles of several major DNA repair systems in effecting survival and mutation in mammalian cells following chemical or radiation insults have been studied extensively. A marked interest in mutant cell lines deficient or proficient in repair now exists in several laboratories [6, 9, 39, 51] and these abnormal cells have been examined to better ascertain the biological impact of processes such as excision-repair, post-replication repair and unscheduled DNA synthesis. Much of this and other work has been useful in clarifying which repair systems are "error-free" or "error-prone" and which are the common major pathways, i.e., those which repair damage resulting from more than one type of insult [23, 38]. Currently, there is a trend towards elucidating roles for processes involving molecules only recently recognized as important in DNA repair. These include, for example, alkyl transferases [23, 26], poly (ADP-ribose) synthetase [3, 4, 50], DNA glycosylases [23, 36] and at least two DNA polymerases [20, 32, 37, 49, 52]. While no role in repair has yet been shown for DNA polymerase γ, a mitochondrical enzyme [16, 32], others have demonstrated important roles for both DNA polymerase α, the normal replicative enzyme, and for DNA polymerase β, thought to facilitate some repair of ionizing-radiation induced lesions [6, 18, 54] and those lesions induced by certain chemotherapy agents [32, 33, 49]. The significance of polymerase α in repair is particularly relevant to the present paper since this enzyme has been implicated in the repair of damage from both alkylating agents and ultraviolet (UV) irradiation [5, 8, 10, 15, 20, 32, 33,

*This paper is based on work supported by United States Department of Energy Contract DE-AC07-76RLO-1830.

42, 52, 53], but not damage resulting from ionizing radiation [20, 37, 53].

Some of the conclusions regarding the roles of DNA polymerases in repair were derived from results with aphidicolin, a diterpene tetraol mycotoxin which uniquely inhibits the replicative activity of DNA polymerase α [18, 19]. Inhibition apparently results when aphidicolin binds to a specific site on the enzyme; this binding is most competitive with deoxycytidine triphosphate (dCTP), slightly competitive with thymidine triphosphate (dTTP) only when the dCTP concentration is low [17, 18] and not competitive with the other deoxynucleoside triphosphates (dNTPs) [18]. Thus there appears to be a common binding site on the polymerase specific for dCTP and aphidicolin - a finding which is intriguing and of potential significance, suggesting a more interesting relationship between DNA polymerase α and dCTP than merely the usual one of enzyme - substrate. Indeed work by Meuth et al. [30] then by Nicander and Reichard [41] led the latter investigators to hypothesize that "aphidicolin might interfere with the binding of dCTP to an allosteric site of the replicase, regulating the affinity of the system for substrates."

While a cell line altered in polymerase α activity would be useful in providing a better understanding of what is putatively a complex interrelationship of dCTP as both regulator and substrate for the replicase, with rare exception [24] few attempts to obtain a mammalian cell mutant with a defective polymerase α by selecting for aphidicolin resistance have been successful. More often, resistance to the inhibition has been due to an enlarged dCTP pool in these mutants [2, 41, 48] probably because of an alteration in the complicated allosteric regulation of ribonucleotide reductase [2, 30, 48]. An abnormal dNTP pool balance of this sort may be expected to lead to increased misincorporation of dCTP during DNA synthesis [22] and a concommitant increase in the frequency of mutations, but not necessarily to a better understanding of the interrelationship of polymerase α activity and cellular dCTP.

Utilizing a dNTP pool-altered, polymerase α-altered mutant of CHO-K1 (see Results) and a permeable cell-DNA synthesis system, the present study examines the complicated interrelationship of dNTP pools as potential regulators as well as substrates for polymerase α. These findings are then related to the commonly observed UV-induced inhibition of DNA synthesis, and, ultimately, to the biological consequences of exposure to radiation and chemical insults.

MATERIALS AND METHODS

Cell Lines, Culture Conditions, Carcinogen Treatment, Aphidicolin Treatment

CHO-K1, its UV-resistant derivative, Mut 8-16, growth conditions in Ham's F-12 medium with 10% fetal calf serum, survival and mutation assays following UV-irradiation have been described earlier [39]. For exposure to dimethylsulfate (DMS), carcinogen was first diluted in growth medium then immediately administered to growing culture for 2 h at 37°C. Carcinogen-containing medium was then removed and replaced with fresh growth medium. Survival assays were as for UV-treated cells. Aphidicolin (graciously supplied by A. H. Todd, Imperial Chemical Industry) was dissolved in DMSO then appropriate volumes were added directly to cultures which were then incubated for seven days prior to staining dishes for clonagenic survival. In some experiments, aphidicolin was added to permeable cells (see below). For X-irradiation cells in culture flasks at cloning densities were exposed to 100 Rad/min of unfiltered radiation provided by a Norelco, Model MG-300 source, which operated at 250 kV, 10 ma, and had a 1/2 value layer of 0.522 mm copper. Clonagenic assays were as for UV and DMS.

Analysis of dNTP Pools

The assay procedure was essentially that described by North et al. [43]. Dishes containing growing monolayers were chilled on ice, washed with cold PBS, then cells were scraped into ice-cold 60% methanol and kept overnight at -20°C. Following centrifugation at 750 × g in the cold, supernatants were lyophilized to dryness overnight while pellets were assayed for DNA content, according to the differential hydrolysis procedure described by Newman and Bockrath [40]. Lyophils were dissolved in cold 0.5 N perchloric acid on ice for 30 min then neutralized with 1.5 N KOH. After an additional 30 min on ice, followed by centrifugation at 12,000 × g in the cold, supernatants were lyophilized as before. These lyophils were then stored at -76°C until assayed for dNTP content.

Dry samples were dissolved in 1 ml of cold distilled water. A 0.02-ml aliquot of each unknown was added to a reaction tube on ice containing in 0.175 ml of buffer (50 mM Tris-HCl, pH 8.3, 1 mM $MgCl_2$, 1 mM mercaptoethanol, 0.22 mg/ml of bovine serum albumin) 8.75×10^{-3} units of copolymer (dI-dC for assay of dCTP or dGTP; dA-dT for assay of dATP or dTTP), and 8×10^{-3} ml of appropriate ^{3}H-dNTP (^{3}H-dATP, 17 Ci/mmol for dTTP assay; ^{3}H-dTTP, 20 Ci/mmol for dATP assay; ^{3}H-dCTP, 25 Ci/mmol for dGTP assay; ^{3}H-GTP, 12 Ci/mmol for dCTP assay). A 0.005 ml portion of purified _E. coli_ polymerase Pol I (from Worthington Biochemicals or Baltimore Research Laboratory) containing 0.375 units of activity was subsequently added to each tube. Tubes were then quickly transferred to a 37°C water bath and periodically

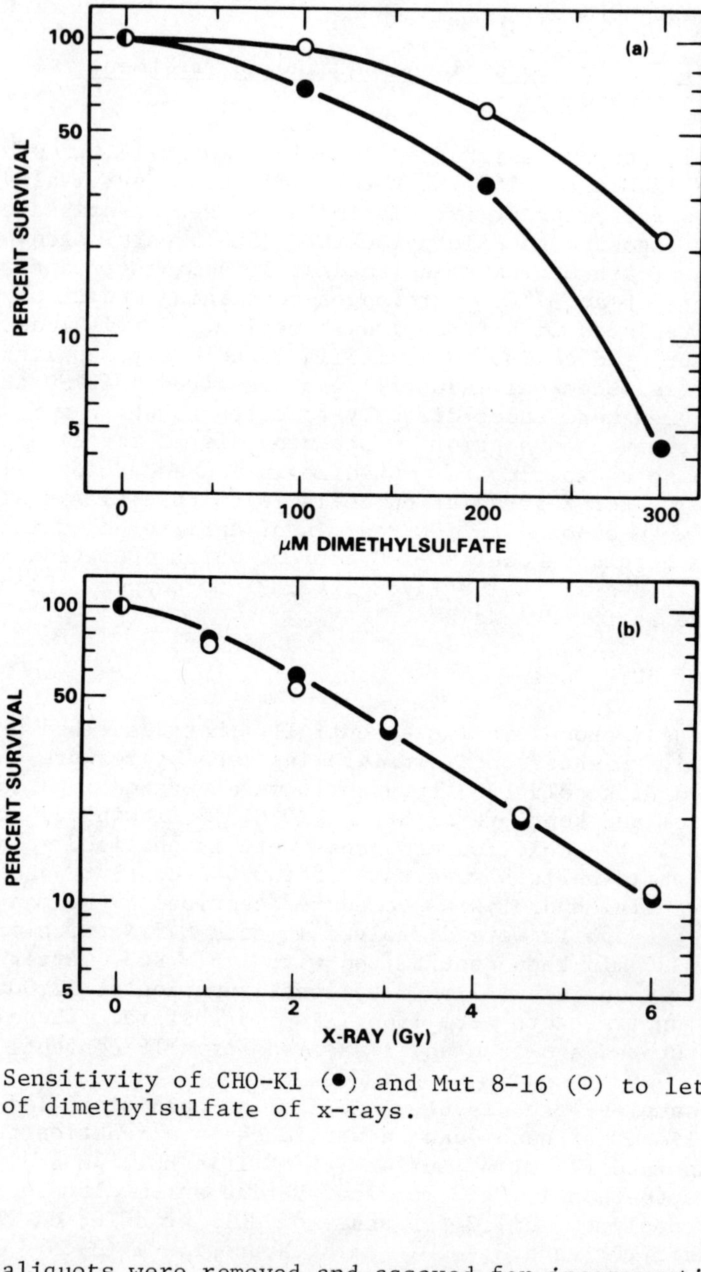

Fig. 1. Sensitivity of CHO-K1 (●) and Mut 8-16 (○) to lethal effects of dimethylsulfate of x-rays.

0.025 ml aliquots were removed and assayed for incorporation of trichloracetic acid insoluble ^{3}H-dpm. To determine the concentrations of each dNTP in unknown samples, incorporation of radioactivity at 20 or 40 min was compared to that obtained in control samples containing known concentrations of respective dNTP substrates (Fig. 3). Results are expressed as pmol dNTP/μg DNA (Table 1).

TABLE 1. dNTP Pools in Exponential Cultures[a]

Cell Line	dATP	dGTP	dCTP	dTTP
CHO-K1	11.1	0.678	18.77 ± 0.62	2.12 ± 0.65
Mut 8-16	9.76 ± 0.21	0.665	12.47 _ 0.61	5.55 ± 1.59

[a]Expressed as pmol/μg DNA, ± S.D.

DNA Synthesis in Permeable CHO Cells

The procedure employed was nearly identical to that described by Berger et al. [3], though in the present study about 10% as many cells were included in each 0.3 ml assay mixture. Cultures were washed with cold PBS, then cells were collected in cold PBS after trypsinization. The number of cells in each suspension was determined, and suspensions were centrifuged as before in the cold. Supernatants were discarded then cells were suspended at 2×10^6 cells/ml for 20 min in cold permeabilization buffer (0.01 M Tris-HCl, pH 7.8, 1 mM EDTA, 4 mM $MgCl_2$, 30 mM mercaptoethanol). Cells were again collected by centrifugation and resuspended in fresh cold buffer at 2×10^6 cells/ml. Judging by dye exclusion, 95% of the cells were made permeable by this protocol. The standard reaction mixture consisted of 0.2 ml of permeable cells and 0.1 ml of a reaction buffer (0.1 M HEPES, pH 7.8, 0.02 M $MgCl_2$, 0.21 M NaCl, 0.015 M ATP, 0.3 mM dATP, 0.3 mM dGTP, 0.3 mM dCTP, and 2.2 μM ^{3}H-dTTP (specific activity 3.4×10^6 dpm/nmol)). In some experiments (Fig. 5), concentrations of individual dNTP substrates were increased but the specific radioactivity of ^{3}H-dTTP was never varied. When indicated aphidicolin was also included in the mixture. Assays were begun when mixtures were transferred to a 37°C water bath. Thereafter, 0.05 ml aliquots were periodically sampled for trichloracetic acid-insoluble ^{3}H-dpm, as described earlier. Rates of incorporation were calculated from slopes which best fitted the data points through the origin, obtained for up to 30 min of incubation (Fig. 4a).

RESULTS

A portion of this work involved studies of a UV-resistant cell line, Mut 8-16, serendipitously selected among surviving UV-irradiated CHO-K1 clones [39]. In order to appreciate what follows, it is first necessary to characterize this unusual cell line. Some of its properties have alrady been described [39], importantly that UV-resistance is most pronounced at low fluences and is conferred via some mechanism not involving conventional repair pathways. Figure 1 reveals that in addition, Mut 8-16 is resistant to the killing

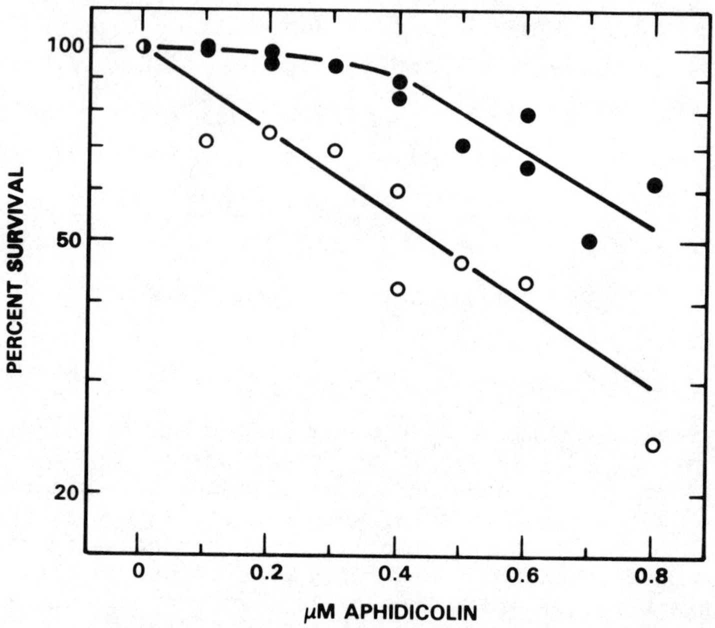

Fig. 2. Sensitivity of CHO-K1 (●) and Mut 8-16 (○) to lethal effects of aphidicolin.

effects of dimethylsulfate (DMS), but not to x-rays. Since the mutant is also resistant to N-methyl-N'-nitro-N-nitrosoguanidine (P. Liu, personal communication), a common mechanism must exist in this cell line which mediates survival after UV- and alkylating agent-induced damage to DNA but not damage induced by ionizing radiation.

Further studies, Fig. 2, reveal that Mut 8-16 is more sensitive than CHO-K1 to killing by aphidicolin. A reduction in the cellular concentration of the principle competitor molecule, dCTP [18] and/or a defect in polymerase α conferring an increased affinity for this inhibitor could explain this finding. Both of these possibilities were investigated, as described in the following experiments.

To determine the concentrations of dNTP pools in exponentially growing cells, monolayers were harvested and dNTP pool extracts were prepared and assayed according to a slight modification of the method of North et al. [43] as described in Materials and Methods. To determine the concentration of each unknown dNTP, incorporated radioactivity in the presence of cell extract was compared to the slope of the line generated by the corresponding known dNTP control which was assayed simultaneously (Fig. 3). Results are shown in Table 1. On the average, there was about 2/3 the steady-state concentration

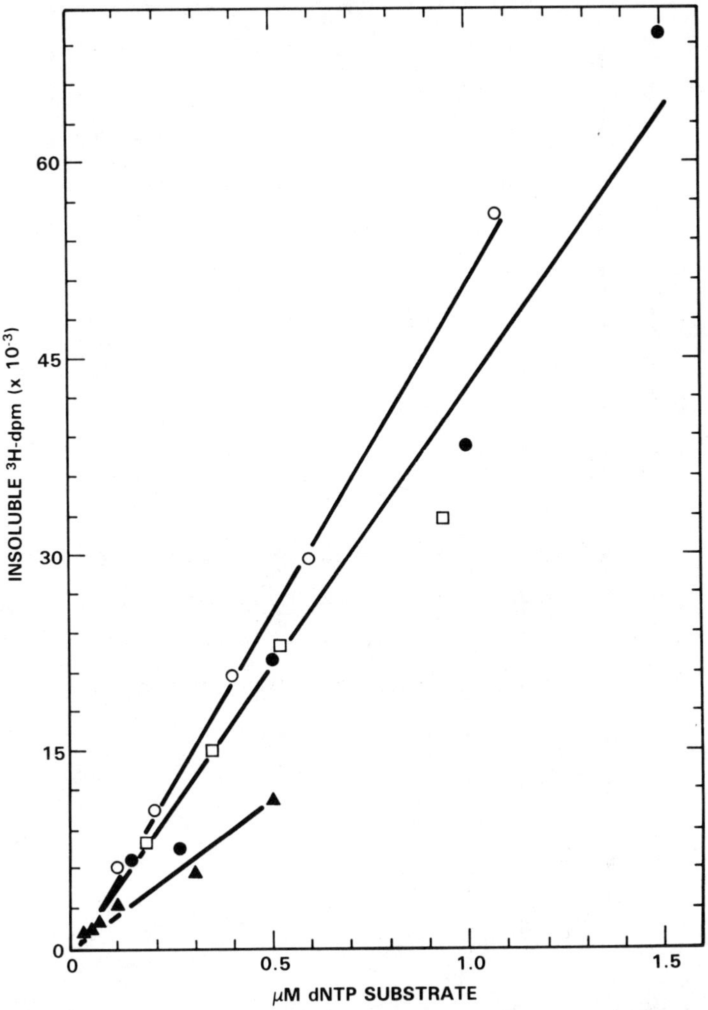

Fig. 3. Calibration of dNTP concentration standards in dNTP pool assay. Appropriate concentrations of dATP (●), dTTP (□), dCTP (○), or dGTP (△) were mixed with the complementary ^{3}H-dNTP, E. coli polymerase I and appropriate copolymer (see Materials and Methods). Acid-insoluble radioactivity was determined after 20 min (dTTP, dGTP) or 40 min (dCTP, dATP) of incubation at 37°C.

of cellular dCTP and about 2- to 3-fold more dTTP in Mut 8-16 than in the parent CHO-K1 cells. No significant differences in corresponding deoxypurine nucleoside triphosphates were apparent.

To compare polymerase α activity on functional chromatin in the presence or absence of aphidicolin and to minimize uncontrolled

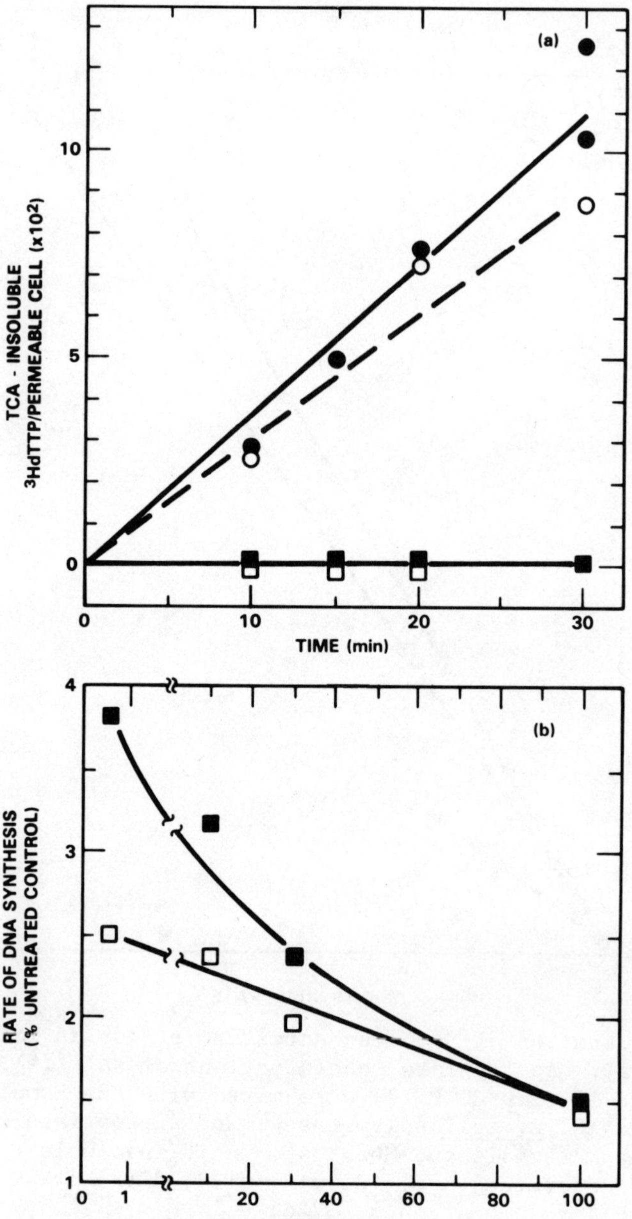

Fig. 4. DNA synthesis in permeable CHO cells. a) CHO-K1 (closed symbols) and Mut 8-16 (open symbols) were made permeable and added to a standard mixture of dNTP substrates (containing ^{3}H-dTTP) in the absence (●, ○) or presence (■, □) or 100 μg aphidicolin/ml. b) Percentage of maximum rate (in absence of inhibitor, from a) vs. concentration of aphidicolin in CHO-K1 (■) and in Mut 8-16 (□).

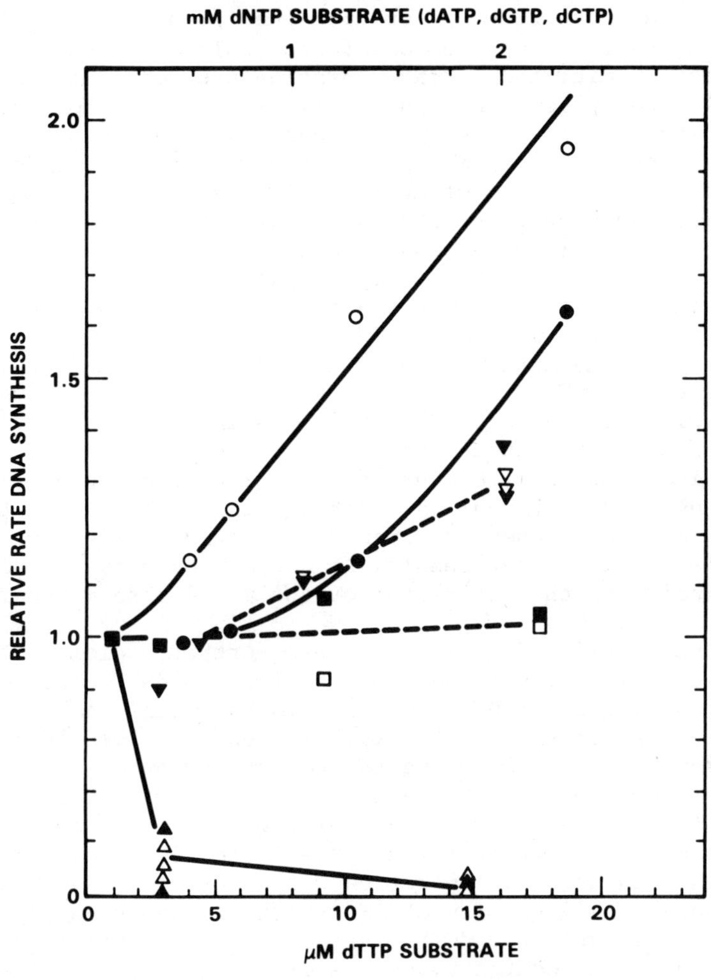

Fig. 5. Effects of dNTP substrate concentrations on rate of DNA synthesis in permeable cells. Rates of DNA synthesis were determined in the standard reaction mixture, as in Fig. 4 (relative rate = 1.0) or in the standard mixture containing excess dCTP (●, ○), dGTP (▽, ▼), dATP (□, ■) or dTTP (△, ▲) at the indicated concentrations. Open symbols are permeable CHO-K1 cells; closed symbols are permeable Mut 8-16 cells.

effects resulting from differences in cellular dNTP substrate concentrations in parent and mutant cell lines (Table 1), cells were first permeabilized and DNA synthesis was in each case measured in the presence of a standard mixture of all four exogenously supplied dNTP substrates including ^{3}H-dTTP, according to the method of Berger et al. [3]. Average rates of tritium incorporation per cell are shown in Fig. 4a. Under the conditions employed, these were linear for at least 30 min, though incorporation was somewhat slower in the mutant. In the presence of aphidicolin, 100 μg/ml, both cell lines were inhibited to about 1.4% of the control rate, indicating the assay measured only DNA polymerase α activity. Figure 4b demonstrates that as the concentration of aphidicolin was decreased inhibition was likewise decreased, but at concentrations below 100 μg/ml, Mut 8-16 was consistently more sensitive to inhibition than CHO-K1. Since exogenous dNTP substrate concentrations and conditions were otherwise identical, this differential sensitivity of the two cell lines must reflect a difference in the relative affinity of their respective DNA polymerase α molecules for aphidicolin. Thus both the low level of dCTP (Table 1) and an abnormal polymerase α in the mutant (Fig. 4b), are consistent with its increased sensitivity to the killing effects of this inhibitor (Fig. 2). However, since dTTP may also compete with aphidicolin for binding to polymerase α [17] the enlarged dTTP pool in Mut 8-16 might be expected to compensate somewhat for the reduction of the chief competitor, dCTP. Thus, it is possible that sensitivity of Mut 8-16 to aphidicolin is due principally to the alteration of the polymerase molecule.

This permeabilized cell system was further exploited to ascertain and compare the effects of varying the dNTP composition of the reaction mixture on DNA synthesis in both cell lines. In Fig. 5, the relative rates of DNA polymerization in permeable cells resulting from singularly varying the concentration of each of the dNTP substrates are given. Rates are normalized to the rate observed in the standard mixture and plotted as a function of the concentration of substrate that was varied. No effect of up to a 22-fold increase in dATP concentration was observed and only a small (but equivalent) stimulation of DNA synthesis was seen in mutant and parent cells when the dGTP concentration was increased by the same magnitude. In contrast, excess dTTP markedly inhibited DNA synthesis in both cell lines. A 4-fold or 20-fold increase in dTTP concentration reduced synthesis to about 20% or 5%, respectively, that observed in the standard reaction mixture. The addition of excess dCTP to permeabilized CHO-K1 cells, on the other hand, resulted in an immediate and dramatic increase in the rate of DNA synthesis over the entire 23-fold concentration range of substrate examined. The response of Mut 8-16 cells to dCTP, though also conspicuous, was significantly less dramatic. Indeed at dCTP concentrations up to 1 mM (10-fold excess), little stimulation of replicative activity was observed compared to an approximately 40% increase in activity in parent cells under the same conditions. These results, therefore, demonstrate that the presence of dCTP, but not dATP or dGTP strongly stimulated DNA polymerase α activity in CHO-K1 cells while excess dTTP inhibited this activity under conditions where, judging from the response to the increased concentrations of the other dNTPs, substrate concentrations were not limiting; more-

TABLE 2. Inhibition of DNA Synthesis in Permeable CHO CHO Cells[a]

Cell Line	Treatment	Reaction Mix	%Relative Rate[b]
	none	standard	100
	UV, 5J/M^2	standard	23.4 ±2.86
	UV, 5J/M^2	standard + 1.1 mM dCTP	32.83
	UV, 5J/M^2	standard + 1.1 mM dATP	25.74
	UV, 5J/M^2	standard + 2.2 mM dGTP	24.99
CHO-K1	UV, 15 J/M^2	standard	20.3 ±6.6
	UV, 15 J/M^2	standard + 1.1 mM dCTP	26.4
	UV, 15 J/M^2	standard + 1.1 mM dATP	26.2
	UV, 15 J/M^2	standard + 2.2 mM dGTP	29.1
	DMS, 100 μM	standard	54.5
	DMS, 200 μM	standard	30.79
	DMS, 300 μM	standard	20
	none	standard	100
	UV, 5J/M^2	standard	31.1 ±3.5
	UV, 5J/M^2	standard + 1.1 mM dCTP	31.9
	UV, 5J/M^2	standard + 1.1 mM dATP	33.9
	UV, 5J/M^2	standard + 2.2 mM dGTP	29.1
Mut 8-16	UV, 15J/M^2	standard	30.9 ±1.1
	UV, 15J/M^2	standard + 1.1 mM dCTP	33.67
	UV, 15J/M^2	standard + 1.1 mM dATP	38.16
	UV, 15J/M^2	standard + 2.2 mM dGTP	30.25
	DMS, 100 μM	standard	70.85
	DMS, 200 μM	standard	59.9
	DMS, 300 μM	standard	37.5

[a]Growing cultures were exposed to UV, then incubated for 2 hr or exposed to DMS for 2 hr before cells were made permeable.

[b]Where indicated, values are ± one standard deviation.

over, the most significant difference in mutant and parent enzyme activities was in response to excess dCTP.

Table 2 shows that the inhibition of DNA synthesis found in whole cells after exposure to UV [11, 13, 14] was also observed when DNA synthesis was assayed as before but in growing cells made permeable 2 h after UV irradiation or after a 2 h exposure to dimethylsulfate (DMS). At equal concentrations of DMS, from 100 μM to 300 μM, inhibition was significantly greater in CHO-K1 cells than in Mut 8-16. Similar results were obtained with UV. Following exposure to 5 J/M^2, the relative synthetic rate in Mut 8-16 was 31.1 of maximum vs. 23.4% of maximum for CHO-K1, or about 1.3-fold greater in the mutant. Addition of 1.1 mM dCTP (but not addition of dATP or dGTP) to the reaction mix significantly increased replicative activity in the parent cell line (by nearly 40%) but no comparable recovery could be induced by dCTP in irradiated Mut 8-16. This cell-specific difference in response to dCTP thus parallels the relative differential response to dCTP observed in unirradiated controls (Fig. 5) and demonstrates that inhibition induced by a low UV fluence is partially reversible by dCTP in normal CHO cells. Inhibition of DNA synthesis at a higher fluence (15 J/M^2) was still greater in CHO-K1 than in Mut 8-16, but could not be differentially relieved by more dCTP as before. This suggests the presence of a secondary source of inhibition at this higher dose (see Discussion).

Implicit in the comparison of Mut 8-16 and CHO-K1 is that the reduced DNA synthesis inhibition as well as the increased survival and mutation frequency in the mutant [39], all of which manifest at low UV fluences, are related to a common defect which resulted in a cellular dTTP/dCTP imbalance and an altered regulation of DNA polymerase α. To better understand the interrelationship of these phenomena in whole cells, cellular dCTP and dTTP pools were also measured 2 h after exposure to UV or after a 2 h exposure to DMS. These results are shown in Fig. 6 where the pool sizes relative to untreated control cells are plotted, for comparison of both mutagens, as a function of the fraction of cells that survive treatment [39] (Fig. 1). Most interestingly, a significant reduction in the dCTP pool to 20-35% of normal resulted from both treatments even at high survival (>85%) levels in both cell lines. In contrast a much enlarged dTTP pool was observed in CHO-K1 after UV (up to 3.7-fold after 15 J/M^2). This increase in dTTP was less pronounced in Mut 8-16 (up to 2.2-fold after 15 J/M^2), though to a first approximation, data points relating relative increases in dTTP and decreases in survival in both cell lines share a common line. DMS-induced increases in dTTP were smaller and were observed only when survival was below 30%. No comparable changes in cellular dATP or dGTP were recorded (not shown).

Figure 7 shows the rate of DNA synthesis (expresssed as percent of untreated controls) as a function of the ratio of the relative (to steady-state) dTTP/dCTP pool imblance induced by mutagens. When plotted in this manner, data from different doses of UV and DMS in both parent and mutant cell lines fall close to a common curve that suggests a rapid decrease in DNA synthesis when the relative dTTP/dCTP pool imbalance falls between 1-5, and a more gradual decrease when this ratio further increases. It is interesting to note that relative dTTP/dCTP ratios at or near 5 correspond approximately to doses of UV [39] or DMS (Fig. 1) at which the transitions in the shapes of the respective survival curves from shouldered regions to non-shouldered (exponential) regions occur.

DISCUSSION

Since Mut 8-16 was selected only for its resistance to UV, it seems likely that the unusual properties of this cell line, namely deviate dTTP and dCTP pools (Table 1) an abnormal polyermase α

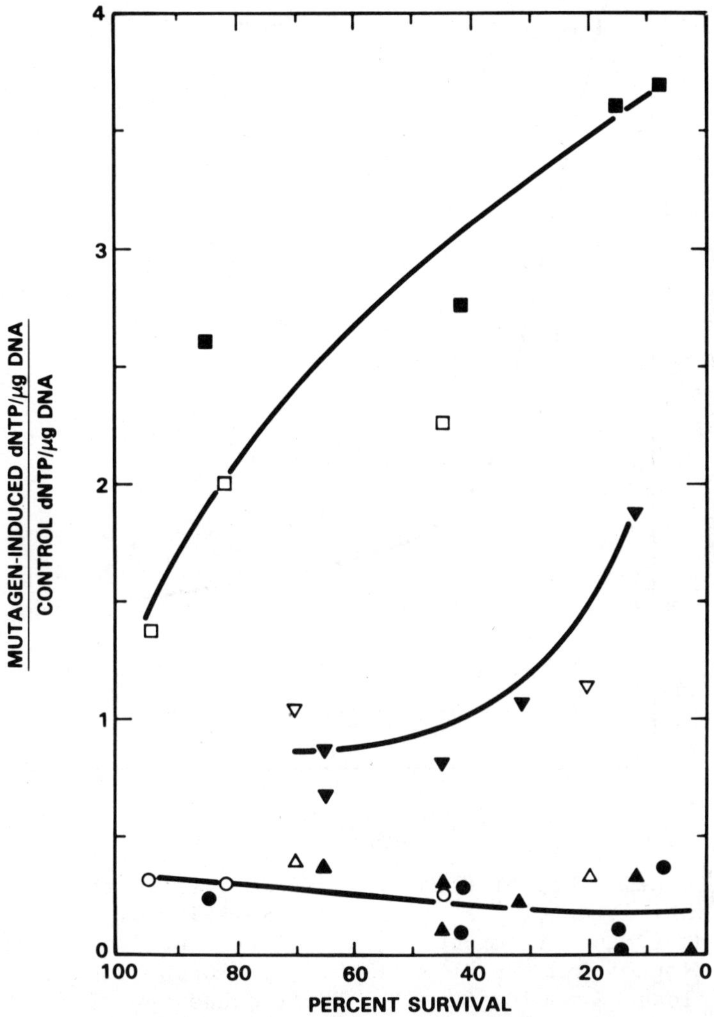

Fig. 6. UV or DMS-induced changes in relative concentrations of dTTP and dCTP. Cell monolayers were exposed to UV then incubated in fresh growth medium for 2 h or cells were grown in the presence of DMS for 2 h. The concentrations of dNTP pools in extracts from treated cells were compared to that from untreated control cultures. Survival values after UV or DMS exposure were obtained earlier [39] or from Fig. 1, respectively. Relative dTTP after UV (■, □); relative dTTP after DMS (▼, ▽); relative dCTP after UV (●, ○) relative dCTP after DMS (▲, △). Closed symbols, CHO-K1; open symbols, Mut 8-16.

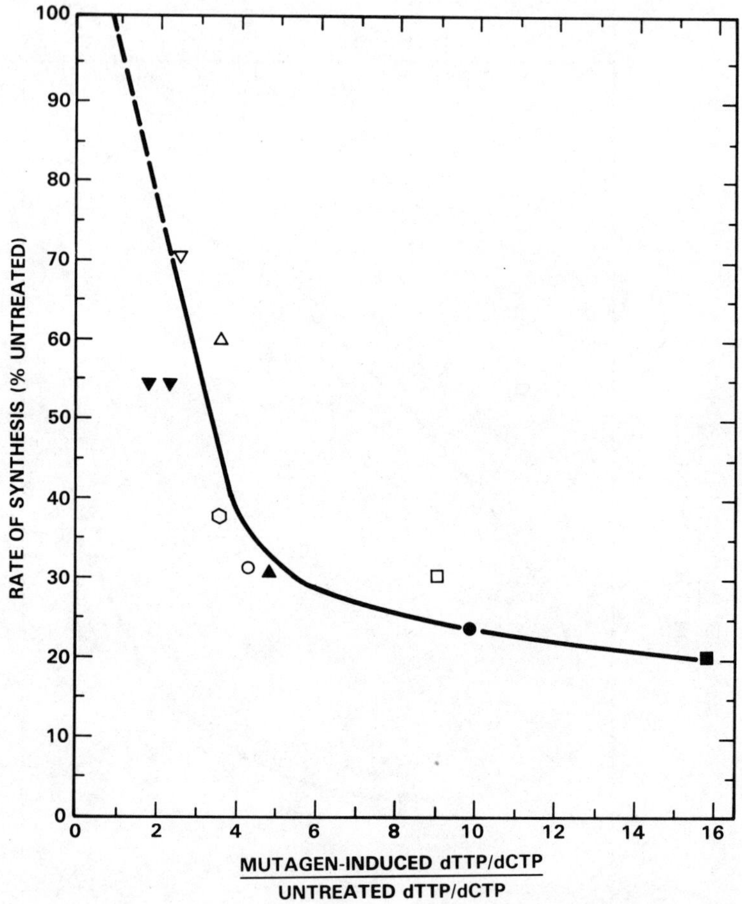

Fig. 7. Relationship of relative dTTP/dCTP ratio to mutagen-induced inhibition of DNA synthesis. Relative dTTP/dCTP values, determined from Fig. 6, are plotted against percent inhibition of DNA synthesis (Table 2). Open symbols, Mut 8-16; closed symbols, CHO-K1. UV fluences were 5 J/M^2 (●, ○) or 15 J/M^2 (■, □); DMS concentrations were 100 μM (▼, ▽), 200 μM (▲, △); 300 μM (⬡).

(Figs. 4 and 5) and a lower induced inhibition of DNA synthesis by UV and by DMS (Table 2) in comparison to the parent cell line, are the result of a single mutation which confers both an increased probability for mutation and resistance to killing by low fluences of UV and by DMS. Based upon this assumption and our other experimental results, we hypothesize that: 1) dTTP and dCTP act antagonistically, as the negative or positive effector, respectively, or DNA polymerase α activity; 2) the ratio of the pools of these regulatory molecules determines the probability that each will bind to the replicase thereby governing its activity; 3) the increase in the relative dTTP/dCTP pool ratio in whole cells after exposure to UV or to DMS is responsible for inhibition of DNA replication; and 4) the greater replicative activity in Mut 8-16 after treatment with low doses of UV or with DMS confers a higher survival probability but a greater frequency of mutations.

Meuth et al. [30] and Nicander and Reichard [41] were first to suggest allosteric regulation of polymerase α activity by dCTP. They examined the correlation of DNA synthesis with alterations in dNTP pools in normal or pool-altered mouse 3T6 fibroblasts caused by addition to cultures of excess thymidine or deoxyadenosine (causing accumulation of all dNTPs but dCTP, which decreased dramatically) or aphidicolin (which also caused a dramatic decrease in dCTP). In all instances, a relatively low dCTP (and normal or elevated dTTP) invariably paralleled a marked decrease in the rate of DNA synthesis. These and other findings led Nicander and Reichard [41] to postulate that dCTP binds to an allosteric site on the replicase which in turn positively regulates the affinity of the system for all the dNTP substrates, in agreement with our findings for enhanced DNA synthesis in permeable cells with excess dCTP (Fig. 5). We further assign a role to dTTP as negative regulator of polymerase α. This conclusion is straight forward and also based upon DNA synthesis in permeable cells but now in the presence of excess dTTP (Fig. 5). Juding by the negative response to this substrate, the absence of response or small response in replicative activity to additional dATP or dGTP, substrate concentrations were not rate limiting.

We further believe that the depression in DNA synthetic rates observed in both mutant and parent cells after exposure to UV or DMS (Table 2) was due to a combination of both the relative increase in cellular dTTP, the inhibitor of polymerase α, and/or a relative decrease in cellular dCTP, the stimulator of polymerase α. We have shown that following exposure, a large imbalance was induced in the dTTP/dCTP ratio; dCTP always decreased several fold from the steady-state value while after UV, but not DMS, dTTP increased several-fold (Fig. 6). As shown in Fig. 7, inhibition of DNA synthesis appeared to correlate with the relative dTTP/dCTP ratio regardless of whether there were marked changes induced in both dTTP and dCTP (UV) or principally in dCTP (DMS). Furthermore, inhibition appeared to be near saturation at relative dTTP/dCTP ratio values >5; at higher values, little additional inhibition was observed. In contrast, an approximately linear relationship of lower relative dTTP/dCTP values and inhibition was obtained. This occurred at doses which correspond to shouldered regions of survival curves generated by UV [39] or DMS (Fig. 1).

It must be noted that inhibition did not correlate with the absolute dTTP/dCTP ratio when comparing both cell lines but only with the deviation from the respective steady-state ratio. The steady-state ratio of dTTP/dCTP in Mut 8-16 was already almost 4-fold larger than that of CHO-K1 (Table 1). Since DNA synthesis in Mut 8-16 was only slightly slower than in CHO-K1 (Fig. 4a), we speculate that the alteration of polymerase α in Mut 8-16 (Fig. 4b) has somewhat conferred resistance to this abnormal steady-state pool imbalance but not to a larger imbalance induced by mutagens.

Inhibition at lower relative dTTP/dCTP ratios could be partially reversed only by excess dCTP in CHO-K1 cells while inhibition at high relative dTTP/dCTP ratios could not be similarly reversed (Table 2). This difference in recovery even in the presence of comparable substrate concentrations, such as that found in our assay system, may reflect a requirement for intermediate protein synthesis [11, 45] and/or a difference in the nature of inhibition at low and high UV fluences. The latter possibility is suggested by the work of Doniger [14]. At 5 J/M^2, initiation is principally (though not exclusively) inhibited while at 10 J/M^2, elongation rates are also significantly inhibited. In conjunction with our data, these findings suggest that polymerase α in the initiation complex is selectively vulnerable to inhibition by small increases in the relative dTTP/dCTP ratio which is partially reversible by dCTP while elongating forks are inhibited by higher relative dTTP/dCTP ratios where inhibition is not likewise reversible, possibly due to the need for an inducible factor [11].

As stated earlier, the correlary to our hypothesis that a high cellular dTTP/dCTP pool ratio mediates inhibition of DNA synthesis is that the latter in turn, contributes to reduced survival and mutation. We propose that cells which synthesize DNA in the presence of damaged template have a selective survival advantage over cells in which DNA synthesis is inhibited. This idea is consistent with the model of Painter [44] for double-strand breaks in DNA resulting from transient inhibition of replication. Painter has suggested that the frequency of double-strand breaks generated by a type II topoisomerase at junctions between regions of supercoiled DNA (replicon clusters) is enhanced when a partially blocked cluster is adjacent to a duplicated cluster. Possibly the enhanced lethality we propose when DNA synthesis is inhibited is associated with this increase in double-strand breaks. However, we believe that continued synthesis under these conditions is also disadvantageous to cells in that it leads to an enhancement of mutation frequency, as observed in Mut 8-16 at low UV fluences [39] possibly because even a mild dNTP pool imbalance, such as that resulting from treatment at low doses of mutagen, leads to base misincorporation [22, 46]. Indeed, Rossman and Stone-Wolff [46] have demonstrated that while DNA synthesis inhibition, alone, has no effect, the simultaneous presence of excess TdR in the medium leads to enhanced mutagenicity (which is prevented by the addition of CdR) in Chinese hamster V-79 cells. Assuming that excess TdR would increase the cellular dTTP/dCTP ratio, their findings further support our idea. The data of Ashman et al.

[1], Meuth et al. [30], Nicander and Reichard [41], and Meuth [29], all of whom investigated the effects of pool alterations in rodent cells are also generally in agreement with our proposal while studies of human fibroblasts [11, 21, 34, 47] are consistent only in so far as reduced DNA synthesis inhibition relates to the enhancement of survival.

The study most amenable to comparison of the results reported in this paper is that of Meuth [30]. By adding exogenous thymidine and deoxyguanosine to pro^- and thy^- CHO-K1 cell lines so as to selectively control ribonucleotide reductase activity, Meuth was able to manipulate the endogenous dNTP pool concentrations in growing cultures, and to examine the resulting effects of such imbalances on mutagen-induced cytoxicity and mutation frequencies. His data reveal that low dTTP/dCTP ratios following exposure to alkylating agents always resulted in higher than normal survival while high dTTP/dCTP ratios were invariably accompanied by higher mutation frequencies and lower survival than normally expected, consistent with our results. However, in contrast to our interpretation, the author suggests that the dTTP/dCTP pool imbalance caused induction of a highly efficient repair-protein and that concommitant with repair misincorporation of dTTP occurred, probably opposite 0^6-alkylguanine. The effect, according to Meuth, was specific for alkylating agents since no similar differential effect of exogenous thymidine and deoxyguanosine was observed in cells after UV. We believe, however, that Meuth did not know that UV, itself, had already induced a marked pool imbalance, as our findings clearly show (Fig. 6), which probably was not further enhanced by exogenous deoxynucleosides. Our approach, comparing the induced dTTP/dCTP pool imbalance with biological responses in Mut 8-16 and in CHO-K1 cells, better illustrates a possible role for a dTTP/dCTP pool imbalance in the enhancement of survival and mutation after UV as well as after exposure to alkylating agents.

The data of Moustacchi et al. [34], Rudé and Friedberg [47], Kaufmann and Cleaver [21], and Cleaver et al. [11] demonstrate the effects of UV on DNA synthesis inhibition in human fibroblasts, both normal and those of the UV-sensitive, excision-repair-proficient xeroderma pigmentosum (XP) variant cells. In both cells, rates of DNA synthesis declined rapidly after exposure to UV but inhibition was more pronounced and recovery slower in the UV-sensitive variant. Cleaver, and coworkers [11] further demonstrated that replication forks in normal but not the variant cells were able to replicate past damaged template, indicating that photolesions were not necessarily responsible for inhibition in normal cells, as implied by our findings (Fig. 7) and which are in agreement with the data of Doniger [14] and Meyn et al. [31]. To a first approximation, increased DNA synthesis inhibition in human cells also paralleled an enhancement of cell killing. However, the variant was more rather than less mutable than normal fibroblasts [25, 35], which appears at first to be a contradiction of our model. Judging from the small decrease in specific activity of radiolabeled thymidine in DNA before and after irradiation of normal human fibroblasts [11], however, the induced increase in dTTP we observed in CHO cells (Fig. 6) did not occur in these human cells (though dCTP may have decreased). As discussed earlier, enhancement of mutation frequency during partial DNA synthesis inhibition occurs only when there is also a dNTP pool imbalance [46]. One possible explanation for the greater mutagenicity of the variant is that dTTP in these cells only (whose specific activity of radiolabeled DNA was not examined by Cleaver et al. [11]) increased after irradiation such that significantly more misincorporation of dTTP occurred even under conditions of greater DNA synthesis inhibition [22]. Alternatively, since the variant was, in essence, clinically selected for its mutable phenotype, there may be a secondary biochemical lesion in these abnormal cells which masked the reduction in mutation frequency we would have otherwise predicted. A defect, for example, in a topoisomerase [12], a terminal nucleotidyl transferase [55] or a DNA gyrase might confer such a phenotype [12] as implied indirectly by Mattern and Painter [26], Painter [45], and by Mattern et al. [28]. Without more data on the properties of variant cells, including measurements of dNTP pools, it is difficult to intelligently speculate on these possibilities in human cells, or to evaluate these findings with respect to our own.

Finally, though we believe that the interdependent relationship of dNTP pools and DNA synthesis ultimately affects survival and mutation, the impact of this interdependence is probably best observed at low doses of mutagen as seen in Mut 8-16 here and earlier [39]. This is because at higher doses, the influence of major repair processes on the biological consequences of damage predominates and may actually obscure the significance of the relationship we postulate.

REFERENCES

1. C. R. Ashman and R. L. Davidson, Bromodeoxyuridine mutagenesis in mammalian cells is related to deoxyribonucleotide pool imbalance, Molec. Cell. Biol., 1:2-4-260 (1981).
2. D. Ayusawa, K. Iwata, and T. Sene, Alteration of ribonucleotide reductase in aphidicolin - resistant mutants of mouse FM3A cells associated with resistance to arabinosyladenine and arabinosylcytosine, Somat. Cell Genet., 7:27-42 (1981).
3. N. A. Berger, A. S. Kaichi, P. G. Steward, R. R. Kleuecz, G. L. Forrest, and S. D. Gross, Synthesis of poly (adenosine diphosphate ribose) in synchronized Chinese hamster cells, Exptl. Cell. Res., 117:127-135 (1978).
4. N. A. Berger and G. W. Sikorski, Nicotinamide stimulates repair of DNA damage in human lymphocytes, Biochem. Biophys. Res. Commun., 95:67-72 (1980).

5. E. Bermúdez, The inhibition of DNA repair in primary rat hepacyte cultures by aphidicolin: evidence for involvement of α-polymerase in the repair process, Biochem. Biophys. Res. Commun., 109:275-281 (1982).
6. C. C. Chang, J. A. Boezi, S. T. Warren, C. L. K. Sabourin, P. K. Liu, L. Glatzer, and J. E. Trosko, Isolation and characterization of a UV-sensitive hypermutable aphidicolin - resistant Chinese hamster cell line, Somat. Cell. Genet., 7:235-253 (1981).
7. L. M. Chang, P. Plevani, and F. J. Bollum, Evolutionary conservation of DNA polymerase β structure, Proc. Natl. Acad. Sci. (USA), 79:758-761 (1982).
8. G. Ciarrocchi, J. G. Jose, and S. Linn, Further characterization of a cell-free system for measuring replicative and repair DNA synthesis with cultured human fibroblasts and evidence for involvement of DNA polymerase α in DNA repair, Nucle. Acids Res., 7:1205-1219 (1979).
9. J. D. Cleaver, DNA repair in Chinese hamster cells of different sensitivities to ultraviolet light, Int. J. Radiat. Biol., 16:277-285 (1969).
10. J. D. Cleaver, Specificity and completeness of inhibition of DNA repair by novobiocin and aphidicolin, Carcinogenesis, 3: 1171-1174 (1982).
11. J. E. Cleaver, G. H. Thomas, and S. D. Park, Xeroderma pigmentosum variants have a slow recovery of DNA synthesis after irradiation with ultraviolet light, Biochem. Biophys. Acta, 564:122-131 (1979).
12. G. C. Crumplin, The involvement of DNA topoisomerases in DNA repair and mutagenesis, Carcinogenesis, 2:157-160 (1981).
13. D. Dahle, T. D. Griffiths, and J. G. Carpenter, Inhibition and recovery of DNA synthesis in UV-irradiated Chinese hamster V-79 cells, Photochem. Photobiol., 32:157-165 (1980).
14. J. Doniger, DNA replication in ultraviolet light irradiated Chinese hamster cells: the nature of replicon inhibition and post-replication repair, J. Mol. Biol., 120:433-446 (1978).
15. S. L. Dresler, J. D. Roberts, and M. W. Lieberman, Characterization of deoxyribonucleic acid repair synthesis in permeable human fibroblasts, Biochem., 21:2557-2564 (1982).
16. L. P. Goscin and J. J. Byrnes, DNA polymerase γ: one peptide, two activities, Biochem., 21:2513-2518 (1982).
17. A. M. Holmes, Studies on the inhibition of highly purified calf thymus 8S and 7.3S DNA polymerase α by aphidicolin, Nucleic Acids Res., 9:161-168 (1981).
18. J. Huberman, New views on the biochemistry of eucaryotic DNA replication revealed by aphidicolin an unusual inhibitor of DNA polymerase α, Cell, 23:647-648 (1981).
19. S. Ikegami, T. Taguchi, and M. Ohashi, Aphidicolin prevents mitotic cell division by interfering with the activity of DNA polymerase α, Nature, 275:458-460 (1978).

20. G. Iliakis, M. Nusee, P. Bryant, Effect of aphidicolin on cell proliferation, repair of potentially lethal damage and repair of DNA strand breaks in Ehrlich ascite tumor cells exposed to x-rays, Int. J. Radiat. Biol., 42:417-434 (1982).
21. W. K. Kaufmann and J. E. Cleaver, Mechanism of inhibition of DNA replication by ultraviolet light in normal human and xeroderma pigmentosum fibroblasts, J. Mol. Biol., 149:171-187 (1981).
22. T. Kunkel and L. Loeb, Fidelity and mammalian DNA polymerases, Science, 213:765-767 (1981).
23. T. Lindahl, DNA repair enzymes, Ann. Rev. Biochem., 51:61-87 (1982).
24. P. K. Liu, C.-C. Chang, J. E. Trosko, D. K. Dube, G. M. Martin, and L. A. Loeb, Mammalian mutator mutant with an aphidicolin-resistant DNA polymerase α, Proc. Natl. Acad. Sci. (USA), 80: 797-801 (1983).
25. V. M. Maher, L. M. Ouellette, R. D. Curren, and J. J. McCormick Frequency of ultraviolet light-induced mutation is higher in xeroderma pigmentosum variant cells than in normal human cells, Nature, 261:593-595 (1976).
26. M. R. Mattern and R. B. Painter, Dependence of mammalian DNA replication on DNA supercoiling. I. Effect of ethidium bromide on DNA synthesis in permeable Chinese hamster ovary cells, Biochem. Biophys. Acta, 563:293-305 (1979).
27. M. R. Mattern and R. B. Painter, Dependence of mammalian DNA replication on DNA supercoiling. II. Effect of novobiocin on DNA synthesis in Chinese hamster ovary cells, Biochim. Biophys. Acta, 563:306-312 (1979).
28. M. R. Mattern, R. F. Paone, and R. S. Day, III, Eukaryotic DNA repair is blocked at different steps by inhibitors of DNA topoisomerases and of DNA polymerases α and β, Biochim. Biophys. Acta, 697:6-13 (1982).
29. M. Meuth, Role of deoxynucleotide triphosphate pools in the cytotoxic and mutagenic effects of DNA alkylating agents, Somat. Cell Genet., 7:89-102 (1981).
30. M. Meuth, E. Aufreiter, and P. Reichard, Deoxyribonucleotide pools in mouse - fibroblast cell lines with altered ribonucleotide reductase, Eur. J. Biochem., 71:39-46 (1976).
31. R. E. Meyn, D. L. Vizard, R. R. Hewitt, and R. M. Humphrey, The fate of pyrimidine dimers in the DNA of ultraviolet-irradiated Chinese hamster cells, Photochem. and Photobiol., 20: 221-226 (1974).
32. M. R. Miller and D. N. Chinault, Evidence that DNA polymerase α and β - participate differentially in DNA repair synthesis induced by different agents, J. Biol. Chem., 257:46-49 (1982).
33. M. R. Miller and D. N. Chinault, The roles of DNA polymerases α, β, and γ in DNA repair synthesis induced in hamster and human cells by different DNA damaging agents, J. Biol. Chem., 257:10204-10209 (1982).

34. E. Moustacchi, J. K. Ehmann, and E. C. Friedberg, Defective recovery of semi-conservative DNA synthesis in xeroderma pigmentosum cells following split-dose ultraviolet irradiation, Mut. Res., 62:159-171 (1979).
35. B. C. Myhr, D. Turnbull, and J. A. DiPaolo, Ultraviolet mutagenesis of normal and xeroderma pigmentosum variant human fibroblasts, Mut. Res., 62:341-353 (1979).
36. Y. Nakabeppu and M. Sekiguchi, Physical association of pyrimidine dimer DNA glycosylase and apurinic/apyrimidinic DNA endonuclease essential for repair of ultraviolet-damaged DNA, Proc. Natl. Acad. Sci. (USA), 78:2742-2746 (1981).
37. H. Nakamura, T. Morita, and S. Yoshida, Differential inhibition of DNA polymerases by X-irradiated DNA, Int. J. Radiat. Biol., 38:449-458 (1980).
38. S. Nakano, H. Yamagami, and R. Takaki, Enhancement of excision-repair efficiency by conditioned medium from density-inhibited cultures in V79 Chinese hamster cells. Evidence for excision repair as an error-free repair process, Mut. Res., 62:369-381 (1979).
39. C. N. Newman, Characterization of a UV-resistant mutant of CHO-K1 with normal repair activity, Mut. Res., 96:99-108 (1982).
40. C. N. Newman and R. C. Bockrath, Growth moderation in slow-growing mutants of Escherichia coli, J. Gen. Microbiol., 85: 203-210 (1974).
41. B. Nicander and P. Reichard, Aphidicolin sensitivity of variant 3T6 cells selected for changes in ribonucleotide reductase, Biochem. Biophys. Res. Commun., 103:148-155 (1981).
42. Y. Nishi, M. M. Hasegawa, N. Inui, S. Ikegami, and M. Yamada, Effect of post-treatment with aphidicolin - a specific inhibitor of DNA polymerase α - on sister-chromatid exchange induced by ethyl methanesulfonate, Mut. Res., 103:155-159 (1982).
43. T. W. North, R. K. Bestwick, and C. K. Mathews, Detection of activities that interfere with enzymatic assay of deoxyribonucleoside 5'-triphosphates, J. Biol. Chem., 255:6640-6645 (1980).
44. R. B. Painter, A replication model for sister-chromatid exchange, Mut. Res., 70:337-341 (1980).
45. R. B. Painter, Response of Chinese hamster ovary cells to DNA damage after a conditioning exposure to ultraviolet light, Biochim. Biophys. Acta, 609:257-263 (1980).
46. T. G. Rossman and S. Stone-Wolff, Inhibition of DNA synthesis is not sufficient to cause mutagenesis in Chinese hamster cells, Biochimie, 64:809-813 (1982).
47. J. M. Rúde and E. C. Friedberg, Semi-conservative deoxyribonucleic acid synthesis in unirradiated and ultraviolet-irradiated xeroderma pigmentosum and normal skin fibroblasts, Mut. Res., 42:433-442 (1977).
48. Sabourin, C. Liki, P. F. Bates, L. Glatzer, C. C. Chang, J. E. Trosko, and J. A. Boezi, Selection of aphidicolin-resistant CHO cells with altered levels of ribonucleotide reductase, Somat. Cell Genet., 7:255-268 (1981).

49. S. Seki and T. Oda, Effects of 2',3'-dideoxy-thymidine triphosphate on replicative DNA synthesis and unscheduled DNA synthesis in permeable mouse sarcoma cells, Biochim. Biophys. Acta, 606:246-250 (1980).
50. C. J. Skidmore, M. I. Davies, P. M. Goodwin, H. Halldorsson, P. J. Lewis, S. Shall, and A. Zia'ee, Involvement of poly (ADP-ribose) polymerase in the degradation of NAD caused by γ-radiation and N-methyl-N-nitrosourea, Eur. J. Biochem., 101: 135-142 (1979).
51. L. H. Thompson, K. W. Brookman, L. E. Dillehay, A. V. Carrano, J. A. Mazrimas, C. L. Mooney, and J. L. Minkler, A CHO-cell strain having hypersensitivity to mutagens, a defect in DNA strand-break repair, and an extraordinary baseline frequency of sister-chromatid exchange, Mut. Res., 95:427-440 (1982).
52. A. A. van Zeeland, C. J. M. Bussmann, F. Degrassi, A. R. Filon, A. C. van Kesteren-van Leeuwen, F. Palitti, and A. T. Natarajan, Effects of aphidicolin on repair replication and induced chromosomal aberrations in mammalian cells, Mut. Res., 92:379-392 (1982).
53. R. Waters, K. Crocombe, and R. Mirzayans, The inhibition of DNA repair by aphidicolin or cytosine arabinoside in x-irradiated normal and xeroderma pigmentosum fibroblasts, Mut. Res., 94:229-234 (1981).
54. A. Weissbach, The functional roles of mammalian DNA polymerase, Arch. Biochem. Biophys., 198:386-396 (1979).
55. S. Yoshida, S. Masaki, H. Nakamura, and T. Morita, Cooperation of terminal deoxynucleotidyl transferase with DNA polymerase α in the replication of ultraviolet-irradiated DNA, Biochim. Biophys. Acta, 652:324-333 (1981).

MISINCORPORATION OF DEOXYURIDINE IN HUMAN CELLS: CONSEQUENCES OF ANTIFOLATE EXPOSURE

Randall G. Richards, Oliver E. Brown, and W. David Sedwick*

Departments of Medicine and Microbiology-Immunology
Duke University Medical Center
Durham, North Carolina 27710

ABSTRACT

Treatment of B and T lymphoblastoid cell lines (SB and MOLT-4, respectively) and a promyelocytic leukemia cell line (HL-60) with the lipid soluble antifolate, 2,4-diamino-5-methyl-6-(2',5'-dimethoxybenzyl)-pyrido(2,4-d) pyrimidine (BW301U), led to drug dose-dependent inhibition of [^{3}H]deoxyuridine (dU) incorporation into DNA as thymidine, and to misincorporation of [^{3}H]dU as dUMP. After a 15 min preincubation with up to 50 μM BW301U and a further 15 min incubation after addition of [^{3}H]dU, the number of alkaline labile apyrimidinic sites increased with increasing drug dose, as demonstrated by alkaline sucrose gradient analysis. Significantly, new replication of DNA was inhibited only ∿50% by 50 μM BW301U when [^{3}H]dU incorporation was ⩾97% inhibited. Additional preliminary findings suggest that newly replicated DNA containing misincorporated dUMP is rapidly degraded <u>in vivo</u> by extensive excision-repair processes.

INTRODUCTION

It has been recognized for some time that antifolates induce perturbations of <u>de novo</u> nucleotide biosynthesis, resulting in nucleotide pool imbalances [4, 14, 25, 26] and fragmentation of DNA under alkaline conditions [5]. However, antifolates such as methotrexate (MTX) or metoprine (DDMP), commonly used in cancer chemo-

*To whom correspondence should be addressed.

therapy, were shown only recently to promote the misincorporation of dUMP into DNA of lymphoblastoid cells and subsequently induce fragmentation of the newly synthesized DNA by postreplication excision-repair [6, 23, 24].

Inhibition of thymidine biosynthesis by antifolates leads to an increase in the intracellular dUMP pool [1, 14, 16, 19-21] with an accompanying fall in the dTTP pool [1, 14, 16, 19, 20]. It has been demonstrated that antifolates induce increases in the dUMP pool of one order [1, 14, 16, 20] to as great as three orders of magnitude [7]. Goulian and coworkers, for example, reported that the dUMP pool size increased from 0.85 pmole per 10^6 cells without methotrexate to 1220 pmol per 10^6 cells with 10 μM methotrexate [7]. Associated with this increase in dUMP pool size was an increase in the dUTP pool size from $<3 \times 10^{-4}$ to 0.189 pmol per 10^6 cells and a corresponding decrease in the dTTP pool from 39 to 0.98 pmol per 10^6 cells. Presumably, these changes in nucleotide pool sizes promoted the utilization of dUTP as substrate for DNA synthesis, resulting in the observed misincorporation of dUMP into DNA.

Previous work in this laboratory showed that dUMP misincorporates into DNA of B lymphoblastoid cells (WIL-2) exposed to metoprine [24]. The experiments presented here further demonstrate that BW301U (2,4-diamino-5-methyl-6-(2',5'-dimethoxybenzyl)-pyrido(2,4-d)pyrimidine) [3], a new lipid soluble antifolate, induces dUMP misincorporation into DNA of not only B lymphoblastoid cells (SB), but also T lymphoblastoid cells (MOLT-4) and promyelocytic leukemia cells (HL-60). Moreover, in the B and T cell lines studied thus far, BW301U rendered newly synthesized DNA alkaline labile in a drug-dose dependent manner. These results strongly suggest that antifolates can universally induce dUMP misincorporation into the DNA of blood cells leading to the formation of apyrimidinic sites resulting from excision of uracil by DNA-glycosylase. Preliminary evidence also suggests that exposure of cells to antifolates perturbs nucleosome assembly at replication forks. Our findings indicate that protein-DNA interactions are abnormal insofar as newly replicated DNA of cells treated with BW301U is more easily degraded than newly replicated DNA of control cells and bulk DNA of both BW301U treated and untreated cells.

MATERIALS AND METHODS

Materials

MOLT-4 and SB cells were obtained from Dr. Richard Metzgar, Department of Microbiology and Immunology, Duke University Medical Center, Durham, North Carolina. HL-60 cells were obtained through Dr. Martina Viegl, Department of Microbiology and Immunology, Duke University Medical Center, Durham, North Carolina. BW301U was obtained from the Burroughs Wellcome Co., Research Triangle Park, North Carolina. RPMI 1640 media, MEM media for suspension culture, fetal calf serum, and penicillin-streptomycin solutions were purchased from Gibco Laboratories, Grand Island, New York. Unlabeled nucelosides were obtained from Sigma; [6-^{3}H]deoxyuridine (dU) (24.5 Ci/mmol) and [methyl-^{3}H]thymidine (dT) (77.2 Ci/mmol) were purchased from New England Nuclear, and [8-^{3}H]deoxyadenosine (dA) (15 Ci/mmol) was purchased from ICN Pharmaceuticals, Inc.

Cell Culture

MOLT-4 and SB cells were maintained at 500,000 cells/ml in RPMI 1640 media supplemented with 10% fetal calf serum and 25 units/ml penicillin-25 μg/ml streptomycin. HL-60 cells were maintained at 300,000 cells/ml in RPMI 1640 media supplemented with 10% fetal calf serum, 10 mM Hepes, and 50 units/ml penicillin-50 μg/ml streptomycin. All experiments with these cell lines were performed at a concentration of 1×10^7 cells/ml in RPMI 1640 media supplemented with 50 mM Hepes and 10% fetal calf serum which had been dialyzed versus 0.15 M NaCl (RPMI 1640 experimental media).

WIL-2 cells were maintained at 200,000 cells/ml in MEM media for suspension culture supplemented with 10% fetal calf serum, 0.15% $NaHCO_3$, 4 mM glutamine and 25 units/ml penicillin-25 μg/ml streptomycin. Experiments with this cell line were performed at a concentration of 1×10^7 cells/ml in MEM media for suspension culture supplemented with 10 mM Hepes and 10% dialyzed fetal calf serum (MEM experimental media).

Inhibition of [^{3}H]dU or [^{3}H]dA Incorporation into DNA

MOLT-4 or SB cells (1×10^7 cells/ml) were preincubated at 37°C for 15 min in RPMI 1640 experimental media with the appropriate concentration of BW301U. Cells which were to be labeled with [^{3}H]dA were preincubated with 0.2 mM adenosine, in addition to the BW301U, in order to inhibit the utilization of metabolized [^{3}H]dA products in RNA synthesis. This concentration of adenosine was found to have no effect on DNA synthesis as measured by [^{3}H]dT incorporation into DNA during a 15 min pulse period (data not shown). After transfer of 1×10^6 cells (0.1 ml) into test tubes containing 10 μCi of either [^{3}H]dU or [^{3}H]dA, the cells were incubated for 15 more min at 37°C, then immediately diluted with ice-cold phosphate buffered saline (PBS) and washed once in the same buffer, centrifugation was at 1000 × g for 2.5 min. The cells were then transferred onto 2.3 cm Whatman No. 3 filter papers and dried under a heat lamp. The filter papers were then prepared by washing in 5% TCA, Hokin's reagent, and 95% ethanol as described previously [17] and counted in minivials with Betafluor (National Diagnostics) in a Packard Tri-Carb 3255 scintillation counter. When [^{3}H]dA was the label employed, the counted paper disks were taken out of the minivials and the Betafluor removed with two 15 min acetone washes. RNA was degraded by saturating the filter disks with 200 μl of a solution 1 N in NaOH, 0.1 N in NaCl, and 0.01 N in Na_2EDTA followed by incubation at 37°C for 1 h in a humidifed incubator. The ribonucleotides were removed by washing in 5% TCA, Hokin's reagent, and 95% ethanol, and the disks were recounted. The effect of the antifolate on [^{3}H]dU or [^{3}H]dA incorporation into DNA was calculated by comparing the percentage of inhibition of label incorporated into the DNA of BW301U treated and untreated cells.

Detection of Misincorporated dUMP in DNA

Cells (1×10^7 cells/ml) which had been pulse labeled for 15 min with [^{3}H]dU in the presence of either 0 or 10 μM BW301U were lysed in 0.15 M NaCl, 1.5 mM $MgCl_2$, 50 mM Tris, pH 7.4 by expression through a 27 gauge needle (1 cc tuberculin syringe) or by homogenization in a Thomas teflon pestle tissue grinder. The lysate was centrifuged at 500 × g for 5 min and the pellet, containing nuclei, was resuspended in 75 mM NaCl, 0.75 mM $MgCl_2$, 5 mM Na_2EDTA, 2% SDS, and 50 mM Tris, pH 7.4, then left at room temperature overnight. After addition of an equal volume of 2 N NaOH, 0.2 N NaCl, and 0.2 M Na_2EDTA, the sample was incubated at room temperature for >30 min and the DNA sedimented through a 5-20% alkaline sucrose gradient to determine the alkaline lability of the DNA (see below). For the SB and HL-60 cells, all the sucrose gradients were pooled. For the MOLT-4 cells, only the bottom fractions (1-16 out of 34) were pooled. The pooled fractions were then dialyzed versus 0.15 M NaCl and concentrated to 0.3-0.4 ml. The DNA in the concentrated samples was then precipitated with two volumes of 95% ethanol at -20°C in the presence of 0.5 mg of sodium acetate. Digestion of the DNA to nucleosides, separation by thin layer chromatography (TLC) on cellulose plates, and preparation for scintillation counting was performed as described previously [24].

Alkaline Sucrose Gradient Analysis

Treated cells were washed in cold PBS, lysed in 10 mM Hepes, 1 mM Na_2EDTA, 2% SDS, pH 7.2, and digested with 100 μg/ml Proteinase K for three hours. [For the detection of dUMP in DNA, this step was modified (see above).] After addition of an equal volume of 2 N NaOH, 0.2 N NaCl, 0.02 M Na_2EDTA, the samples were incubated at room temperature for >30 min and poured onto 11.0 ml linear 5-20% alkaline sucrose gradients [5% sucrose solution: 0.5 mM Na_2EDTA, 0.2 N NaOH, and 0.8 N NaCl, 20% sucrose solution: 0.5 mM Na_2EDTA, 0.8 N NaOH, and 0.2 N NaCl [2]]. Separation by centrifugation was carried out in a Beckman SW-41 rotor for 3.5 h at 38,000 rpm (20°C). Gradients were then eluted into fractions of 1/3 ml volume. DNA in an aliquot of each fraction was TCA precipitated onto 2.3 cm Whatman No. 3 filter paper disks and prepared for scintillation counting as described above for the inhibition of [^{3}H]dU or [^{3}H]dA incorporation into DNA.

Nuclei Isolation and Micrococcal Nuclease Digestion

Replication of DNA in B lymphoblastoid cells (WIL-2 or SB) was stopped by adding an equal volume of ice-cold 40 mM sodium azide in PBS, and immediately placing the cells in an ice sludge-bath. The cells were then pelleted and washed once with cold 20 mM sodium azide in PBS, centrifugation was for 2.5 min at 1000 × g. The final pellets were resuspended in buffer A (10 mM Hepes, 0.1 mM Na_2EDTA, pH 7.2), allowed to swell for 5 min on ice, and lysed by expression through a 25 gauge needle into an equal volume of buffer B (10 mM Hepes, 2 mM $MgCl_2$, pH 7.2) plus 1.0 mM phenylmethylsulfonylfluoride (PMSF). After centrifugation at 1100 × g for 10 min, the nuclear pellet was resuspended in 10 mM Hepes, 1 mM $MgCl_2$, pH 7.2, 0.01% Triton X-100 and allowed to stand for 5 min in an ice bath. Nuclei were pelleted by centrifugation at 1100 × g for 10 min and washed in the same buffer without Triton X-100, centrifugation as before. Digestion with micrococcal nuclease was carried out at 37°C in 10 mM Hepes, 1 mM $CaCl_2$, pH 7.2 (3.3-3.6×10^7 nuclei/ml). At the indicated times, an aliquot of the digestion mixture was taken. Digestion was stopped by adjusting the sample to 5 mM Na_2EDTA and putting the sample in an ice bath. Digestion kinetics were determined by monitoring the percentage of soluble radiolabel or A_{260} absorbing material in cold 5% perchloric acid (PCA)-0.5 M NaCl after digestion for three hours at 37°C with 1 mg/ml RNAse A. The values reported reflect correction for the presence of RNA determined by comparison of t = 0 samples treated with and without RNAse A.

RESULTS

The lipid soluble antifolate, BW301U, dramatically inhibits the incorporation of exogenous [^{3}H]deoxyuridine (dU) into DNA before it severely inhibits total DNA synthesis, as measured by incorporation of [^{3}H]deoxyadenosine (dA). This is shown in Fig. 1, where the degree of inhibition of [^{3}H]dU or [^{3}H]dA incorporation into trichloroacetic acid (TCA) precipitatable DNA fragments of SB and MOLT-4 cells were compared at various concentrations of BW301U from 0 to 50 μM. Both the SB cells (a B lymphoblastoid cell line) and the MOLT-4 cells (a T lymphoblastoid cell line) responded about equally to BW301U. After a 15 min preincubation at 37°C, as little as 0.5 μM BW301U inhbited [^{3}H]dU incorporation during a 15 min pulse by >60%. Under the same conditions, 20 μM BW301U inhibited [^{3}H]dU incorporation by at least 97% (as seen in the SB cells), and additional drug up to 50 μM increased the observed inhibition level only slightly. In contrast, total DNA synthesis was inhibited to only about 50% by 50 μM BW301U. Similar experiments employing [^{32}P]phosphate have also shown that total DNA synthesis is inhibited by only about 50% under conditions where the incorporation of exogenous dU is inhibited by ⩾97% (data not shown).

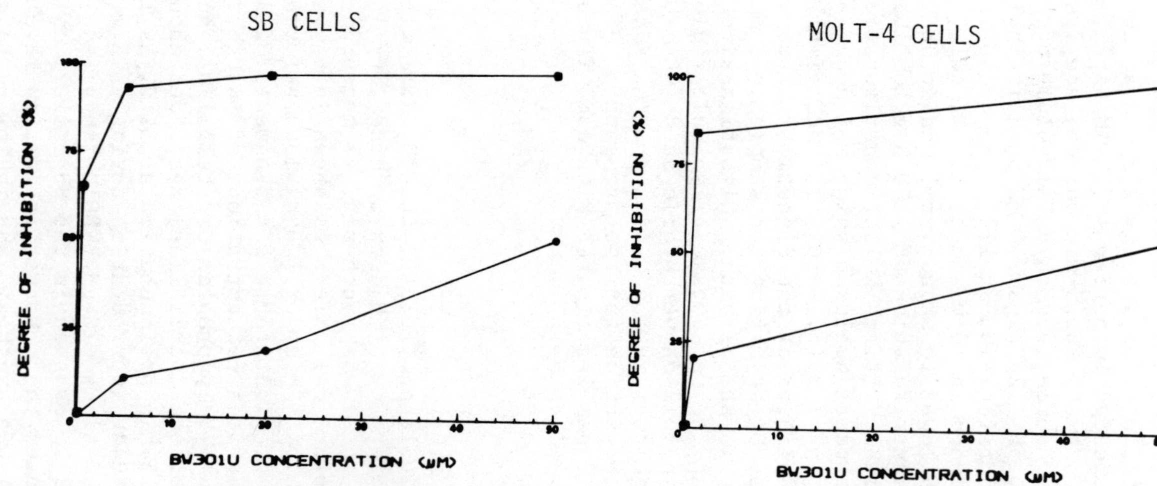

Fig. 1. Inhibition of $[^3H]$dU (squares) or $[^3H]$dA (circles) incorporation into DNA of MOLT-4 and SB cells exposed to BW301U.

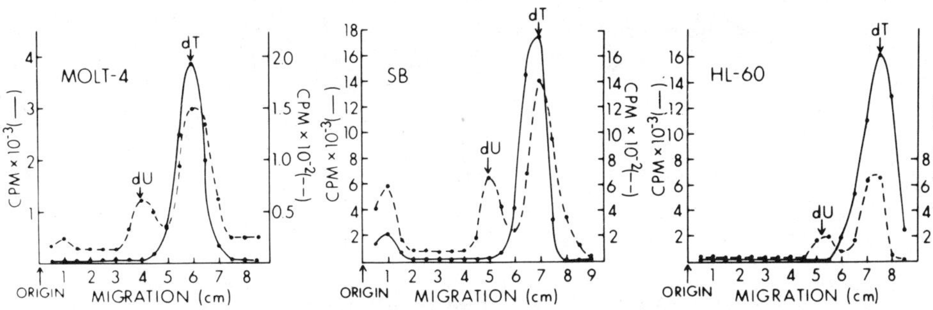

Fig. 2. Detection of dUMP in DNA. MOLT-4, SB, or HL-60 cells were exposed to either 0 or 10 μM BW301U for 30 min at 37°C with a pulse of 100 μCi/ml [^{3}H]dU for the last 15 min. After sedimentation through alkaline sucrose gradients, the DNA was pooled, isolated, and digested to nucleosides which were then separated by TLC on cellulose plates (see Materials and Methods). Solid lines: nucleoside profile of DNA from control cells. Dotted lines: nucleoside profile of DNA from 10 μM BW301U treated cells.

Misincorporation of dUMP into DNA was previously shown to occur in replicating DNA of B lymphoblastoid cells exposed to either methotrexate [6] or metoprine [23, 24]. Recently, we have extended these results to show that misincorporation of dUMP into DNA is not specific to B lymphoblastoid cells, and is likely to occur universally in blood cells treated with antifolates. Human B lymphoblastoid cells (SB), human T lymphoblastoid cells (MOLT-4), or promyelocytic leukemia cells (HL-60) were exposed to a 15 min pulse of exogenous [^{3}H]dU and either 0 or 10 μM BW301U. Following sedimentation through alkaline sucrose gradients, ethanol precipitation of the pooled fractions, and digestion of the DNA with nucleases and bacterial alkaline phosphatase, the resulting nucleosides were separated by TLC on cellulose plates (Fig. 2). Whereas none of these cell lines incorporated dUMP into DNA in the absence of BW301U, at a BW301U concentration which inhibits [^{3}H]dU incorporation by >90%, (10 μM, see SB cells in Fig. 1) an appreciable amount of [^{3}H]dUMP was detected in nuclear DNA. Since SB and MOLT-4 cells are B and T lymphoblastoid cell lines, respectively, and HL-60 cells are a promyelocytic leukemia cell line, these results strongly suggest that antifolate-induced dUMP misincorporation occurs in blood cells, generally, and is not limited to B lymphoblastoid cells.

One consequence of dUMP misincorporation into DNA is the creation of alkaline labile apyrimidinic sites resulting from the excision of uracil by uracil-DNA-glycosylase, the first enzymatic step in excision-repair. Quantification of fragmented DNA in alkaline sucrose gradients as a function of BW301U concentration (or

TABLE 1. Alkaline Sucrose Gradient Distribution of [^{3}H]-Labeled DNA from Cells Treated with BW301U

	MOLT-4 CELLS			SB CELLS			
BW301U Concentration	0 uM	1 uM	50 uM	0 uM	0.5 uM	5 uM	50 uM
% Inhibition*	0%	84%	99%	0%	84%	97%	99%
Small DNA# (Percent of Total)	5.2%	10.8%	28.8%	4.7%	7.5%	27.8%	30.4%

*Percent inhibition pertains to the amount of exogenously added [^{3}H]deoxyuridine which incorporated into TCA precipitatable DNA compared with the control to which no BW301U was added.

#Small DNA pertains to the [^{3}H]-labeled DNA which sedimented in the top 1/6th of an alkaline sucrose gradient (fractions 28-34 out of 34 total fractions).

inhibition of [^{3}H]dU incorporation) indicated that, at about the concentration of BW301U necessary for the direct detection of dUMP IN DNA, a significant increase occurred in the detectable amount of alkaline fragmentable DNA. This can be seen in Table 1, which shows the distribution of tritium in alkaline sucrose gradients after treatment of cells with exogenous [^{3}H]dU and increasing concentrations of BW301U. For both SB and MOLT-4 cells, BW301U concentrations which led to 84% inhibition of exogenous [^{3}H]dU incorporation into DNA, resulted in only about 10% of the label remaining in the top one-sixth of the gradient. Compared to the controls, this represented an increase by a factor of two. However, at BW301U concentrations which led to ≥97% inhibition of exogenous [^{3}H]dU incorporation (i.e., 5-50 μM BW301U), approximately 30% of the label remained in the top one-sixth of the gradient - an increase by a factor of six over controls. Thus, the degree of alkaline fragmentation of DNA induced by BW301U demonstrated a drug-dose dependency. The rapid increase in alkaline labile sites at ∿5 μM BW301U suggests that the function is logarithmic rather than linear, but documentation of this point will require further experimentation.

In order to more fully understand the potential cytotoxic effect of antifolates, it is important to know if dU misincorporation is inhibitory to normal protein-DNA interactions leading to a mature chromatin structure. Since chromatin proteins (histones) protect DNA from nucleases, chromatin maturation can be evaluated by monitoring the sensitivity of DNA to micrococcal nuclease digestion. Figure 3A shows the micrococcal nuclease sensitivity of bulk and newly replicated DNA, demonstrating that DNA synthesized during a 30 sec [^{3}H]dT pulse was much more sensitive to micrococcal nuclease than bulk DNA protected by chromatin proteins. In contrast, when

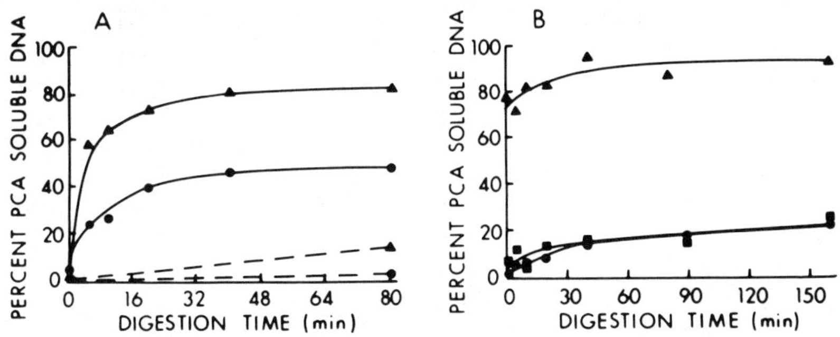

Fig. 3. Chromatin maturation is inhibited in BW301U treated cells. A) Human lymphoblastoid cells, WIL-2 3.6 ml at 1 × 10^7 cells/ml) were preincubated at 37°C for 15 min in MEM experimental media, then pulsed for 30 sec with 80 μCi/ml [^{3}H]dT. DNA replication was stopped by adding an equal volume of ice-cold 40 mM sodium azide in PBS and putting the sample in an ice sludge-bath. Nuclei isolation, digestion with 650 units/ml micrococcal nuclease (2.8 × 10^5 nuclei/unit) and PCA digestiom kinetics were performed as described in Materials and Methods. Triangles: newly replicated DNA ([^{3}H]dT). Circles: bulk DNA (absorbance at 260 nm with correction for RNA). Dotted lines: controls for endogenous nuclease activity (no micrococcal nuclease added). B) Human lymphoblastoid cells, SB (0.5-1.0 × 10^6 cells/ml), were labeled for ∿20 h with 0.1 μCi/ml [^{14}C]dT to label bulk DNA. These were then concentrated to a volume of 2 ml at 1 × 10^7 cells/ml and preincubated at 37°C for 15 min in RPMI 1640 experimental media plus 0.2 mM adenosine (to inhibit the utilization of metabolized [^{3}H]dA products in RNA synthesis) with or without 50 μM BW301U. DNA was labeled for 15 min at 37°C with [^{3}H]dU and [^{3}H]dA. Cells without drug treatment were given 5 μCi/ml [^{3}H]dU and 5 μCi/ml [^{3}H]dA supplemented with enough cold nucleosides to be equivalent to the 100 μCi/ml [^{3}H]dU and 100 μCi/ml [^{3}H]dA which was given to the drug treated cells. DNA replication was stopped by adding an equal volume of ice-cold 40 mM sodium azide in PBS and putting the samples in an ice sludge-bath. Nuclei isolation, digestion with 5 units/ml micrococcal nuclease (5 × 10^6 nuclei/unit), and PCA digestion kinetics were performed as described in Materials and Methods. Triangles: newly replicated DNA from control cells ([^{3}H]dU and [^{3}H]dA with correction for RNA). Circles: bulk DNA ([^{14}C]dT; BW301U treated and untreated cells were identical).

newly replicated DNA was pulse labeled with [^{3}H]dU and [^{3}H]dA for 15 min (i.e., 30 times longer), it was protected from micrococcal nuclease to about the same extent as the [^{14}C]dT-labeled bulk DNA (Fig. 3B), indicating the association of chromatin proteins with the newly replicated DNA. Significantly, DNA synthesized in the presence of 50 μM BW301U was found largely in a degraded state subsequent to isolation - approximately 75% was PCA soluble, even before the addition of micrococcal nuclease. Since DNA replication was inhibited by only ∿50% under the conditions employed in this experiment (Fig. 1), the possibility that the observed degradation resulted from slow progression of DNA synthesis was eliminated. Moreover, the [^{14}C]dT-labeled bulk DNA of this experiment (Fig. 3B) as well as both the bulk and newly replicated DNA of the control experiment (Fig. 3A) were not degraded before addition of micrococcal nuclease, suggesting that the degradation of DNA from the BW301U treated cells was not due to digestion by endogenous nucleases during nuclei isolation. We interpret these results to indicate that high levels of dUMP misincorporation induced by the antifolate led to the degradation of newly synthesized DNA by excision-repair processes. Thus, at high antifolate concentrations which promote misincorporation of dUMP into DNA, progression of DNA into a mature chromatin structure appears to be inhibited.

DISCUSSION

Antifolates are potent inhibitors of dihydrofolate reductase. It has been shown that addition of folate and tetrahydrofolate derivatives can reverse the effects of antifolate exposure [18, 22]. Similarly, products of tetrahydrofolate-dependent reactions, such as thymidine and purines for nucleic acid synthesis, and glycine for protein synthesis can effect partial rescue of cells from antifolate toxicity when given alone, or complete rescue of cells when given in combination [9-11, 15, 22, 26]. Thus, the cytotoxic consequences of antifolates is dependent on the complex interrelationship of independent metabolic pathways possessing a common dependence on tetrahydrofolate. The experiments presented here have concentrated on the inhibition of thymidine biosynthesis - the first of the three metabolic pathways to be affected significantly by antifolate exposure, and the only one that depletes tetrahydrofolate pool levels in the cell by oxidation of this cofactor to dihydrofolate.

The antifolate-induced increase in dUTP pool and corresponding decrease in dTTP pool [7] leads to misincorporation of dUMP in DNA and excision of uracil by DNA-glycosylase [6, 23, 24]. This effect is drug-dose dependent and appears to be universal in blood cells (this work). dUMP has been detected directly in DNA of B lymphoblastoid cells (WIL-2 and SB), T lymphoblastoid cells (MOLT-4), and promyelocytic leukemia cells (HL-60), and the number of alkaline labile sites in newly synthesized DNA increases with increasing antifolate concentration.

Since the primary site for dUMP misincorporation is at replication forks, we have attempted to characterize the fate of newly replicated DNA with respect to chromatin structure. A toxic effect of antifolates at the DNA level is supported by several observations showing that high concentrations of deoxyuridine in DNA affect normal DNA-protein interactions in chromatin. It has been reported that low levels of dUMP (0.0021 mole %) in DNA of nuclei, chromain, and reconstituted chromatin lead to excision of only about half of the uracil residues when exposed to cell-free extracts of human lymphoblastoid cells, contrasting to the excision of >90% of all uracils from the same DNA when free of protein [13]. Analogous to the well studied protection of DNA from micrococcal nuclease (for a recent review, see [12]) these results are consistent with the concept that nucleosomal proteins can also protect the DNA from glycosylases. <u>B. subtilis</u> phage PBS1 DNA, on the other hand, contains a high level of dUMP (35.9 mole %), and uracil excision from chromatin made by reconstituting histones with this DNA was reported to be equal to that of free DNA (i.e., >90% excision of uracil) [13]. Thus, the protection of deoxyuridine residues observed at low dUMP substitution levels was lost at high dUMP substitution levels. Presumably, a high level of enzymatic excision and repair at accessible regions of the chromatin led to DNA strand breakage and annihilation of nucleosomal structures.

Ths interpretation is also supported by the observations of Grafstrom and coworkers [8]. Using a whole cell lysate of lymphoblastoid cells, DNA was synthesized <u>in situ</u> in the presence of [^{3}H]dCTP and a 10:1 ratio of dUTP-dTTP (in addition to appropriate amounts of the remaining nucleoside triphosphates). After a two minute pulse, the cells were chased with an excess of dTTP and unlabeled dCTP and the amount of TCA precipitatable tritium labeled DNA was determined. Within two minutes, $\sim$50% of the labeled DNA was degraded to acid soluble material; by 10 minutes, a plateau had been reached at $\sim$70% acid soluble material. Again, these results suggested that high levels of dUMP misincorporation led to the degradation of DNA.

Our experiments were originally designed to determine if high-dose BW301U exposure prevented nucleosome assembly at replication forks. These experiments were based on the knowledge that 'free' DNA at replication forks is more sensitive to micrococcal nuclease digestion that bulk DNA in a mature chromatin structure (Fig. 3A). The fact that newly replicated DNA in cells treated with BW301U was highly degraded after nuclei isolation (Fig. 3B), even without exposure to micrococcal nuclease, suggested that there was active degradation occurring either during the nuclei isolation procedure or during the 15 min pulse period, itself.

The experiments of Grafstrom and coworkers, above, would tend to indicate that at least some, if not all, of the degradation occurred during the 15 min pulse period. The value of ∿75% which we found for the amount of predigestion PCA soluble material when DNA was labeled in vivo corresponds remarkably well to the ∿70% degradation of in situ-labeled dUMP-containing DNA which they reported. Moreover, the value of ∿50% inhibition which we determined for overall DNA synthesis (Fig. 1) may reflect, in part, the degradation of newly synthesized DNA. Thus, our results strongly suggest that high-dose antifolate exposure leads to elevated levels of dUMP misincorporation and rapid degradation of newly synthesized DNA in vivo by cellular attempts to excise and repair DNA synthesized under these conditions. Experiments are now in progress to more fully characterize this potentially important cytotoxic lesion induced by antifolates.

ACKNOWLEDGMENTS

This research was supported by an NCI Research Grant (R01 CA 31110), by an NCI Research Training Grant in Cancer Chemotherapy (NCI 5 T 32 CA09307-05), by an NIH National Research Service Award (NCI 5 T 32 CA09111), and by a grant from The Samuel and Rae Eckman Charitable Foundation.

REFERENCES

1. B. Ardalan, M. D. Buscaglia, and P. S. Schein, Tumor 5-fluorodeoxyuridylate concentration as a determinant of 5-fluorouracil response, Biochem. Pharmacol., 27:2009-2013 (1978).
2. M. L. DePamphilis, P. Beard, and P. Berg, Synthesis of superhelical simian virus 40 deoxyribonucleic acid in cell lysates, J. Biol. Chem., 250:4340-4347 (1975).
3. D. S. Duch, C. W. Sigel, S. W. Bowers, M. P. Edelstein, J. C. Cavallito, R. G. Foss, and C. A. Nichol, Lipid-soluble inhibitors of dihydrofolate reductase: Selection and evaluation of the 2,4-diaminopyridopyrimidine BW301U and related compounds as anticancer agents, Current Chemotherap. Infect. Dis. (J. D. Nelson and C. Grassi, eds.), American Society for Microbiology, Washington, D.C., 1597-1599 (1980).
4. A. Fridland, Effect of methotrexate on deoxyribonucleotide pools and DNA synthesis in human lymphocytic cells, Cancer Res., 34:1883-1888 (1974).
5. A. Fridland and T. P. Brent, DNA replication in methotrexate-treated human lymphoblasts, Eur. J. Biochem., 57:379-385 (1975).
6. M. Goulian, B. Bleile, and B. Y. Tseng, Methotrexate-induced misincorporation of uracil into DNA, Proc. Natl. Acad. Sci. U.S.A., 77:1956-1960 (1980).
7. M. Goulian, B. Bleile, and B. Y. Tseng, The effect of methotrexate on levels of dUTP in animal cells, J. Biol. Chem., 255: 10630-10637 (1980).

8. R. H. Grafstrom, B. Y. Tseng, and M. Goulian, The incorporation of uracil into animal cell DNA in vitro, Cell, 15:131-140 (1978).
9. M. T. Hakala and E. Taylor, The ability of purine and thymine derivatives and of glycine to support the growth of mammalian cells in culture, J. Biol. Chem., 234:126-128 (1959).
10. K. R. Harrap, G. A. Taylor, and G. P. Browman, Enhancement of the therapeutic effectiveness of methotrexate and protection of normal proliferating tissues with purines and pyrimidines, Chem.-Biol. Interact., 18:119-128 (1977).
11. W. M. Hryniuk, L. W. Brox, J. F. Henderson, and T. Tamaoki, Consequences of methotrexate inhibition of purine biosynthesis in L5178Y cells, Cancer Res., 35:1427-1432 (1975).
12. T. Igo-Kemenes, W. Horz, and H. G. Zachau, Chromatin, Ann. Rev. Biochem., 51:89-121 (1982).
13. K. Ishiwata and A. Oikawa, Chromatin structure interferes with excision of abnormal bases from DNA, Biochim. Biophys. Acta, 698:15-21 (1982).
14. R. C. Jackson, The regulation of thymidylate biosynthesis in Novikoff hepatoma cells and the effects of amethopterin, 5-fluorodeoxyuridine, and 3-deazauridine, J. Biol. Chem., 253: 7440-7446 (1978).
15. J. J. Kinahan, M. Otten, and G. B. Grindley, Evaluation of ribonucleoside and deoxyribonucleoside triphosphate pools in cultured leukemia cells during exposure to methotrexate or methotrexate plus thymidine, Cancer Res., 39:3531-3539 (1979).
16. P. Klubes, K. Connelly, I. Cerna, and H. G. Mandel, Effects of 5-fluorouracil on 5-fluorodeoxyuridine-5'-monophoshate and 2-deoxyuridine-5'-monophosphate pools, and DNA synthesis in solid mouse L1210 and rat Walker 256 tumors, Cancer Res., 38: 2325-2331 (1978).
17. J. Laszlo, M. J. Fyfe, D. Sedwick, L. Lee, and O. E. Brown, Comparison of metoprine (DDMP) and etoprine (DDEP) by measuring the inhibition of deoxyuridine incorporation into DNA of human leukemia cells, Cancer Treatment Reports, 62:341-344 (1978).
18. J. A. R. Mead, J. M. Venditti, A. W. Schrecker, A. Goldin, and J. C. Keresztesy, The effect of reduced derivatives of folic acid on toxicity and antileukemic effect of methotrexate in mice, Biochem. Pharmacol., 12:371-383 (1963).
19. R. G. Moran, C. P. Spears, and C. Heidelberger, Biochemical determinants of tumor sensitivity to 5-fluorouracil: Ultrasensitive methods for the determination of 5-fluoro-2'-deoxyuridylate, 2'-deoxyuridylate, and thymidylate synthetase, Proc. Natl. Acad. Sci., U.S.A., 76:1456-1460 (1979).
20. C. E. Myers, R. C. Young, and B. A. Chabner, Biochemical determinants of 5-fluorouracil response in vivo, J. Clin. Invest., 56:1231-1238 (1975).

21. C. E. Myers, R. C. Young, D. G. Johns, and B. A. Chabner, Assay of 5-fluorodeoxyuridine-5'-monophosphate and deoxyuridine-5'-monophosphate pools following 5-fluorouracil, Cancer Res., 34:2682-2688 (1974).
22. D. Roberts and E. V. Warmath, Methotrexate inhibition of CCRF-CEM cultures of human lymphoblasts, Europ. J. Cancer, 11:771-782 (1975).
23. W. D. Sedwick and J. Laszlo, An antifolate-induced lesion in newly synthesized DNA, pp. 295-308, in: "Advances in Enzyme Regulation," Vol. 19 (G. Weber, ed.), Pergamon Press, New York (1981).
24. W. D. Sedwick, M. Kutler, and O. E. Brown, Antifolate-induced misincorporation of deoxyuridine monophosphates into DNA: Inhibition of high molecular weight DNA synthesis in human lymphoblastoid cells, Proc. Natl. Acad. Sci. U.S.A., 78:917-921 (1981).
25. M. H. N. Tattersall and K. R. Harrap, Changes in the deoxyribonucleoside triphosphate pools of mouse 5178Y lymphoma cells following exposure to methotrexate or 5-fluorouracil, Cancer Res., 33:3086-3090 (1973).
26. M. H. N. Tattersall, R. C. Jackson, S. T. M. Jackson, and K. R. Harrap, Factors determining cell sensitivity to methotrexate: Studies of folate and deoxyribonucleoside triphosphate pools in five mammalian cell lines, Eur. J. Cancer, 10:819-826 (1974).

EFFECTS OF NUCLEOTIDE POOL IMBALANCES ON THE EXCISION REPAIR OF ULTRAVIOLET-INDUCED DAMAGE IN THE DNA OF HUMAN DIPLOID FIBROBLASTS

Ronald D. Snyder

Stauffer Chemical Company
400 Farmington Avenue
Farmington, Connecticut 06032

ABSTRACT

Recent studies in our laboratory and others have demonstrated that DNA polymerase inhibitors such as the ara nucleosides, aphicolin and dideoxythymidine are potent inhibitors of the DNA excision repair process in confluent human fibroblasts as evidenced by the agent-dependent accumulation of single-strand interruptions in the DNA of UV-irradiated, but not in unirradiated, cellular DNA. In rapidly cycling cells, on the other hand, these agents are weak inhibitors at best but when used in combination with the ribonucleotide reductase inhibitor, hydroxyurea, a significant enhancement of inhibitory capacity is seen. In an attempt to better understand the mechanism of repair inhibition by DNA polymerase inhibitors, and the nature of this hydroxyurea enhancement, experiments were initiated in which the effects of a series of ribonucleotide reductase inhibitors on dNTP pools and on the DNA repair process were determined in both quiescent cultures and log-phase cultures of human fibroblasts. It was determined that hydroxyurea, deoxyadenosine, pyridine-2-carboxaldehyde thiosemicarbazone (TSC), pyrozoloimidazole (IMPY), 3,5-diamino-1,2,4-triazole (guanazole), 3,4,5-trihydroxy benzohydroxamic acid (THBA) and 3,4-dihydroxy benzohydroxamic acid (DHBA) are all effective inhibitors of the DNA repair process in confluent cells but not in log-phase cells. Moreover, the effects of these inhibitors can be reversed by the addition of certain combinations of deoxynucleosides. These reversal studies and the direct analysis of dNTP pool modulation by these compounds in log phase and confluent cultures support the notion that specific pool depletions rather than general imbalance of pools gives rise to the inhibition of the DNA excision repair process.

1. INTRODUCTION

Hydroxyurea (HU) has long been recognized as a potent inhibitor of the enzyme ribonucleotide reductase [8, 16, 17, 25, 26] and the modulation of dNTP pools by this agent has since been demonstrated convincingly in many procaryotic and eucaryotic systems. As would be predicted, HU is also a strong inhibitor of DNA replication in rapidly cycling cells but paradoxically, does not appear to inhibit incorporation of DNA precursors during DNA repair processes in confluent cells (UDS). In fact, HU is widely used in UDS assays in order to reduce the background incorporation resulting from cycling cells. This apparent resistance of repair synthesis to HU inhibition has lead to the general belief that HU does not inhibit DNA repair in mammalian cells. However, it has been clearly demonstrated that HU treatment of UV-irradiated mammalian cells leads to single-strand DNA breaks [1, 5, 12] and inhibition of the removal of pyrimidine dimers from normal human fibroblasts [4]. In addition, it has been shown that HU enhances the repair inhibitory effects of the aranucleosides [7, 23, 24]. It became of interest, therefore, to determine what the relationship was between the inhibition of ribonucleotide reductase, the subsequent perturbation of dNTP pools, and the apparent inhibition of DNA synthesis. This paper describes the preliminary results of one appraoch at better understanding this relationship through the examination of the effects of a number of ribonucleotide reductase inhibitors on the DNA repair process.

2. MATERIALS AND METHODS

2.1. Cell Culturing and Labeling

Normal human foreskin fibroblasts (HSBP) were grown in modified Eagles' medium (MEM) supplemented with 10% fetal bovine serum, and were maintained at 37°C in humidity- and carbon-dioxide controlled incubators. All experiments were conducted by inoculating $\sim 5.0 \times 10^4$ cells into 60 mm plastic tissue culture dishes, allowing 48 h for growth then labeling for 24 h with 0.7 μCi/ml [^{3}H]thymidine (6.7 Ci/mmol) or 0.7 μCi/ml [^{14}C]thmidine (543 mCi/mmol) in MEM with 10% newborn calf serum.

2.2. Chemicals

Hydroxyurea, guanazole, and the deoxynucleosides were purchased from Sigma Chemical Company. IMPY was obtained from Dr. Van Narayanan, Drug Synthesis and Chemistry Branch, NCI. The thiosemicarbazone derivative was the generous gift of Dr. R. Wallace Brockman, Southern Research Institute, Birmingham, Alabama. The benzohydroxamic acids were the generous gifts of Dr. B. van't Riet, Medical College of Virginia, Richmond. [^{3}H]thymidine was purchased from ICN Pharmaceuticals and [^{14}C]thymidine was purchased from Amersham.

2.3. dNTP Pool Determinations

dNTP pool determinations were made using modifications of the DNA polymerase method of Hunting and Henderson [14] described in detail elsewhere [18].

2.4. Inhibitor Treatment and Alkaline Sucrose Sedimentation

The details of the repair inhibition assay have been presented elsewhere [7, 22]. Briefly, cells in the appropriate growth stage prelabeled with [^{3}H]- or [^{14}C]thymidine were irradiated with 20 J/M^2 UV light and then incubated for 3 h in the presence [^{3}H] or absence [^{14}C] of a ribonucleotide reductase inhibitor. The differentially labeled cells were then mixed together in cold saline-0.12% EDTA and 50 μl (10^4 cells) were lysed for 1 h at room temperature in 200 μl of 1 N NaOH overlayed on 5-20%, 4.3 ml alkaline sucrose gradients containing 2 M NaCl. Centrifugation was carried out in a Beckman SW 60 Ti rotor for 150 min or in a Beckman SW 50.1 rotor for 105 min at 30,000 rpm. Six drop fractions were collected on filter paper strips which were then given a 15 min wash with 5% TCA and two-15 min washes in 95% ethanol. Radioactivity in acid insoluble material was assayed in a Beckman liquid scintillation spectrometer. Molecular weight analyses were done by computer program. The difference in the reciprocal weight average molecular weight ($\Delta 1/M_w$) of the [^{3}H]- and [^{14}C]-labeled DNA is a direct measure of the excess DNA strand breaks formed due to the repair process ($2 \times \Delta 1/M_w$).

3. RESULTS AND DISCUSSIONS

The rate-limiting step in DNA excision repair is the initial endonucleolytic incision, for once this nick is made in the DNA the damaged site is excised and the resultant gap is very rapidly refilled and ligated [11]. For this reason, normal DNA isolation procedures and sedimentation analyses are incapable of "capturing" repair intermediates (i.e., single-strand DNA interruptions). It has been demonstrated, however, that if repair is allowed to proceed in the presence of certain inhibitors of DNA polymerase such as the aranucleosides or aphidicolin, stable single-strand interruptions occur at each inhibited site [7, 20-24] and that these strand breaks are easily quantitated with alkaline sucrose velocity sedimentation. That repair can be similarly inhibited by means other than direct inhibition of the DNA polymerases was suggested by the finding that HU, a potent inhibitor of ribonucleotide reductase also induces strand breaks in UV-irradiated mammalian cells [1, 5, 12]. To determine if this phenomenon was actually due to interaction at the level of the reductase (i.e., dNTP pool perturbations) a series of ribonucleotide reductase inhibitors was examined for their ability to inhibit DNA excision repair.

TABLE 1. DNA Strand Breaks Induced in Human Fibroblasts by Ribonucleotide Reductase Inhibitors

Inhibitor	Conc.	Culture State	UV	DNA Strand Breaks 10^8 Daltons
HU	500 uM	confluent	+	0.4
	1 mM	confluent	+	0.9
	2 mM	confluent	+	1.9
		confluent	-	0
		log	+	0
dA	500 uM	confluent	+	0.6
	1 mM	confluent	+	1.1
		confluent	-	0
		log	+	0
TSC	1 mM	confluent	+	3.0
		log	+	0
		confluent	-	0
IMPY	500 uM	confluent	+	1.3
		confluent	-	0
		log	+	0
Guanazole	1 mM	confluent	+	0.5
		confluent	-	0
		log	+	0
THBA	500 uM	confluent	+	1.0
		confluent	-	0
		log	+	0
DHBA	500 uM	confluent	+	1.4
		confluent	-	0
		log	+	0

Strand breaks were determined from alkaline sucrose gradients as described in Materials and Methods.

It is shown in Table 1 and Figs. 1 and 2 that the reductase inhibitors HU [1-5], guanazole [3], IMPY [2, 6, 10], TSC [13], dA [15, for review], THBA and DHBA [9] all cause an accumulation of single-strand DNA breaks in UV-irradiated confluent human fibroblasts and that no such breaks are seen in unirradiated cells. Moreover, the inhibition is completely absent in rapidly cycling log phase cells. This same differential activity of repair inhibitors in confluent vs. log phase cells has been seen with the aranucleosides [23] and with aphidicolin [21]. It is likely that at least a partial explanation for this differential inhibition in the case of the DNA polymerase inhibitors is that competing dNTP pools are much lower

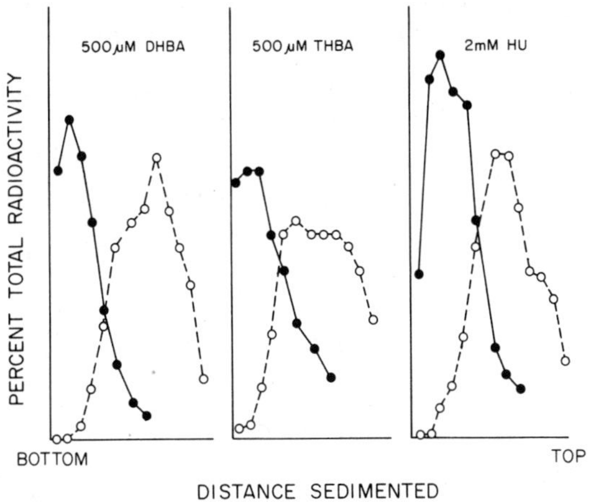

Fig. 1. DNA strand breaks in UV-irradiated confluent human fibroblasts incubated in the presence of ribonucleotide reductase inhibitors. Alkaline sucrose gradients were run after 3 h of inhibitor treatment as described in Materials and Methods. ○-○-○, 20 J/M^2 UV + inhibitor; ●-●-●, 20 J/M^2 UV, no inhibitor.

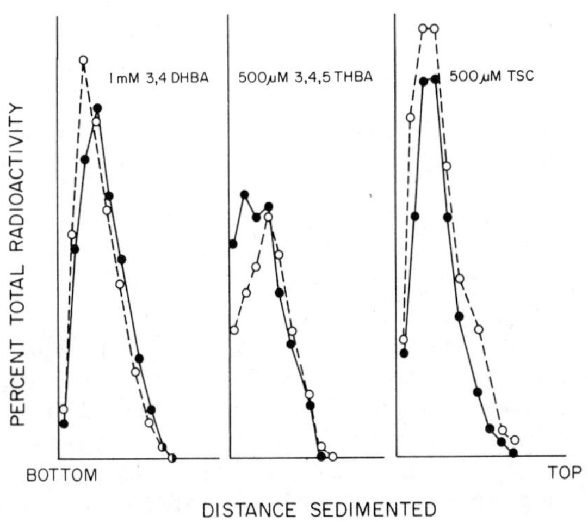

Fig. 2. Effects of ribonucleotide reductase inhibitors on unirradiated human fibroblast DNA. Confluent cultures were treated for 3 h with inhibitor prior to harvesting and alkaline sucrose sedimentation. ○-○-○, inhibitor treatment; ●-●-●, no inhibitor treatment.

TABLE 2. Effects of Deoxynucleosides and Ribonucleotide Reductase Inhibitors on dNTP Pools in Confluent and Log Phase Human Fibroblasts

Agent	Culture State	% Control dATP	dTTP	dCTP	dGTP
2 mM HU	log	32	120	120	48
	confluent	40	156	143	10
500 uM dA	log	2640	115	61	15
	confluent	230	100	57	56
1 mM Guanazole	log	76	107	41	90
	confluent	84	181	39	51
500 uM TSC	log	46	100	37	88
	confluent	80	249	74	20
500 uM IMPY	log	86	85	80	114
	confluent	60	163	89	39
500 uM DHBA	confluent	37	206	76	17
500 uM THBA	confluent	15	263	52	21
500 uM dT	log	178	470	88	225
	confluent	110	332	129	150
1 mM dC	log	93	183	176	183
	confluent	84	137	900	134
1 mM dG	log	120	119	149	900
	confluent	75	100	100	250

dNTP pool determinations were made after 3 hrs incubation of appropriately treated cell cultures at 37°C. The numbers shown were calculated as an average of 2-6 experiments. dNTP levels in untreated log phase and confluent cultures were, respectively, dATP - 23.6, 2.6; dTTP - 49.4, 4.4; dCTP - 11.4, 9.5; dGTP - 6.3, 5.1 pmoles/106 cells.

in confluent cells an thus that the inhibition would be stronger in those cells. An examination of dNTP pools, however, showed that while pools were lower in confluent cultures [18], the magnitude of the reduction was not great enough to account solely for the differential effects of the inhibitors. It seems most likely in those cases that an important additional factor is the lowered amount of DNA polymerase in confluent cells which would make repair more sensitive to inhibition.

Since the agents used in this study are fairly specific for ribonucleotide reductase, perturbations in dNTP pools are likely to be the major case of DNA repair inhibition. An examination of pools

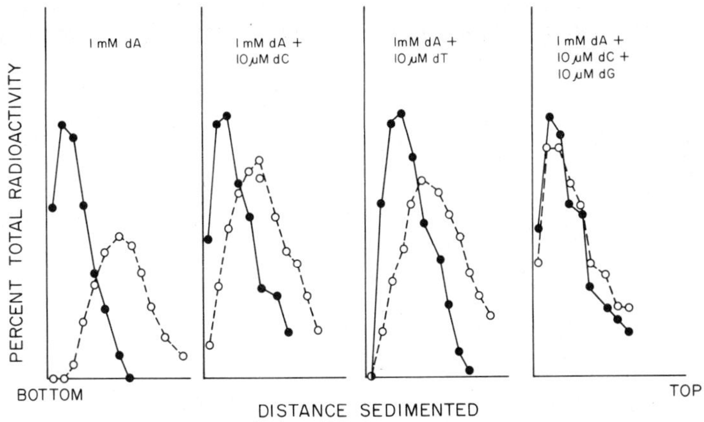

Fig. 3. Antagonism of deoxyadenosine inhibition of DNA repair by deoxynucleosides in confluent cultures of human fibroblasts. Deoxynucleosides were administered simultaneously with the inhibitor, cells were incubated for 3 h at 37°C and then sedimented through alkaline sucrose gradients. o-o-o, 20 J/M^2 UV + dA + deoxynucleoside(s); ●-●-●, 20 J/M^2 UV.

was made following a 3-h treatment of confluent or log phase cultures with inhibitors or deoxynucleosides. These data are presented in Table 2. It is shown that all of the agents elevate dTTP pools in confluent cells, some by 200-300%. This increase in dTTP pools probably results from the slow turnover rate of dTTP pools in cells in which DNA synthesis has been blocked. Of all of the reductase inhibitors only HU increases dCTP pools; the others decrease dCTP pools by 10-60%. dATP pools are decreased by all of the reductase inhibitors, except dA, but the level of the effect varies quite widely from an 85% reduction seen with THBA to a 20% reduction seen with guanazole and TSC. It must be pointed out that the doses chosen for these pool studies are the doses used for the initial repair inhibition studies and thus may not be optimal for detecting maximal pool effects. A more complete dose analysis will be presented elsewhere. The dGTP pools were decreased most consistently and effectively by all of the reductase inhibitors, being reduced by at least 50% and in some cases 80-90% as compared to controls. For the most part, the reduction in all pools was more marked in confluent than in log phase cells; thus, due to the inherently larger pools in log cells, the absolute pool sizes on a pmole/cell basis were always more dramatically lowered in confluent cells. It seems apparent, then, that the inability of the reductase inhibitors to inhibit the DNA repair process in log-phase cells is directly attributable to their failure to lower pools to a critical level. Moreover, it seems most likely that the purine dNTP pools

TABLE 3. Deoxynucleoside Antagonism of DNA Strand-Break Induction by Ribonucleotide Reductase Inhibitors

	% Reversal of Strand Breaks		
	dA + dG + dC	dC + dG	dG
2 mM HU	100	48	35
1 mM dA	100	95	48
500 uM IMPY	92	6	45
500 uM TSC	96	47	16
500 uM THBA	100	80	--
500 uM DHBA	100	67	--

In these experiments, deoxynucleosides (10 uM) were added simultaneously with the inhibitor following irradiation of cultures with 20 J/M^2 254 nm light. Strand breaks were assessed on alkaline sucrose gradients after 3 hours incubation of cultures at 37°C. The numbers represent the averages of 2-3 determinations. The degree of antagonism of breaks is related to the number of strand breaks made, i.e., dose of inhibitor used (data not shown).

because they are most lowered by inhibitor treatment, are rate-limiting in the repair process. The data are also consistent with the notion that it is specific pool depletions rather than pool imbalances which lead to the inhibition of DNA repair since treatment of cells with up to 1 mM dT, dC, or dG dramatically changes pool ratios but leads to no inhibition of DNA repair at least as measured in the strand-break assay (data not shown).

That the reduction in dNTP pools is actually the cause of the inhibition of DNA repair may be demonstrated by studies in which deoxynucleosides are added simultaneously with the inhibitor. Figure 3 shows the resultant effects of deoxynucleosides on DNA strand breaks induced by dA treatment of UV-irradiated cells. Simultaneous treatments with dC or dT are shown to partially reverse the inhibition while a combination of dC and dG completely reverses it. This is the predicted pattern since it is shown in Table 2 that the primary effect of dA treatment is to lower the dCTP and dGTP pools. The partial effectiveness of dT is most likely due to the fact that dT (dTTP) is a positive effector for GDP reduction and thus increases dGTP pools (Table 2). Table 3 shows that a combination of dC, dG, and dA completely reverses the formation of strand breaks induced by all of the inhibitors tested and that combinations of dC and dG, or dG alone are also partially effective antagonists. It is clear, then, that pool perturbations, or more precisely, pool depletions lead to inhibition of the DNA excision repair process. It is to be expected that these periods of pool depletion would be opportunities for mutagenesis as incorrect nucleotides would be inserted in place of the missing correct one. Studies are currently underway to assess the mutagenicity of the ribonucleotide reductase inhibitors in human fibroblasts.

In conclusion it has been demonstrated that hydroxyurea as well as a number of other ribonucleotide reductase inhibitors inhibit the DNA excision repair process and that they most likely do this through depletion of purine dNTP pools. It will be shown in more detail elsewhere [19] that this inhibition is dose-dependent and that it is potentiated by pyrimidines. It will also be shown that all of the ribonucleotide reductase inhibitors increase the uptake of tritiated DNA precursors into cells resulting in inflated estimates of DNA synthesis and most likely accounting for the observations that HU does not appear to inhibit UDS in cultured cells.

REFERENCES

1. E. Ben-Hur and R. Ben-Ishai, DNA repair in ultraviolet light irradiated HeLa cells and its reversible inhibition by hydroxyurea, Photochem. Photobiol., 13:337-345 (1971).
2. R. W. Brockman, S. C. Shaddix, J. W. Carpenter, N. F. Dubois, and R. F. Struck, Studies on the mechanism of action of 2,3-dihydro-1-H pyrazolo [2,3-a] imidazole (NSC-51143), Proc. Am. Assoc. Cancer Res., 19:21 (1978).
3. R. W. Brockman, S. Shaddix, W. R. Laster, Jr., and F. M. Schabel, Jr., Inhibition of ribonucleotide reductase, DNA synthesis and L1210 leukemia by guanazole, Cancer Res., 30:2358-2368 (1970).
4. A. R. S. Collins, DNA damage in ultraviolet-irradiated HeLa and CHO K1 cells examined by alkaline lysis and hydroxyapatite chromatography, Biochim. Biophys. Acta, 478:461-473 (1977).
5. A. R. S. Collins, S. L. Schor, and R. T. Johnson, The inhibition of repair in ultraviolet-irradiated human cells, Mut. Res., 42:413-432 (1977).
6. J. G. Cory and A. E. Fleischer, Mode of inhibition of tumor cell ribonucleotide reductase by 2,3-dihydro-1-H-pyrazolo [2,3-A] imidazole (NSC 51143), Cancer Res., 40:3891-3894 (1980).
7. W. C. Dunn and J. D. Regan, Inhibition of DNA excision repair in human cells by arabinofuranosyl cytosine: Effect on normal and xeroderma pigmentosum cells, Molec. Pharmacol., 15:367-374 (1979).

8. H. Elford, Effect of hydroxyurea on ribonucleotide reductase, Biochem. Biophys. Res. Commun., 33:129-135 (1968).
9. H. L. Elford, L. Wampler, and B. van't Riet, New ribonucleotide reductase inhibitors with antineoplastic activity, Cancer Res., 39:844-851 (1979).
10. H. L. Ennis, L. Moller, J. J. Wang, and O. S. Selawry, 2,3-dihydro-1-H-imidazole [1,2-b] pyrazole. A new inhibitor of deoxyribonucleic acid synthesis, Biochem. Pharmacol., 20:2639-2646 (1971).
11. K. Erixon and G. Ahnstsrom, Single-strand breaks in DNA during repair of UV-induced damage in normal human and XP cells as determined by alkaline DNA unwinding and hydroxylapatite chromatography. Effects of hydroxyurea, 5-fluoro-deoxyuridine and 1-β-D-arabinofuranosyl-cytosine on the kinetics of repair, Mut. Res., 59:257-271 (1979).
12. A. A. Francis, R. D. Blevins, W. L. Carrier, D. P. Smith, and J. D. Regan, Inhibition of DNA repair in ultraviolet-irradiated human cells by hydroxyurea, Biochim. Biophys. Acta, 563:385-392 (1979).
13. F. A. French, E. J. Blanz, S. C. Shaddix, and R. W. Brockman, α-(N)-Formyl-heteroaromatic thiosemicarbazones. Inhibition of tumor-derived ribonucleotide diphosphate reductase and correlation with in vivo antitumor activity, J. Med. Chem., 17:172-181 (1974).
14. D. Hunting and J. F. Henderson, Determination of deoxyribonucleotide triphosphates using DNA polymerase: a critical evaluation, Can. J. Biochem., 59:723-727 (1981).
15. D. Hunting and J. F. Henderson, Models of the regulation of ribonucleotide reductase and their evaluation in intact mammalian cells, CRC Critical Rev. Biochem., 13:325-348 (1982).
16. I. H. Krakoff, N. C. Brown, and P. Reichard, Inhibition of ribonucleotide diphosphate reductase by hydroxyurea, Cancer Res., 28:1559-1568 (1968).
17. J. Neuhard, Studies on the acid-soluble nucleotide pool in Escherichia coli, IV. Effects of hydroxyurea, Biochim. Biophys. Acta, 145:1-6 (1967).
18. R. D. Snyder, Deoxyribonucleotide triphosphate pools in human diploid fibroblasts and their modulation by hydroxyurea. Submitted to Biochem. Pharmacol.
19. R. D. Snyder, The effects of ribonucleotide reductase inhibitors on DNA excision repair in human fibroblasts and modulation of their effects by deoxynucleosides. Manuscript in preparation.
20. R. D. Snyder and J. D. Regan, Aphidicolin inhibits repair of DNS in ultraviolet-irradiated human fibroblasts, Biochem. Biophys. Res. Commun., 99:1088-1094 (1981).
21. R. D. Snyder and J. D. Regan, Differential responses of log and stationary phase human fibroblasts to inhibition of DNA repair of aphidicolin, Biochim. Biophys. Acta, 697:229-234 (1982).

22. R. D. Snyder, W. L. Carrier, and J. D. Regan, Application of arabinofuranosylcytosine in the kinetic analysis and quantitation of DNA repair in human cells after ultraviolet irradiation, Biophys. J., 35:339-350 (1981).
23. R. D. Snyder, B. Van Houten, and J. D. Regan, The accumulation of DNA breaks due to incision; comparative studies with various inhibitors, in: "DNA Repair and Its Inhibition" (A. Collins, R. Johnson, and S. Downes, eds.), in press, IRL Press (1984).
24. R. D. Snyder, B. Van Houten, and J. D. Regan, The inhibition of ultraviolet radiation-induced DNA repair in human diploid fibroblasts by arabinofuranosyl nucleosides. In Press, Chem.-Biol. Interactions (1984).
25. W. G. Thurman (ed.), Symposium on hydroxyurea. A discussion of pharmacology, mechanism of action, animal screening studies and chemical studies, Cancer Chemother. Reports, 40:1-75 (1964).
26. M. K. Turner, R. Abrams, and J. Lieberman, Meso-α,β-diphenyl succinate and hydroxyurea as inhibitors of deoxycytidylate synthesis in extracts of Ehrlich ascites and L cells, J. Biol. Chem., 241:5777-5780 (1966).

EFFECT OF VARIATIONS IN THE CONDITIONS OF DNA SYNTHESIS UPON THE ACCURACY OF DNA REPLICATION

Tashmeem Razzaki* and Navin K. Sinha†

*Department of Genetics
University of Albera
Edmonton, Alberta, Canada

†Waksman Institute of Microbiology
Rutgers University
Piscataway, New Jersey 08854

INTRODUCTION

Recent refinements in the methods for measurement of accuracy of DNA replication have led to accurate estimates for a number of mispairs during the duplication of natural DNA for bacterial and mammalian DNA polymerases and for E. coli and bacteriophage T4 multienzyme replication complexes (see Refs. 17 and 25 for recent reviews). Using a combination of seven purified replication proteins of phage T4 (the products of genes 32, 41, 43, 44, 45, 61, and 62) to replicate DNA *in vitro*, it has been shown that the accuracy of this complex is very similar to that seen during replication of ϕX174 DNA *in vivo* [12, 16, 23-25]. DNA sequences (such as am33) that are replicated with a low accuracy *in vivo* show high error rates during replication *in vitro*. Other sequences (such as am16 and am9), which have a low reversion rate *in vivo*, are replicated with comparably high accuracy *in vitro* [23-25]. It has also been shown that by the appropriate use of deoxynucleotide pool biases, it is possible to induce reversions of a given amber mutant by specific transition and transversion pathways [12, 23, 24]. We have examined the influence of a variety of perturbations in the conditions of DNA replication *in vitro*, on the rates of formation of mispairs leading to transition and transversion mutations. These results are summarized here.

MATERIALS AND METHODS

Materials

The procedures used for the preparation of replication proteins, Pstl restriction enzyme, DNA templates and spheroplasts have been described elsewhere [24].

Methods

Standard conditions for DNA synthesis were 0.5 mM dithiothritol, 10 mM magnesium acetate, 33 mM Tris-HCl pH 7.8, 67 mm potassium acetate, 1 mM rATP, 0.2 mM CTP, GTP and UTP, 0.2 mM each dNTPs with one of the 4 dNTPs in a radiolabeled form (specific activity 50-200 cpm per picomole), p32 (200 μg/ml), p44/62 (50 μg/ml), p45 (25 μg/ml), p41/61 (15 μg/ml), p43 (6 μg/ml) and ϕX174 replicative form DNA template (1-2 μg/ml). Reactions were incubated at 37° for 60 min and the accuracy of DNA replication was measured by an infectivity assay as previously described. The exact variations in the condition of DNA synthesis used are described in the legens to the tables.

Rationale

The assay we have used in these studies [12, 16, 23] measures the accuracy of duplication of phage ϕX174 DNA *in vitro* with a seven enzyme complex (the products of phage T4 genes 32, 41, 43, 44, 45, 61, and 62). By upsetting the accuracy of DNA replication with appropriate deoxunucleotide pool biases, this assay allows for the determination of rates of formation of base mispairs leading to transition and transversion mutations. We wanted to be able to detect both increases as well as decreases in the accuracy of replication. Therefore, in these studies we used deoxynucleotide pool imbalances that were sufficiently mutagenic so that the frequency of revertants formed during DNA replication *in vitro* was slightly higher than the background frequency of mutants already present in the DNA template. If a particular variation in the condition of DNA replication *in vitro* increased the accuracy of replication, then the revertant frequency would be observed to go down. Conditions that decrease the accuracy of DNA replication, on the other hand, would increase the frequency of reverants observed. We have used the ϕX174 mutant *am*3 (gene E^-) to detect effects of transition mutations (this mutant reverts primarily by an AT → GC transition; TAG → TGG) and ϕX174 mutant *am*16 (gene B^-) to detect effects on transversion mutations (the reversions detected were AT → TA and AT → CG; TAG → AAG, GAG, TCG, and possibly TTG). In a few experiments we have also used the mutants *am*18, *am*33, and *am*86. These mutants map in the gene A of ϕX174 and all three revert to wild type by the transiton TAG → CAG.

RESULTS

1. Factors that Affect the Efficiency of Proofreading

1.1. Variations in the Concentration of the Next Correct Deoxynucleotide

As shown by Clayton et al. [3], for phage T4 DNA polymerase, by Fersht et al., for *E. coli* DNA replication apparatus [5, 6] and by Kunkel et al. [15] for *E. coli* DNA polymerase I, the error rate during DNA replication can depend not only upon the relative ratio of the incorrect and correct substrates but also upon the absolute concentration of the deoxynucleotides. This is because once a misinsertion has occurred, the absolute concentration of the next correct dNTP determines the time that will be available for the 3' → 5' exonuclease proofreading function of the DNA polymerase to edit out the misinserted nucleotide. In contrast to the results of these workers, results obtained by Hibner and Alberts [12] and by Sinha and Haimes [23, 24] using the T4 mutlienzyme complex indicated that the error rate was not affected by the concentration of the next correct nucleotide (at least for the formation of G:T mispairs at the *am*3 site on ϕX174 DNA).

We wanted to be sure that the results at the *am*3 site did not represent an exception. Also the interpretation of the results at the *am*3 site was complex because variations in the concentration of the next correct dNTP (dGTP) also changed the ratio of the incorrect/correct substrates. The conclusion that the concentration of the next correct dNTP was not significant was based on the observation that a slope of 1 was obtained when the frequency of revertants was plotted against the ratio of incorrect/correct dNTPs [12, 23]. This suggested that the error rate was a function of G/A rather than G^2/A.

In an attempt to get results that were easier to interpret, we have examined the reversion of the mutants *am*33 and *am*18 of ϕX174. Both of these mutants can be forced to revert to the true wild type (TAG → CAG) by AT → GC transitions. These mutations occur during DNA replication *in vitro* if a high G/A pool bias is present. We raised the error rate approximately 20-fold over the frequency of background revertants (3×10^{-5} for *am*33 and 1×10^{-5} for *am*18) by using a pool bias of G/A = 100 during DNA synthesis catalyzed by the T4 apparatus *in vitro* (Fig. 1). The next correct deoxynucleotide after the misinsertion of a G at these 2 sites is a dCTP. Since dCTP is not involved in the misinsertion event, its concentration can be varied without affecting the ratio of the two competing nucleotides (G and A) at the misinsertion step. We varied the concentration of dCTP from 20 μM (the lowest concentration at which sufficient DNA synthesis was obtained to permit measurement of accuracy of replication) to 200 μM (a concentration at which V_{max} is

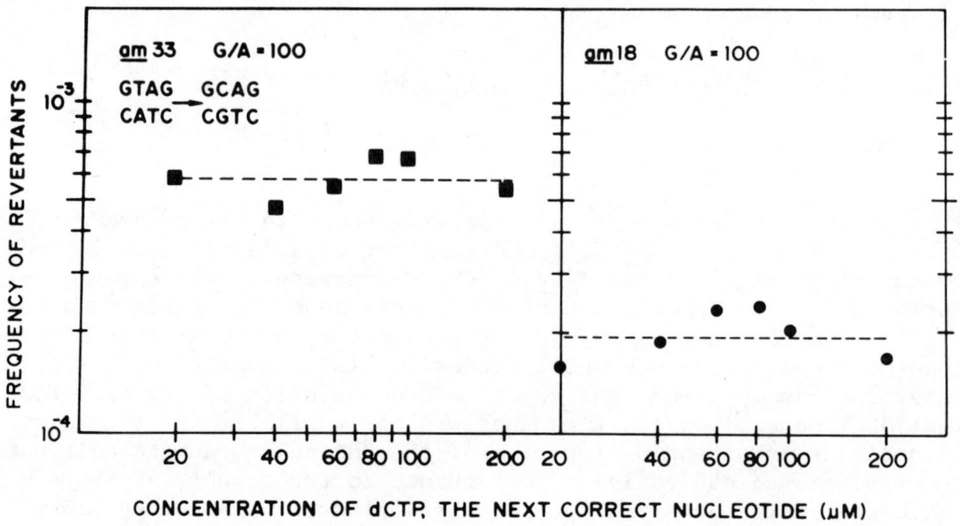

Fig. 1. Effect of variations in the concentration of the next correct deoxynucleotide on the fidelity of DNA replication. DNA was synthesized in vitro using ϕX174 am33 or am18 replicative form DNA as template. Both of these mutants revert by the transition GTAG → GCAG. A pool bias of dGTP/dATP = 100 was used to induce G:T mispairs during replication in vitro. The concentration of the next correct nucleotide (dCTP) was varied from 20 μM to 200 μM and its effect on accuracy of replication was measured by infectivity assays.

reached). As can be seen from the results in Fig. 1, this 10-fold variation in the concentration of the next correct deoxynucleotide had no effect on the error rate.

1.2. Effect of the Addition of Pyrophosphate or dNMPs

The substrate for DNA synthesis by a DNA polymerase is a dNTP. During polymerization, the dNMP portion is added on to the primer chain resulting in the release of a pyrophosphate molecule. Therefore, it is conceivable that a high concentration of pyrophosphate might result in slowing of the rate of polymerization, which could result in an increase in accuracy. Editing by the 3' → 5' exonuclease function of a DNA polymerase results in the release of dNMPs. Therefore, one might reason that the addition of a high concentration of dNMPs might result in some reversal of the proofreading exonuclease activity yielding decreased accuracy. In fact, some evidence for such a reversal has recently been obtained in studies with E. coli replication enzymes [7]. As shown in Table 1, the

TABLE 1. Effect of the Addition of Sodium Pyrophosphate or Deoxynucleoside Monophosphates on the Accuracy of DNA Replication

Variable	Transition Pathway[1] AT → GC (am3 mutant)		Transversion Pathway[2] AT → TA and AT → CG (am16 mutant)	
	DNA Synthesis[3]	Frequency of Revertants[4]	DNA Synthesis[3]	Frequency of Revertants[4]
NaPPi				
0 mM	16.0	5.2×10^{-5}	6.3	2.3×10^{-5}
1 mM	11.9	5.4×10^{-5}	3.5	2.0×10^{-5}
1.5 mM	11.5	3.0×10^{-5}	---	--------
2 mM	10.2	2.1×10^{-5}	1.1	1.7×10^{-5}*
2.5 mM	8.6	2.4×10^{-5}	---	--------
3 mM	4.3	1.9×10^{-5}	0.3	6.5×10^{-5}*
4 mM	0.6	3.1×10^{-5}*	0.2	4.2×10^{-5}*
dNMPs				
0 mM	16.0	5.2×10^{-5}	16.7	3.5×10^{-5}
0.5 mM each	17.1	5.1×10^{-5}	15.3	3.5×10^{-5}
1 mM each	17.9	6.7×10^{-5}	10.4	3.0×10^{-5}
1.5 mM each	18.8	5.5×10^{-5}	----	--------
2 mM each	15.2	4.1×10^{-5}	2.8	2.7×10^{-5}
2.5 mM each	20.1	3.4×10^{-5}	----	--------
3 mM each	----	--------	1.3	1.7×10^{-5}*

DNA was synthesized on ϕX174 am3 or am16 RF molecules using T4 replication proteins *in vitro*. The standard conditions of synthesis are described in the Methods section.

1. For measurements of effects on transition mutations, ϕX174 am3 (E^-) mutant was used. DNA was synthesized in the presence of a pool bias of dGTP/dATP=25 or 50. This stimulates the formation of G:T mispairs leading the the reversion of the am3 codon by an AT → GC transition (TAG → TGG).

2. For measurements of effects on transversion mutations, ϕX174 am16 (B^-) was used. DNA was synthesized in the presence of a pool bias of dGTP and dATP/dTTP=50. This stimulates A:A and G:A mispairs leading to AT → TA and AT → CG transversions (TAG → AAG, GAG, TCG and possibly TTG).

3. DNA synthesis is expressed as number of duplications of input DNA template.

4. Frequency of revertants was measured by an infective center assay using *E. coli* W3350 (Su^+) spheroplasts for am3. With am16 DNA template, *E. coli* W3110 (Su^-) spheroplasts were used to measure revertants and *E. coli* CR63 (Su^+) for total infective centers.

* Whenever DNA synthesis *in vitro* was less than 2 duplications of the input DNA template, a correction was made for the contribution of template DNA to the frequency of revertants.

TABLE 2. Effect of Temperature and pH Variations on the Accuracy of DNA Replication

Variable	Transition Pathway AT → GC (am3 mutant)		Transversion Pathway AT → TA and AT → CG (am16 mutant)	
	DNA Synthesis	Frequency of Revertants	DNA Synthesis	Frequency of Revertants
Temperature				
20°C	----	--------	1.5	2.2×10^{-5}
25°C	13.6	2.9×10^{-5}	3.5	1.0×10^{-5}
30°C	16.8	4.1×10^{-5}	5.5	2.6×10^{-5}
35°C	17.9	5.4×10^{-5}	7.9	1.8×10^{-5}
40°C	13.1	7.9×10^{-5}	8.2	1.3×10^{-5}
45°C	5.1	4.1×10^{-5}	1.8	2.2×10^{-5}
pH				
6.4	less than 0.1	3.3×10^{-6}	0.5	2.5×10^{-5}*
7.5	9.8	2×10^{-4}	3.3	2.5×10^{-5}
8.5	7.1	1.4×10^{-4}	4.6	1.9×10^{-5}
8.9	3.6	1.5×10^{-4}	4.7	2.2×10^{-5}
9.2	6.2	0.9×10^{-4}	---	--------
9.7	---	--------	1.4	2.4×10^{-5}*

* Corrected for the contribution of template DNA to frequency of revertants. See legend to Table I for details.

TABLE 3. Effect of Variations in the Concentrations of Mg^{++} and K^{+} Ions

Variable	Transition Pathway AT → GC (am3 mutant)		Transversion Pathway AT → TA and AT → CG (am16 mutant)	
	DNA Synthesis	Frequency of Revertants	DNA Synthesis	Frequency of Revertants
Mg^{++}				
5 mM	20.2	9.2×10^{-5}	3.7	1.3×10^{-5}
10 mM	23.1	1.2×10^{-4}	7.5	1.6×10^{-5}
15 mM	21.9	8.6×10^{-5}	5.8	1.8×10^{-5}
20 mM	21.5	9.4×10^{-5}	4.9	1.7×10^{-5}
25 mM	22.4	6.2×10^{-5}	2.3	2.1×10^{-5}
30 mM	----	--------	0.8	1.9×10^{-5}*
K^{+}				
0 mM	7.2	2×10^{-4}	12.1	0.74×10^{-5}
25 mM	----	------	19.0	1.9×10^{-5}
33 mM	17.6	1.3×10^{-4}	----	---------
50 mM	----	------	23.4	3.8×10^{-5}
67 mM	15.9	0.94×10^{-4}	----	---------
100 mM	14.6	0.77×10^{-4}	21.4	4.0×10^{-5}
125 mM	----	------	11.9	2.3×10^{-5}
133 mM	21.4	4.6×10^{-4}	----	---------
167 mM	17.8	3.2×10^{-4}	----	---------

TABLE 4. Effect of the Addition of Spermidine on the Accuracy of DNA Replication *in vitro*

Variable	Transition Pathway AT → GC (am3 mutant)		Transversion Pathway AT → TA and AT → CG (am16 mutant)	
	DNA Synthesis	Frequency of Revertants	DNA Synthesis	Frequency of Revertants
Spermidine				
0 mM	17.9	5.4×10^{-5}	11.4	1.5×10^{-5}
1 mM	15.4	9.7×10^{-5}	12.4	1.8×10^{-5}
2 mM	----	------	12.9	1.5×10^{-5}
3 mM	19.0	5.0×10^{-5}	13.1	1.5×10^{-5}
4 mM	20.3	3.9×10^{-5}	12.5	1.6×10^{-5}
5 mM	20.1	6.0×10^{-5}	10.1	1.6×10^{-5}
6 mM	22.7	3.1×10^{-5}	5.7	0.6×10^{-5}
7 mM	18.6	7.1×10^{-5}	----	------
8 mM	3.0	2.7×10^{-5}	----	------

addition of sodium pyrophosphate did not result in enhanced accuracy for the T4 replication apparatus for either the transition or the transversion pathway. The addition of dNMPs (up to 3 mM each) did not significantly reduce the accuracy of the T4 apparatus either (Table 1).

2. Effect of Perturbations in the Conditions of DNA Synthesis

2.1. Variations in Temperature, pH and the Concentration of Magnesium and Potassium Ions

We have varied the rates of DNA replication by a variety of perturbations and measured the rates of transition and transversion mutations during this replication. The data in Tables 2 and 3 show that variations in temperature (from 20 to 45°C), pH (from 6.4 to >9), the concentrations of potassium acetate (from 0 to 167 mM) and magnesium acetate (from 5 to 30 mM) had no effect on the accuracy of T4 DNA replication apparatus *in vitro*.

2.2. Addition of Spermidine

The addition of the polycation, spermidine (up to 8 mM), did not affect the accuracy of DNA replication (Table 4).

2.3. Addition of Manganese

Manganese has long been known to be a mutagen *in vivo* [4, 20]. Several workers have shown that replacement of the divalent ion magnesium by manganese during DNA synthesis *in vitro* by purified DNA

TABLE 5. Effect of the Addition of Manganese on Accuracy of DNA Replication

DIVALENT METAL	TRANSITION PATHWAY Mutant am3 (TAG → TGG)		TRANSVERSION PATHWAYS Mutant aml6 (TAG → AGC, GAG, TCG and TTC)	
	DNA Synthesis	Frequency of Revertants	DNA Synthesis	Frequency of Revertants
10 mM Mg^{++}	16.5	$1.2x10^{-4}$	17.0	$2.6x10^{-5}$
" " " + 1 mM Mn^{++}	19.8	$4.4x10^{-4}$	14.2	$6.6x10^{-5}$
" " " + 1.5 mM Mn^{++}	17.3	$5.5x10^{-4}$	11.4	$1.8x10^{-4}$
" " " + 2.0 mM Mn^{++}	4.3	$5.4x10^{-4}$	4.1	$1.8x10^{-4}$
" " " + 2.5 mM Mn^{++}	1.8	$4.2x10^{-4}$(Corrected)	2.0	$5.3x10^{-4}$
" " " + 3 mM Mn^{++}	0.4	$4.3x10^{-4}$ "	----	------

With am3, a pool bias of G/A = 25 was used to force G:T mispairs leading to AT → GC transitions. With aml6, a pool bias of GA/T = 50 was used to force AT → TA and AT → CG transversions. Viable revertant through transition pathways are also possible at the aml6 site. However, their formation was blocked by keeping the ratio of G/A = 1. The freqency of revertants in the template DNA was $3.3x10^{-6}$ (am3) and $1x10^{-6}$ (aml6).

TABLE 6. Effect of 5-Bromouracil on the Fidelity of DNA Replication

Mutant	Reversion Pathway	Deoxynucleotide	DNA Synthesis	Frequency of Revertants	Relative Error Rate
am3	Transition	dTTP	31.3	$5.1x10^{-5}$	1.0
		5-BrdUTP	13.8	$5.8x10^{-4}$	11.4
aml8	Transition	dTTP	42.8	$1.35x10^{-4}$	1.0
		5-BrdTUP	35.6	$1.33x10^{-3}$	9.9
am86	Transition	dTTP	9.5	$9.1x10^{-5}$	1.0
		5-BrdUTP	7.4	$1.3x10^{-3}$	14.3
aml6	Transversion	dTTP	39.3	$4.6x10^{-6}$	1.0
		5-BrdUTP	24.8	$6.2x10^{-6}$	1.35

DNA was synthesized under standard conditions (see Methods section) with 200 µM dATP, dCTP, dGTP and either 200 µM dTTP or 200 µM 5-bromodeoxyuridine triphosphate. All manipulations were done in dim yellow light. The mutants am3, aml8 and am86 revert primarily by transitions (TAG → TGG, CAG and CAG respectively). The mutant aml6 (B^-) can revert by both transitions and transversions. The predominant transition pathway at this site leads to the formation of TGG codon. In our hands, this revertant forms pinpoint plaques at 37° and fails to form plaques at 40°C. All platings involving the aml6 mutant were done at 40°C so as to insure that revertants could arise by transversions only.

TABLE 7. Effect of the Addition of Intercalating Agents

Variable	Transition Pathway AT → GC (am3 mutant)		Transversion Pathway AT → TA and AT → CG (am16 mutant)	
	DNA Synthesis	Frequency of Revertants	DNA Synthesis	Frequency of Revertants
9-Aminoacridine				
0 μM	16.0	$5.2x10^{-5}$	11.4	$2.9x10^{-5}$
5 μM	15.6	$5.4x10^{-5}$	8.9	$3.3x10^{-5}$
10 μM	16.9	$4.3x10^{-5}$	7.0	$3.5x10^{-5}$
15 μM	15.3	$2.1x10^{-5}$	----	------
20 μM	14.4	$2.8x10^{-5}$	3.4	$3.1x10^{-5}$
30 μM	12.1	$2.2x10^{-5}$	1.7	$4.3x10^{-5}$
40 μM	8.7	$3.0x10^{-5}$	0.52	$2.2x10^{-5}$*
Adriamycin (Doxorubicin)				
0 μM	16.0	$5.2x10^{-5}$	11.4	$2.9x10^{-5}$
0.16 μM	----	------	11.9	$2.8x10^{-5}$
0.32 μM	----	------	10.9	$4.3x10^{-5}$
0.40 μM	15.1	$5.9x10^{-5}$	----	------
0.48 μM	----	------	9.3	$3.2x10^{-5}$
0.64 μM	----	------	6.7	$1.9x10^{-5}$
0.80 μM	15.3	$4.1x10^{-5}$	5.8	$3.8x10^{-5}$
1.2 μM	15.4	$1.9x10^{-4}$	----	------
1.6 μM	13.7	$4.8x10^{-5}$	----	------
2.0 μM	12.8	$3.6x10^{-5}$	----	------
2.5 μM	4.7	$7.1x10^{-5}$	----	------

polymerase is also mutagenic [11, 14, 26]. We find that a total replacement of magnesium by manganese (as was done in the above studies with DNA polymerase alone) resulted in such a drastic inhibition of DNA replication reaction catalyzed by the T4 replication complex *in vitro* (to less than 1 duplication of the input template) that the accuracy of this reaction could not be measured. Because of this limitation, we were only able to measure the effect of the addition of manganese to reactions where magnesium was also present. Even under these conditions, the presence of manganese resulted in a pronounced inhibition. The rate of DNA synthesis decreased by more than a factor of 20 as the concentration of manganese increased from 0 to 3 mM despite the presence of 10 mM magnesium (Table 5). Both transitions and transversions were seen to increase. In the presence of 2.5 mM manganese, the error rate for the transition pathway went up by a factor of 3.5 only. The effect on the transversion pathway was much more pronounced. The addition of 2.5 mM manganese increased the error rate more than 20-fold (Table 5).

2.4. Replacement of Thymidine Triphosphate by 5-Bromodeoxyuridine Triphosphate

Thymine analog, 5-bromouracil, is known to be a mutagen *in vivo* [1, 9, 10]. A complete replacement of thymine by the analog

5-bromouracil results in only a 2-3-fold reduction in the rate of DNA synthesis of the T4 phage replication apparatus in vitro (Table 6). We find that this complete substitution of TTP by BrdUTP increases the error rate approximately 10-fold but only for those mutants (am3, am18, and am86) which can revert to wild type by a transition pathway (Table 6). No increase in reversion is observed for the mutant am16 in these experiments because under the conditions of plating used here, the formation of a viable revertant requires a transversion mutation (see Table 6 for details).

2.5. Addition of Intercalating Agents

Planar dyes capable of intercalating between the base pairs in a DNA double helix are strong mutagens in vivo resulting in the formation of frameshift mutations exclusively. Recently, it has been reported, however, that these compounds can also produce base substitution mutations, at least during the copying of ϕX174 by E. coli DNA polymerase I in vitro [21]. In contrast to the results of these workers, we find that the addition of the intercalating agents 9-aminoacridine (up to 40 μM) or adriamycin (up to 2.5 μM) had no effect on rates of formation of substitution mutations by the T4 replication apparatus in vitro. Both the transition and the transversion pathways were unaffected (Table 7). It is worth pointing out that the infectivity assay used in this study is incapable of detecting any frameshift mutants that might have been induced by these mutagens.

DISCUSSION

A combination of 7 replication proteins of phage T4 can replicate natural DNA templates very efficiently with a rate and accuracy comparable to that seen during DNA replication in vivo [12, 16, 22-25]. The accuracy of this replication process can, however, be decreased by imbalances in the concentrations of the dNTP substrates. This allows one to easily force in mutations by any substitution pathway that can lead to a viable revertant [23]. Using this approach, we have examined the effects of a number of variations, in the conditions of DNA replications in vitro, on the rates of base substitutions leading to either transitions or transversions. We find that most variations do not affect the accuracy of DNA replications even though the rate of DNA replication can readily be varied by greater than 10-fold by most of these variations. Changes in temperature, pH, or the concentration of potassium or magnesium salts, or the addition of spermidine, sodium pyrophosphate or deoxynucleoside monophosphates have no effect. Also the addition of intercalating agents adriamycin or 9-aminoacridine has no effect. The addition of manganese is mutagenic with transversion mutations being stimulated about 5 times more than transition mutations. Substitution of thymine by the base analog 5-bromouracil produces transition mutations exclusively.

Changes in the concentration of the next correct deoxynucleotide do not affect the error rate of the T4 phage replication apparatus *in vitro*. Is it because there is no proofreading nuclease activity under these conditions of DNA replication or does it mean that the proofreading function is so efficient that variations in the concentration of the next correct deoxynucleotide do not significantly affect its level? Data presented here do not allow us to distinguish between these 2 possibilities. Results of other experiments (not shown here) indicate that the latter of the above two possibilities is correct. Even at saturating concentrations of deoxynucleotide substrates, approximately 1/3 of all correctly polymerized nucleotides are edited out by the 3' → 5' exonuclease activity. Due to the presence of such a high level of proofreading function, it is not surprising that attempts to alter the accuracy of DNA replication by limiting the concentration of the next correct dNTP or by the addition of a high level of dNMPs have no effect. Since *E. coli* DNA polymerases have a much lower level of the 3' → 5' exonuclease activity [13], it also becomes intelligible why variations in the concentration of the next correct dNTPs or the addition of high concentrations of dNMPs result in altered accuracy. The exact reasons why only certain misinsertions show next nucleotide effects are not yet understood.

ACKNOWLEDGMENTS

We gratefully acknowledge the expert technical assistance of Sandy Maimone and Royann Blakely. This work was supported by grants from the National Institutes of Health (GM24391) and the Charles and Johanna Busch Foundation to Navin Sinha. Tashmeem Razzaki was supported by a strategic grant from the Natural Sciences and Engineering Research Council of Canada (to Profs. R. C. Von Borstel and P. J. Hastings).

REFERENCES

1. S. Benzer and E. Freese, Induction of specific mutations with 5-bromouracil, Proc. Natl. Acad. Sci., U.S.A., 44:112-119 (1958).
2. M. Bittner, R. L. Burke, and B. Alberts, Purification of the T4 gene 32 protein free from detectable deoxyribonuclease activities, J. Biol. Chem., 254:9565-9572 (1979).
3. L. K. Clayton, M. F. Goodman, E. W. Branscomb, and D. J. Galas, Error induction and correction by mutant and wild type T4 DNA polymerases. Kinetic error discrimination mechanisms, J. Biol. Chem., 253:1902-1912 (1979).
4. M. Demerec and J. Hanson, Mutagenic action of manganous chloride, Cold Spring Harbor Symp. Quant. Biol., 16:215-227 (1951).

5. A. R. Fersht, Fidelity of replication of phage φX174 by DNA polymerase III holoenzyme: spontaneous mutation by misincorporation, Proc. Natl. Acad. Sci., U.S.A., 76:4946-4950 (1979).
6. A. R. Fersht and J. W. Knill-Jones, DNA polymerase accuracy and spontaneous mutation rates: frequencies of purine:purine, purine-pyrimidine, and pyrimidine-pyrimidine mismatches during DNA replication, Proc. Natl. Acad. Sci., U.S.A., 78:4251-4255 (1981).
7. A. R. Fersht, J. W. Knill-Jones, and W. C. Tsui, Kinetic basis for spontaneous mutation. Misinsertion frequencies, proofreading specificities and cost of proofreading by DNA polymerases of E. coli, J. Mol. Biol., 156:37-52 (1982).
8. A. R. Fersht and J. W. Knill-Jones, Fidelity of replication of bacteriophage φX174 DNA in vitro and in vivo, J. Mol. Biol., 165:633-654 (1983).
9. E. Freese, The difference between spontaneous and base analog induced mutations in phage T4, Proc. Natl. Acad. Sci., U.S.A., 45:622-633 (1959).
10. E. Freese, The specific mutagenic effect of base analogs on phage T4, J. Mol. Biol., 1:87-105 (1959).
11. Z. W. Hall and I. R. Lehman, An in vitro transversion by a mutationally altered T4-induced DNA polymerase, J. Mol. Biol., 36:321-333 (1968).
12. U. Hibner and B. Alberts, Fidelity of DNA replication catalyzed in vitro on natural DNA template by the T4 bacteriophage multienzyme complex, Nature, 285:300-305 (1980).
13. A. Kornberg, DNA Replication, W. H. Freeman, San Francisco (1980).
14. T. Kunkel and L. A. Loeb, On the fidelity of DNA replication. Effect of divalent metal ion activators and deoxynucleoside triphosphate pools on in vitro mutagenesis, J. Biol. Chem., 255:9961-9966 (1979).
15. T. A. Kunkel, R. M. Schaaper, R. Beckman, and L. A. Loeb, On the fidelity of DNA replication. Effect of the next nucleotide on proofreading, J. Biol. Chem., 256:9883-9889 (1981).
16. C. C. Liu, R. L. Burke, J. Barry, U. Hibner, and B. Alberts, Probing DNA replication mechanisms with the T4 bacteriophage in vitro, Cold Spring Harbor Symp. Quant. Biol., 43:469-487 (1979).
17. L. A. Loeb and T. A. Kunkel, Fidelity of DNA synthesis, Ann. Rev. Biochem., 52:429-457 (1982).
18. C. F. Morris, L. A. Moran, and B. M. Alberts, Purification of gene 41 protein of bacteriophage T4, J. Biol. Chem., 254:6797-6802 (1979).
19. C. F. Morris, D. C. Mace, H. Hama-Inaba, N. Sinha, and B. Alberts, Purification of the gene 43, 44, 45, and 62 proteins of the bacteriophage T4 DNA replication apparatus, J. Biol. Chem., 254:6787-6796 (1979).
20. A. Orgel and L. E. Orgel, Induction of mutations in bacteriophage T4 with divalent manganese, J. Mol. Biol., 14:453-457 (1965).

21. C. W. Shearman and L. A. Loeb, On the fidelity of DNA replication. Mechanisms of misincorporation by intercalating agents, J. Biol. Chem., 258:4485-4491 (1983).
22. N. K. Sinha, C. F. Morris, and B. M. Alberts, Efficient in vitro replication of double-stranded DNA templates by a purified T4 bacteriophage system, J. Biol. Chem., 255:4290-4303 (1980).
23. N. K. Sinha and M. Haimes, Probing the mechanism of transition and transversion mutagenesis using the phage T4 DNA replication apparatus, in: "Mechanistic Studies of DNA Replication and Genetic Recombination" (B. Alberts, ed.), ICN-UCLA Symp. Molec. Cell Biol., 19:707-723 (1980).
24. N. K. Sinha and M. Haimes, Molecular mechanism of substitution mutagenesis. An experimental test of the Watson-Crick and Topal-Fresco models of base mispairings, J. Biol. Chem., 256: 10671-10683 (1981).
25. N. K. Sinha and M. F. Goodman, Fidelity of DNA replication, in: "The Bacteriophage T4" (C. Mathews, E. Kutter, G. Mosig, and P. Berget, eds.), ASM Press, Washington, D.C., 131-137 (1983).
26. M. A. Sirover and L. A. Loeb, Infidelity of DNA synthesis in vitro: screening for potential metal mutagens or carcinogens, Science, 194:1434-1436 (1976).

THYMINELESS MUTAGENESIS IN BACTERIA

Bernard A. Kunz*

Laboratory of Genetics
Mutagenesis Section
National Institute of Environmental Health Sciences
Research Triangle Park, North Carolina

SUMMARY

Experimental evidence indicates that while thymine starvation induces primarily A:T → G:C transitions in bacteria, it also may cause other uncharacterized base substitutions as well as frameshifts and deletions. However, models have been proposed to explain only the induction of point mutations by thymine deprivation. In this study, we demonstrate that thymine nucleotide depletion induces both point mutations in the *his-4* and *lacI* genes of *Escherichia coli* and reversion of the frameshift mutations *trpE9777*, *trpA21*, *trpA540*, and *trpA9813*. Analysis of the *lacI* amber spectrum revealed that thymine starvation resulted in G:C → A:T transitions and all possible transversions. A defect in uracil-DNA glycosylase has little effect on the induction of *lacI*$^-$ mutations but reduces substantially the induction of *trpE9777* revertants. These data show that the mutagenic specificity of thymine nucleotide depletion is not limited to A:T → G:C transitions and suggest that removal of uracil from DNA plays a role in the generation of frameshift mutations by thymine deprivation. A model that involves nucleotide misincorporation into DNA and induction of error-prone repair functions in response to thymine starvation is proposed to account for these findings.

---

*Present Address: Biology Dept.
York University
4700 Keele St.
Toronto, Ontario,
Canada M3J 1P3

INTRODUCTION

Three decades ago, Cohen and Barner [10] demonstrated that incubation of a thymine auxotroph of Escherichia coli in the absence of thymine, under conditions otherwise conducive to growth, resulted in cell death. Subsequently, loss of cell viability during thymine nucleotide depletion was reported for a variety of bacterial species [7, 15, 24, 28, 36, 44, 46, 67]. In addition to cell killing, thymineless conditions have been found to cause mutation [4-6, 11, 16, 31, 36, 37, 55, 70], recombination [27, 54], prophage induction [40], sensitization to ultraviolet light [28, 44], induction of recA-dependent SOS functions [4, 71, 72], and nonconservative DNA replication [56]. These discoveries provided evidence for the occurrence of DNA damage during thymine starvation and it has been established that this damage includes single and double-strand DNA breaks and DNA degradation [2, 3, 8, 25, 35, 52, 58, 60, 61, 68]. Despite extensive genetic and biochemical characterization of thymineless phenomena, the underlying molecular mechanisms have not been fully elucidated. However, the available data suggest that nucleotide pool imbalances play an intrinsic role in these processes.

Thymineless mutagenesis in bacteria was described initially for E. coli by Coughlin and Adelberg [11, who reported a 100-fold increase in the frequency of reversion to histidine prototrophy. Forward mutation to suppression, auxotrophy and drug resistance and F-specific insertion mutations also have been found [4, 5, 31, 36, 51, 70]. Several studies were directed at determining the mutagenic specificity of thymine nucleotide depletion. The responses of the induced mutations to base analogs, alkylating agents and frameshift agents led to the conclusion that thymine deprivation caused primarily A:T → G:C transitions although it was evident that other uncharacterized base substitutions as well as frameshifts and deletions also might be induced [5, 6, 36, 55]. Bridges et al. [6] proposed that point mutations occurred upon error-prone 'filling in' of gaps produced as a consequence of the DNA strand breakage provoked by thymine starvation. This hypothesis was based on the finding that thymineless mutagenesis was prevented by a defect in a gene (exr = lexA) involved in the control of error-prone repair. In apparent contradiction, Bresler et al. [5] demonstrated that mutations were induced in a lexA$^-$ thymine auxotroph incubated on solid medium supplemented with sub-optimal concentrations of thymine. However, this latter finding might be misleading and reflect a different mechanism for mutageneis since total omission of thymine was not mutagenic for wild-type or lexA$^-$ strains in this system. Thymine starvation also causes a decrease in the dTTP pool and an increase in the dCTP pool [53] and Pauling [55] suggested that such an imbalance might enhance base-pairing errors resulting in misincorporation of cytosine in place of thymine. This hypothesis is consistent with the predominance of A:T → G:C transitions among the mutations induced by thymine nucleotide depletion. Neither of these proposed models readily predicted the induction of deletions or frameshifts.

Thymine starvation of E. coli causes an increase in the dUTP pool [33]. This finding led Breitman et al. [3] to suggest that dUTP might be incorporated into DNA during replicative or repair synthesis under thymineless conditions and so provoke enzyme attack resulting in DNA breakage and degradation. Such damage might be expected to give rise to frameshifts and/or deletions. With the use of appropriate mutants, it was demonstrated that uracil was incorporated into bacterial DNA during thymine deprivation and that this could lead to DNA strand breakage [45, 65, 66, 69]. However, the relation between uracil incorporation and the induction of mutation by thymine starvation was not investigated.

In the present study, we examined the effects of a mutation in uracil-DNA glycosylase on thymineless mutagenesis and we sought to characterize the base substitutions, other than A:T → G:C transitions, that might be induced by thymine nucleotide depletion. To determine the mutagenic specificity of thymine deprivation, we employed the E. coli lacI system which permits the detection of amber and ochre mutations at 65 independent sites within the lacI gene [12, 13, 49, 63]. Since the lacI DNA sequence and the location of the nonsense mutations have been established, each mutation can be attributed to a specific base change [12, 22, 48]. We found that thymine starvation induced $lacI^-$ mutations and reversion of the trpE9777, trpA21, trpA540, and trpA9813 frameshift alleles. Nonsense mutations occurred in the lacI gene as a consequence of G:C → A:T transitions and all possible transversions but the amber and ochre mutations constituted less than 2% of the total $lacI^-$ mutations analyzed. Confirmation that thymine deprivation provokes A:T → G:C transitions was obtained by a separate analysis of induced his-4 revertants. A defect in uracil-DNA glycosylase had little effect on the production of $lacI^-$ mutations but reduced reversion of the trpE9777 frameshift allele by 90%. These results indicate that thymine nucleotide depletion can induce various types of base substitution and suggest that removal of uracil incorporated into DNA plays a role in the generation of frameshift mutations by thymine starvation. The data also suggest that both misincorporation of cytosine and error-prone repair of thymineless damage cause point mutations. A model to account for these findings is presented.

MATERIALS AND METHODS

Strains and Media

Unless otherwise stated, suppressor and deletion strains and media and techniques for the lacI system were the same as described by Coulondre and Miller [12, 13]. Strain KMBL3835 [F'pro-lac/ara, Δ(pro-lac), thi^-, trpE9777] has been described previously [32]. A thymine-requiring derivative of this strain (LMG8008) was selected using trimethoprim as described by Miller [47]. Strain LMG8009 [F'pro-lac/ara, Δ(pro-lac), thi^-, trpE9777, thyA, Tn10:tyrA, ung-1, nadB] was constructed by P1-vir-mediated transduction using BD1467 [Hfr P045, thi⁻, relA1, λ⁻, Tn10:tyrA, ung-1, nadB] (provided by Dr. B. K. Duncan) as the donor and LMG8008 as the recipient. Tetracycline-resistant transductants were selected and screened for tyrosine and nicotinic acid auxotrophy as the tyrA and nadB genes flank the ung-1 allele [19]. To verify the presence of ung-1, suspected Ung⁻ isolates were tested for the ability to support the growth of T2 bacteriophage containing uracil in their DNA (bacteriophage provided by Dr. D. Fix). Such bacteriophage are unable to grow on Ung⁺ bacteria [19]. Strains JC3890trpE9777, JC3890trpA21, JC3890trpA540, and JC3890trpA9813 were provided by Dr. T. Kato. In addition to the trp mutations indicated, each strain also carries uvrB301, argE3, thr-1, leu-6, proA2, his-4, thi-1, lacY1, galK2, ara-14, xyl-5, mtl-1, tsx-33, strA31, supE44. Thymine-requiring derivatives of these strains (JCT9777, JCT21, JCT540, and JCT9813) were selected as described above. The media used to assay histidine and tryptophan prototrophs has been described [38].

lacI System

The production of a mutational spectrum requires the isolation and characterization of thousands of mutants. To facilitate these procedures, the lacI⁻ mutations are studied on the transferable F'pro-lac episome in a host strain having a deletion of the homologous chromosomal region. lacI⁻ mutants are selected on medium containing phenyl-β-D-galactoside, a non-inducer of the lac operon, as the sole carbon source. Since the only copy of lacI resides on the F', no contribution of lesions from homologous chromosomal DNA is possible. Thus, any mutations arising within the lacI gene must be the result of DNA synthesis or repair processes acting on F' DNA. To identify nonsense mutations, the F' lacI⁻ mutations are transferred by replica-mating into a series of suppressor strains that are, in turn, replicated onto indicator medium containing 5-bromo-4-chloro-3-indolyl-β-D-galactoside. On this medium, nonsuppressed lacI⁻ muatants form blue colonies whereas suppression of the mutations leads to the emergence of white colonies. Deletion mapping and suppression pattern analysis of the nonsense mutants, carried out by replicamating transfer of the F' into deletion strains and suppressor strains, are used to identify particular sites within the lacI gene. Finally, the lacI DNA sequence [22] can be used to determine the specific base substitutions that gave rise to the nonsense mutations at these sites.

Thymine Starvation

Thymine-requiring strains were grown at 37°C to exponential phase in nutrient medium supplemented with thymidine (final thymidine concentration for all media: 50 μg/ml). Cells were then harvested by centrifugation, washed 3 times in VB solution [41], resuspended in appropriately supplemented minimal broth lacking thymidine and incubated with shaking at 37°C. At regular intervals, samples were withdrawn, washed one time in VB solution, diluted in VB solution if necessary and plated onto appropriately supplemented media to detect survivors, lacI⁻ mutants, tryptophan prototrophs or histidine prototrophs.

RESULTS

Thymine Starvation of Ung$^+$ and Ung$^-$ Strains: Effects on Survival and Mutation

Incubation of exponential phase cells of strains LMG8008 (Ung$^+$) and LMG8009 (Ung$^-$) in appropriately supplemented minimal medium lacking thymine led to cell death (Fig. 1). For both strains survival began to decrease within two hours of transfer to thymine omission medium and the survival curves are very similar. This result is in agreement with previous findings for thymine starvation of Ung$^-$ *E. coli* strains [19]. However, it has been reported that mutations in the gene coding for uracil-DNA glycosylase in *Bacillus subtilis* confer a modest degree of resistance to thymineless death [45].

In LMG8008 and LMG8009, base substitutions, deletions and frameshifts can give rise to *lacI*$^-$ mutations whereas frameshifts can re-

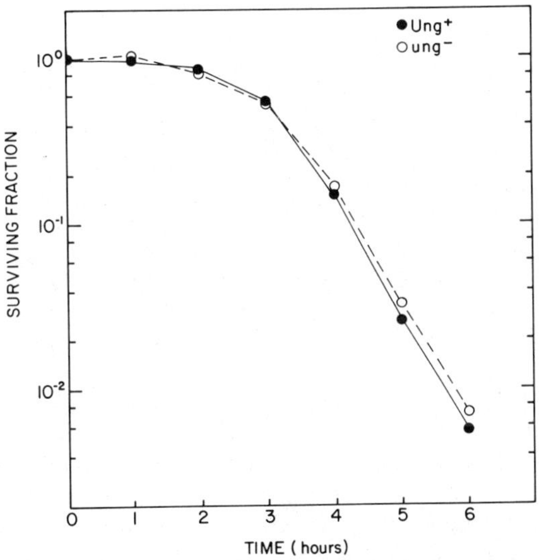

Fig. 1. Thymineless death in strains wild-type or mutant for uracil-DNA glycosylase. (●——●): LMG8008 (Ung$^+$), (○---○): LMG8009 (Ung$^-$).

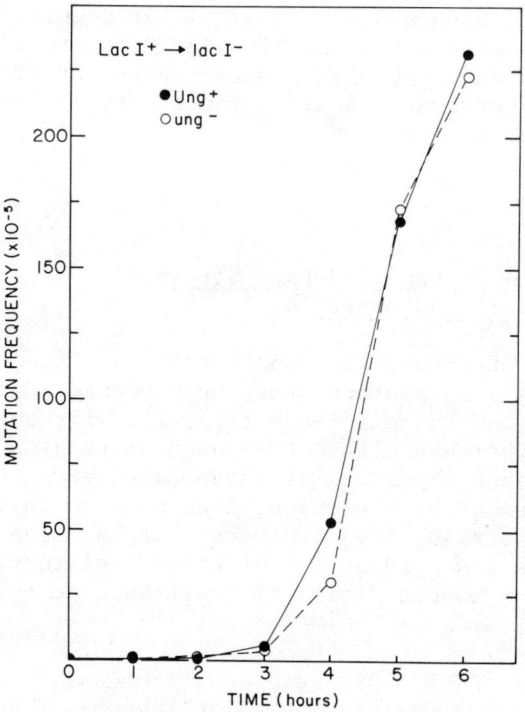

Fig. 2. Induction of lacI$^-$ mutations by thymine starvation. Symbols as for Fig. 1.

vert the trpE9777 mutation. As the survival of the Ung$^+$ strain decreased, the frequencies of lacI$^-$ mutants and tryptophan prototrophs increased and after 6 h of thymine starvation were 500 and 300-fold greater, respectively, than the spontaneous frequencies (Figs. 2 and 3). Since these increases in frequency were accompanied by 7 and 14-fold enhancements in the numbers of lacI$^-$ and Trp$^+$ cells, respectively, per ml of culture, and the total cell population underwent less than one doubling, the increased frequencies are not due to selection for pre-existing mutants. The Ung$^-$ strain showed the same induction kinetics for lacI$^-$ mutation as did the Ung$^+$ strain. However, the induction of tryptophan prototrophs was reduced by 90%.

The trpE9777 mutation resulted from the addition of an A:T base pair to a run of 5 A:T base pairs [64] and reversion is expected to result from loss of one of the 6 A:T base pairs. Targets for the induction of frameshift mutagenesis by thymine starvation are not restricted to additions in A:T runs. We have found that the trpA21 and trpA540 mutations, which are base pair deletions in sequences of mainly G:C pairs, and the trpA9813 mutation, which is a base pair deletion in a run of primarily A:T pairs [64], also are reverted by thymine deprivation (Table 1).

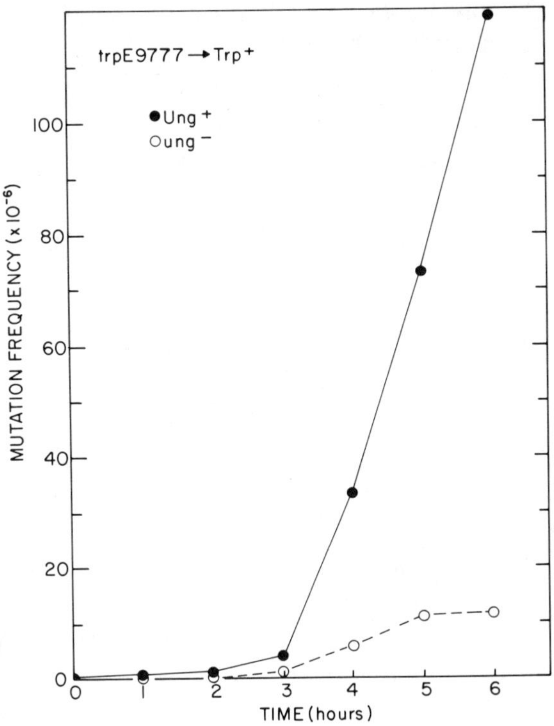

Fig. 3. Induction of tryptophan prototrophs by thymine starvation. Symbols as for Fig. 1.

These results demonstrate that thymine nucleotide depletion induces frameshift mutations. Moreover, the data suggest that removal of uracil incorporated into DNA during thymine starvation gives rise to the frameshifts. It cannot be ascertained from our data whether frameshift mutation following thymine deprivation is totally dependent on uracil excision. The low level of trpE9777 reversion in the Ung⁻ strain may reflect leakiness of the ung-1 allele since strains carrying this mutation have been found to possess residual uracil-DNA glycosylase activity [43].

Analysis of lacI⁻ Mutations

The lacI mutational responses are summarized in Table 2. Nonsense mutations comprised 2% of the spontaneous mutations in the Ung+ strain. In contrast, 8.8% of the mutations that arose spontaneously in the Ung⁻ strain were nonsense mutations. This difference can be attributed to the fact that defects in uracil-DNA glycosylase result in a specific enhancement of G:C → A:T transitions presumably due to the failure to repair DNA cytosine residues that have under-

TABLE 1. Induction of Frameshift Revertants by Thymine Starvation

Strain	Period of Starvation (hours)	Viable cells (%)	Trp^+ frequency*	
JCT9777	0	100	3.1	(1)**
(trpE9777)	6	0.3	1443	(465)
JCT21	0	100	0.8	(1)
(trpA21)	6	0.4	247	(309)
JCT540	0	100	0.7	(1)
(trpA540)	6	0.4	11	(15)
JCT9813	0	100	0.2	(1)
(trpA9813)	6	0.8	89	(445)

* Frequencies are per 10^7 viable cells

** Numbers in parentheses are the relative increases

gone spontaneous deamination to uracil [17, 20, 21]. Indeed, 97% of the spontaneous amber mutations occurred in the Ung^- strain as a consequence of G:C → AT transitions (data not shown). Following starvation for 6 h, similar increases in the frequencies of $lacI^-$ and nonsense mutations were found for the Ung^+ and Ung^- strains. Clearly, the ung-1 allele has no marked effect on thymineless induction of mutations in the lacI gene.

The amber mutations recovered during the screening of the $lacI^-$ mutations induced in the Ung^+ strain were analyzed to determine the base substitutions responsible. An important feature of the lacI system is that A:T → G:C transitions cannot generate nonsense codons in the lacI gene [48]. This fortuitously eliminates the high background of A:T → G:C transition events expected for thymine deprivation and so facilitates detection of other base substitutions that may be induced less frequently.

TABLE 2. Induction of lacI$^-$ Mutations by Thymine Starvation

Strain	Period of starvation (hours)	lacI$^-$ frequency*	Total nonsense mutants %	Total nonsense mutants frequency	Total lacI$^-$ mutants screened
LMG8008 (Ung$^+$)	0	4.8	2.0	0.1	1,000
	6	2312	2.1	47	16,600
LMG8009 (Ung$^-$)	0	8.1	8.8	0.7	1,000
	6	2224	1.2	27	11,450

* Frequencies are per 10^6 viable cells

TABLE 3. Distribution of Base Substitutions Leading to lacI Amber Mutations

Base substitution	Sites available	Sites found: Spontaneous	Sites found: Thymine starvation
G:C → A:T	14	14 (71.8)*	12 (30.4)
G:C → T:A	10	10 (16.4)	7 (35.8)
A:T → T:A	4	3 (6.2)	4 (18.9)
A:T → C:G	5	5 (4.7)	4 (14.2)
G:C → C:G	3	1 (0.9)	1 (0.7)
Total ambers mapped		464	148

Amber mutations arose spontaneously in KMBL3835 or were induced by thymine starvation of LMG8008 for 6 hours

* Numbers in parentheses are the percents of total occurrences

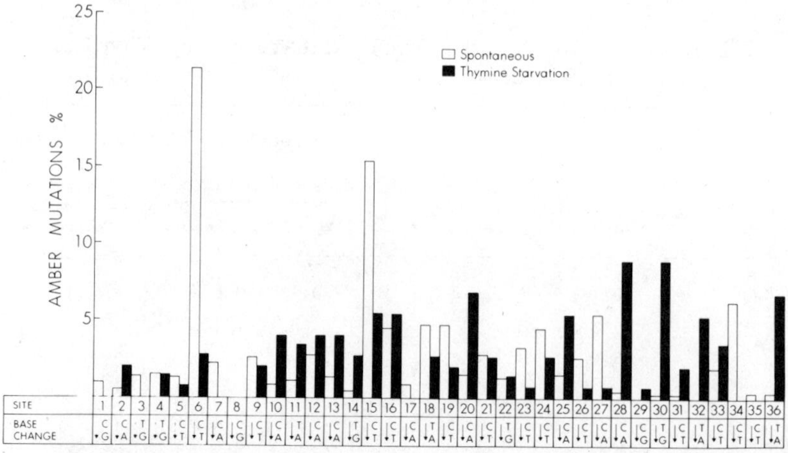

Fig. 4. Spectra for lacI amber mutations. The open bars represent spontaneous mutations obtained in the parental strain KMBL3835 (464 amber mutants were analyzed). The solid bars represent mutations obtained following thymine starvation of strain LMG8008 for 6 h (148 amber mutants were analyzed). Amber mutations occur at 36 sites within the lacI gene. The base change that gives rise to a particular mutation is indicated below the site number for that mutation. Data are presented as the percent of the total occurrences. It should be noted that the mutation frequencies at individual sites are much greater following thymine starvation. For example, induced amber mutation frequencies at sites 4, 9, 16, and 21 are all 1000-fold greater than the spontaneous frequencies.

The mutational spectrum of lacI amber mutations induced by thymine starvation is given in Fig. 4 where it is compared to the spectrum for amber mutations arising spontaneously in the parental strain KMBL3835. Table 3 summarizes the mutational specificity of thymineless mutagenesis in terms of the base substitutions required to generate the amber mutations. Although the data are expressed as percent of total occurrences, the frequencies of individual amber mutations are as much as 45,000-fold greater after thymine deprivation. It is apparent that the spectra are quite different. The spontaneous spectrum has hotspots at the G:C → A:T transition sites, ambers 6 and 15. Each of these sites contains a 5-methylcytosine residue and it has been proposed that the hotspots result from spontaneous deamination of the 5-methylcytosine to thymine [14]. While the mutation frequencies for ambers 6 and 15 are 1000-fold greater following thymine starvation, the percent of amber mutations induced at these sites is markedly lower than the spontaneous fraction. In fact, as a percentage of the total, fewer mutations are recovered

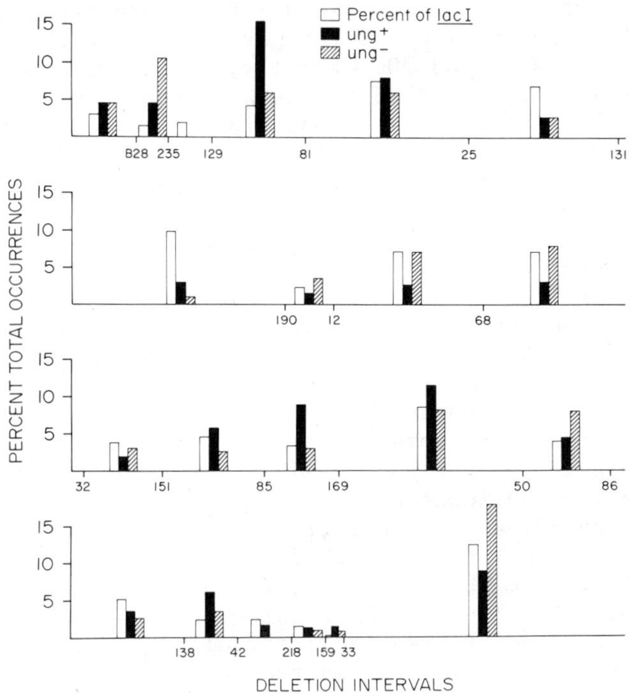

Fig. 5. Distribution of non-suppressible (by amber and ochre suppressors) lacI$^-$ mutations among 21 deletion intervals covering the lacI gene. Mutations were obtained following thymine starvation for 6 h. The open bars represent the percent of the lacI gene occupied by particular intervals. The closed bars represent the percent of total mutations mapped to particular intervals for strain LMG8008 (Ung$^+$) (226 mutations were mapped). The cross-hatched bars represent the percent of total mutations mapped to particular intervals for strain LMG8009 (Ung$^-$) (226 mutations were mapped).

at the majority of G:C → A:T transition sites. In contrast to the spontaneous spectrum, there is a large increase in the percentage of events occurring at A:T sites. Such events account for only 11% of the spontaneous mutations but for fully one-third of the mutations induced by thymine deprivation (Table 3). However, not all A:T sites behave identically - of the 9 sites where base substitutions involve an initial A:T pair, the relative proportion of mutations increased at only five: ambers 11, 14, 30, 32, and 36. Moreover, notable increases were found at 8 of the 27 sites where the base substitution involves an initial G:C pair and this indicates that thymineless mutagenesis is not restricted to A:T sites. It also can be seen that thymineless mutagenesis is not limited to misincorporation of cytosine. There are 8 sites where cytosine mispairing can lead to transversion but the percent of mutation was greater at only two: ambers 14 and 30. Finally, it is interesting to note that increases were found at 13 of the 28 sites where adenine can be substituted to give a nonsense mutation. This may reflect a heightened misincorporation of adenine related to the swelling of the dATP pool [53] and the induction of SOS functions [34, 71, 72] during thymine starvation.

Nonsense mutations constitute only a small fraction of the mutations induced in the lacI gene by thymine deprivation (Table 2). Presumably, the remainder includes missense mutations [50], frameshifts [9], and deletions [23]. In order to determine whether these mutations are localized to specific regions or are scattered throughout the lacI gene, 452 non-suppressible (by amber and ochre suppressors) $lacI^-$ mutations were mapped using 20 deletion strains. Figure 5 shows that the non-suppressible mutations occurred in 20 of the 21 intervals delineated by the lacI deletions. In general, the percent of mutations that occurred in a given interval seems to be roughly proportional to the fraction of the lacI gene occupied by that interval. For the most part, the distribution of non-suppressible mutations is not affected notably by a defect in uracil-DNA glycosylase. However, several intervals contain more mutations selected in the Ung^+ strain than in the Ung^- strain. Perhaps the reduced recovery of mutations in such intervals reflects the presence of sequences sensitive to frameshifting in response to thymine deprivation. While this may be the case, the recovery of nearly identical numbers of $lacI^-$ mutants in the Ung^+ and Ung^- strains suggests that if frameshifts result from removal of uracil incorporated into DNA, then frameshifts compose only a small fraction of the total $lacI^-$ mutations.

Analysis of his-4 Revertants Induced by Thymine Starvation

Locus revertants of his-4 occur by A:T → G:C transitions whereas other base substitutions can give rise to suppressor mutations that suppress his-4, an ochre allele [39]. The base substitutions responsible for the his-4 revertants can be characterized by screening the revertants against a series of amber and ochre mutant strains of bacteriophage T4. The pattern of suppression of the T4 nonsense mutants, obtained by infecting the revertants with the bacteriophage, reveals whether the his-4 revertants are due to transitions or transversions [39]. We used this system to determine the relative proportions of A:T → G:C transitions versus other base substitutions induced by thymine starvation of a single bacterial strain. To carry out this analysis, His^+ revertants were induced by thymine deprivation of JCT9777 for 6 h. This treatment increased the his-4 reversion frequency by in excess of 1000-fold. Several hundred spontaneous and induced revertants were selected and characterized. Table 4 shows that spontaneous mutations arose as a consequence of transversion

TABLE 4. Distribution of Base Substitutions Leading to his-4 Reversion

Base substitution	Period of starvation (hours) 0		6	
A:T → G:C	194	(74.0)*	200	(100)
G:C → A:T	37	(15.0)	0	
Transversion	17	(11.0)	0	
Total mutants tested	248		200	

Mutations were induced by thymine starvation of JCT9777. Spontaneous events contribute to less than 1% of the total induced mutations

* Numbers are the percents of total occurrences

and A:T → G:C and G:C → A:T transitions. However, the bacteriophage suppression patterns indicated that all of the mutations induced by thymine starvation were locus revertants due to A:T → G:C transitions. This result is consistent with previous reports that thymine nucleotide depletion induces predominantly A:T → G:C transitions.

DISCUSSION

Investigations of the specificity of thymineless mutagenesis have led to the conclusion that primarily A:T → G:C transitions are induced [5, 6, 36, 55]. In certain of these studies, indirect evidence based on the responses of the induced mutations to chemical mutagens indicated that thymine starvation also might provoke other base substitutions as well as frameshifts and deletions. It was proposed that the point mutations arose either by error-prone repair of gaps generated during thymine nucleotide depletion [6] or by misincorporation of cytosine in place of thymine due to an elevated dCMP/dTMP ratio [55]. These models did not readily predict thymineless induction of frameshifts. However, it was suggested, and subsequently demonstrated, that uracil incorporation into DNA during thymine starvation could lead to DNA strand breakage [3, 45, 65, 69]. It seemed possible that such damage might cause frameshifts.

In this report we have shown that thymine deprivation of E. coli induces forward mutation in the lacI gene and reversion of several frameshift mutations (Tables 1 and 2, Figs. 2 and 3). While a majority of lacI⁻ mutations are expected to result from A:T → G:C transitions, an analysis of the lacI amber mutations revealed that G:C → A:T transitions and all possible transversions could be recovered (Fig. 4, Table 3). However, nonsense mutations constituted only 2% of the total induced lacI mutations examined (Table 1). Presumably this is due to the fact that the A:T → G:C transition, the major base substitution induced by thymine starvation, does not produce nonsense codons in the lacI gene [48]. This seems to be a likely explanation since characterization of 200 his-4 revertants induced by thymine deprivation indicated that all appeared to be due to A:T → G:C transitions (Table 4). A defect in uracil-DNA glycosylase had little or no effect on cell killing or the induction of lacI⁻ mutations by thymine starvation but reduced the induction of trpE9777 revertants by 90% (Figs. 1, 2, and 3). These results demonstrate that in E. coli thymine nucleotide depletion can induce all types of base substitutions and suggest that removal of uracil from DNA may play an important role in the generation of frameshift mutations by thymine deprivation.

The induction of some frameshift mutagenesis in the Ung⁻ strain may reflect leakiness of the ung-1 allele [43]. However, it is possible that other mechanisms exist for the removal of uracil from DNA. For example, endonuclease V of E. coli has been found to have increased activity on uracil-containing DNA in vitro, although an in vivo role of the enzyme in uracil excision repair has not been established [30, 66]. Recently, an endonuclease that acts specifically on DNA containing uracil was isolated from third instar larvae of Drosophila [17]. In view of the fact that we demonstrated a reduction of thymineless frameshift mutagenesis by the ung-1 mutation for only the trpE9777 allele, it may be that alternative mechanisms, independent of uracil excision from DNA, can cause frameshift mutations following thymine deprivation.

Normally, permanent inclusion of uracil in DNA is prevented by at least two mechanisms. First, the enzyme deoxyuridine 5'-triphosphate nucleotidohydrolase (dUTPase) hydrolyzes dUTP to dUMP and pyrophosphate [1] thereby limiting the availability of dUTP as a substrate for DNA synthesis. Second, the enzyme uracil-DNA glycosylase cleaves the uracil-deoxyribose linkage in the first step of a base excision repair system that removes uracil from DNA [42]. Defects in these enzymes confer a mutator phenotype [19, 26]. As described above, mutations in uracil-DNA glycosylase result in a specific enhancement of G:C → A:T transitions [20, 21]. Defects in dUTPase have no effect on this base substitution but may induce frameshifts and deletions [21, 26]. These findings suggest that mere substitution of uracil for thymine in DNA cannot account for the predominance of A:T → G:C transitions observed as a consequence of thymine starvation. Furthermore, the fact that lacI⁻ mutations are induced to a similar extent in Ung⁺ and Ung⁻ strains argues that removal of incorporated uracil does not give rise to the majority of lacI⁻ mutations.

What then might be the mechanism for the induction of point mutations by thymine starvation? If the majority of lacI⁻ mutations are caused by A:T → G:C transitions, then the hypothesis that these events result from misincorporation of cytosine in place of thymine [55] is attractive. However, this cannot be the sole mechanism for generating point mutations by thymine deprivation since G:C → A:T transitions and all possible transversions were detected (Table 3). In accord with the proposal of Bridges et al. [6] experimental evidence suggests that the recA-dependent SOS repair system [57, 71] also may play a role in thymineless mutagenesis. First, it is known that thymine starvation induces SOS functions [34, 71, 72]. Second, Bridges et al. [6] found that a mutation in a controlling gene (lexA) of the SOS system prevented thymineless reversion to leucine prototrophy. Similarly, a defect in the umuC gene, which is required for recA-dependent mutagenesis, reduces reversion of his-4 induced by thymine nucleotide depletion (unpublished observations). Third, the dATP pool increases under thymineless conditions [53] and in response to SOS induction (L. A. Loeb, personal communication) and it appears that SOS-induced cells preferentially incorporate adenine residues during DNA synthesis past apurinic sites [62]. Taken collectively, these findings suggest that base substitutions other than A:T → G:C transitions may arise in response to thymine starvation as a consequence of SOS-dependent DNA synthesis in the presence of elevated dCTP and dATP pools.

A model to account for the known features of thymineless mutagenesis is presented in Fig. 6. Both a heightened dCTP/dTTP ratio [55] and error-prone repair [6] play essential roles in the model and these components are linked by the induction of DNA damage due to uracil incorporation during thymine starvation. Initially, inhibition of dTMP biosynthesis by withholding of thymine leads to a swelling of the dCTP and dUTP pools presumably through lack of feedback inhibition by dTTP on ribonucleotide reductase. This would result in misincorporation of cytosine and uracil nucleotides in place of dTTP during DNA synthesis. Cytosine mispairing could give rise directly to A:T → G:C transitions. Incorporation of uracil would lead to daughter strand gaps formed as a result of uracil excision by uracil-DNA glycosylase and subsequent AP endonuclease attack. Such gaps might play a role in the induction of frameshifts and deletions and in the activation of SOS functions. Repair replication to fill in the gaps, under conditions of continued thymine starvation and SOS induction, could result in misincorporation of dUTP, dCTP, and possibly dATP in the repair patches. In the case of dUTP incorporation, this would give rise to a process of reiterative uracil insertion and excision with the result that few repair patches would be completed. Frequent incomplete episodes of repair could lead to enlarged gaps in daughter strand DNA and such damage also

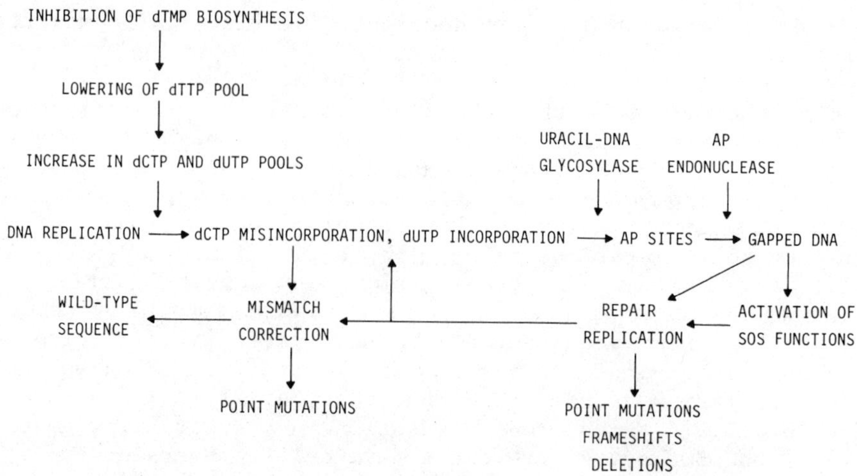

Fig. 6. Model for induction of of mutation in bacteria by thymine nucleotide depletion. See text for details.

might act as a substrate for the production of frameshifts or deletions. Misincorporation of dCTP and dATP during repair or vegetative replication could result in additional base substitution events not restricted to A:T → G:C transitions. At the various stages where base mispairing could occur, mismatch correction might act to restore the wild-type DNA sequence or to fix the base pairing errors.

A number of thymineless phenomena have been described in bacteria including induction of mutation and recombination, stimulation of repair replication, sensitization to UV and activation of SOS functions (see Introduction). These features have in common the involvement of DNA damage and not unexpectedly, thymine starvation also has been found to cause breakage and degradation of DNA. The model proposed here for thymineless mutagenesis is based on the creation of nucleotide pool imbalances in response to thymine deprivation. It seems likely that other thymineless effects in bacteria also may be accounted for by such imbalances.

ACKNOWLEDGEMENTS

The author thanks Mary Skrzynski, Ronnie Dunn, and Don Halderman for expert technical assistance, Drs. B. K. Duncan, D. Fix, and T. Kato for providing bacterial and bacteriophage strains and Drs. B. W. Glickman and T. A. Kunkel for their critical reading of the manuscript.

REFERENCES

1. L. E. Bertani, A. Häggmark, and P. Reichard, Enzymatic synthesis of deoxyribonucleotides. II. Formation and interconversion of deoxyuridine phosphates, J. Biol. Chem., 238:3407-3413 (1963).
2. J. L. Bousque and N. Sicard, Size and transforming activity of deoxyribonucleic acid in Diplococcus pneumonia during thymine starvation, J. Bacteriol., 128:540-548 (1976).
3. T. R. Breitman, P. B. Maury, and J. N. Toal, Loss of deoxyribonucleic acid-thymine during thymine starvation of Escherichia coli, J. Bacteriol., 112:646-648 (1972).
4. S. E. Bresler, M. I. Mosevitsky, and L. Vyacheslavov, Complete mutagenesis in a bacterial population induced by thymine starvation on solid media, Nature, 225:764-766 (1970).
5. S. Bresler, M. Mosevitsky, and L. Vyacheslavov, Mutations as possible replication errors in bacteria growing under conditions of thymine deficiency, Mutat. Res., 19:281-293 (1973).
6. B. A. Bridges, J. Law, and R. J. Munson, Mutagenesis in Escherichia coli. II. Evidence for a common pathway for mutagenesis by ultraviolet light, ionizing radiation and thymine deprivation, Molec. Gen. Genet., 103:266-273 (1968).
7. F. Brunel, A. M. Sicard, and N. Sicard, Transformaing activity of bacterial deoxyribonucleic acid in relation to marker efficiencies in Diplococcus pneumoniae during thymine starvation, J. Bacteriol., 106:904-907 (1971).
8. R. N. Buick and W. J. Harris, Thymineless death in Bacillus subtilis, J. Gen. Microbiol., 88:115-122 (1975).
9. M. P. Calos and J. H. Miller, Genetic and sequence analysis of frameshift mutations induced by ICR-191, J. Molec. Biol., 153:39-66 (1981).
10. S. S. Cohen and H. D. Barner, Studies on unbalanced growth in Escherichia coli, Proc. Natl. Acad. Sci. U.S.A., 40:885-893 (1954).
11. C. A. Coughlin and E. A. Adelberg, Bacterial mutation induced by thymine starvation, Nature, 178:531-532 (1956).
12. C. Coulondre and J. H. Miller, Genetic studies of the lac repressor. III. Additional correlation of mutational sites with specific amino acid residues, J. Molec. Biol., 117:525-575 (1977).
13. C. Coulondre and J. H. Miller, Genetic studies of the lac repressor. IV. Mutagenic specificity in the lacI gene of Escherichia coli, J. Molec. Biol., 117:577-606 (1977).
14. C. Coulondre, J. H. Miller, P. J. Farabaugh, and W. Gilbert, Molecular basis of base substitution hotspots in Escherichia coli, Nature, 274:775-780 (1978).
15. H. DeBarjac, Thymine-requiring mutants of the insect pathogen Bacillus thuringiensis, J. Invert. Pathol., 16:321-324 (1970).
16. C. E. Deutch and C. Pauling, Thyminelss mutagenesis in Escherichia coli, J. Bacteriol., 119:861-867 (1974).

17. W. A. Deutsch and A. L. Spiering, A new pathway expressed during a distinct stage of Drosophila development for the removal of dUMP residues in DNA, J. Biol. Chem., 257:3366-3368 (1982).
18. B. K. Duncan and J. H. Miller, Mutagenic deamination of cytosine residues in DNA, Nature, 287:560-561 (1980).
19. B. K. Duncan, P. A. Rockstroh, and H. R. Warner, Escherichia coli K12 mutants deficient in uracil-DNA glycosylase, J. Bacteriol., 134:1039-1045 (1978).
20. B. K. Duncan and B. Weiss, in: "DNA Repair Mechanisms" (P. C. Hanawalt, E. C. Friedberg, and C. F. Fox, eds.), pp. 183-186, Academic Press, Inc., New York (1978).
21. B. K. Duncan and B. Weiss, Specific mutator effects of ung (uracil-DNA glycosylase) mutations in Escherichia coli, J. Bacteriol., 151:750-755 (1982).
22. P. J. Farabaugh, Sequence of the lacI gene, Nature, 274:765-769 (1978).
23. P. J. Farabaugh, U. Schmeissner, M. Hofer, and J. H. Miller, Genetic studies of the lac repressor. VII. On the molecular nature of spontaneous hotspots in the lacI gene of Escherichia coli, J. Molec. Biol., 126:847-863 (1978).
24. J. L. Farmer and F. Rothman, Transformable thymine-requiring mutant of Bacillus subtilis, J. Bacteriol., 89:262-263 (1964).
25. D. Freifelder, Single strand breaks in bacterial DNA associated with thymine starvation, J. Molec. Biol., 45:1-7 (1969).
26. S. M. Frisch, J. L. Couch, and D. A. Glaser, Mutator activity of a short Okazaki fragment mutant of Escherichia coli, J. Bacteriol., 134:1192-1194 (1978).
27. J. Gallant and T. Spottswood, Measurement of the stability of the repressor of alkaline phosphatase synthesis in Escherichia coli, Proc. Natl. Acad. Sci., U.S.A., 52:1591-1598 (1964).
28. J. Gallant and S. R. Suskind, Relation between thymineless death and ultraviolet inactivation in Escherichia coli, J. Bacteriol., 82:187-194 (1961).
29. A. M. Gasc, N. Sicard, J. P. Claverys, and A. M. Sicard, Lack of SOS repair in Streptococcus pneumoniae, Mutat. Res., 70:157-165 (1980).
30. F. T. Gates and S. Linn, Endonuclease V of Escherichia coli, J. Biol. Chem., 252:1647-1653 (1977).
31. C. S. Genther, R. S. Schoeny, J. C. Loper, and C. C. Smith, Mutagenic studies of folic acid antagonists, Antimicrob. Agents Chemother., 12:84-92 (1977).
32. B. W. Glickman, Spontaneous mutagenesis in Escherichia coli strains lacking 6-methyladenine residues in their DNA. An altered mutational spectrum in dam^- mutants, Mutat. Res., 61:153-162 (1979).
33. M. Goulian and W. Beck, Variations of intracellular deoxyribosyl compounds in deficiencies of vitamin B_{12}, folic acid, and thymine, Biochim. Biophys. Acta, 129:336-349 (1966).
34. L. J. Gudas and A. B. Pardee, DNA synthesis inhibition and the induction of protein X in Escherichia coli, J. Molec. Biol., 101: 459-477 (1976).

35. W. E. Hill and W. L. Fangman, Single-strand breaks in deoxyribonucleic acid and viability loss during deoxyribonucleic acid synthesis inhibition in Escherichia coli, J. Bacteriol., 116:1329-1335 (1973).
36. A. J. Holmes and A. Eisenstark, The mutagenic effect of thymine starvation on Salmonella typhimurium, Mutat. Res., 5:15-21 (1968).
37. D. Kanazir, The apparent mutagenicity of thymine deficiency, Biochim. Biophys. Acta, 30:20-23 (1958).
38. T. Kato and E. Nakano, Effects of the umuC36 mutation on ultraviolet-radiation-induced base-change and frameshift mutations in Escherichia coli, Mutat. Res., 83:307-319 (1981).
39. T. Kato, Y. Shinoura, A. Templin, and A. J. Clark, Analysis of ultraviolet light-induced suppressor mutations in the strain of Escherichia coli K12 AB1157: an implication for molecular mechanisms of mutagenesis, Molec. Gen. Genet., 180:283-291 (1980).
40. D. Korn and A. Weissbach, Thymineless induction in Escherichia coli K12 (λ), Biochim. Biophys. Acta, 61:775-790 (1962).
41. B. A. Kunz and B. W. Glickman, The infidelity of conjugal DNA replication in Escherichia coli, Genetics, 105:489-500 (1983).
42. T. Lindahl, An N-glycosidase from Escherichia coli that releases free uracil from DNA containing deaminated cytosine residues, Proc. Natl. Acad. Sci., U.S.A., 71:3649-3653 (1974).
43. T. Lindahl, DNA glycosylases, endonucleases for apurinic/apyrimidinic sites, and base excision repair, Progr. Nucl. Acid Res. Molec. Biol., 22:135-191 (1979).
44. J. G. Little and P. C. Hanawalt, Thymineless death and ultraviolet sensitivity in Micrococcus radiodurans, J. Bacteriol., 113:233-240 (1973).
45. F. Makino and N. Munakata, Deoxyuridine residues in DNA of thymine-requiring Bacillus subtilis strains with defective N-glycosidase activity for uracil-containing DNA, J. Bacteriol., 134:24-29 (1978).
46. L. G. Mathieu, J. DeRepentigny, J. Turgeon, and S. Sonea, Thymineless death of Staphylococcus aureus and formation of its alpha toxin, Can. J. Microbiol., 14:983-987 (1968).
47. J. H. Miller, Experiments in Molecular Genetics, Cold Spring Harbor Laboratory, Cold Spring Harbor (1972).
48. J. H. Miller, C. Coulondre, and P. J. Farabaugh, Correlation of nonsense sites in the lacI gene with specific codons in the nucleotide sequence, Nature, 274:770-775 (1978).
49. J. H. Miller, D. Ganem, P. Lu, and A. Schmitz, Genetic studies of the lac repressor. I. Correlation of mutational sites with specific amino acid residues: construction of a co-linear gene-protein map, J. Molec. Biol., 109:275-301 (1977).
50. J. H. Miller and U. Schmeissner, Genetic studies of the lac repressor. X. Analysis of missense mutations in the lacI gene, J. Molec. Biol., 131:223-248 (1979).

51. P. Mukopadhyay and S. Palchaudhuri, Trimethoprim-produced F-specific insertion mutations in Escherichia coli K12, J. Bacteriol., 150:755-761 (1975).
52. H. Nakayama and P. C. Hanawalt, Sedimentation analysis of deoxyribonucleic acid from thymine-starved Escherichia coli, J. Bacteriol., 121:537-547 (1975).
53. J. Neuhard and A. Munch-Peterson, Studies on the acid soluble nucleotide pool in thymine requiring mutants of Escherichia coli during thymine starvation. II. Changes in the amounts of deoxycytidine triphosphate and deoxyadenosine triphosphate in Escherichia coli $T^-A^-U^-$, Biochim. Biophys. Acta, 114:61-71 (1966).
54. A. J. Norin and E. P. Goldschmidt, Effect of mutagens, chematherapeutic agents and defects in DNA repair genes on recombination in F' partial diploid Escherichia coli, Mutat. Res., 59:15-26 (1979).
55. C. Pauling, in: "Structural Chemistry and Molecular Biology" (A. Rick and N. Davidson, eds.), pp. 393-398, W. H. Freeman Co., San Francisco (1968).
56. C. Pauling and P. C. Hanawalt, Non-conservative DNA replication in bacteria after thymine starvation, Proc. Natl. Acad. Sci., U.S.A., 54:1728-1735 (1965).
57. M. Radman, in: "Molecular and Environmental Aspects of Mutagenesis" (L. Prakash, F. Sherman, M. W. Miller, C. W. Lawrence, and H. W. Taber, eds.), pp. 128-142, Charles C. Thomas, Publisher, Springfield (1974).
58. G. Ramareddy and H. Reiter, Sequential loss of loci in thymine-starved Bacillus subtilis 168 cells: evidence for a circular chromosome, J. Molec. Biol., 50:525-532 (1970).
59. J. Reich and J. Soska, Thymineless death and deoxyribonucleotideless death in Lactobacillus acidophilis R-26, Biochem. Biophys. Res. Commun., 29:62-67 (1967).
60. D. L. Reichenbach, G. E. Schaiberger, and B. Sallman, The effect of thymine starvation on the chromosomal structure of Escherichia coli JG-151, Biochem. Biophys. Res. Commun., 42: 23-30 (1971).
61. H. Reiter and G. Ramareddy, The loss of DNA behind the growing point of thymine-starved Bacillus subtilis 168, J. Molec. Biol., 50:533-548 (1970).
62. R. M. Schaaper, T. A. Kunkel, and L. A. Loeb, Infidelity of DNA synthesis associated with bypass of apurinic sites, Proc. Natl. Acad. Sci., U.S.A., 80:487-491 (1983).
63. U. Schmeissner, D. Ganem, and J. H. Miller, Genetic studies of the lac repressor. II. Fine structure deletion map of the lacI gene, and its correlation with the physical map, J. Molec. Biol., 109:303-326 (1977).
64. E. C. Siegel and K. K. Vaccaro, The reversion of trp frameshift mutations in mut, polA, lig, and dnaE mutant strains of Escherichia coli, Mutat. Res., 50:9-17 (1978).

65. F. Tamanoi and T. Okazaki, Uracil incorporation into the nascent DNA of thymine-requiring mutant of Bacillus subtilis 168, Proc. Natl. Acad. Sci., U.S.A., 75:2195-2199 (1978).
66. B.-K. Tye, J. Chien, I. R. Lehman, B. K. Duncan, and H. Warner, Uracil incorporation as a source of pulse-labelled DNA fragments in the replication of the Escherichia coli chromosome, Proc. Natl. Acad. Sci., U.S.A., 75:233-237 (1978).
67. J. T. Wachsman, S. Kemp, and L. Hogg, Thyminelss death in Bacillus megaterium, J. Bacteriol., 87:1079-1086 (1964).
68. J. R. Walker, Thymine starvation and single strand breaks in chromosomal deoxyribonucleic acid of Escherichia coli, 104: 1391-1392 (1970).
69. H. R. Warner, B. K. Duncan, C. Garett, and J. Neuhard, Synthesis and metabolism of uracil-containing deoxyribonucleic acid in Escherichia coli, J. Bacteriol., 145:687-695 (1981).
70. R. Weinberg and A. B. Latham, Apparent mutagenic effect of thymine deficiency for a thymine-requiring strain of Escherichia coli, J. Bacteriol., 72:570-572 (1956).
71. E. M. Witkin, Ultraviolet mutagenesis and inducible repair in Escherichia coli, Bacteriol. Rev., 40:869-907 (1976).
72. E. M. Witkin and I. E. Wermundsen, Targeted and untargeted mutagenesis by various inducers of SOS functions in Escherichia coli, Cold Spring Harbor Symp. Quant. Biol., 43:881-886 (1978).
73. K. Yoshinaga, Double strand scission of DNA involved in thymineless death of Escherichia coli 15 TUA, Biochim. Biophys. Acta, 294:204-213 (1973).

GENETIC AND BIOCHEMICAL EFFECTS OF THYMIDYLATE STRESS IN YEAST

J. G. Little

Department of Biology
York University
Toronto, Ontario M3J1P3
Canada

1. INTRODUCTION

In addition to the *de novo* pathway of deoxynucleotide biosynthesis, most organisms possess salvage pathways whereby nucleosides, and sometimes bases, can be channeled into the deoxynucleotide pathway. The yeast *Saccharomyces cerevisiae* lacks the salvage enzymes thymidine kinase [17] and deoxycytidine kinase (McIntosh and Haynes, unpublished results). It probably also lacks purine deoxynucleoside salvage enzymes, although the evidence for this is, at present, indirect. Yeast can thus be viewed as having a "minimal" DNA precursor pathway (Fig. 1).

The absence of thymidine kinase impeded for a time the study of DNA metabolism in yeast and stimulated a search for mutant strains capable of incorporating exogenous thymidine 5'-monophosphate (dTMP). Strains capable of dTMP uptake (*tup* mutants) were isolated in various laboratories [12, 18, 22, 25, 35]. Such strains can also incorporate nucleotides other than dTMP. From these *tup* strains, mutants blocked in *de novo* thymidylate synthesis were isolated: these are auxotrophic for dTMP. One class of auxotrophs (*tmp1*) [10, 13, 26] has been shown to be defective in thymidylate synthetase [8]; the other class (*fol* mutants) is auxotrophic for dTMP, and adenine, histidine and methionine, because of defects in folate metabolism [26, 27]. These strains facilitate pool manipulations in yeast and can incorporate various nucleotide analogs which are useful in studies of DNA replication and precursor metabolism.

To explore the effects of deoxynucleotide pool imbalance we focussed on the thymidine nucleotide pathway both for theoretical

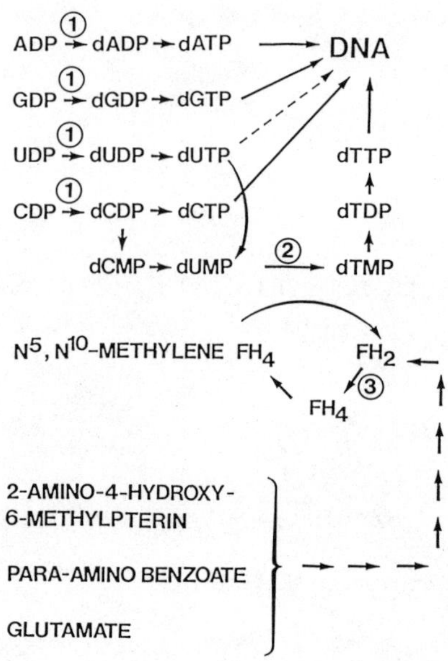

Fig. 1. DNA precursor pathway in yeast. The enzymes indicated are: (1) ribonucleoside diphosphate reductase; (2) thymidylate synthetase; and (3) dihydrofolate reductase.

and practical reasons. DNA precursor pathways are inherently asymmetrical (Fig. 1). The terminal product of the thymidylate pathway, dTTP, is synthesized in a series of steps longer, more complex and more energetically costly than those of other DNA precursors. Synthesis of dTTP involves the intermediate dUTP which can itself function as a DNA precursor [7]. The two purine triphosphates are terminal products of their biosynthetic pathways, whereas of the three pyrimidine triphosphates only dTTP is terminal: dCTP and dUTP can be converted to dTTP. The thymidylate pathway is coupled to the major carbon transfer reactions in cells by its linkage to folate metabolism; N^5,N^{10}-methylenetetrahydrofolate donates methyl groups in the conversion of dUMP to dTMP, and this reaction can significantly drain intracellular folate pools [31], which are implicated in several vital biochemical pathways. This relative complexity of the thymidylate pathway suggests that it may serve some significant regulatory function. In practice, it is much easier to manipulate the pools of thymidine nucleotides than pools of other precursors, and a variety of thymidylate analogs, with differing metabolic potential, can be incorporated by yeast tup and tmp1 strains.

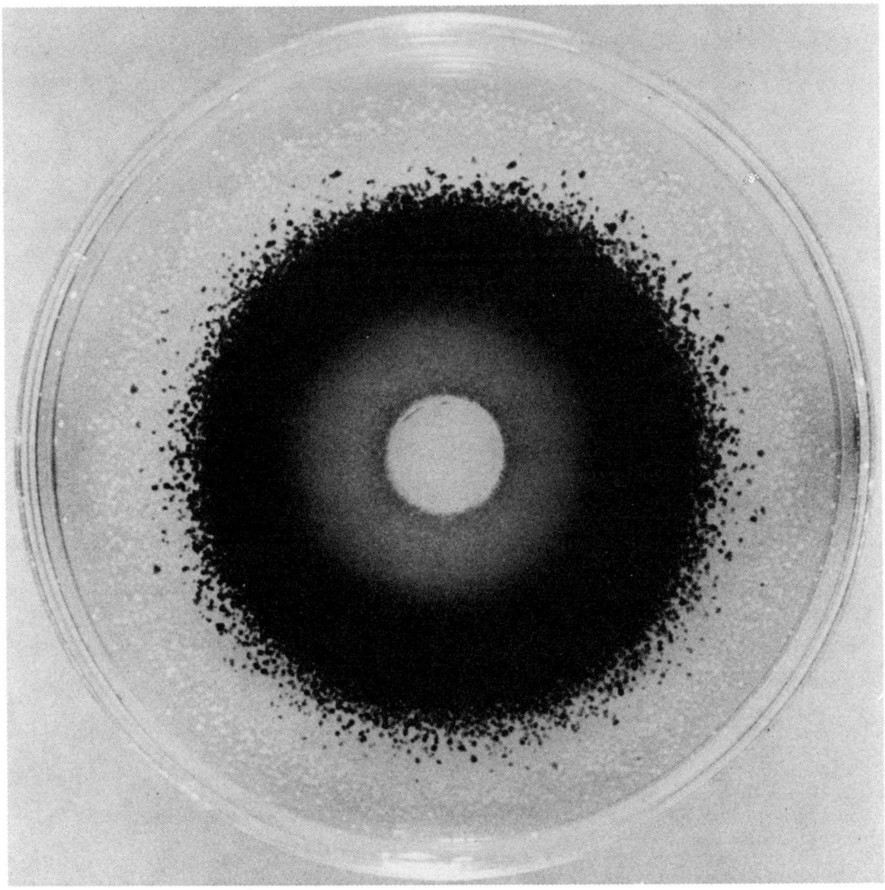

Fig. 2. Thymidylate stress effects in yeast. In a plate overlaid with a uniform lawn of yeast cells auxotrophic for thymidylate, a central well containing dTMP (2 mg/ml) was used to generate a radial gradient of concentration. After two days incubation at 34°C, the plate was overlaid with tetrazolium dye. Respiratory-competent grande colonies turn red (black in this photograph). Petite colonies remain white. Inhibition of cell growth is evident at the extremes of thymidylate concentration. (Photograph supplied by Dr. B. J. Barclay.)

2. THYMIDYLATE STRESS

In yeast tup strains which are blocked in de novo synthesis of dTMP, cell growth is inhibited when the exogenous concentration of dTMP is either limiting or excessive [26]. This can be easily visualized by generating a radial gradient of dTMP concentration in

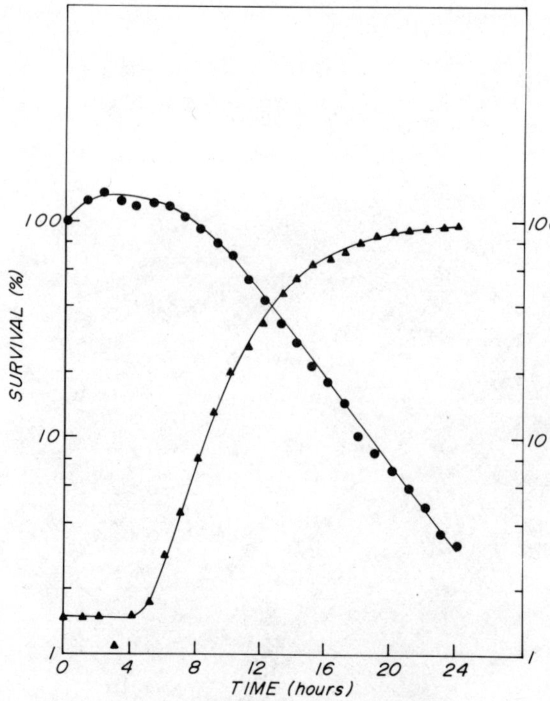

Fig. 3. Induction of cytoplasmic petites during thymidylate starvation. Exponentially growing cells of a grande strain auxotrophic for dTMP were incubated in medium without dTMP. Aliquots were taken at the times indicated and plated to determine cell viability (●). After 4 days growth, petite induction was scored (▲).

a plate containing a uniform lawn of yeast tmpl cells (Fig. 2). Growth inhibition is seen at the two extremes of thymidylate concentration. The magnitude of growth inhibition and the optimial concentration of dTMP needed for normal growth can be markedly influenced by the genetic background and by the medium used to grow dTMP auxotrophs [26]. In addition to growth inhibition, genetic changes occur in cells exposed to thymidylate starvation or excess.

3. THYMIDYLATE STARVATION

It is now possible to induce a thymineless condition in cultures of Saccharomyces cerevisiae by a variety of methods. When strains blocked in de novo thymidylate synthesis (tmpl or foll) are incubated in the absence of dTMP, cell growth is arrested and thymineless death ensues [26]. Cells assume the dumbbell morphology typical of yeast cells in which DNA synthesis has been arrested [26]. In strains conditionally defective in thymidylate synthetase [8, 13, 26], thymidine nucleotide starvation can be effected by incubating cells at the restrictive temperature [13]. Drugs can also be used to abolish dTMP biosynthesis. In yeast, the combination of antifolate drugs aminopterin (or methotrexate) plus sulfonamide is a potent inhibitor. Cells thus treated require thymidylate, adenine, histidine and methionine for growth [3, 26]. If dTMP is omitted, thymineless death ensues. In tup yeast strains fluorodeoxyuridine-5'-monophosphate (FdUMP), which inhibits thymidylate synthetase, can be used to induce a thymineless condition [20]. Thymidylate deprivation provokes characteristic genetic alterations in yeast.

3.1. Mitochondrial Effects

We find that the mitochondrial genome of yeast is especially vulnerable to thymidylate starvation. Point mutations, deletions, and, eventually, progressive erosion of mitochondrial DNA occur under conditions of dTMP deprivation [4].

In *Saccharomyces cerevisiae*, the induction of respiratory-incompetent cytoplasmic petites is associated with the deletion of some or all of the mitochondrial genome [28]. With the onset of thymineless death in yeast, the fraction of petites in the surviving population began to increase. After 24 h of starvation almost all survivors were petite (Fig. 3).

In yeast, resistance to the antibiotics erythromycin and chloramphenicol is conferred by point mutations in mitochondrial genes [33, 36]. Mutation to erythromycin resistance (ERY^R) or chloramphenicol resistance (CAP^R) was therefore scored in a population of yeast cells undergoing dTMP starvation. Figure 4 shows the induction of antibiotic-resistant mutants. Control experiments, described elsewhere [4], showed that the increase in the number of antibiotic-resistant mutants was not the result of selection of these mutants during the starvation episode. These tests also revealed that the resistant mutants arose as a consequence of mitochondrial gene mutation and not by nuclear gene mutation. This finding - that organellar DNA is susceptible to thymineless mutagenesis - was subsequently confirmed by other workers. Thus, thymidylate inhibition by FUdr in *Chlamydomonas* induces chloroplast gene mutations [37]; mitochondrial mutations are induced in mammalian cells by the folate antagonist methotrexate (D. Hoar - this volume).

The induction of cytoplasmic petites during thymidylate starvation is correlated with a progressive degradation of the mitochondrial DNA. This was shown by cesium chloride gradient analysis of DNA extracted from starved cells (Fig. 5). This experiment depicts the DNA profiles of starved cells under two sets of condi-

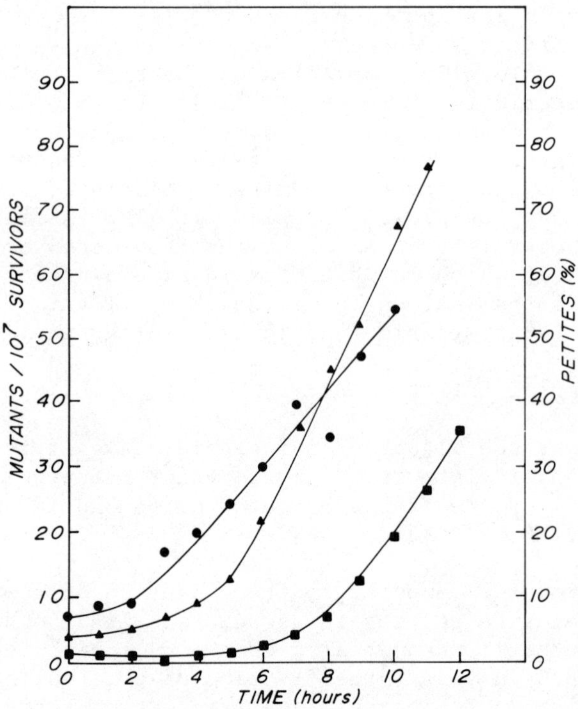

Fig. 4. Induction of mitochondrial point mutations during thymidylate starvation. The experimental conditions are similar to those described in Fig. 3. Mutation to erythromycin resistance (▲) or chloramphenicol resistance (●) was scored in a population of yeast cells undergoing dTMP starvation. The induction of petite colonies (■) was also scored.

tions: incubation at low cell densities (a-d) or high cell densities (e-h). Incubation at high cell densities conferred protection against thymineless death whereas induction of petites was approximately equivalent under both sets of conditions [4]. Incubation in the absence of dTMP resulted in the progressive erosion of mitochondrial DNA and the pattern of this degradation was essentially the same in the low and high cell density cultures. This confirms the interpretation that thymineless induction of petites is due to loss of mitochondrial DNA.

As shown, an interesting distinction emerged in the nuclear DNA profiles of the two cultures. The low cell density culture - where thymineless death is rapid - reveals a significant broadening of the nuclear DNA band, suggesting a correlation between chromosome degradation and thymineless death. This was subsequently studied by the more sensitive method of sucrose gradient analysis of nuclear DNA (see below).

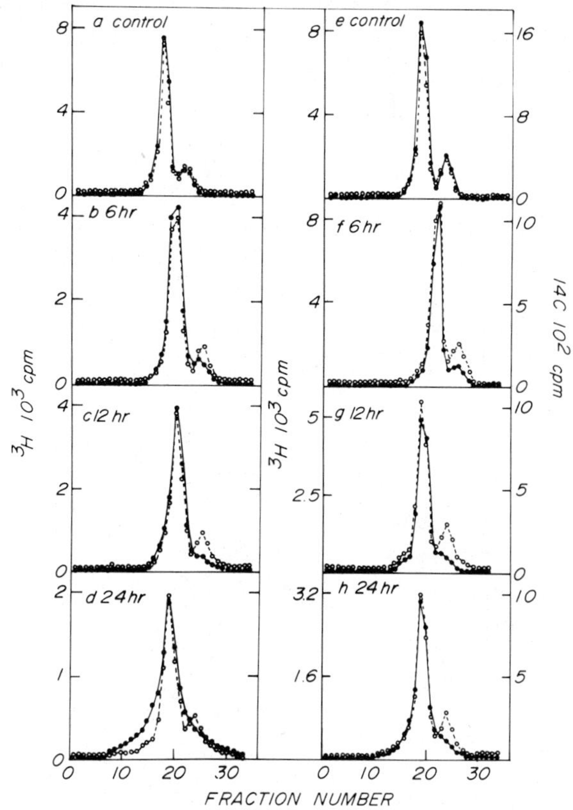

Fig. 5. CsCl gradient analysis of yeast DNA during dTMP starvation. Cells of a grande strain auxotrophic for dTMP were grown in medium in the presence of either ^{3}H-dTMP or ^{14}C-dTMP. ^{14}C-labelled cells were used as a source of marker DNA, and were not starved. ^{3}H-labelled cells were incubated in medium lacking dTMP for the times indicated. Cells were starved at either low cell density (1.5×10^6/ml) or high cell density (3.0×10^7/ml). ^{3}H-labelled and ^{14}C-labelled cells were mixed, and lysed spheroplasts were centrifuged in neutral CsCl gradients. (a), (b), (c), and (d) are neutral CsCl density profiles of DNA from cells incubated at low density. (e), (f), (g), and (h) are the corresponding profiles from cells starved at high density. ^{14}C-label (○), ^{3}H-label (●).

3.2. Nuclear Effects

Hints that thymidylate starvation induced strand breaks in yeast nuclear DNA were confirmed and extended by alkaline sucrose gradient analysis of DNA isolated by gentle lysis of spheroplasts [2].

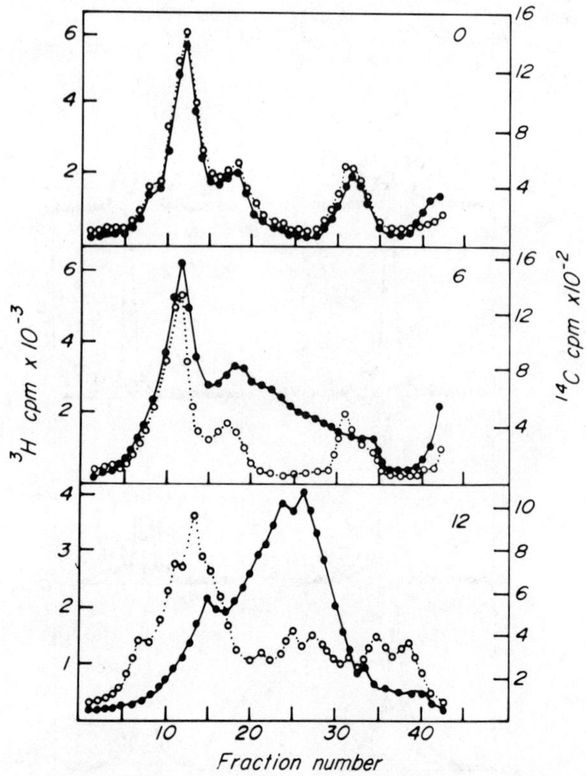

Fig. 6. Alkaline sucrose gradient analysis of DNA from thymidylate-starved dTMP auxotrophs. Cells prelabehled with ^3H-dTMP (●) were starved for periods of 0, 6, or 12 h, and the DNA subjected to alkaline sucrose gradient analysis. ^{14}C-labelled (○) unstarved cells served as a source of control DNA. The shift observed in the DNA profiles demonstrates the production of DNA single-strand breaks during dTMP starvation.

A typical experiment is shown in Fig. 6. During dTMP depletion, there was a decline in the amount of DNA sedimenting at the control molecular weight. In this experiment the shift in the DNA profile indicates the induction of approximately 50 strand breaks per haploid cell. Very similar data were obtained in different experiments in which the dTMP starvation was induced either by genetic lesion or by the drug regimens previously described which block thymidylate biosynthesis [2]. Thus it is evident that thymidylate starvation, no matter how provoked, leads to the accumulation of single-strand breaks in nuclear DNA.

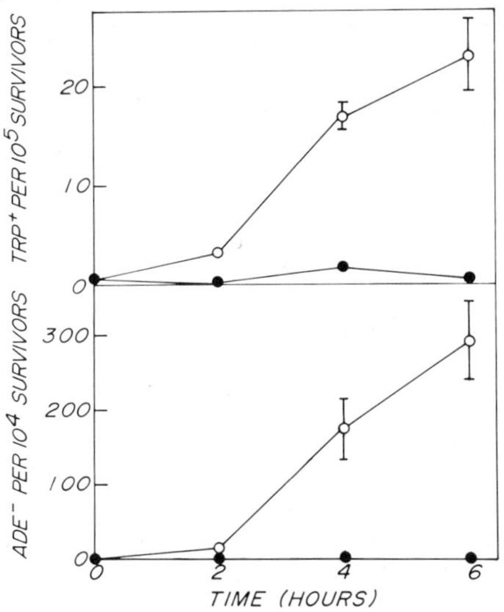

Fig. 7. Mitotic recombination induction by thymidylate starvation. Diploid yeast cells containing appropriate genetic markers were treated with aminopterin plus sulfanilamide in the presence (●) or absence (○) of dTMP for the times indicated. The emergence of tryptophan-independent and of adenine-requiring colonies signifies the induction of mitotic recombination (see text).

Persistent DNA strand breaks can promote genetic recombination. In diploid yeast strains containing appropriate genetic markers, mitotic crossing-over and gene conversion events can be readily scored. Such strains were constructed and subjected to thymidylate starvation by using FdUMP or antifolate drugs or by incubating a strain temperature-sensitive for thymidylate synthetase at the restrictive temperature [20]. Thymidylate deprivation was found to be highly recombinagenic (Fig. 7). In this experiment, tryptophan-independent colonies emerge by gene conversion events; mitotic crossing-over produces adenine-requiring colonies (for complete experimental details see Ref. 20). Provision of exogenous dTMP eliminated induced mitotic recombination and cell killing. Similar tests showed that several sulfa drugs, which impede de novo folate synthesis, were recombinagenic [2]. Again it was found that thymidylate supplementation prevented recombination. It is evident that the recombinagenic action of these various drugs is due to depletion of intracellular thymidylate pools.

Since thymineless conditions induce mutations in mitochondrial genes, we sought evidence of thymineless induction of nuclear gene

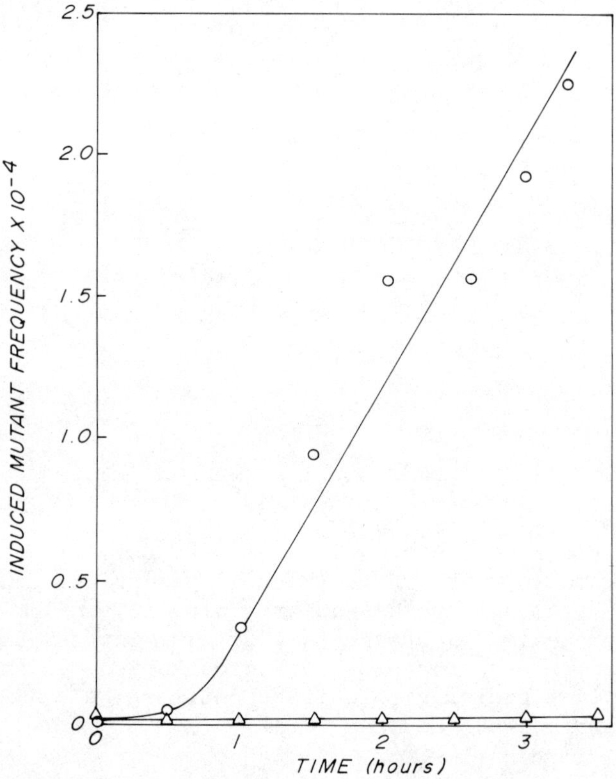

Fig. 8. RAD6-dependent mutagenesis by high concentrations of thymidylate. Nucleotide-permeable tup strains of yeast (see text) which differed at the RAD6 locus were incubated in medium containing dTMP (5 mg/ml) for the times indicated. Induction of mutations in RAD6 (○) or rad6-1 (△) strains are shown.

mutations. A variety of test systems were employed, capable of detecting either forward mutations of any one of several genes in the biosynthetic pathway of adenine or reversion to prototrophy of several auxotrophic markers [2, 4, 20]. Mutation induction by thymidylate deprivation was not observed. The fact that yeast is apparently refractory to mutation induction by thymidine nucleotide starvation whereas several prokaryotes are readily mutated [2] suggests that eukaryotes and prokaryotes may differ in their response to thymidylate stress.

TABLE 1. Genetic Effects of Thymidine Nucleotide Pool Perturbations in Yeast

dTMP Deprivation
Thymineless death
Mutagenesis of mitochondrial DNA
Petite formation
Mitochondrial DNA degradation
Recombination of nuclear genes
Absence of nuclear gene mutation
Nuclear DNA strand breaks
Sensitization to UV and X-rays
Formation of nuclear dense body
dTMP Excess
Mutagenesis of nuclear genes. RAD6 dependent.

4. THYMIDYLATE EXCESS

In yeast strains auxotrophic for dTMP, or in prototrophic tup strains, excessive concentrations of exogenous dTMP inhibit cell growth. These inhibitory concentrations of thymidylate can also induce mutations in nuclear genes [5]. The manifestation of dTMP-induced mutagenesis required the RAD6 gene (Fig. 8), thought to be essential for error-prone DNA repair in yeast [11, 23]. This finding suggested that the simplest hypothesis for thymidylate mutagenesis - that swollen dTTP pools may lead to erroneous insertion of thymine residues in place of cytosine during DNA replication - was probably inadequate, since this model does not invoke error-prone repair.

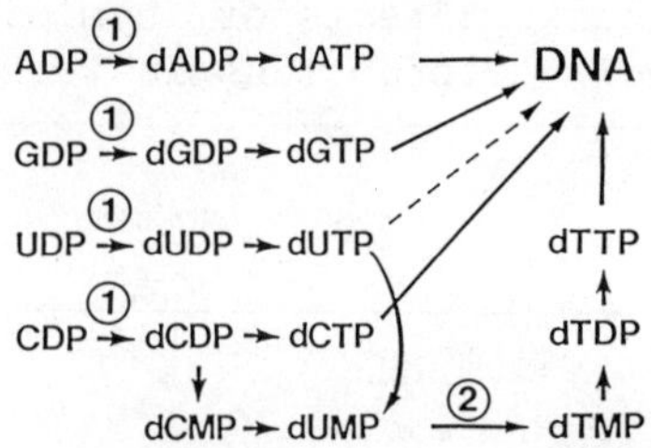

Fig. 9. DNA precursor pathway. Under conditions of thymidylate depletion, dUTP can serve as a DNA precursor.

5. THMIDYLATE STRESS: SUMMARY OF EFFECTS

Table 1 summarizes the principal effects induced by perturbations of the thymidine nucleotide pools in yeast. In addition to the effects already described we have found that brief non-lethal episodes of thymidylate starvation sensitize yeast cells to UV and X-rays. Thymineless conditions also induce the appearance of a nuclear dense body [30], which is normally a symptom of yeast cells undergoing meiosis. The metabolic basis for this effect is not understood.

6. POSSIBLE MECHANISMS

As the observations concerning the genetic effects of thymidine pool imbalance accumulate in this and other laboratories, there is a need to formulate working hypothesis in an attempt to provide comprehensive explanations, and to ask if there may exist additional components of nucleotide metabolism other than those contained in the current orthodoxy. We have discussed possible mechanisms in detail elsewhere [2] and Kunz has recently published a pertinent review [19]. Here I will describe briefly the hypotheses we favor.

The difference in the genetic effects induced at the two extremes of thymidylate stress suggests that quite different biochemical events at the level of DNA replication or repair are involved in these phenomena. However, it has been found that the contrasting mutagenic and recombinagenic effects observed at the extremes of dTMP concentration can both be abolished by a single mutation at the RAD6 locus [5, 21]. This perplexing finding suggests that there exists some common element of control in the effects engendered at the extremes of thymidylate stress, even though the genetic consequences of these perturbations are radically different. The RAD6 gene product is required for error-prone DNA repair in yeast [23] but its specific biochemical function is not understood; at present, therefore, we cannot speculate about its precise role in the phenomena described here, other than to indicate that it will be important to a full understanding of these processes.

In considering the possible causes of the genetic effects which arise from thymidylate deprivation, the most parsimonious approach is to look for a common molecular mechanism which underlies the various effects. The most plausible hypothesis is that under conditions of dTMP starvation, uracil can be incorporated into nascent DNA (Fig. 9), and the observed effects arise as a consequence of this incorporation.

In the *de novo* synthesis of thymidylate, dUTP is an intermediate and it can serve as a substrate in place of dTTP for DNA polyerase [7]. The permanent inclusion of uracil in DNA is prevented usually by two mechanisms. First, the enzyme dUTPase rapidly hydrolyses dUTP to dUMP [6, 16], thus limiting available dUTP as a substrate for DNA replication. Secondly, the enzyme uracil-DNA glycosylase recognizes and removes uracil residues in DNA [24]. The resulting apyrimidinic site is then subject to endonuclease attack and excision repair [34].

Goulian and co-workers have shown that inhibition of thymidylate synthetase in mammalian cells results in increased dUTP levels and enhanced misincorporation of uracil into DNA [14, 15]. Makino and Munakata [29] have also demonstrated uracil incorporation into the DNA of thymine-starved bacteria. In that study, thymine-requiring *B. subtilis* cells defective in uracil-DNA glycosylase were incubated in the presence or absence of thymine. Under conditions of thymine starvation extensive uracil incorporation into DNA was observed. A direct link between thymineless death and uracil excision was established by these authors since strains defective in uracil-DNA glycosylase were resistant to thymineless killing.

These observations suggested that uracil entry into nascent DNA and its subsequent excision may be central to thymineless phenomena. Removal of uracil residues and subsequent attempted excision repair in the absence of thymidylate could result in reiterated episodes of uracil insertion and excision, leading to persistent DNA strand breaks or gaps. These could, in turn, stimulate recombinational events, leading to the introduction of breaks in parental DNA strands. Overlapping gaps may lead to chromosome aberrations or cell death. Thymineless induction of point mutations in the mitochondrial genome and the corresponding absence of nuclear gene mutations could be ascribed to differing efficiencies in uracil excision, i.e., highly efficient removal of uracil moieties (nuclear genes) prevents mutation but promotes recombination whereas less efficient removal of uracil (mitochondrial genes?) permits mutation induction.

We have attempted to test this hypothesis in yeast but so far we have not detected uracil incorporation into DNA under conditions of thymidylate starvation. This may either indicate that the hypothesis is incorrect or that removal of the incorporated uracil occurred during the DNA extraction procedures. A definitive understanding of the relevance of uracil incorporation and excision as an explanation for the yeast thymineless phenomena will require the isolation of mutants defective in uracil-DNA glycosylase.

Our findings that high concentrations of thymidylate induce mutations in yeast nuclear genes is consonant with the observations that excess thymidine concentrations can induce mutations in Drosophila melanogaster [32] and in animal cells [1, 9]. The molecular nature of the mutagenic lesions caused by excessive doses of thymidine or dTMP remains unclear. Since it is known that high levels of dTTP can deplete endogenous pools of dCTP via allosteric inhibition of ribonucleotide reductase (reviewed in Kunz: Ref. 19), the simplest explanation for this mode of mutagenesis is to suppose that the low dCTP: dTTP ratio may promote misincorporation of thymine residues in place of cytosine during DNA replication. GC → AT transitions would occur. This is an attractive model but our finding that dTMP excess is not mutagenic in strains deficient in the RAD6 gene product casts doubt upon its adequacy.

In a previous paper [5] we suggested an alternative possibility: that thymidine nucleotides - or metabolic derivatives of such nucleotides - could operate as signals which effect an alteration in the DNA replication complex leading to diminished accuracy of template copying. The asymmetric architecture of DNA precursor metabolism suggests ways in which transiently high fluxes of thymidine nucleotides can be generated naturally by DNA lesions (of any nature) which impede the progress of chromosome replication. When replication forks are arrested, precursor pools increase. Purine triphosphates are terminal products of their biosynthesis pathways; pyrimidine triphosphates are not. Thus, when DNA lesions engender a swelling of precursor pools, dCTP and dUTP can be converted to thymidine nucleotides. These nucleotides, or metabolites derived from them, or the bias generated in the ratio of pyrimidine precursor pools could signal an alteration in replicative fidelity [5], permitting the replicative bypass of damaged DNA.

This model therefore postulates that experimental manipulations which increase the thymidylate pools mimic the circumstances which would usually arise when DNA replication is blocked, and elicit a similar diminution in replication fidelity, with consequent mutagenesis. Our observation that RAD6 strains are refractory to dTMP-induced mutation is explained by this model if the RAD6 gene product is required for the alteration of the DNA replication complex.

In considering possible mechanisms which might be implicated in the genetic effects induced by thymidylate stress we wondered if nucleotide interconversions other than those represented in the familiar metabolic charts (e.g., Fig. 9) might exist. We conducted a test in the yeast system by asking if the DNA thymine requirements of yeast strains blocked in de novo thymidylate metabolism could be satisfied by nucleotides other than 5'-dTMP.

7. NOVEL NUCLEOTIDE INTERCONVERSIONS

Yeast tup strains blocked in de novo synthesis of thymidylate will grow if exogenous 5'-dTMP (pT) is supplied. We conducted tests (Orna Landman and J. G. Little; unpublished results) to ask if 3'-dTMP (Tp) or 3',5'-cyclic dTMP (cT) could substitute for pT. The results of a typical experiment are shown in Fig. 10, which shows that Tp and cT can support cell growth. Thymidine cannot substitute for pT since yeast lacks thymidine kinase. We have obtained similar results in a wide variety of yeast tup strains blocked in thymidylate biosynthesis by mutation or drugs. The nucleotides are utilized by yeast in the following order of preferences: pT > Tp > cT. Cells supplied with exogenous ^{3}H-3'-dTMP incorporated label into both nuclear and mitochondrial DNA.

We envisaged two possible pathways whereby these nucleotides could be converted to the orthodox DNA precursor 5'-dTTP (pppT). These are depicted in Fig. 11. Pathway 1 involves phosphorylation at the 5'-position to yield pTp, followed eventually by dephosphorylation of the 3'-phosphate to yield pppT. Dephosphorylation need not occur at the step shown in Fig. 11 (see below) but should, as a minimum, require the production of pTp. Alternatively, Tp might undergo intramolecular migration of phosphate via the cyclic intermediate to yield ultimately the 5'-monophosphate as shown. Similarly, cT could be converted either by pathway 2 or pathway 3 to generate pppT.

The following experiment was used to decide which of these mechanisms was operative. Yeast auxotrophs (tmp1) were grown in medium containing inorganic ^{32}p plus one of the following nucleotides: unlabelled pT or Tp or cT. DNA was isolated and digested to 5'-nucleotides. These nucleotides were separated and the radioactivity of each was counted. Full details of these experiments will be published elsewhere; here I summarize the principal findings. Cells grown in pT yielded DNA with three labelled nucleotides as expected, ^{32}pA, ^{32}pG, and ^{32}pC; cells grown in cT had identical patterns; cells grown in Tp yielded DNA with four labelled nucleotides, ^{32}pA, ^{32}pG, ^{32}pC, and ^{32}pT and these were present in the correct Chargaff ratios.

From these results and the rationale depicted in Fig. 11, cyclic nucleotide cT is cleaved exclusively to pT, presumably by a cyclic phosphodiesterase. The 3'-nucleotide Tp must be phosphorylated in vivo at the 5'-position by a route similar to path-

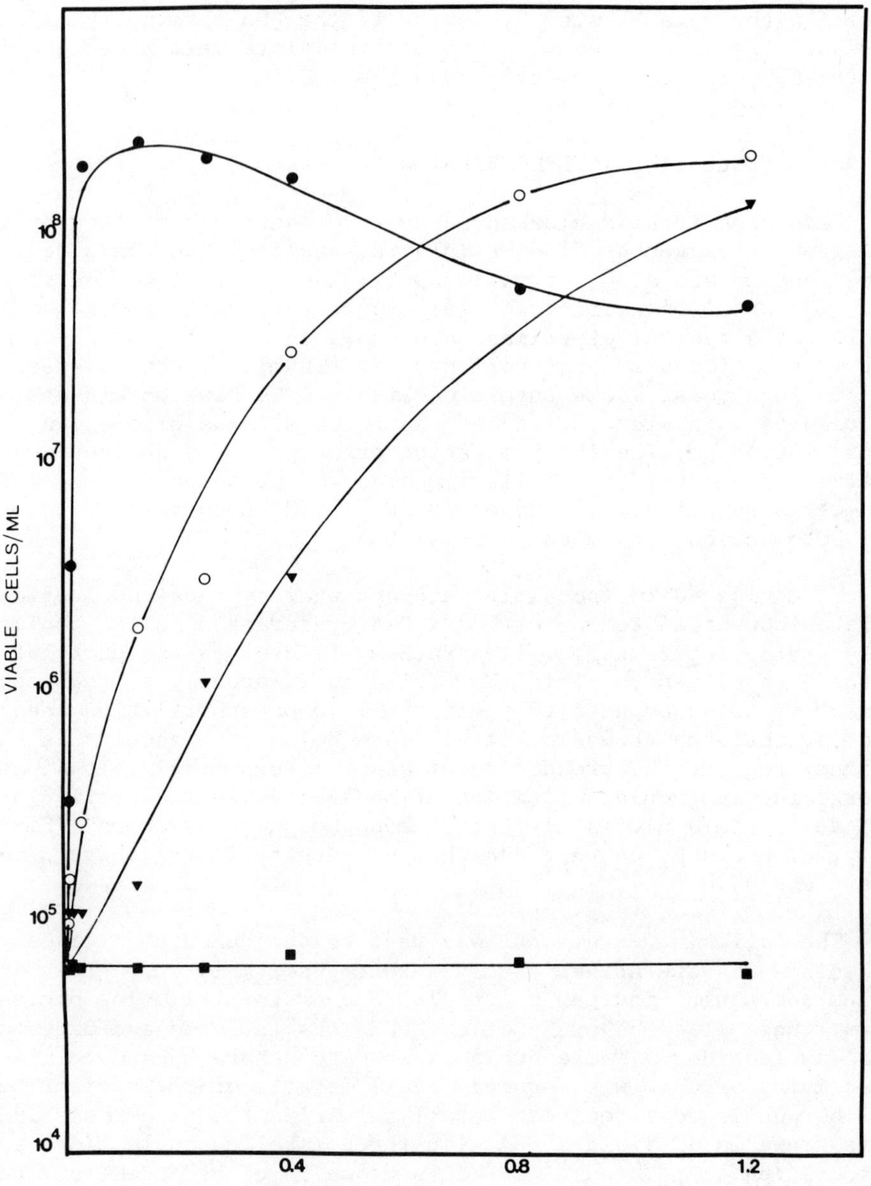

Fig. 10. Growth of a thymidylate auxotroph in the presence of various thymidine nucleotides. Cells a yeast tmp1 strain were incubated for 24 h in media containing one of the following supplements at the concentrations indicated, and viable cell counts were determined. (●) 5'-dTMP; (○) 3'-dTMP; (▲) 3'5'-cyclic dTMP; (■) thymidine.

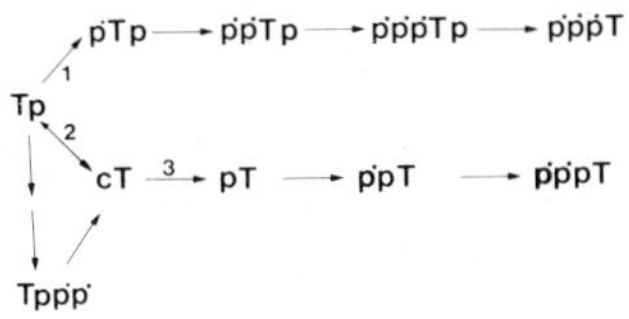

Fig. 11. Two pathways of thymidine nucleotide inverconversions. The 3'-nucleotide Tp could undergo phosphorylation at the 5'-position (pathway 1), leading ultimately to the DNA precursor pppT. Alternatively, intramolecular migration of phosphate (pathway 2 and 3) could occur. The cyclic nucleotide could enter pathway 2 or 3. In cells grown in ^{32}P, utilization of T_p by pathway 1 would lead to DNA-thymine residues flanked by label at the 5'-side. Pathway 2, 3 would lead to unlabelled pT residues in DNA.

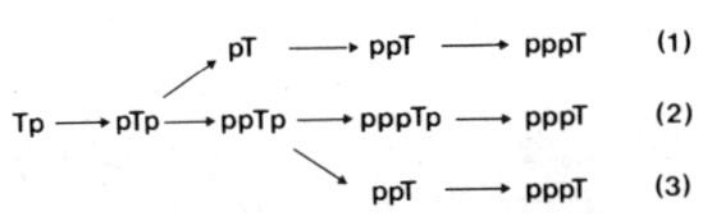

Fig. 12. Dephosphorylation of 3'-phosphate during thymidylate interconversions. In the conversion of Tp to pppT, dephosphorylation could occur at any one of the indicated steps.

way 1 (Fig. 11). The pathway from Tp to pppT involves a dephosphorylation step at the 3'-position. Three routes are possible (Fig. 12) and we are at present investigating which of these routes actually occurs. All routes involve the intermediate pTp.

A minimal inference from these observations is that yeast possesses enzymes which can catalyze nucleotide interconversions other than those known to exist in the conventional pathways of DNA precursor metabolism. It seems likely that this metabolic capacity makes possible some useful biological process(es) not presently understood. Tentative hypotheses include the following: (a) the capacity to channel Tp into the normal pathway may reflect a nucleotide salvage pathway, whereby DNA hydrolysis products can be recycled or (b) nucleotide interconversions generating unusual products (such as pTp) may have regulatory potential.

Since precursor metabolism has been extensively investigated in various organisms, it may, at first, seem surprising that analogous nucleotide interconversions have not been reported previously. However, meaningful detection of such occurrences in cells requires special conditions. Nucleotide-permeable strains, such as the yeast tup mutants, are required. In addition, the cells must be devoid of thymidine kinase, since otherwise dephosphorylation of incorporated nucleotides, such as Tp, could yield thymidine capable of phosphorylation to pT - a trivial result. It seems unlikely that these nucleotide interconversion capacities are unique to yeast. We suspect that in other eukaryotic cells mutants permeable to nucleotides and deficient in thymidine kinase would reveal similar findings.

In studies of the genetic consequences of precursor pool derangements in organisms, it is often tacitly assumed that our present knowledge of the components of precursor pathways is essentially complete, and explanations are based upon this assumption. Although the biological significance of the thymidylate interconversions which we have detected in yeast is unclear, these findings reveal inadequacies in our understanding, and suggest that one should be alert to the possibility of unorthodox nucleotide production.

8. CONCLUSIONS

Thymidylate stress in the yeast Saccharomyces cerevisiae ensues when the concentrations of thymidine nucleotides are either limiting or excessive. Cell growth inhibition, cytological aberrations and a gradual decline in cell viability are induced by this stress. In addition, a constellation of genetic changes occur, which are different in kind at the two extremes of thymidylate concentration.

Different genetic alterations are induced in the mitochondrial and nuclear genomes by thymidylate starvation. Mitochondrial point mutations, formation of respiratory-incompetent cytoplasmic petites and gradual erosion of mitochondrial DNA are observed during thymineless deprivation. The same conditions induce nuclear DNA strand breaks and greatly enhance the levels of mitotic recombination events; nuclear gene mutations are not induced. Thymidine nucleotide starvation sensitizes yeast cells to radiation (UV and X-rays).

Excessive concentrations of thymidylate induce mutations in nuclear genes and the manifestation of this mutagenesis requires the RAD6 gene, thought to be pivotal for error-prone DNA repair in yeast. Preliminary observations suggest that this mode of thymidylate stress may impede respiratory competence in yeast but mitochondrial genetic effects under these conditions have not been systematically investigated.

The genetic consequences which are engendered by thymidylate stress almost certainly involve aberrations in DNA replication and repair, but comprehensive explanations are lacking. Work in other systems strongly suggests that uracil entry into replicating DNA and its subsequent fate - retention or excision - may be centrally involved in thymineless phenomena. Definitive evidence for this mechanism has not so far emerged in these yeast studies. Similarly, the exact nature of the genetic lesions caused by excessive thymidine nucleotide concentrations and the mechanisms responsible for these changes require more intensive investigation. One perplexing feature is that the RAD6 gene product appears to be involved in the mechanisms which provoke genetic alterations at the two extremes of thymidylate stress, although these alterations differ radically.

Finally, it is important to recognize that in studies of pool derangements and their genetic consequences, components of nucleotide metabolism other than those presently known may be significant. We have detected unorthodox interconversions of thymidylate nucleotides in yeast. Similar reactions are likely to occur in other organisms and may have regulatory potential for nucleotide orchestration.

9. REFERENCES

1. D. Anderson, C. R. Richardson, and P. J. Davies, The genotoxic potential of bases and nucleosides, Mutat. Res., 91:265-272 (1981).
2. B. J. Barclay, B. A. Kunz, J. G. Little, and R. H. Haynes, Genetic and biochemical consequences of thymidylate stress, Can. J. Biochem., 60:172-194 (1982).
3. B. J. Barclay and J. G. Little, Selection of yeast auxotrophs by thymidylate starvation, J. Bacteriol., 132:1036-1037 (1977).
4. B. J. Barclay and J. G. Little, Genetic damage during thymidylate starvation in *Saccharomyces cerevisiae*, Mol. Gen. Genet., 160:33-40 (1978).
5. B. J. Barclay and J. G. Little, Mutation induction by deoxythymidine monophosphate: a model, Mol. Gen. Genet., 181:279-281 (1981).
6. L. F. Bertani, A. Haggmark, and P. Reichard, Enzymatic synthesis of deoxyribonucleotides. Formation and interconversion of deoxyuridine phosphates, J. Biol. Chem., 238:3407-3413 (1963).
7. M. J. Bessman, I. R. Lehman, J. Adler, S. B. Zimmerman, E. S. Simms, and A. Kornberg, Enzymatic synthesis of deoxyribonucleic acid; the incorporation of pyrimidine and purine analogs into deoxyribonucleic acid, Proc. Natl. Acad. Sci., U.S.A., 44:633-640 (1958).
8. L. Bisson and J. Thorner, Thymidine 5'-monophosphate-requiring mutants of *Saccharomyces cerevisiae* are deficient in thymidylate synthetase, J. Bacteriol., 132:44-50 (1977).
9. M. O. Bradley and N. A. Sharkey, Mutagenicity of thymidine to cultured Chinese hamster cells, Nature, 274:607-608 (1978).
10. M. Brendel and W. W. Fath, Isolation and characterization of mutants of *Saccharomyces cerevisiae* auxotrophic and conditionally auxotrophic for 5'-dTMP, Z. Naturforsch., 29c:733-738 (1974).

11. B. Cox and J. C. Game, Repair systems in Saccharomyces, Mutat. Res., 26:257-264 (1974).
12. W. W. Fath, M. Brendel, W. Laskowski, and E. Lehmann-Brauns, Economizing DNA-specific labelling by deoxythymidine-5'-monophosphate in Saccharomyces cerevisiae, Mol. Gen. Genet., 132: 335-345 (1974).
13. J. C. Game, Yeast cell cycle mutant cdc21 is a temperature-sensitive thymidylate auxotroph, Mol. Gen. Genet., 146:313-315 (1976).
14. M. Goulian, B. Bleile, and B. Y. Tseng, The effects of methotrexate of levels of dUTP in animal cells, J. Biol. Chem., 255: 10630-10637 (1980a).
15. M. Goulian, B. Bleile, and B. Y. Tseng, Methotrexate-induced misincorporation of uracil into DNA, Proc. Natl. Acad. Sci., U.S.A., 77:1956-1960 (1980b).
16. G. R. Greenberg and R. L. Somerville, Deoxyuridylate kinase activity and deoxyuridine triphosphatase in Escherichia coli, Proc. Natl. Acad. Sci., U.S.A., 71:249-257 (1962).
17. A. R. Grivell and J. F. Jackson, Thymidine kinase: evidence for its absence from Neurospora crassa and some other microorganisms, and the relevance of this to the specific labelling of deoxyribonucleic acid, J. Gen. Microbiol., 54:307-317 (1968).
18. S. Jannsen, I. Witte, and R. Megnet, Mutants for the specific labelling of DNA in Saccharomyces cerevisiae, Biochim. Biophys. Acta, 299:681-685 (1973).
19. B. A. Kunz, Genetic effects of deoxyribonucleotide pool imbalances, Environmental Mutagenesis, 4:695-725 (1982).
20. B. A. Kunz, B. J. Barclay, J. C. Game, J. G. Little, and R. H. Haynes, Induction of mitotic recombination in yeast by starvation for thymine nucleotides, Proc. Natl. Acad. Sci., U.S.A., 77:6057-6061 (1980).
21. B. A. Kunz and R. H. Haynes, DNA repair and the genetic effects of thymidylate stress in yeast, Mutat. Res., 93:353-375 (1982).
22. W. Laskowski and E. Lehmann-Brauns, Mutants of Saccharomyces able to grow after inhibition of thymidine phosphate synthesis, Mol. Gen. Genet., 125:275-277 (1973).
23. C. W. Lawrence and R. Christensen, UV mutagenesis in radiation sensitive strains of yeast, Genetics, 83:207-232 (1976).
24. T. Lindahl, An N-glycosidase from Escherichia coli that releases free uracil from DNA containing deaminated cytosine residues, Proc. Natl. Acad. Sci., U.S.A., 71:3649-3653 (1974).
25. J. G. Little and R. H. Haynes, DNA-specific labelling in yeast mutants, Genetics, 74:S161 (1973).
26. J. G. Little and R. H. Haynes, Isolation and characterization of yeast mutants auxotrophic for 2'-deoxythymidine-5'-monophosphate, Mol. Gen. Genet., 168:141-151 (1979).
27. M. Luzzati, Isolation and properties of a thymidylate-less mutant in Saccharomyces cerevisiae, Eur. J. Biochem., 56:533-538 (1975).

28. H. Mahler, Biogenetic autonomy of mitochondria, C.R.C. Crit. Rev. Biochem., 1:381-460 (1973).
29. E. Makino and N. Munakata, Deoxyuridine residues in DNA of thymine-requiring Bacillus subtilis strains with defective N-glycosidase activity for uracil-containing DNA, J. Bacteriol., 134:24-29 (1978).
30. P. B. Moens, B. J. Barclay, and J. G. Little, Nuclear morphology of yeast under thymidylate starvation, Chromosoma, 82:333-340 (1981).
31. G. A. O'Donovan and J. Neuhard, Pyrimidine metabolism in microorganisms, Bacteriol. Rev., 34:278-343 (1970).
32. O. M. Parkash, Thymidine teratogenesis and mutagenesis in Drosophila melanogaster, Experientia, 23:859-861 (1967).
33. D. Y. Thomas and D. Wilkie, Inhibition of mitochondrial synthesis in yeast by erythromycin: cytoplasmic and nuclear factors controlling resistance, Genet. Res., 11:33-41 (1968).
34. W. G. Verly and E. Rassart, Purification of Escherichia coli endonuclease specific for apurinic sites in DNA, J. Biol. Chem., 250:8214-8219 (1975).
35. R. B. Wickner, Mutants of Saccharomyces cerevisiae that incorporate deoxythymidine-5'-monophosphate into deoxyribonucleic acid in vivo, J. Bacteriol., 117:252-260 (1974).
36. D. Wilkie, G. Saunders, and A. W. Linnane, Inhibition of respiratory enzyme synthesis in yeast by chloramphenicol: relationship between chloramphenicol tolerance and resistance to other antibacterial antibiotics, Genet. Res., 10:105-109 (1967).
37. E. A. Wurtz, B. B. Sears, D. K. Rabert, H. S. Sheperd, N. W. Gillham, and J. E. Boynton, A specific increase in chloroplast gene mutations following growth of Chlamydomonas in 5-fluorodeoxyuridine, Mol. Gen. Genet., 170:235-242 (1979).

BIOLOGICAL CONSEQUENCES OF GUANINE STARVATION

Wolfgang Sadée and Binh Trong Nguyen

School of Pharmacy
University of California
San Francisco, California 94143

1. INTRODUCTION

One of the first effects of guanine starvation is a drastic inhibition of DNA synthesis. The DNA inhibition is associated with guanine, rather than adenine, nucleotide depletion [2]. Moreover, the DNA inhibition was found to be associated with GTP, rather than dGTP, depletion [1]. This was demonstrated with the use of mouse lymphoma S49 wild-type and HGPRT-ase negative mutants: DNA synthesis inhibition by mycophenolic acid (resulting in guanine nucleotide starvation by IMP dehydrogenase inhibition) in the mutant cells was not reversed by selectively establishing normal or elevated dGTP levels with exogenous deoxyguanosine (dGuo) (see metabolic pathways in Fig. 1). The same phenomenon was observed in two other cell lines (mouse L1210 and human CCRF-CEM, wild type and HGPRT-ase negative) [3], which suggests a more general applicability of the finding that DNA synthesis inhibition by guanine starvation is independent of the total dGTP cellular pool size.

There are two general hypotheses that can be proposed to account for the apparent relationship between DNA synthesis inhibition and guanine ribonucleotide depletion. 1) Guanine ribonucleotides may regulate some functions that are essential for DNA replication (e.g., GTP could represent a regulatory factor of proteins associated with DNA replication). 2) The incorporation of guanine into DNA could represent an extreme case of DNA precursor channeling, such that the obligatory pathway for the immediate DNA precursor is GDP → dGDP → dGTP → DNA, while dGTP generated from salvaged dGuo has no access to DNA.

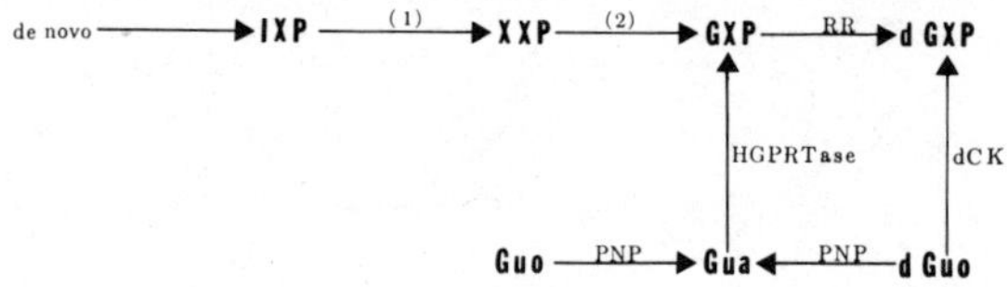

Fig. 1. Pathways of guanine metabolism. 1) IMP dehydrogenase; 2) guanylate synthetase. HGPRT-ase) hypoxanthine-guanine phosphoribosyl transferase; PNP) purine nucleoside phosphorylase; dCK) deoxycytidine kinase.

There is growing evidence that the DNA replicase system indeed exists of multiple enzymes that operate in concert to make distant DNA precursors directly available to DNA replication (for a general review of this topic, see [3]). However, no such pathway as yet was found to be exclusive for a single metabolic precursor channel; rather, deoxyribonucleotide triphosphates provided from the general pool in the cell can also be utilized for DNA replication, however, with a reduced efficiency. If the hypothesis is correct that guanine incorporation into DNA occurs exclusively or very predominantly via the ribonucleotide reductase precursor channel, then this finding would be significant on two accounts. 1) Guanine incorporation into DNA could serve as a primary tool to study DNA precursor channeling. 2) An important function of DNA-guanine precursor channeling would be to prevent DNA replication under unfavorable conditions (purine starvation).

2. APPROACH AND RATIONALE

In order to address the question of guanine precursor channeling, it is necessary to biochemically isolate the two pathways of possible guanine incorporation, i.e., GMP $\rightarrow$ GDP $\rightarrow$ dGDP $\rightarrow$ dGTP $\rightarrow$ DNA and dGMP $\rightarrow$ dGDP $\rightarrow$ dGTP $\rightarrow$ DNA. Figure 1 illustrates that the S49 wild type cells are not useful, since they also can convert dGuo to Gua, thereby feeding dGuo tracer into both metabolic pathways. Similarly, the HGPRT-ase negative mutant is not useful since GMP can not be labeled directly with Gua tracer. We therefore selected a nucleoside phosphorylase negative (PNP-negative) mutant of S49 cells [6]. We expected that in the presence of mycophenolic acid to block de novo guanine formation, ^{14}C-Gua and ^{3}H-dGuo tracers would selectively label GMP and dGMP, respectively (Fig. 2). Furthermore, by diluting the tracers with large concentrations of unlabeled Gua and dGuo, respectively, one would provide exclusive sources of guanine nucleotide with specific ^{3}H or ^{14}C activities that are largely determined by the externally added tracer.

The following questions then need to be addressed: 1) What is the concentration of Gua that fully rescues the cells from the myco-

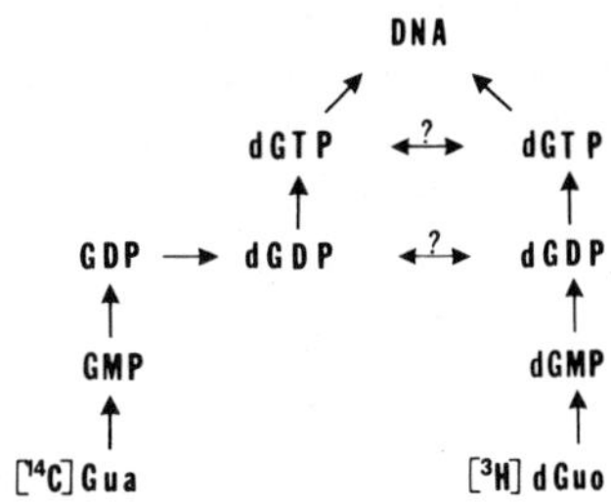

Fig. 2. Two potentially independent pathways of guanine incorporation into DNA.

phenolic acid block? 2) What is a suitable concentration of dGuo to significantly increase the intracellular dGTP pool? 3) What is the mechanism of toxicity of dGuo at higher doses? And finally, 4) which is the predominant precursor pathway for the incorporation of guanine into DNA?

3. EXPERIMENTAL

Mouse lymphoma S49 wild-type and PNP negative mutants [6] were grown as described previously [1, 4]. DNA synthesis was measured either with ^{3}H-Thd (80 Ci/mmole, 0.5 μCi/ml incubation medium, 10^5 cells/ml, 20 min exposure to the tracer 4 h after drug exposure) or with ^{14}C-Gua (51 mCi/mmole, ∿0.5 μCi/ml incubate) or ^{3}H-dGuo (18 Ci/mmole, 2 μCi/ml incubate). In the case of ^{14}C-Gua and ^{3}H-dGuo, tracer was added together with the drug and nucleosides at the beginning of the incubations. DNA incorporation was then measured at 2, 4, and 6 h, after digestion of RNA with 0.3 N KOH [4]. Complete DNA synthesis inhibition (^{3}H-Thd incorporation) by 0.5 μM mycophenolic acid was obtained within 2 h in wild-type cells and the PNP negative mutant. Cell growth was measured in a Coulter Counter 24 h after drug exposure.

4. RESULTS AND DISCUSSION

4.1. Mycophenolic Acid Toxicity and Guanine/Deoxyguanosine Rescue

Wild-type and PNP negative S49 cells were equally sensitive to mycophenolic acid (Fig. 3), which suggests that the mutation had no effect on the mode of mycophenolic acid toxicity. Moreover, a dose of 400 μM Gua was capable of fully preventing mycophenolic acid (1.2 μM) toxicity of both wild-type and mutant cells. In contrast, PNP negative mutants are more sensitive towards toxic effects of

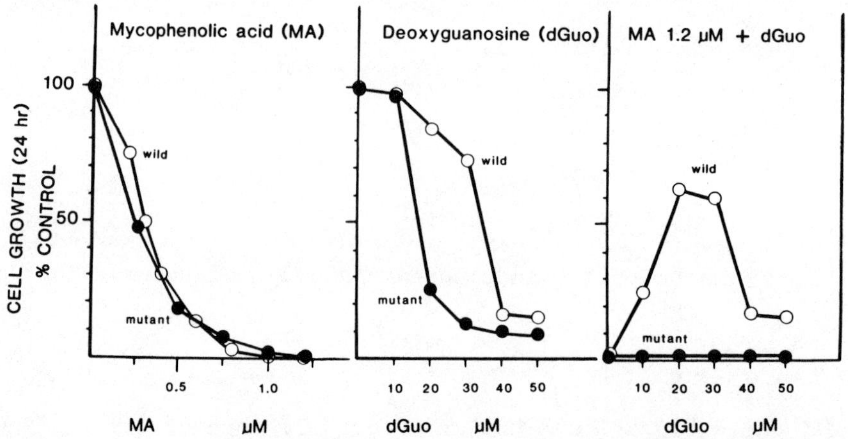

Fig. 3. Cell growth inhibition in wild-type and PNP negative mutants in the present of mycophenolic acid or deoxyguanosine added alone or in combination. There was no rescue from mycophenolic acid by dGuo in the mutant cells, although dGuo was toxic to these cells at higher levels (evidence for efficient cell uptake).

dGuo (Fig. 3, middle panel) than are wild-type cells. This is to be expected, since the absence of PNP allows more of the nucleoside to accumulate within the cell. A dramatic difference between wild-type and PNP negative cells is observed with the rescue from mycophenolic acid toxicity by dGuo. dGuo is capable of partially restoring cell growth in the wild-type cells that allow salvage via dGuo → Gua → GMP; the toxicity of dGuo at higher levels is then caused by dGuo → dGMP → dGDP → dGTP. However, dGuo is completely ineffective in reversing mycophenolic acid toxicity in the PNP negative mutant. This result is identical to that observed previously in HGPRT-ase negative mutants [1]; both mutant types are incapable of salvaging dGuo into GMP, but they do convert dGuo to dGMP.

We conclude that the PNP negative cells provide the desired opportunity to study DNA precursor uptake by two distinct pathways via GMP and dGMP, respectively. The results from the cell growth experiments shown above are compatible with both the guanine precursor channeling and the GTP regulatory role hypotheses.

In order to appreciate the subsequent experiments, it is first necessary to discuss the mechanism of dGuo cell toxicity. Ullman et al. [6] have provided strong evidence that dGuo is toxic because of the feedback inhibition by its metabolite dGTP of the ribonucleotide reductase catalyzed conversion of CDP to dCDP (and presumably also UDP to dUDP). Cell mutants that either failed to ac-

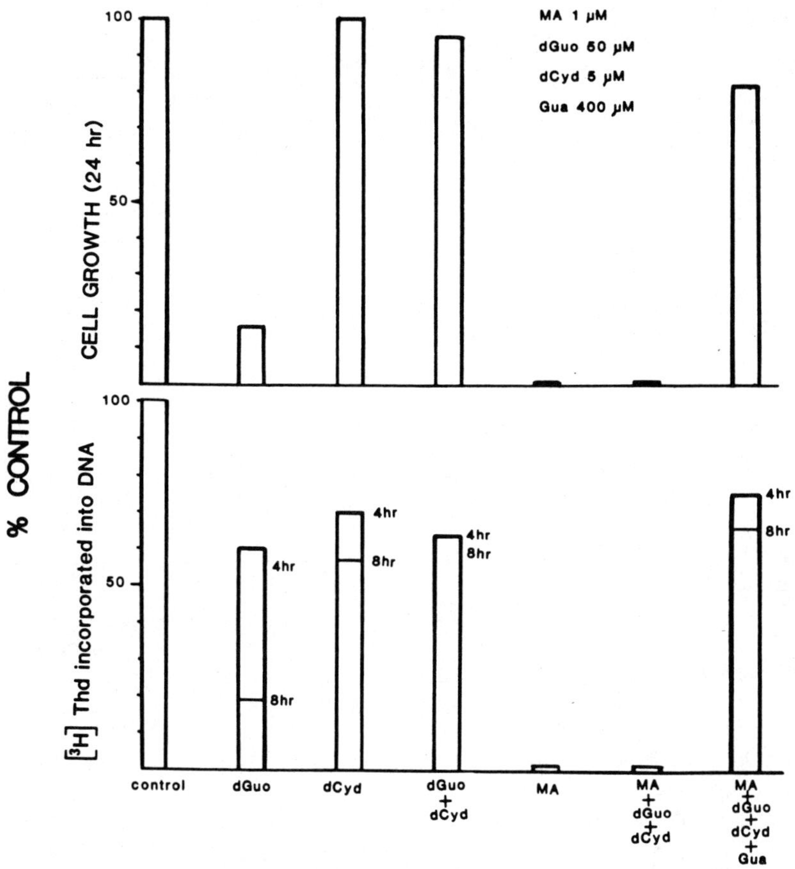

Fig. 4. Rescue by guanine (Gua) of DNA synthesis and cell growth in PNP negative mutants exposed to mycophenolic acid (MA), deoxyguanosine (dGuo), and deoxycytidine (dCyd). Deoxycytidine (dCyd) was added to reverse dGuo toxicity. The ability of dCyd to reverse dGuo toxicity confirms earlier reports [6] that dGuo is toxic by feedback inhibition of ribonucleotide reductase. Therefore, at the concentrations used in this study, dGuo should produce high total cellular dGTP pools. However, only guanine was capable of reversing mycophenolic acid toxicity.

tivate dGuo to dGTP or that expressed an altered ribonucleotide reductase activity insensitive to dGTP were resistant to the toxic effects of dGuo [5]. Moreover, we have observed that as little as 2.5 μM dCyd completely reverses dGuo (50 μM) toxicity. The dCyd effect was apparently not caused by an inhibition of dGuo uptake and incorporation into DNA, since there was no difference of ^{3}H-dGuo

TABLE 1. Incorporation of ^{14}C-Gua and ^{3}H-dGuo into DNA, in the Presence of Mycophenolic Acid (MA) and Various Combinations of Gua, dGuo, and dCyd. DNA Incorporation Rates were Corrected for the Specific Activity of ^{14}C-Gua and ^{3}H-dGuo in the Reaction. Results were repeated at least once

Expt. (#)	MA (μM)	Gua (μM)	dGuo	dCyd	^{14}C-Gua (μCi)	^{3}H-dGuo (μCi)	n Moles / 10^8 cells/30 min
1	-0-	-0-	50	2.5	-0-	2	1.35±0.50
2	1	-0-	50	2.5	-0-	2	2.55±0.37
3	1	400	50	2.5	-0-	2	1.94±0.37
4	-0-	400	50	2.5	0.5	-0-	9.34±3.12
5	1	400	50	2.5	0.5	-0-	9.60±0.65
6	1	400	-0-	-0-	0.5	-0-	9.51±0.71

uptake into acid insoluble cell material in the presence of either 2.5 or 10 μM dCyd. Rather, dCyd is an effective salvage precursor for dCTP and dTTP, thereby overcoming dGuo toxicity. These results suggest that dGuo, at least at concentrations of 50 μM, acts solely via dGTP and resultant ribonucleotide reductase inhibition. Assuming that the exclusive guanine precursor channeling hypothesis is correct, then one must postulate that dGTP generated from dGuo has not access to DNA synthesis, although it readily reaches the regulatory site of ribonucleotide reductase, even in toxic concentrations.

The subsequent DNA precursor uptake studies generally employed concentrations of 400 μM Gua and 50 μM dGuo. In order to overcome the intrinsic dGuo toxicity, between 2.5 to 10 μM dCyd was added, which allowed normal cell growth and DNA synthesis to occur in the presence of 50 μM dGuo.

4.2. DNA Precursor Uptake

Figure 4 summarizes results of studies on cell growth and ^{3}H-Thd uptake into DNA of PNP negative mutants in the presence of various combinations of mycophenolic acid, Gua, dGuo, and dCyd. It can be seen that dCyd (5 μM) fully prevents dGuo (50 μM) toxicity. However, dGuo (50 μM) and dCyd (5 μM) are ineffective against mycophenolic acid toxicity. Cell growth and DNA synthesis resume only in the presence of 400 μM Gua.

It is now of principal interest to determine which source of guanine is taken up into DNA under the condition of mycophenolic acid + Gua + dGuo + dCyd (last bar, Fig. 4). This question was addressed by adding ^{14}C-Gua and ^{3}H-dGuo, respectively, to the incubation mixture in order to determine the relative roles of Gua and dGuo in delivering the DNA precursor. Since massive concentrations of exogenous guanine sources are employed and since de novo guanine synthesis is blocked by mycophenolic acid, it is assumed that precursor dilution in both pathways (Gua → GMP → GDP → dGDP → dGTP → DNA and dGuo → dGMP → dGDP → dGTP → DNA) is negligible.

The results of this experiment are shown in Table 1. Clearly, dGuo contributed only ∿20% to guanine incorporation into DNA, while Gua provided by far the predominant share of guanine in DNA. Our previous estimate of DNA synthesis rates in S49 cells [4] are similar to the figure obtained with ^{14}C-guanine labeling of DNA. When ^{3}H-dGuo incorporation was measured in the absence of Gua, there was an increase of the uptake of ^{3}H-dGuo into DNA, suggesting that some competition occurs between Gua and dGuo. However, it is clear that dGuo was incapable of providing a sufficient dGTP-DNA precursor pool to allow DNA synthesis at a substantial rate.

It could be argued that the effect of mycophenolic acid on DNA synthesis is a result of protein synthesis inhibition by a lack of GTP. However, DNA synthesis is much more drastically affected by mycophenolic acid than are protein and RNA synthesis. Moreover, the onset of DNA synthesis inhibition is rather rapid (< 2 h) which argues against de novo protein synthesis as the responsible target. Finally, addition of actinomycin D in concentrations that prevented RNA, but not DNA synthesis, to cell incubations containing mycophenolic acid actually reversed the DNA synthesis inhibition presumably by making more guanine precursor available [1].

5. CONCLUSIONS

The results presented here provide suggestive evidence that the incorporation of guanine into DNA may represent an extreme case of DNA precursor channeling. Thus, dGuo may be ineffective in delivering dGTP into the immediate DNA precursor pool, although the dGTP from dGuo has ready access to the regulatory site of ribonucleotide reductase. Moreover, the functional guanine precursor pool may be GMP or GDP, the concentrations of which could assume an important role in regulating the replication of DNA synthesis. However, these studies must be expanded to include measurements of the specific activities of guanine ribo- and deoxyribonucleotide pools afer ^{14}C-Gua and ^{3}H-dGuo labeling. Careful measurements of the metabolic flux through these cellular nucleotide pools should reveal the extent of guanine-DNA precursor channeling and thereby its biological significance.

5. REFERENCES

1. M. B. Cohen, J. Maybaum, and W. Sadée, Guanine nucleotide depletion and toxicity in mouse T lymphoma (S-49) cells, J. Biol. Chem., 256:8713-8717 (1981).
2. M. B. Cohen and W. Sadée, Contributions of the depletions of guanine and adenine nucleotide to the toxicity of guanine starvation in the mouse T lymphoma cell line, Cancer Res., 43: 1587-1591 (1983).
3. C. K. Mathews, Enzymatic channeling of DNA precursors, Chapter in this volume (1983).
4. J. Maybaum, M. B. Cohen, and W. Sadée, In vivo rates of pyrimidine nucleotide metabolism in intact mouse T-lymphoma (S-49) cells treated with 5-fluorouracil, J. Biol. Chem., 25: 2126-2130 (1981).
5. B. T. Nguyen, M. B. Cohen, and W. Sadée, Guanine ribonucleotide depletion in mammalian cells: a target of purine antimetabolites, Cancer Chemother. Pharmacol., 11:117-119 (1983).
6. B. Ullman, L. J. Gudas, S. M. Clift, and D. W. Martin, Jr., Isolation and characterization of purine-nucleoside phosphorylase-deficient T-lymphoma cells and secondary mutants with altered ribonucleotide reductase: genetic model for immunodeficiency disease, Proc. Natl. Acad. Sci., U.S.A., 76:1074-1078 (1979).

THYMINELESS DEATH AND GENETIC EVENTS IN MAMMALIAN CELLS

Takeshi Seno,* Dai Ayusawa,*
Kimiko Shimizu,* Hideki Koyama,†
Keiichi Takeishi,*, and Tada-aki Hori‡

*Department of Immunology and Virology
Saitama Cancer Center Research Institute
Ina-machi, Saitama-Ken 362, Japan

†Department of Biochemistry
Cancer Institute
Toshima-ku, Tokyo 170, Japan

‡Division of Genetics
National Institute of Radiological Sciences
Chiba 280, Japan

1. SUMMARY

Thymidylate synthase-negative mutants of cultured mouse FM3A cells were immediately committed to cell death upon thymidine deprivation especially when the cells were synchronized in the S-phase. Thymidine deprivation induced single strand breaks in parental DNA strands, as measured by alkaline sucrose gradient sedimentation, giving rise to two peaks, one with large and the other with short fragments. Increase in the short DNA fragments paralleled that of thymineless death. Thymidine deprivation also accumulated double strand DNA fragments as determined by a method of neutral filter elution, and their extent paralleled that of cell death. Double-strand DNA eluted through the filter sedimented as a single peak both in a neutral and in an alkaline sucrose gradient that coincided with that of the above short DNA fragments. Therefore, the double strand breaks seemed to occur in some defined portions of the genome and in some specific manners in contrast to those induced by X-ray, which occurred rather randomly. Cycloheximide blocked thymineless death and accumulation of the double stranded DNA fragments in parallel. The double strand breaks induced by thymidine starvation were not repaired, but instead advanced on subsequent incubation of the cells in growth medium containing thymidine. Cytogenetically, thymidine deprivation induced chromosome aberrations such as chromatid breaks, chromatid interchanges, and chromosome fragmentation. Also, 5-bromodeoxyuridine deprivation induced sister chromatid exchange. Thymidylate stress also induced loss of a stably integrated human gene in mouse cells, possibly by DNA rearrangements, under the conditions where no point mutations were induced.

2. INTRODUCTION

It is well known that growing cells die if de novo synthesis of thymidylate is blocked. This block occurs in auxotrophic thy (thymidylate synthase deficient) mutants or in normal cells by use of drugs that block thymidylate metabolism. The phenomenon "thymineless death" (TLD) was first observed in *Escherichia coli* by Cohen and Barner in 1954 [14]. Since that time, thymine starvation has been studied in many organisms ranging from bacteriophages to mammalian cells.

As to a mechanism of TLD, Makino and Munakata proposed a possible link between excision of uracil-containing DNA and TLD by the finding that *B. subtilis* mutants defective in uracil-DNA glycosylase (ung^-) were slightly resistant to thymineless killing [32]. In mammalian cells, Goulian and co-workers have shown that thymidylate stress resulted in marked misincorporation of uracil into DNA [20] as a consequence of expansion of the intracellular level of dUTP [19].

Thymine starvation provokes physical alteration of DNA in bacteria in terms of, e.g., the transforming ability and viscosity [33], substrate capacity for DNA methylase [18], and template activity for RNA polymerase [31]. Alkaline sucrose gradient sedimentation analysis showed that breaks occurred in the parental DNA strands during thymine starvation [11, 15, 21, 34, 36]. Moreover, double strand breaks of DNA, detected by decrease in sedimentation velocity in a neutral sucrose gradient, have also been proposed by Yoshinaga for TLD in *E. coli* $15T^-$ carrying unknown prophages [41]. These findings imply that a certain type of DNA degradation could underlie the mechanism of TLD. At present, however, an exact molecular mechanism of TLD is still unknown in prokaryotes.

In mammalian cells, inhibition of thymidylate synthesis by drugs such as methotrexate and 5-fluoro-2'-deoxyuridine 5'-monophosphate derived from 5-fluorouracil is known to result in immediate cell death. Studies on the molecular events occurring in association with such cell death are important not only because thymidylate synthase is a target enzyme in cancer chemotherapy but also because limitation of thymidylate supply is known to induce mitotic recombination in yeast [29, 30] and mutations in the mouse FM3A thy mutants [28]. But owing to lack of genetically defined thymidine auxotrophs in mammalian cells, the molecular events occurring in TLD of mammalian cells have not been studied in detail.

Recently, we isolated absolute thymidylate synthase-negative mutants [5, 6] and temperature-sensitive enzyme mutants [1, 4] from mouse FM3A cells. Here we describe the molecular and genetic events induced in these mutants by thymidylate stress.

3. CYTOCIDAL EFFECT OF THYMIDINE DEPRIVATION ON MOUSE MUTANT CELLS DEFICIENT IN THYMIDYLATE SYNTHASE

3.1. S-Phase Cells as Target of Thymineless Death

Thymidine (dThd) auxotrophic mutant cells were immediately committed to cell death under conditions of dThd starvation, as shown in Fig. 1. Microscopic observation of the cells, which had been exposed to dThd-free medium for a short time and then grown in the presence of dThd, revealed that there were two morphologically different populations: One with a normal cell shape growing normally and the other enormously swollen and unable to divide on subsequent culture with dThd. To test whether this morphological heterogeneity was due to asynchrony of the cells, we synchronized the cells in the S-phase before dThd starvation by treatment with aphidicolin. This antibiotic is known to be a specific inhibitor of DNA polymerase α and its cytokinetic effect, which has only slight cytotoxicity, is reversible [26]. The portions of cells at various stages after aphidicolin-treatment were monitored by flow cytofluorometry. When asynchronously growing cells were treated with aphidicolin for 12 h, most of the cells were arrested in the S phase with a minor population in the G1 phase. On further incubation of the treated cells for 90 min in medium containing dThd, the minor cell population at the G1/S boundary was shown to enter the S-phase. By these treatments almost all the cells could be synchronized in the S-phase. The synchronized cells did not show the above morphological heterogeneity, consisting entirely of the swollen cell type unable to divide, and the survival curve of dThd-starved cells followed approximately first order kinetics (Fig. 1).

Taken together, these results show that the S-phase cells are more susceptible to TLD than cells in the other phases, indicating that DNA replicative capacity is necessary for TLD.

3.2. Effects of Antimetabolites on Thymineless Death

In prokaryotes, chloramphenicol [16] and sometimes actinomycin D are known to block TLD [37]. We thus examined the effects of several agents on TLD in mouse cells synchronized in the S-phase (Fig.

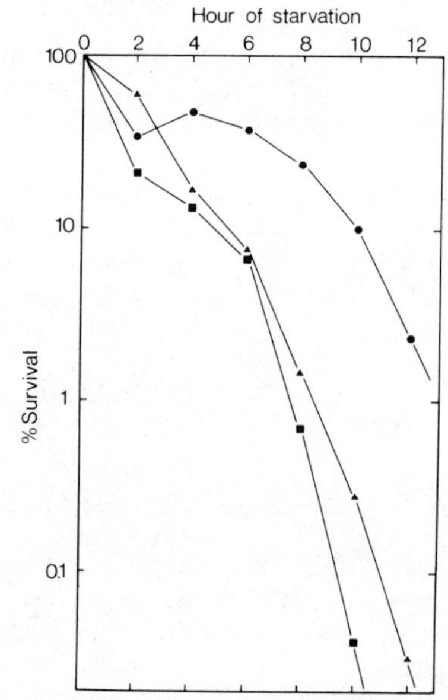

Fig. 1. Effect on survival curves of FSthy-21 cells of aphidicolin pretreatment before dThd starvation. Asynchronously growing cells (●), and cells pretreated with aphidicolin at 0.2 μg/ml (▲) or 1.0 μg/ml (■) were washed extensively and seeded on agar-medium containing no dThd. After incubation for the indicated time, the plates were overlaid with one hundredth volume of medium containing 1 mM dThd. The number of visible colonies was counted and expressed as a fraction of the control number. The control value was for cells starved for a few min. The plating efficiency of these control cells was nearly 100%. For isolation and characterization of the mutants, and other details, see Refs. 6, 9.

2). Cycloheximide clearly blocked TLD under conditions where overall protein synthesis was inhibited more than 95%. Because cycloheximide is known to inhibit the initiation and elongation of DNA replication in animal cells, TLD may be associated with DNA replication. However, it should be noted that amino acids deprivation also blocked TLD and that aphidicolin did not block TLD (data to be published). Alternative explanation of the effect of cycloheximide is that the drug blocked the synthesis of an inducible factor involved in TLD. Actinomycin D at the various concentrations tested did not block TLD but rather stimulated it slightly. Uracil, which is reported to inhibit the action of uracil-DNA glycosylase in vivo [20] and in vitro [33] in human cells, had no effect.

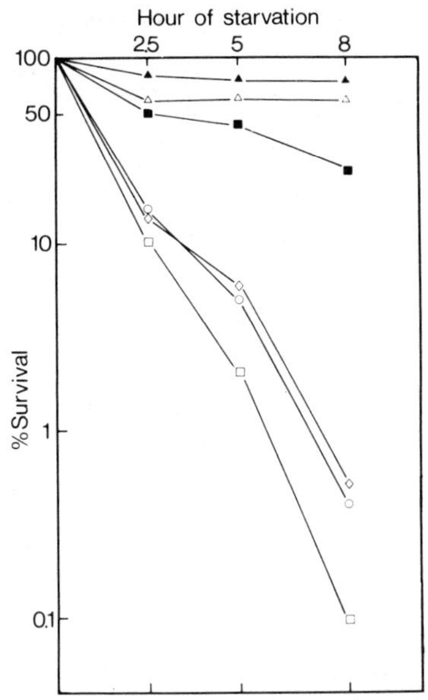

Fig. 2. Effects of cycloheximide and other agents on the survival curves of FS*thy*-21 cells during dThd starvation. The cells synchronized in the S-phase by aphidicolin (1 μg/ml) treatment for 12 h were cultured in the presence (closed symbols) or absence (open symbols) of 10 μM dThd for the time indicated with none (○), 3 μg/ml of cycloheximide (△, ▲), 1 μg/ml actinomycin D (□, ■), or 1 mM uracil (◊). The cells were washed and seeded on agar-medium containing dThd to determine viable cells. For other details, see legend to Fig. 1.

In our mouse *thy* mutant cells the extent of uracil incorporation was only slightly above a background level in contrast to the significant levels as reported for bacteria [38] and human lymphocytes cells [20].

4. INDUCTION OF DNA STRAND BREAKS ASSOCIATED WITH THYMINELESS DEATH

4.1. Analysis of Single Strand DNA Breaks

First, asynchronously growing FS*thy*-21 cells in which DNA has been labeled with [^{14}C]dThd for 24 h followed by 12 h chase were starved of dThd for the indicated times and induction of single strand breaks in the DNA was monitored by sedimentation in an alkaline sucrose gradient. In normally growing FS*thy*-21 cells, as shown

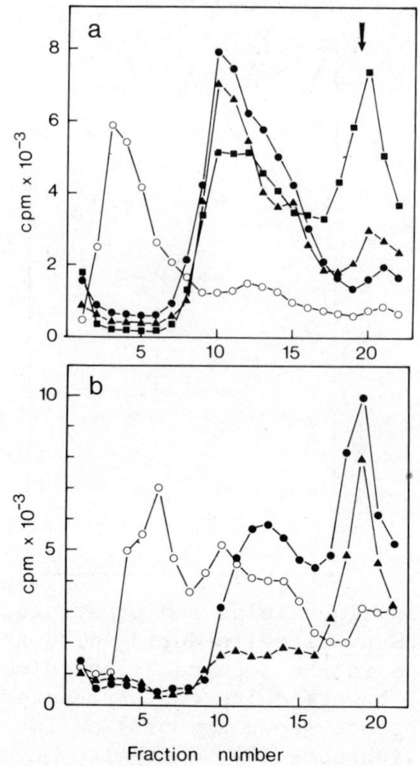

Fig. 3. Determination of single strand breaks of parental DNA strands in FS<u>thy</u>-21 cells during dThd starvation by sedimentation in an alkaline sucrose gradient. Asynchronous cells (a) or aphidicolin-synchronized cells (b) in which parental DNA strands had been labeled by incubation for 24 h with 10 μM [^{14}C]dThd (50 mCi/mmol, New England Nuclear) and chased for more than 12 h in the absence of labeled dThd. They were then starved of dThd for 0 h (○), 4 h (●), 8 h (▲), and 12 h (■). They were then collected, chilled in ice and suspended in cold buffer [10 mM Tris-HCl (pH 7.4), 0.15 M NaCl and 1 mM EDTA] after being washed once with the same buffer. A sample of 0.5 ml of cell suspension (about 10^6 cells) was layered on the top of 0.5 ml of lysing solution [0.45 M NaOH, 0.55 M NaCl, 10 mM EDTA, and 1% sodium N-lauroylsarcosinate] which in turn had been layered on top of 10 ml of a linear gradient of 5-20% sucrose in buffer [0.1 M NaOH, 0.9 M NaCl, and 10 mM EDTA] over 1.0 ml of a cusion of 60% sucrose in 0.1 M NaOH as described elsewhere [3, 35]. The samples were stood overnight at 2°C, and then centrifuged at 17,000 rpm for 3 h at 2°C in a Hitachi RPS-40T-2 rotor. Then 0.5 ml fractions were collected from the bottom of the tube. To each fraction 300 μg of carrier DNA

in Fig. 3a, most of the labeled DNA sedimented to the bottom of the gradient as nearly 400 S chromosome-size DNA [35]. After dThd starvation for 4 h, almost all of this high molecular weight DNA was broken down to sizes of 200-300 S, the postulated size of the intermediates of chromosome replication in human cells [35]. On further starvation of dThd, this major peak of DNA did not shift toward the top but smaller DNA fragments, close to the postulated size of the replication unit in mammalian cells [23], emerged gradually with increase in the time of starvation. Since production of DNA fragments of this smaller size seemed to be correlated with TLD, we carried out the same experiments using cells synchronized in the S-phase which were shown to be targets of TLD (see Fig. 1). As expected, accumulation of the smaller DNA fragments were more obvious in association with more rapid cell death as compared to those with asynchronous cells (Fig. 3b).

The effect of cycloheximide on the induction of single strand breaks to DNA in asynchronized cells was tested. The drug rescued more than 60% of the cells from TLD during dThd starvation for 12 h (see Fig. 2), while the drug suppressed the production of smaller DNA fragments but not the larger DNA (data not shown).

4.2. Analysis of Double Strand DNA Breaks

We employed the method of neutral filter elution for determining double strand breaks of DNA, because this technique gave reproducible results. We could not employ sedimentation analysis in neutral sucrose gradient, because it did not give reproducible results in any conditions we tried. Figure 4 shows the elution profiles of double strand DNA fragments produced upon dThd starvation. Asynchronous FS<u>thy</u>-21 cells prelabeled with [^{14}C]dThd as described in the previous section were starved of dThd for the time indicated, placed on a filter and eluted as described in the legend to Fig. 4. In the figure, the amount of DNA remaining on the filter was plotted against the elution time. The extent of DNA eluted were found to be proportional to the time of dThd deprivation. The DNA elution curves of dThd-starved cells, however, leveled off within a few hours. This is in marked contrast to results on cells irradiated with X-ray: X-ray irradation causes random double strand breaks on DNA and thus the elution curves followed typical first-

---

and cold trichloroacetic acid at a final concentration of 5% were added. The resulting precipitates were transferred to glass filters (Whatman, CF/C). The filters were washed with 5% trichloroacetic acid and then ethanol, and DNA was determined as radioactivity. For other details, see Ref. 9. Sedimentation was from right to left. The arrow indicates a peak position of T4 phage DNA (Mr, 0.5 × 10^8).

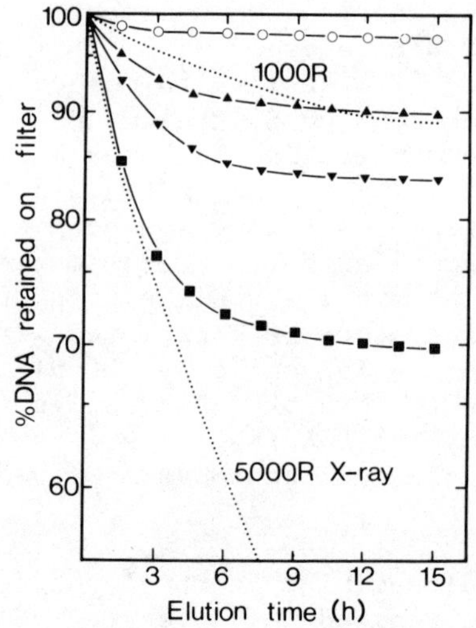

Fig. 4. DNA double strand breaks induced by dThd deprivation in FS<u>thy</u>-21 cells as detected by neutral filter elution. Cells labeled with [^{14}C]dThd as described in Fig. 3 caption were starved of dThd for 0 (○), 4 (▲), 8 (▼), or 12 h (■). As described by Bradley and Kohn [12], 1-5 × 10^5 cells starved of thymidine were placed on a Unipore filter (Bio-Rad Laboratories, 25 mm in diameter; pore size 0.8 μm; polycarbonate filter) and tubing connected to a pump was attached to the outlet of the filter. Then the cells were lysed with 3 ml of a solution [0.05 M Tris base, 0.05 M glycine, 0.025 M EDTA, and 2% sodium lauroyl sulfate, adjusted to pH 9.6] containing 1 mg/ml of proteinase K (Merck). The filter was stored for 45 min at room temperature, and then pumping was started to obtain a flow rate of 2 ml per hour. One hour later, 30 ml of the above solution but without proteinase was layered on top of the remaining lysing solution and 3 ml fraction (corresponding to 90 min elution) were collected. DNA's eluted from the filter and remaining on the filter were each determined. The upper and lower dotted lines represent DNA double strand breaks in cells irradiated with X-ray at doses of 1 krad and 5 krad, respectively. The cells were irradiated on ice to prevent repair reaction [3]. For other details, see Ref. 9.

order kinetics and the slopes were dependent on the doses, indicating the reliability of this elution technique [12].

The effect of cycloheximide on the production of double strand DNA fragments in dThd-starved cells was then tested because this compound blocked TLD (see Fig. 2). As expected, cycloheximide pre-

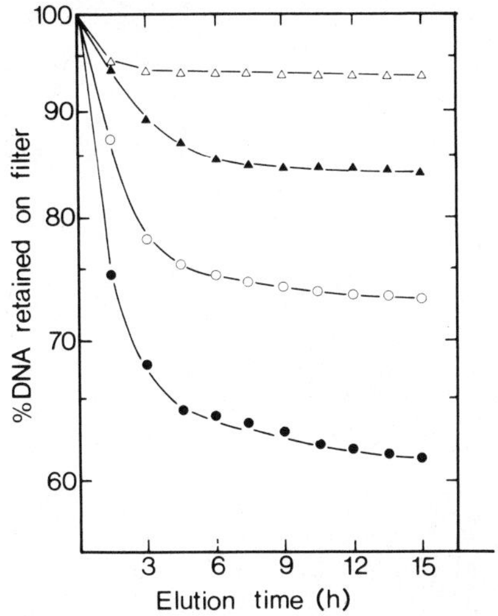

Fig. 5. Effect of cycloheximide on the induction of DNA double strand breaks in FS<u>thy</u>-21 cells by dThd deprivation and subsequent changes during further incubation in the presence of dThd. The cells prelabeled as described in Fig. 3 were starved of dThd for 12 h in the presence (○, ●) or absence (△, ▲) of 3 μg/ml of cycloheximide. Aliquots of these starved cells were washed and allowed to recover for 4 h in medium containing dThd (closed symbols). The samples were analyzed by the method of neutral filter elution as for Fig. 4. For other details, see Ref. 9.

vented production of double strand DNA fragments in cells deprived of dThd for 12 h (Fig. 5).

Next, portions of cells starved of dThd for 12 h in the presence or absence of cycloheximide were further incubated for 4 h in the fresh medium containing dThd and then the fate of the double strand DNA fragments was examined. The production of the fragments advanced rather than being reversed (Fig. 5). In this context, it is noteworthy that double strand breaks of DNA induced by X-ray irradiation are largely rejoined immediately in prokaryotes and animal cells [12, 39, 40], including mouse FM3A cells (data not shown).

4.3. Nature of Double Strand DNA Fragments Eluted from the Neutral Filtration

Since the exten of neutral elution (Figs. 4 and 5) correlated well with those of TLD (Figs. 1 and 2), it was tempting to characterize the nature of DNA fragments that were eluted through the filter. It has been demonstrated that DNAs eluted through the filter

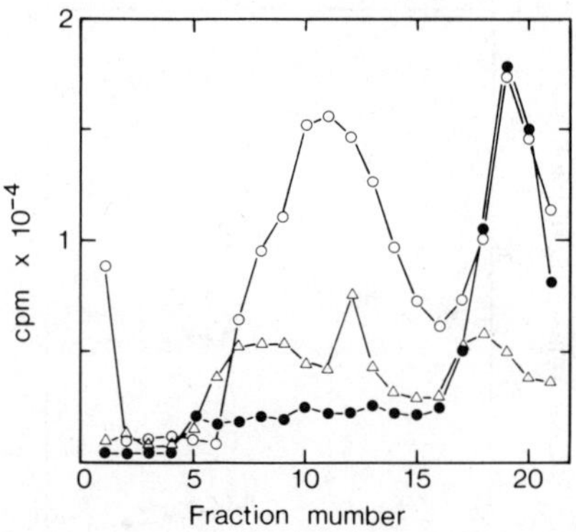

Fig. 6. Sedimentation in an alkaline sucrose gradient of the DNA fragments eluted through the neutral filter with cells starved of thymidine. FS*thy*-21 cells labeled and starved of thymidine for 12 h were eluted in triplicates through filters as described in Fig. 4. Eluents of first 6 h were pooled and combined. DNA was precipitated with two volume of ethanol at -40°C after addition of *E. coli* tRNA at a final concentration of 10 μg/ml, collected by centrifugation at 15,000 × g for 15 min, and dissolved in 1 ml of a buffer [1 mM Tris-HCl (pH 7.9) and 1 mM EDTA]. An aliquot of 0.5 ml was subjected to sedimentation analysis in an alkaline sucrose gradient (●) as described in Fig. 3. As controls, the starved cells were directly subjected to an alkaline sucrose gradient sedimentation (○), or normally growing cells in the presence of dThd were subjected to neutral filter elution steps except that a filter was omitted and eluted DNA was processed similarly as above to see to what extent the DNAs were artificially degraded in the elution steps (△). For other details, see Ref. 9.

in the present method are double stranded in nature [12]. DNAs from the 12 h-starved cells were eluted through the filter and eluents of first 6 h were pooled and combined. A portion of the sample was analyzed in an alkaline sucrose gradient. As shown in Fig. 6, the DNAs (closed circles) sedimented as a single peak which coincided with that of the smaller size DNA fragments (see Fig. 3) observed when the starved cells were directly subjected to an alkaline sucrose gradient sedimentation (open circles in Fig. 6). The results showed nicely that the smaller size DNA in Fig. 3 represents quantitatively the fraction of DNA eluted through the filter at each time point of dThd starvation (compare Fig. 4 with Figs. 3 and 6).

Next, a portion of the same DNA sample eluted through the filter was sedimented in a neutral sucrose gradient in order to see whether

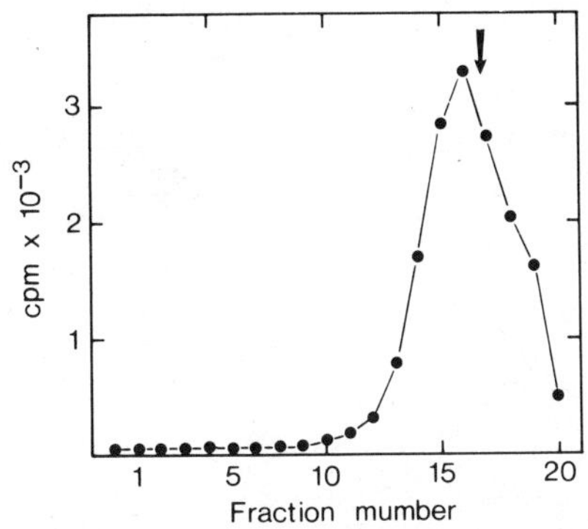

Fig. 7. A neutral sucrose gradient sedimentation of DNA fragments eluted through the neutral filter with FS<u>thy</u>-21 cells starved of dThd. The neutral filter eluents obtained as described in the caption to Fig. 6 were incubated for 6 h in the presence of 1% sodium lauroyl sulfate and 1 mg/ml of proteinase K and overlayed on top of 10 ml of a 5-20% sucrose gradient in a buffer [25 mM Tris-HCl (pH 7.5), 1 M NaCl, 2 mM EDTA, and 0.015% N-lauroyl sarcosinate] over 1 ml of a cushion of 60% sucrose in the same buffer. The gradient was centrifuged at 8500 rpm for 12 h in a Hitachi RPS-40T-2 rotor. Then 0.5 ml fractions were collected from the bottom of the tube and counted for their radioactivities in the aqueous scintillation cocktail. The arrow indicates sedimentation of T4 phage DNA as an internal marker (Mr. 1.1×10^8) which was extracted carefully from the phage with phenol. Our extraction procedure usually yields DNA fragments as long as 150 kilobase pairs. For other details, see Ref. 9.

or not the double strand DNA fragments contained nicks. If they do, molecular size of the fragments in a neutral sucrose should be much longer than that in an alkaline sucrose. In both sedimentations, phage T4 DNA was used as an internal marker. As shown in Fig. 7, the DNA fragments sedimented in a neutral sucrose gradient as a single peak and appeared to be a little bigger than T4 DNA, whereas they sedimented as a peak slightly smaller than T4 DNA in an alkaline sucrose gradient (see Fig. 3). In the alkaline sedimentation, T4 phages were applied to a gradient without prior DNA extraction, whereas DNA was extracted from the phage particles with phenol before the neutral sedimentation. Since T4 DNA (around 200 kilobase pairs) is too large to be extracted intact, it is probable that majority of the phage DNA sedimented as partially cleaved genome. Therefore, we tend to conclude that the double strand DNA fragments obtained during TLD by the neutral filter elution were of rather a unique size and might not possess a considerable number of single

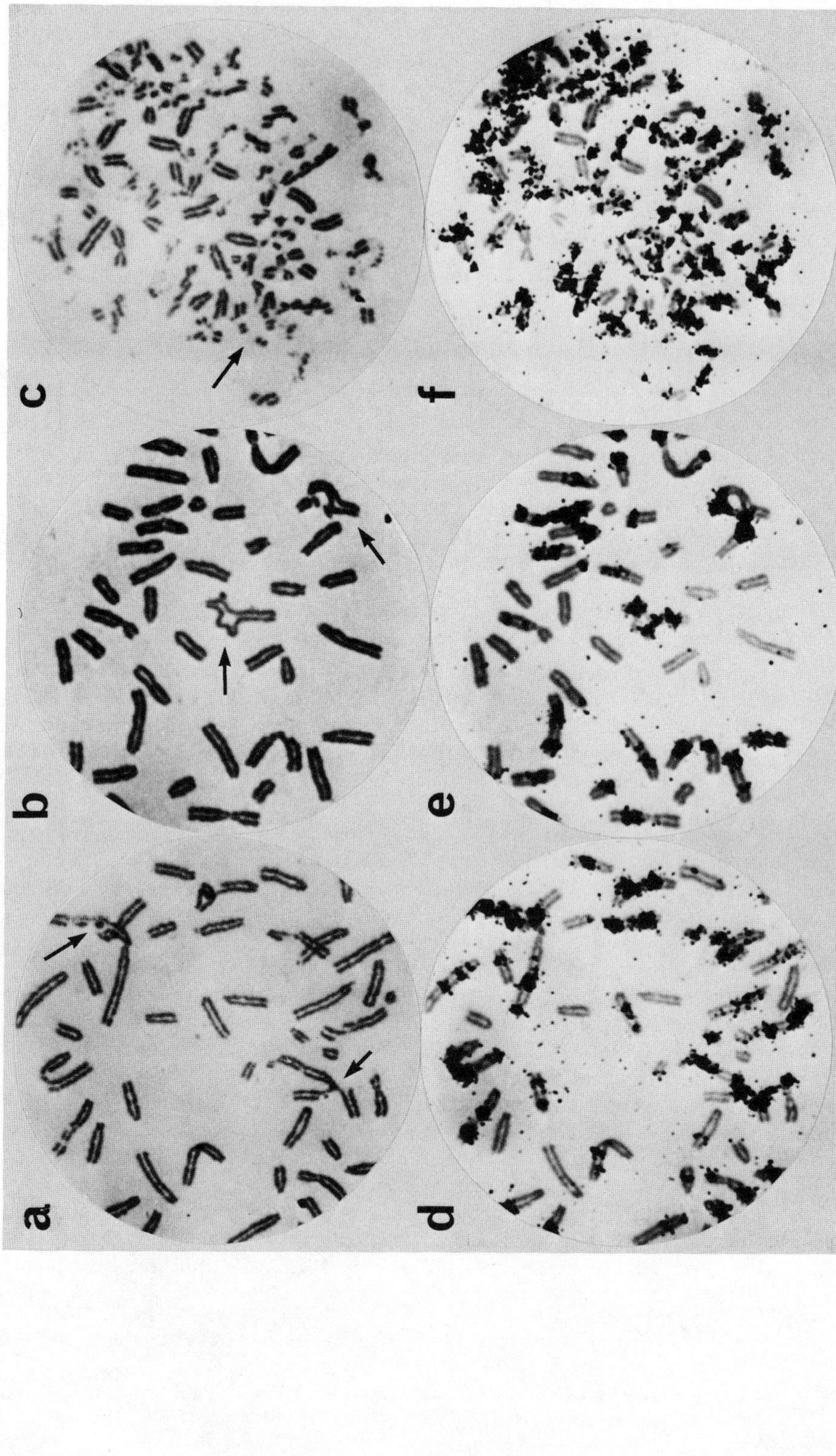
a
b
c
d
e
f

Fig. 8. Examples of chromosome aberrations induced in FSthy-21 by complete thymidine starvation. Cells starved for 12 h were recovered by dThd (10 μM) and grown for 6 h. a, b, and c represent metaphases containing chromatid breaks, chromatid interchanges and fragmentations, forming double minutes-like segments, respectively (arrows); d, e, and f are autoradiographs of the same metaphase spread of a, b, and c, respectively. For other details, see Ref. 25.

strand nicks, if any. These results are consistent with, but do not prove that double strand breaks of DNA are present in the dThd-starved cells.

5. CHROMOSOME INSTABILITY INDUCED BY THYMIDYLATE STRESS IN MOUSE MUTANT CELLS DEFICIENT IN THYMIDYLATE SYNTHASE

5.1. Mitotic Inhibition of dThd Starvation

Effect of thymidine starvation on mitotic activity was examined with exponentially growing FS*thy*-21 cells which had been prelabeled with ^{3}H-dThd for 10 min prior to the complete starvation of thymidine. Metaphase index (percent metaphase) was determined on the preparations made after 1 h-colchicine treatment. The majority of cells labeled with ^{3}H-dThd in the S-phase failed to enter mitosis during the starvation.

When dThd (10 μM) was added to the culture after 12 h of thymidine starvation, the mitotic activity recovered and resulted in a burst of mitosis 5-8 h later. In addition, autoradiographic analysis indicated that in such burst of mitosis, labeled mitoses appeared first and nonlabeled mitoses followed (data not shown). Taken together, these results suggest that cells synthesizing DNA were arrested in the S-phase, while the other cells were accumulated at the G_1/S boundary during thymidine starvation.

5.2. Chromosome Aberrations at Sites of DNA Replication

Chromosome damaging effect of thymidine starvation was examined on the metaphase spreads prepared by 2 h-colchicine treatment of the cells which had been starved completely for 12 h and then recovered by an addition of dThd (10 μM). Metaphases recovered by dThd addition contained extensive chromosome damages, exhibiting a variety of chromosome aberrations such as chromatid breaks, chromatid interchanges and chromosome pulverization, forming double minutes-like segments, as shown in Fig. 8a, b, c, respectively. The results are consistent with the biochemical results in the earlier section that double strand breaks of DNA were induced in the dThd-starved cells. The extent of chromosome damage seemed to be different in the cells which are arrested at different stages in the S-phase starvation. Cells arrested at early S-phase exhibited chromosome aberrations in less extent, whereas the majority of cells arrested at middle or late S-phase had more extensive chromosome damages.

From an inspection of the autoradiographs shown in Fig. 8d, e, and f, it seems likely that the sensitive sites to chromosome damaging effect of dThd starvation are localized at the chromosomal sites which had replicated just before the initiation of starvation. The majority of break points and interchange sites were labeled,

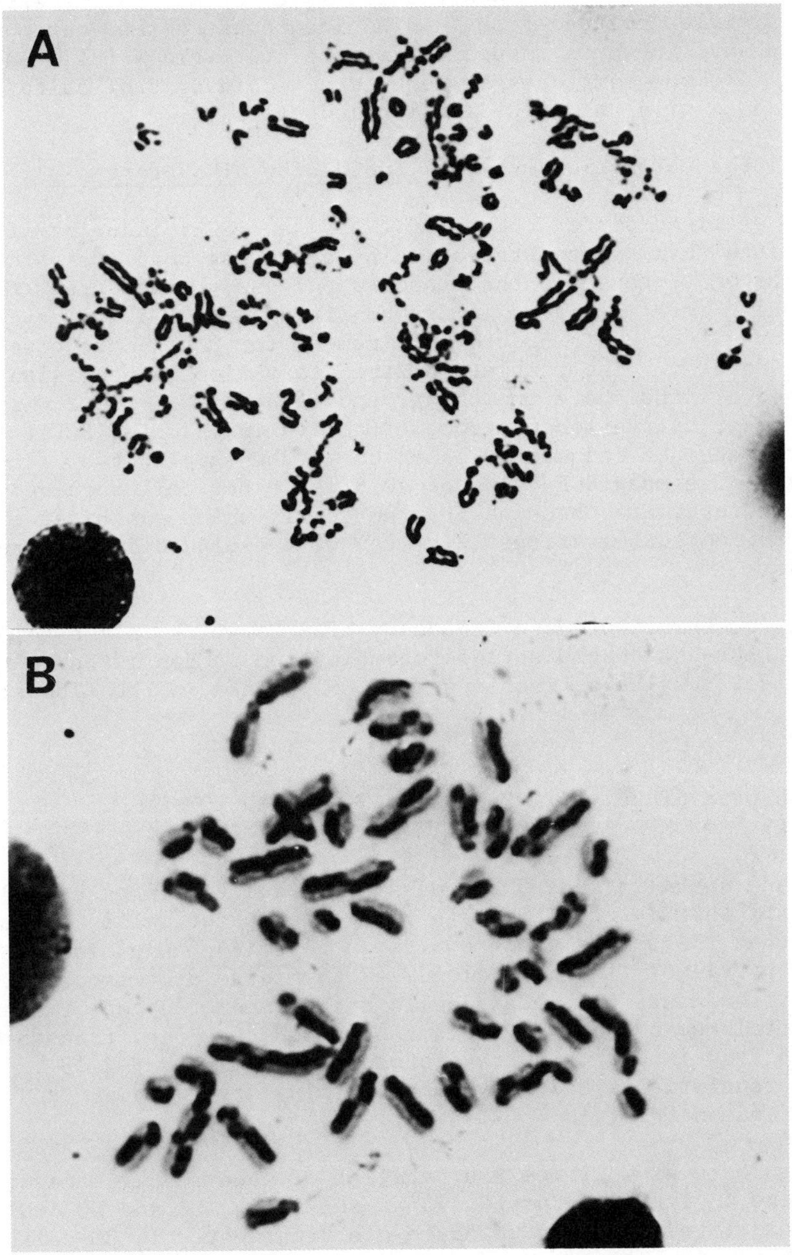

Fig. 9. Sister chromatid exchanges (B) induced in FS*thy*-21 cells by complete BrdUrd starvation for 6 h as described in the text. Under this condition the cell population contained a fraction which exhibited extensive chromosome aberrations (A). For other details, see Ref. 25.

whereas intact whole or parts of chromosomes were not labeled. It must be added here that these results were not due to the effect of the β-particles emitted from ^{3}H-dThd incorporated into chromosomal DNA which are known to induce chromosome aberrations [24], because no sign of chromosome damages were detected in control culture in the presence of dThd of ^{3}H-dThd-labeled cells.

5.3. Sister Chromatid Exchanges Induced by dThd Starvation

We examined sister chromatid exchanges (SCE) inducing effect of complete thymidylate starvation by depriving completely for BrdUrd at the second S-phase in the standard SCE assay procedure. FS*thy*-21 cells which were incubated in BrdUrd (10 μM) for 1.5 cell cycle were starved for their BrdUrd requirement for 6 h and reversed by BrdUrd addition. The complete BrdUrd starvation induced significant increases of SCEs above the background value of 8.5 ± 0.2 per cell and a longer starvation was more effective in inducing SCEs. Figure 9b represents SCE inductions observed in the sample at 6 h after the reversal. The mean SCE value of 40.8 ± 0.9 per cell obtained at 6 h was underestimated because aberrant cells which exhibited extensive chromosome aberrations (Fig. 9a) were excluded in the present analysis.

Again, this increased frequency of SCEs which is a cytological evidence for DNA double strand recombination is consistent with the biochemical results of double strand DNA breaks in the dThd-starved cells.

6. LOSS OF A STABLY INTEGRATED HUMAN GENE IN MOUSE CELLS PROVOKED BY THYMIDYLATE STRESS

Since double strand DNA breaks and chromatid interchanges are induced by thymidylate stress in our mouse mutant cells, one might expect that the stress also results in genetic changes in such cells. We have previously isolated clones of thymidine prototrophic transformant from mouse *thy* mutant cells by DNA mediated gene transfer using total human DNA. We further obtained secondary transformants with DNA from the primary transformant FS*thy*-11/*thy*H$^+$-2 [8]. The stable transformants had only human thymidylate synthase and segments of human DNA [8].

Figure 10 exemplifies the behavior of such stably integrated human gene in the secondary transformant lines, CO and D3, following thymidylate stress by methotrexate treatment. In CO cells thymidylate stress induced loss of the human gene as a function of time of methotrexate treatment when measured by the appearance of thymidine auxotrophic colonies [6]. In contrast, D3 cells did not give rise to thymidine auxotrophs under the identical conditions. In both transformant lines mutations as measured by ouabain re-

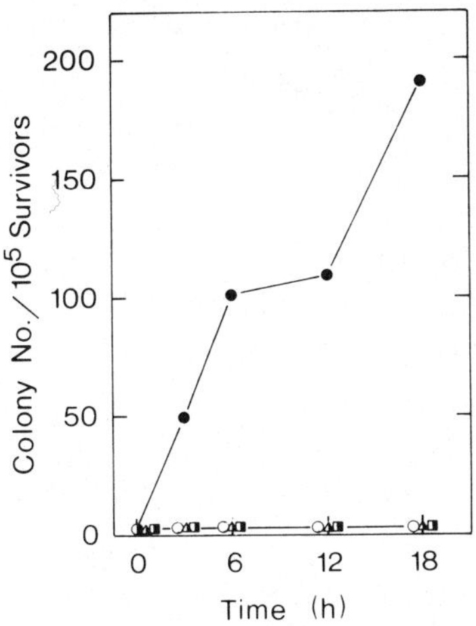

Fig. 10. Induction of the loss of human thy gene in mouse cell transformants by thymidylate stress. The secondary transformant lines, CO (closed symbols) and D3 (open symbols), as described in the text, were subjected to thymidylate stress by incubation in the presence of 10 μM methotrexate and 20 μM hypoxanthine. After each time indicated, portions of the cells were withdrawn and recovered by incubation in medium containing 1 μM 5-methyltetrahydrofolate, 20 μM hypoxanthine and 10 μM dThd, and then in medium containing 10 μM dThd for a total of 5 days. They were then assayed on soft agar plate for colonies resistant to 1 μM 6-thioguanine (triangles), resistant to 2 mM ouabain (squares). Colonies auxotrophic for thymidine (circles) were assayed for growth on the plate containing 0.3 μM methotrexate, 10 μM 5-methyltetrahydrofolate, and 10 μM dThd as described previously [6]. Addition of thymidine together with methotrexate in the above thymidylate starving incubation as control abolished the appearance of thymidine auxotrophs in CO line. For further details, see Ref. 9.

sistance and 6-thioguanine resistance were not induced under thymidylate stress. As a control, Fig. 11 shows that mutations were induced in CO cells by BrdUrd [7]. These results are consistent with the results reported by Kunz et al. [29] that thymidylate stress in yeast is recombinagenic, but not mutagenic. It should also be mentioned that thymidylate stress induced adenine phosphoribosyltransferase-negative segregants from the mouse cells presumably heterozygous for the alleles, but not 6-thioguanine or ouabain resistant clones [7].

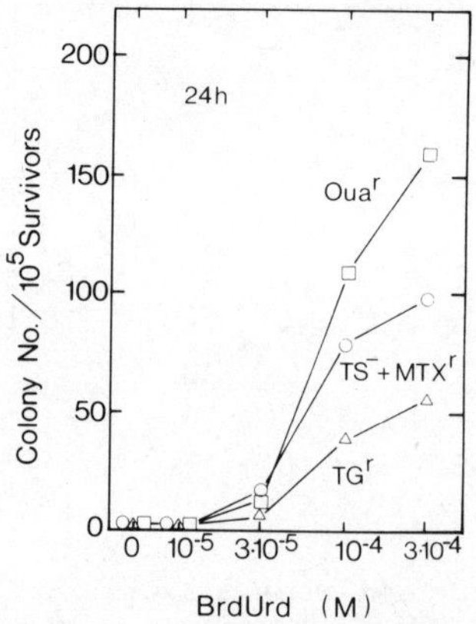

Fig. 11. Mutation induction by BrdUrd in clone CO cells. The secondary transformant line CO cells were cultured for 24 h in medium containing BrdUrd as indicated. Cells were recovered in medium containing dThd for 5 days before plating for colonies resistant to ouabain (□), 6-thioguanine (△), and methotrexate in the presence of 5-methyltetrahydrofolate and dThd (○) which should allow both thymidine auxotrophic and methotrexate-resistant cells to grow (see Ref. 6).

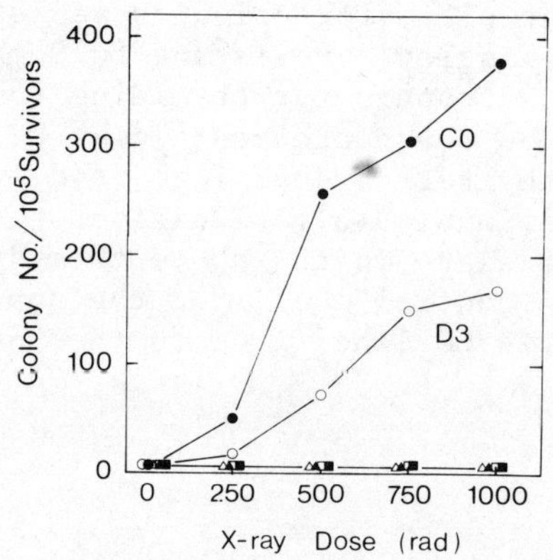

Fig. 12. Induction of the loss of human thy gene in mouse cell transformant by X-ray irradiation. Experiments identical to Fig. 10 were performed except that thymidylate stress was replaced by X-ray irradiation. For cell lines and figure symbols, see captions to Fig. 10. After irradiations, the cells were recovered by incubation for 5 days in medium containing dThd, and then assayed on the selection plates as specified in Fig. 10.

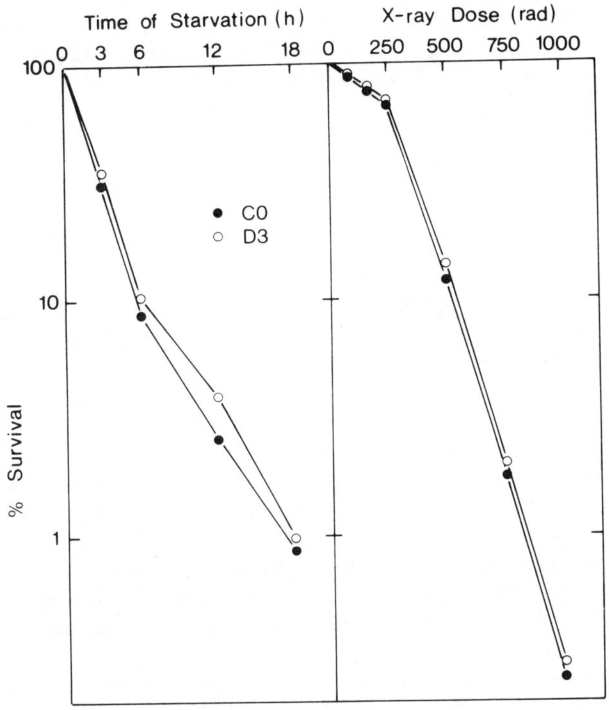

Fig. 13. Effects of thymidylate stress and X-ray irradiations on the survivals of CO and D3 cell lines. The secondary transformant lines, CO (●) and D3 (○), were subjected to thymidylate stress or X-ray irradiations and assayed for survival colonies on agar plate. Conditions for thymidylate stress and X-ray irradiations were identical to those in Fig. 10 and in Fig. 12, respectively.

Although it is not clear at the present moment why the behavior of the human gene in D3 line was different from that in CO line, the results in Fig. 12 may give a clue to this subject: X-ray did induce loss of the human gene in both transformant lines. The survival curves of the two lines as a function of methotrexate treatment and that of X-ray doses are shown for comparison in Fig. 13. The results in Figs. 4 and 5 suggest that the DNA breaks in the dThd starved cells occurred in a specific fashion and localized in limited regions of nuclear genome in contrast to the random DNA breaks in X-ray irradiated cells. Figure 8 also shows that the breaks are not randomly induced in the chromosomes of the starved cells. Probably, the human gene in D3 line may have been integrated in an early replicating region of genomes which were shown to be less susceptible to thymidylate stress (see the preceding section), whereas the gene in CO line may have located in the more susceptible middle or late replicating genomes. It is also probable that more copies of the human gene are present in D3 line than in CO line, since transferred gene is usually amplified in tandem before integration in the nuclear chromosomes.

7. CONCLUSIONS

Thymineless death occurs in mammalian cells as it does in bacteria. Cells in the S-phase are the target of thymineless death, in which double strand breaks of DNA correlate with the cell death. The DNA breaks occur at sites of DNA replication resulting in the production of DNA fragments approximately of a replicon size and fragmentations in the metaphase chromosomes. Cycloheximide treatment or amino acids deprivation blocks both thymineless death and the DNA strand breaks. Moreover, the DNA breaks are not repaired, but rather advanced following incubation of the starved cells in the presence of thymidine. This is in contrast to the repairable DNA breaks induced by X-ray. These results together suggest that the DNA strand breaks are tightly associated with the replicative capacity of DNA and with some specific enzyme system.

Thymidylate stress induced DNA rearrangements, but not mutations, in mouse cells as evidenced by the genetic changes and by the aberrations of metaphase chromosomes. The genetic changes as assayed by the loss of the marker gene could represent a type of recombination in mammalian nuclear genomes, since in yeast thymidylate stress also induces mitotic recombination, but not mutations, while it induces both the genetic events in bacteria.

The DNA arrangements reported in this paper is implicative by the fact that they were induced by the physiological stress, but not by the direct interactions of DNA with the chemical or physical agents. Capability of somatic cells to give rise to such genetic rearrangements as initiated by the double strand DNA breaks could underlies the mechanisms of differentiation processes as well as carcinogenesis. More specifically, further study on the molecular mechanisms of thymineless death in mammalian cells may shed light on the nature of heritable fragile sites of human chromosomes some of which are associated with mental retardations and severe inborn errors [17, 27].

ACKNOWLEDGMENTS

This work was supported in part by grants for cancer research from the Ministry of Education, Science, and Culture of Japan, and by the Society for Promotion of Cancer Research of Japan.

REFERENCES

1. D. Ayusawa, K. Iwata, and T. Seno, Isolation of mouse FM3A cell mutants with thermolabile thymidylate synthetase by resistance to aphidicolin, Biochem. Biophys. Res. Commun., 96: 1654-1661 (1980).
2. D. Ayusawa, K. Iwata, and T. Seno, Alteration of ribonucleotide reductase in aphidicolin-resistant mutants of mouse FM3A cells with associated resistance to arabinosyladenine and arabinosylcytosine, Somat. Cell Genet., 7:27-42 (1981).
3. D. Ayusawa, K. Iwata, and T. Seno, Unusual sensitivity to bleomycin and joint resistance to 9-β-D-arabinofuranosyladenine and 1-β-D-arabinofuranosylcytosine of mouse FM3A cell mutants with altered ribonucleotide reductase and thymidylate synthase, Cancer Res., 43:814-818 (1983).
4. D. Ayusawa, K. Iwata, H. Koyama, and T. Seno, Conditional thymidine auxotrophic mutants of mouse FM3A cells due to thermosensitive thymidylate synthase and their prototrophic revertants, J. Biol. Chem., 256:12005-12012 (1981).
5. D. Ayusawa, H. Koyama, K. Iwata, and T. Seno, Single-step selection of mouse FM3A cell mutants defective in thymidylate synthetase, Somat. Cell Genet., 6:261-270 (1980).
6. D. Ayusawa, H. Koyama, K. Iwata, and T. Seno, Selection of mammalian thymidine auxotrophic cell mutants defective in thymidylate synthase by their reduced sensitivity to methotrexate, Somat. Cell Genet., 7:523-534 (1981).
7. D. Ayusawa, H. Koyama, K. Shimizu, and T. Seno, Genetic consequences of methotrexate-induced dTMP starvation in mouse FM3A cells, Cell Struct. Funct., 7:413 (1982).
8. D. Ayusawa, K. Shimizu, H. Koyama, K. Takeishi, and T. Seno, Unusual aspects of human thymidylate synthase in mouse cells introduced by DNA-mediated gene transfer, J. Biol. Chem., 258: 48-53 (1983).
9. D. Ayusawa, K. Shimizu, H. Koyama, K. Takeishi, and T. Seno, Accumulation of DNA strand breaks during thymineless death in thymidylate synthase-negative mutants of mouse FM3A cells, J. Biol. Chem., 258:12448-12454 (1983).
10. D. Ayusawa, H. Koyama, K. Shimizu, and T. Seno, Thymidylate starvation induces DNA rearrangement as evidenced by genetic analysis of the loci introduced by gene transfer in mouse cells, in preparation.
11. J. L. Bousque and N. Sicard, Size and transforming activity of deoxyribonucleic acid in Diplococcus pneumoniae during thymidine starvation, J. Bacteriol., 128:540-548 (1976).
12. M. O. Bradley and K. W. Kohn, X-ray induced DNA double strand break production and repair in mammalian cells as measured by neutral filter elution, Nucleic Acids Res., 7:793-804 (1979).
13. S. J. Caradonna and Y.-C. Cheng, Uracil DNA-glycosylase, J. Biol. Chem., 255:2293-2300 (1980).

14. S. S. Cohen and H. D. Barner, Studies on unbalanced growth in Escherichia coli, Proc. Natl. Acad. Sci., U.S.A., 40:885-893 (1954).
15. D. Freifelder, Single-strand breaks in bacterial DNA associated with thymine starvation, J. Mol. Biol., 45:1-7 (1969).
16. N. Ganguli and S. B. Bhattacharjee, Influence of chloramphenicol on thymineless death in bacteria, Biochim. Biophys. Acta, 169:545-547 (1968).
17. T. W. Glover, FUdR induction of the X-chromosome fragile site: evidence for the mechanisms of folic acid and thymidine inhibition, Am. J. Hum. Genet., 33:234-242 (1981).
18. M. Gold and J. Hurwitz, The enzymatic methylation of the nucleic acids, Cold Spring Harbor Symp. Quant. Biol., 28:149-156 (1963).
19. M. Goulian, B. Bleibe, and B. Y. Tseng, The effect of methotrexate on levels of dUTP in animal cells, J. Biol. Chem., 255:10630-10637 (1980).
20. M. Goulian, B. Bleibe, and B. Y. Tseng, Methotrexate-induced misincorporation of uracil into DNA, Proc. Natl. Acad. Sci., U.S.A., 77:1956-1960 (1980).
21. W. E. Hill and W. L. Fangman, Single-strand breaks in deoxyribonucleic acid and viability loss during deoxyribonucleic acid synthesis inhibition in Escherichia coli, J. Bacteriol., 116:1329-1335 (1973).
22. T. Hori, High incidence of sister chromatid exchanges and chromatid interchanges in the conditions of lowered activity of poly(ADP-ribose) polymerase, Biochem. Biophys. Res. Commun., 102:38-45 (1981).
23. T. Hori and K. G. Lark, Effect of puromycin on DNA replication in Chinese hamster cells, J. Mol. Biol., 77:391-404 (1973).
24. T. Hori and S. Nakai, Unusual dose-response of chromosome aberrations induced in human lymphocytes by very low dose exposures to tritium, Mutat. Res., 50:101-110 (1981).
25. T. Hori, D. Ayusawa, K. Shimizu, H. Koyama, and T. Seno, Chromosome instability induced under the conditions of thymidylate stress in thymidylate synthease-negative mutants of mouse FM3A cells, Cancer Res., 44:703-709 (1984).
26. S. Ikegami, T. Taguchi, M. Ohashi, M. Oguro, H. Nagano, and Y. Mano, Aphidicolin prevents mitotic cell division by interfering with the activity of DNA polymerase-α, Nature (London), 275:458-460 (1978).
27. P. B. Jacky, B. Beek, and G. R. Sutherland, Fragile sites in chromosomes: Possible model for the study of spontaneous chromosome breakage, Science, 220:69-70 (1983).
28. H. Koyama, D. Ayusawa, M. Tsuji, and T. Seno, Thymidineless death and mutation induction in cultured mouse FM3A cell mutants deficient in thymidylate synthease, Mutat. Res., 105:433-438 (1982).
29. B. A. Kunz, B. J. Barclay, J. C. Game, J. G. Little, and R. H. Haynes, Induction of mitotic recombination in yeast by starva-

tion for thymine nucleotide, Proc. Natl. Acad. Sci., U.S.A., 77:6057-6061 (1980).

30. J. G. Little and R. H. Haynes, Isolation and characterization of yeast mutants auxotrophic for 2'-deoxythymidine 5'-monophosphate, Mol. Gen. Genet., 168:141-151 (1979).

31. M. Luzzati, Effect of thymine starvation on messenger ribonucleic acid synthesis in Escherichia coli, J. Bacteriol., 92: 1435-1446 (1966).

32. F. Makino and N. Munakata, Deoxyuridine residues in DNA of thymine-requiring Bacillus subtilis strains with defective N-glycosidase activity for uracil-containing DNA, J. Bacteriol., 134:24-29 (1978).

33. H. D. Mennigmann and W. Szybalski, Molecular mechanism of thymineless death, Biochem. Biophys. Res. Commun., 9:398-404 (1962).

34. H. Nakayama and P. Hanawalt, Sedimentation analysis of deoxyribonucleic acid from thymine-starved Escherichia coli, J. Bacteriol., 121:537-547 (1975).

35. J. W. Rawles and J. M. Collins, Deoxyribonucleic acid replication in human diploid fibroblasts stimulated to proliferate, J. Biol. Chem., 252:4762-4766 (1977).

36. D. L. Reichenbach, G. E. Schaiberger, and B. Sallman, The effect of thymine starvation on chromosomal structure of Escherichia coli JG-151, Biochem. Biophys. Res. Commun., 42: 23-30 (1971).

37. R. Rolfe, On the mechanism of thymineless death in Bacillus subtilis, Proc. Natl. Acad. Sci., U.S.A., 57:114-121 (1967).

38. F. Tamanoi and T. Okazaki, Uracil incorporation into nascent DNA of thymine-requiring mutant of Bacillus subtilis 168, Proc. Natl. Acad. Sci., U.S.A., 75:2195-2199 (1978).

39. K. F. Weibezahn and T. Coquerelle, Radiation induced DNA double strand breaks are joined by ligation and recombination processes, Nucleic Acids Res., 9:3139-3150 (1981).

40. G. W. Woods, Quantitation of the repair of gamma-radiation-induced double-strand DNA breaks in human fibroblasts, Biochim. Biophys. Acta, 655:342-348 (1981).

41. K. Yoshinaga, Double strand scission of DNA involved in thymineless death of Escherichia coli 15 TAU, Biochim. Biophys. Acta, 294:204-213 (1973).

INDUCTION OF MITOCHONDRIAL MUTATIONS IN HUMAN CELLS BY METHOTREXATE

David I. Hoar and Leo S. Dimnik

Departments of Medical Biochemistry and Pediatrics
The University of Calgary
Calgary, Alberta
Canada

ABSTRACT

Inhibition of dihydrofolate reductase by the folate analog, methotrexate (MTX) results in a depletion of tetrahydrofolate dependent one carbon transfer reactions in amino acid and nucleic acid biosynthesis. When human cells (either HeLa or normal skin fibroblasts) are exposed to MTX in a defined medium containing dialyzed fetal calf serum, essential and non-essential amino acids, and purine source, the thymidylate pools alone are depleted. Under these conditions exposure to 10^{-6} M MTX induces mitochondrial mutagenesis, measured as an increase in the frequency of chloramphenicol resistant (CAP^R) colonies, without altering the rate of nuclear mutation monitored by determining the frequency of 6-thioguanine resistance (TG^r). The occurrence of CAP^R mutations is time, and MTX concentration dependent and the frequency of CAP^R can be decreased quantitatively by adding thymidine to the culture medium. This mitochondrial specific mutagenesis can also be achieved using the thymidylate synthetase inhibitor, 5-fluorodeoxyuridine further implicating thymidylate pools as the mediator of this effect. During the course of exposure to 10^{-6} M MTX the thymidine kinase deficient HeLa BU25 cell line exhibits a progressive depletion and degradation of mitochondrial DNA suggesting that the mutagenesis and DNA degradation represent portions of a progressive process. The basis for the selective sensitivity of the mitochondrial genome to thymidylate depletion mutagenesis may be the consequence of its differences from the nuclear genome in mechanisms of DNA replication or repair.

1. INTRODUCTION

Methotrexate (MTX) is an effective chemotherapeutic drug used in the treatment of various human cancers [2]. Little is known about the basis for its cytotoxicity [13], although the specific site of its action in mammalian cells is well characterized [27]. Through the inhibition of dihydrofolate reductase (DHFR) (5,6,7,9-tetrahydrofolate: NADP oxidoreductase; EC 1.5.1.3), MTX depletes tetrahydrofolate pools and thus interferes with one-carbon transfer reactions requiring this co-factor. One such reaction is that carried out by thymidylate synthetase (TS) (5,10-methylenetetrahydrofolate: dUMP C-methyltransferase; EC 2.1.1.45). Depletion of co-factor, and thus inhibition of the enzyme, leads to a depletion of dTMP (and hence dTTP) pools with a concomitant rise in dUMP. As levels of dUMP rise, this is phosphorylated to dUTP and is incorporated into DNA, in place of dTTP [12]. Under normal circumstances, dUTP is degraded by the enzyme dUTPase; however, with a block in thymidylate synthetase, levels of dUTP rise dramatically and appear to be sufficient to oversaturate this degradative system [12]. The abnormal insertion of uracil (Ura) into DNA initiates an excision repair cycle triggered by the Ura-DNA glycohydrolase [12, 16, 17] and where dUTP/dTTP ratios are high, uracil has a high probability of being reinserted during excision repair and a self defeating cycle will ensue. A cyclical excision repair cycle such as this would result in a high turnover of strand breaks particularly in replicating segments of a genome. In support of these predictions <u>in vivo</u> studies on the effects of MTX on Ura incorporation into DNA of human lymphoblasts have shown that within six hours there is a substantial increase in the dUTP/dTTP ratio ($\sim 10^5$-fold) and measurable quantities of Ura can be found in DNA [12]. On the basis of these observations it could be predicted that other inhibitors of TS would boost the dUTP/dTTP ratio and permit Ura incorporation into DNA. One such inhibitor is 5-fluorodeoxyuridine (FUdR). On entry into mammalian cells, FUdR is converted to its deoxyribonucleotide FdUMP (a non-competitive inhibitor of TS) by thymidine kinase (TK) (ATP: thymidine 5'-phosphotransferase; EC.2.7.1.21). The inhibition of TS leads to increased dUTP pools which favor higher level phosphorylation of FdUMP to FdUTP and thus its incorporation into DNA [16, 17]. As both dUTP and FdUTP would enter DNA and both are substrates of Ura-DNA glycohydrolase, a futile excision repair cycle would again be initiated.

In the lower eukaryotes <u>Saccharomyces</u> and <u>Chlamydomonas</u> studies of the effects of dTTP depletion by mutation, antifolate drugs or FUdR have revealed a most remarkable organelle specific mutagenesis [3, 22, 38]. These studies were undertaken to determine whether a similar organelle specific (mitochondrial) mutagenesis occurs in human cells in culture. Our results indicate that, in HeLa cells or normal human skin fibroblasts, treatment with the folate antagonist MTX or the TS inhibitor FUdR causes a significant elevation in chloramphenicol resistant (CAP^R) variants (referred to hereforth as mutants) without any change in the frequency of chromosomal mutants. As CAP^R in mammalian cells is cytoplasmically inherited and is associated with point mutations in the mitochondrial DNA [4, 18, 31] this is interpreted as mitochondrial specific mutageneis. The mitochondrial mutagenesis is accompanied by progressive degradation and a quantitative depletion of the mitochondrial DNA. The mutagenesis is reversed by exogenous supplementation of thymidine; hence, we believe that the specific mutagenesis is a consequence of the depletion of the intracellular dTTP pools.

2. MATERIALS AND METHODS

2.1. Cells and Culture Conditions

The HeLa BU25 cell line resistant to 5-bromodeoxyuridine and deficient in TK [19] was used in the majority of experiments as the utilization of exogenous TdR was essentially blocked and intracellular dTTP pools were primarily dependent on the activity of TS (and hence folate dependent). Three separate human fibroblast cell strains (HSC172CTO, CSH233CTO and CSH234CTO) and the parental TK^+ HeLa cell line were used in experiments where the levels of dTTP could be manipulated through the supplementation of exogenous TdR.

Six independent clones of HeLa BU25 recently cloned in 5-bromodeoxyuridine were grown to confluency in 75 cm^2 flasks (Lux), split into four equal portions and two of each set subject to MTX treatment. The second two flasks were carried through without MTX treatment and selected in parallel for chloramphenicol resistance (CAP^R) to determine whether clonal variation would influence mutant recovery. In the remainder of the experiments with HeLa BU25, a single clonal population of cells was employed.

The V79 Chinese hamster line VEOT resistant to emetine (emt^r, a recessive marker), ouabain (oua^R, a dominant marker) and 6-thioguanine (-TG^r, and X-linked recessive marker) was used in cybrid selection experiments.

Cells were grown routinely in alpha medium (Flow Laboratories), supplemented with 10% fetal calf serum and 20 mM HEPES buffer pH 7.15 (referred to as medium); however, during treatments, dialyzed serum was employed and adenine was added at 10 μg/ml.

2.2. Treatments and Mutant Selections

We have previously demonstrated that the recovery of CAP^R mutants following MTX treatment is dependent upon the cell density during the treatment interval [10, 11]. For these reasons a standardized treatment density was employed. Twenty four hours prior to the treatment, late log phase cells were harvested with trypsin, counted and plated in 75 cm^2 or 175 cm^2 flasks (Lux) at a density of 3×10^4 $cell/cm^2$ in fresh medium. Treatment medium (20 mls for the 75 cm^2 flask) containing MTX or FUdR was added following a brief rinse with phosphate buffered saline. At various time intervals, the cells were harvested with trypsin counted, plated for viability and selected in either 25 μg/ml chloramphenicol or 2.5 μg/ml 6-thioguanine. Cell densities for selection experiments were at 1 to 5×10^5 cells/100 mm petri disch (Lux). All incubations were at 37°C in a 5% CO_2 atmosphere for 12-16 days after which colonies were visualized by staining with methylene blue [15]. Representative mutant colonies from the experiments with HeLa BU25 were isolated and cloned once in non-selective media using a specially designed punch press. Such clones were then tested for CAP^R by determining their relative plating efficiencies in 0, 25, and 50 μg/ml CAP.

2.3. Cytoplast Isolation and Hybridizations

Ten independent cloned CAP^R mutants were grown in duplicate in non-selective media in 25 cm^2 flasks and, while in late log phase, used for the isolation of cytoplasts using methods previously described [29, 31]. The highly enriched cytoplast populations were washed twice with medium and incubated for 2 h. At this time a total of 2×10^6 log phase VEOT cells were added to one flask of each set, co-cultivated for a further $2^1/_2$ h and both flasks hybridized using the PEG (44% PEG 6000, 10% DMSO in serum free medium) technique [28]. After overnight culture at 37°C, in fresh medium, fusogen treated cultures were harvested with trypsin, counted and plated at 5×10^5 cells/100 mm petri dish in medium containing 25 μg/ml CAP. Seventy-two hours later, 6-thioguanine was added to 2.5 μg/ml and incubation continued for 2 weeks. Representative clones were isolated and tested for CAP^R, TG^r (and HAT sensitive) emt^r and oua^R.

2.4. DNA Analysis

2.4.1. DNA Preparation

One or two 175 cm^2 flasks of HeLa BU25 untreated or treated for increasing times with 1 μM MTX were used for preparation of DNA. Washed cells were resuspended in 100 mM Tris 40 mM EDTA (TE) and immediately lysed by injecting an equal volume of 0.2% SDS in TE. The lysate was phenol and chloroform:isoamyl alcohol (24:1) extracted before precipitation with ethanol. The nucleic acids were dissolved, RNase digested and reextracted prior to determining the DNA concentration spectrophotometrically.

2.4.2. DNA Dot Quantitative Analysis

Serially diluted samples were placed in a boiling waterbath for 5 min, dotted in two duplicate sets at 1.0, 0.1, and 0.01 μg/5 μl

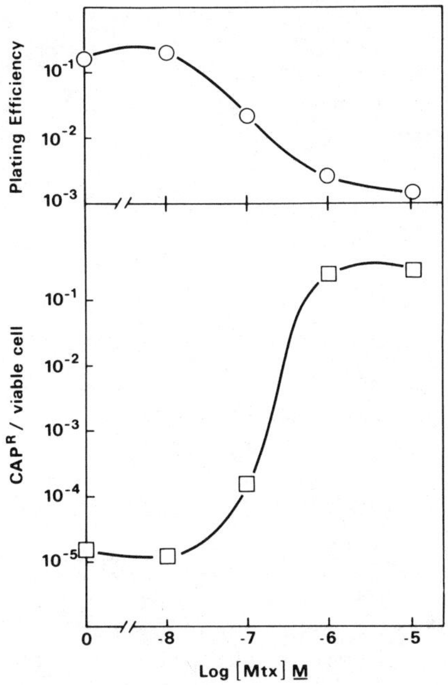

Fig. 1. Plating efficiency (upper) and chloramphenicol resistant colonies per viable cell (lower) for HeLa BU25 treated 72 h with methotrexate at the concentrations shown. These points are the average of triplicate samples for three independent experiments.

dot on nitrocellulose (BA83 Schleicher Schuell) previously soaked with 20X SSC and backed at 80°C for 2 h prior to hybridization.

2.4.3. Agarose Gel Analysis

Five μg aliquots of the DNA samples were elctrophoresed in native gels or denatured by glyoxalation, electrophoresed in 1% agarose according to the procedure of Carmichael and McMaster [7] and Southern transferred [30] to BA83 nitrocellulose and baked as above.

2.4.4. Hybridization Analysis

All filters were prehybridized at least 2 h at 43°C in 5X SSC 0.1% SDS, 0.2% polyvinyl pyrrolidone (PVP-10 Sigma), 250 μg/ml denatured salmon sperm DNA and 50% formammide. One set of the DNA dots, the native and the glyoxal treated Southern transfers were then hybridized with the 850 bp Hind III human mitochondrial DNA fragment (nick translated with ^{32}P dCTP to 2.5×10^8 cpm/μg). The remaining set of dots were hybridized to a 4.4 kb human Pst I β globin probe similarly labelled. Hybridization was for 6 h at 43°C in fresh prehybridization solution to which 7.5% (wt/vol) PVP-360

(Sigma) and probe (10^5 cpm/ml for the mitochondrial probe and 10^6 cpm/ml for the β globin probe) had been added. Washes were 4 × 10 min in 2X SSC at 43°C followed by 4 × 10 min in 0.2X SSC at 43°C. Autoradiography was carried out using preflashed XAR-5 film and Chronex Quanta III screens.

3. RESULTS

3.1. MTX Mutagenesis

Treatment of the HeLa BU25 TK deficient cell line for 72 h with increasing concentrations of MTX in the presence of all amino acids and adenine lead to a dramatic dose dependent increase in the recovery of mutants capable of growth in 25 μg/ml of chloramphenicol (Fig. 1). Selection of duplicate treated and untreated cultures in thioguanine resulted in recovery of 1.8×10^{-5} to 2×10^{-6} mutants per viable cell plated, without any evidence of an effect of the MTX treatment (data not shown). These findings suggested MTX induced mitochondrial specific mutagenesis and a simple fluctuations analysis was carried by treating 6 dependent clones of HeLa BU25 with increasing concentrations of MTX. From each independent set a total of 10 flasks were plated at 10^5 cells/flask in 25 μg/ml chloramphenicol and after 12 days incubation colonies were evident. Half of each set was stained and colonies were isolated and cloned from the 1 μM and 10 μM MTX treatments for further testing. A total of 6 very poorly defined small colonies were evident amongst 2.4×10^6 untreated cells (3 from clone 1 and 3 from clone 6), whereas well defined healthy colonies were present amongst populations treated at 1, 10, and 100 μM MTX. At 10 nM and 100 nM MTX, colonies were sparse and poorly defined ($2/10^5$ and $130/10^5$ respectively), hence it would appear that the condition favoring induction of CAP^R mutants also improve their recovery. Two additional experiments conducted on a single clone of HeLa BU25 gave essentially identical results and these are pooled in Fig. 1.

To ascertain whether the observed mitochondrial mutagenesis was peculiar to the TK^- cell line or a general phenomona we carried out similar experiments with three normal skin fibroblast cell strains. As these cell strains could utilize exogenous supplies of thymidine, the MTX treatments were done in the presence or absence of thymidine. Some variations in the absolute frequency of CAP^R mutants/surviving cell were observed (five fold lower in HSC172 CTO vs CSH233 CTO), although the relative MTX dose related relationships were identical. The results shown in Fig. 2 were obtained with CSH233 CTO and here the effect of adding back 1 μg/ml or 10 μg/ml TdR during the MTX treatment period can be observed. The relative plating efficiencies obtained following MTX treatment (upper graph in Fig. 2) did not reveal a marked concentration dependent cytotoxicity; however, when 10 μg/ml or 1 μg/ml TdR was added during the treatment interval, all

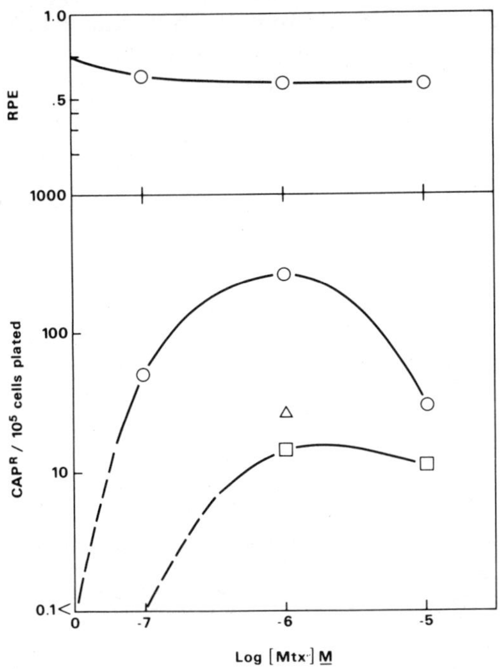

Fig. 2. The relative plating efficiency of human diploid fibroblast CSH233 CTO) (upper) and recovery of chloramphenicol resistant colonies (lower) following 72 h exposure to methotrexate (open circles). The open triangle and squares show the effect of adding respectively 1 μg/ml for 10 μg/ml thymidine to the medium during methotrexate treatment. Each point is the average of 4 replicate cultures.

survivals exceeded controls by an average of 50% and 15% respectively (data not shown).

3.2. Cytoplasmic Inheritance of CAP^R

Chloramphenicol resistance in eukaryotic organisms normally has a mitochondrial basis [4, 18]. As the recovery of CAP^R occurs at such a dramatic frequency we undertook cybrid analysis of 10 independent CAP^R clones of HeLa BU25 to check on the nature of these mutants.

All the independently isolated HeLa Bu25 CAP^R clones were capable of growing in high levels (50 μg/ml) of CAP, a characteristic of cytoplasmic CAP^R mutants [34]. The cytoplast fusions to the hamster VEOT cell line yielded cybrids at a frequency of between 3.7 and $4.5 \times 10^4/10^6$ recipient VEOT cells. Platings of 10^6 fusogen treated cytoplasts or VEOT cells alone failed to yield a single CAP^R, TG^r colony. Individual cybrid colonies were cloned in selective

TABLE 1. FUdR Induction of CAP^R Mutants[a]

FUdR (M)	HeLa[b]	HeLa BU25[b]
0	0	0
10^{-8}	1.006 ± .213	NT[c]
10^{-7}	0.953 ± .123	0
10^{-6}	0.980 ± .184	0
10^{-5}	1.220 ± .389	0
10^{-4}	0.732 ± .248	0.820 ± .110

[a]The induction of CAP^R mitochondrial mutants following a 72 hour exposure to varying concentrations of FUdR.

[b]Frequency of CAP^R colonies per viable cell plated ± one standard deviation.

[c]Not tested

media and tested for the presence of recipient and not donor genomic markers. As the cytoplast donor was TG^+, TK^-, emt^+, oua^+, and the recipient TG^r, TK^+, emt^r, oua^R, true hybrids would be TG^+, TK^+, emt^+, and oua^R. All tested cybrids, in addition to being CAP^R, were TG^r, TK^+, emt^r, and oua^R indicating the donor CAP^R was transferred in the absence of genomic markers and hence the CAP^R mutation is most probably cytoplasmically inherited [4, 31, 34]. It is important to note that the persistence of CAP^R human mitochondria in hamster cells is selected for and in the absence of such selection the interspecific maintenance of CAP^R would not persist.

3.3. Time Course of MTX Induced CAP^R

The HeLa BU25 cell line was subject to treatment with 1 μM MTX and samples tested for CAP^R at various intervals from 0 to 72 h. Following an initial 12 h when no CAP^R mutants were recovered, a maximal increase occurred by 24 h. The recovery of CAP^R was associated with a rapid fall in viability which appeared to plateau by 36 h (Fig. 2). In repeated experiments the CAP^R mutants first appear between 10 and 14 h following exposure to 1 μM MTX [10, 11].

3.4. Thymidylate Pool Effects on CAP^R Recovery

The treatment conditions employed in the above experiments favor the argument that depression of the dTMP pools leads to the mitochondrial mutagenesis. To test this further we supplied thymid-

TABLE 2. Thymidine Reversal of Mitochondrial Mutagenesis[a]

Time of Thymidine Addition[b]	Cell Viability[c]	Relative Frequency of CAP^R/v [d]
4	0.929 ±.027	0
8	0.736 ±.108	0
24	0.233 ±.008	0
72	0.002 ±.0002	1

[a]The effect of adding 100 μM thymidine after varying exposure to 1 μM MTX on cell viability and CAP^R mutant recovery in HeLa.

[b]Time = 0 is the time of addition of MTX; measured in hours.

[c]Plating efficiency ± one standard deviation

[d]A relative frequency of 1 = 0.875 ±.154

ine exogenously and employed the thymidylate synthetase inhibitor 5-fluorodeoxyuridine (FUdR). As FUdR is taken up and phosphorylated to FdUMP by TK, we utilized the TK^+ HeLa line. The TK^- HeLa BU25 has a residual TK activity believed to be predominantly mitochondrial in origin [20] and therefore it was also used. A standard 72 h exposure in treatment medium containing various FUdR concentrations instead of MTX was used here. The results of this study on the two HeLa cell lines are presented in Table 1.

A separate series of experiments were conducted on the TK^+ HeLa cell line where thymidine was added to the treatment medium at various times during exposure to 1 μM MTX. These findings are presented in Table 2. These results and the effects of thymidine on normal fibroblasts treated with MTX (Fig. 2) indicate that a reversal of the MTX induction of CAP^R occurs when thymidine is supplied exogenously to cells having a functional TK.

3.5. Mitochondrial DNA Analysis

Quantitative mitochondrial DNA dot hybridization analysis performed on samples taken at increasing intervals of exposure of HeLa BU25 to 1 μM MTX are shown in Fig. 4. After an initial 8 h period when the mitochondrial DNA content is apparently unchanged, there is a decrease in the 24 h sample. As total cellular DNA was used for these studies and hybridization analysis with a genomic DNA

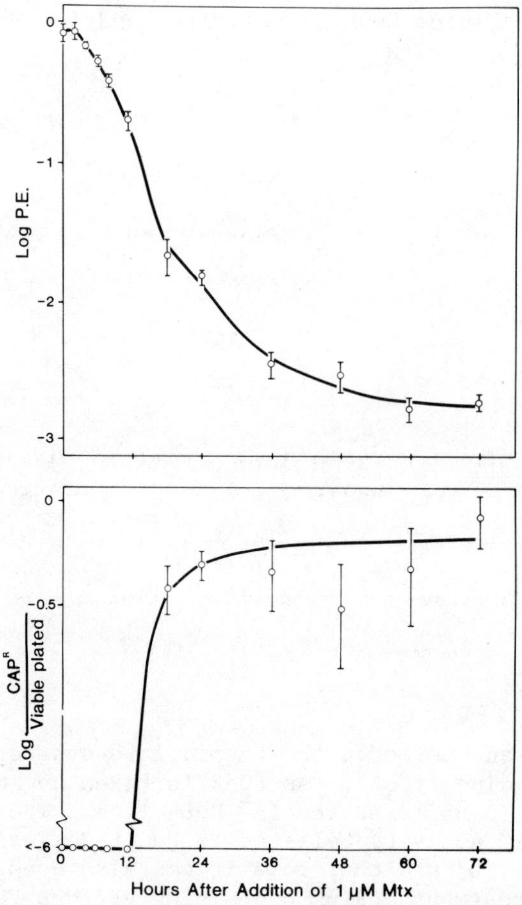

Fig. 3. Time dependent recovery of CAP^R mutants (0 to 72 h) following exposure of HeLa BU25 to 1 μM methotrexate. Cell viability (upper) and CAP^R (lower) are shown. Error bars represent one standard deviation.

sequence indicated the concentrations are similar with the exception of the 48 and 72 h points that are about 40-60% of the others, this is interpreted as a decline in the amount of mitochondrial DNA per cell.

The qualitative analysis of the same DNA was undertaken using native and glyoxal treated samples, denaturing agarose gels, and Southern hybridization analysis with the same mitochondrial DNA probe. The results shown in Fig. 5 indicate that during the time interval when there is a quantitative decrease in mitochondrial DNA (Fig. 4) and a dramatic increase in CAP^R mutants (Fig. 3), there is increasing strand breakage with a decrease in the proportion of intact mitochondrial DNA forms (compare Fig. 5, lane 8 to lanes 10-12).

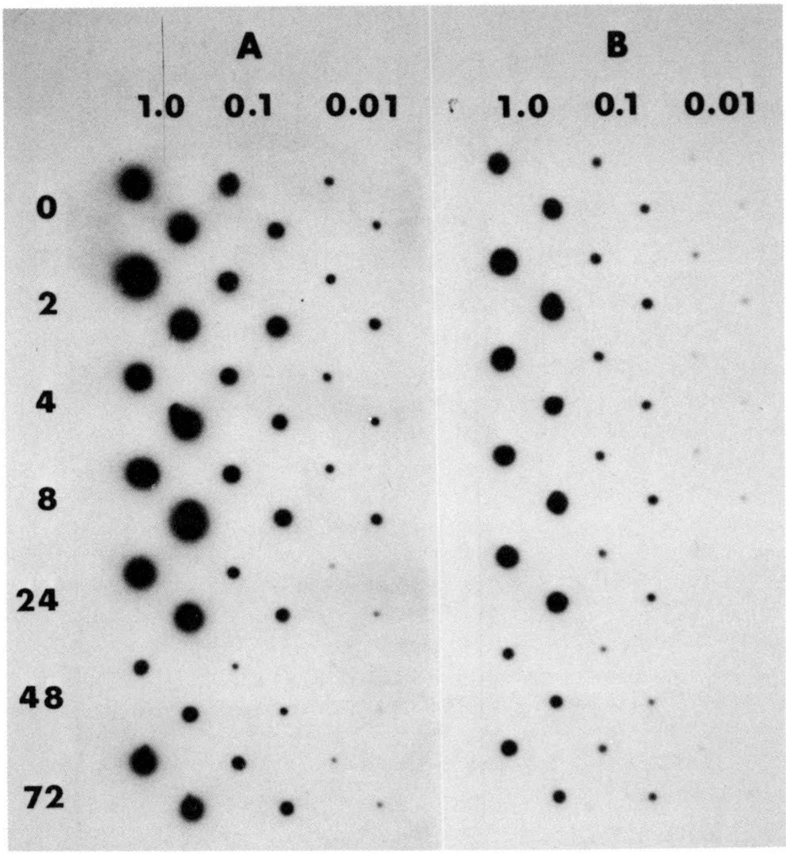

Fig. 4. Duplicate dots of 1.0, 0.1, and 0.01 μg of total DNA (left to right) isolated from HeLa BU25 treated with 1 μM MTX for 0, 2, 4, 8, 24, 48, and 72 h (top to bottom). The filter on the left (A) was probed with the human mitochondrial 850 bp Hind III fragment and the filter on the right (B) with the globin probe.

In the denaturing glyoxal treated samples, there is an increase in the amount of hybridization to material smaller than the major mitochondrial form (compare Fig. 5, lane 1 to Fig. 5, lane 7) indicating an accumulation of single strand nicks has occurred during this time.

The average molecular weight of the genomic DNA analyzed by agarose gel electrophoresis does not apparently change however, at this time analysis of genomic DNA single strand lengths is incomplete and we are unable to state this is exclusively a mitochondrial effect.

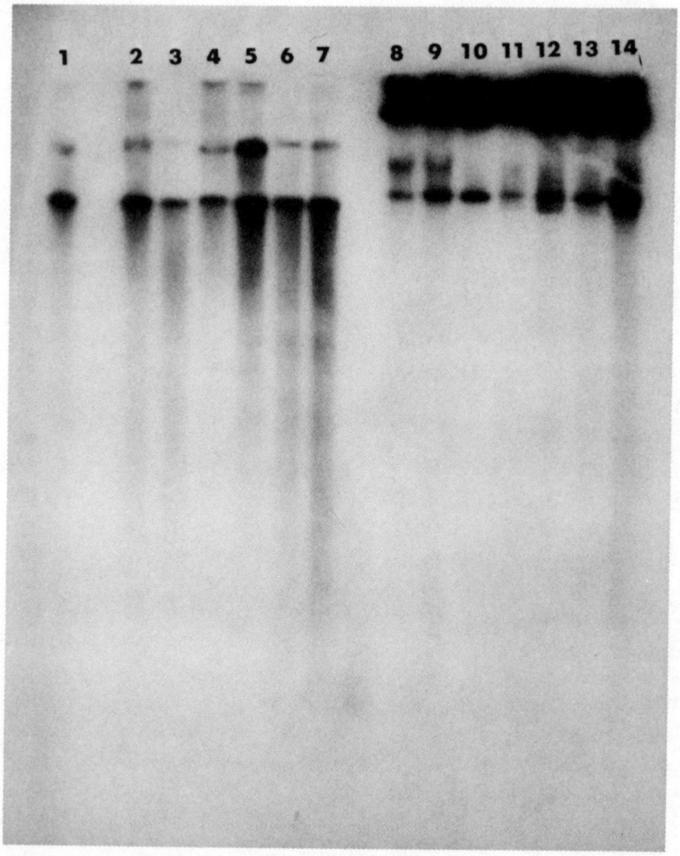

Fig. 5. Southern transfer of glyoxal treated (lanes 1-7) and native (lanes 8-14) DNA samples obtained from HeLa BU25 cells treated with 1 μM MTX for 0, 4, 8, 24, 48, 72, and 96 h (left to right). Hybridization was with the mitochondrial fragment used in Fig. 4.

4. DISCUSSION

4.1. Recovery of CAP^R Mutations

The intracellular concentrations of the DNA precursor dTTP can be modulated experimentally by restricting exogenous supplies (mutationally or by precursor limitation) and inhibiting endogenous biosynthesis. The biosynthesis of dTMP can be effectively blocked by inhibiting the enzyme TS either directly, using the non-competitive inhibitor FUdR (FdUMP is its active form), or by co-factor depletion through inhibition by MTX of the co-factor regenerating enxyme DHFR [17, 27]. As the intracellular pools of dTTP are depleted in cultured human cells by MTX inhibition of DHFR or FUdR inhibition of TS and precursor limitation, there is a marked, dose dependent increase in the occurrence of CAP^R mutants without any change in the frequency of occurrence of TG^r mutants. Since the cytoplast fusion reported here support the cytoplasmic nature of our CAP^R mutants and the site of action of CAP inhibition is localized to the mitochondrion [4, 31, 34], and as the TG^r ($HGPRT^-$) mutants are predominantly X chromosomal mutants, it is assumed that a mitochondrial specific mutagenesis is occurring. When thymidine is supplied exogenously to normal fibroblasts or HeLa TK^+ during MTX treatments, there is a thymidine concentration dependent depression in the frequency of CAP^R mutants (Fig. 2 and Table 2). These studies support the prediction that the mutagenesis is being mediated through a manipulation of the dTTP pool.

A high FUdR concentration is also able to elevate the frequency of CAP^R mutants in the TK^- HeLa BU25 cell line (Table 1). FUdR itself is not known to be a direct inhibitor of TS. There are; however, two TK activities in mammalian cells, one soluble (coded for by a gene on chromosome 19 of man) and one mitochondrial (coded for by a gene on chromosome 16 of man [26]). It is possible that residual levels of either enzyme are responsible for formation of FdUMP, the active inhibitor of TS. As the effect of FUdR is also to reduce the dTTP pools, this finding further supports a role for low dTTP in inducing mitochondrial specific mutagenesis.

When yeast auxotrophic for dTMP are starved for dTMP for prolonged periods depleting their dTTP pools, mitochondrial specific mutagenesis and progressive erosion of the mitochondrial DNA occurs [3]. This leads to a marked increase in respiratory deficient petit mutants [3, 23]. In the unicellular algae Chlamydomonas, FUdR induced dTMP (hence dTTP) depletion significantly elevates chloroplast mutagenesis without affecting the mutation frequency for chromosomal markers [33]. Hence, it would appear that similar mitochondrial specific mutagenic mechanisms dependent upon the levels of dTTP within the cell are operating in these lower eukaryotes and cultured human cells.

The recovery of mitochondrial mutants resistant to CAP in mammalian cells has been accomplished utilizing a variety of mutagenesis protocols [29, 31, 34, 37]. In general, these mutants are exceedingly difficult to recover using standard mutagenesis protocols effective in the isolation of genomic mutants. Growth intervals of one to several months were required [29, 31]. However, when pregrowth in ethidium bromide was employed, recovery of mitochondrial mutants could be enhanced [36]. This is believed to be related to the ethidium bromide induced depletion of mitochondrial DNA [37] as there are normally several thousand identical genomes per cell distributed in at the most a few hundred mitochondria [5].

In these studies we were able to induce and recover CAP^R mutants in either HeLa or normal fibroblasts in less than two weeks. Those mutants isolated under the optimal conditions grew exceedingly well from the onset and remained CAP^R after cloning in non-selective growth medium. As the number of mitochondria per cell appears to be regulated [25], these characteristics suggest that the mitochondrial population was drastically depleted and then repopulated by a surviving CAP^R mutants (or several such mutants). It has been estimated that only a small proportion of the total mitochondrial population needs to be CAP^R to exhibit the drug resistant phenotype [34]. Hence, a total depletion of the normal mitochondrial population would not be necessary to provide the CAP^R mitochondria with a selective advantage. The results of mitochondrial DNA analysis shown in Figs. 4 and 5 provide evidence that there is a quantitative decrease and qualitative destruction of the mitochondrial genome during MTX induced thymidylate starvation. These factors would set the stage for the rare CAP^R surviving mitochondrial DNA to repopulate a mitochondria and this resistant mitochondria could then repopulate the depleted population in the cell.

4.2. The Nature of the Mutagenesis by Thymidylate Deprivation

In cultured human cells, one of the consequences of TS inhibition (by MTX depletion of co-factor or presumably FdUMP inhibition) is a fall in dTTP levels and a rise in dUTP (and FdUTP) which is incorporated into DNA [12, 17]. The Ura or FUra containing DNA would be substrates for Ura-DNA-glycohydrolase, and excision repair in high dUTP conditions could result in reinsertion of Ura and thus a futile repair cycle would occur [12, 16, 17]. This process should lead to accumulating nicks in DNA and provides the starting point for a mutagenic process. As the excision repair process is believed to be highly accurate (not error prone) [9], genomic mutagenesis should not occur as a consequence of these repair cycles. On the other hand, little is known about the repair fidelity or capacity of mammalian mitochondrial DNA. In yeast, although efficient photoreactivation of UV dimers does occur in the mitochondrion, there is no excision repair as measured by UV endonuclease activity [35]. Human cells do, however, have an active mitochondrial Ura-DNA-glycohydrolase [1] suggesting some mitochondrial repair enzymes exist. The conditions leading to mitochondrial mutagenesis clearly produce significant nicks in the mitochondrial DNA. Such conditions are speculated to be promoters of mitotic recombination in the nuclear genome and hence may stimulate mitochondrial recombination [21, 23]. If there is an error prone recombinational repair system operating this could produce the observed effects.

Replication of the mitochondrial DNA unlike the nuclear DNA is dependent upon DNA polymerase gamma. This polymerase is known to have a higher rate of misincorporation and, hence, in the high dUTP situation produced during MTX or FUdR treatment, the potential for U-G base pairs could be increased [21]. This could lead to base pair substitution mutagenesis. The results presented here suggest that the organelle genome and not the nuclear genome is exquisitely sensitive to mutagenesis resulting from depletion of the DNA precursor dTTP. Whether this is a consequence of differences in DNA repair capabilities or differences in the error rate of the DNA polymerases operating in these two cellular compartments remains to be determined. As organelle specific mutagenesis is markedly enhanced at a time when, organelle and not nuclear, DNA would be replicating in *Chlamydomonas* [38], it could be related to differences in modes of DNA replication between the two genomes.

4.3. Methotrexate and Cancer Chemotherapy

There have been many studies demonstrating that MTX is much more effective as a cytotoxic agent against rapidly dividing cells [13, 27]. There are two major theories that can be used to explain these observations. In the first, it is assumed that rapidly cycling cells are exposed more frequently to the sensitive phase of the cell cycle. This is suggested to be S-phase as MTX is only toxic to cells undergoing DNA synthesis [14] and not cells in stationary phase or those blocked in the cell cycle by amino acid starvation [6].

An alternate explanation that is also dependent upon cells cycling is favored by these studies. Rapidly dividing cells and particularly some tumor cells have higher net energy requirements and rates of aerobic glycolysis [8, 32]. Thus, an agent that causes selective mitochondrial destruction would be expected to have a more pronounced cytotoxicity towards those cells demanding greater mitochondrial activity.

The viability curve in Fig. 3 demonstrates two phases. The first, a rapid decline at which point (10-16 h) the CAP^R mutants begin to appear. The cell viability curve then takes on a more gradual decline during which the frequency of CAP^R remains unchanged. Over the time interval from 24 to 72 h, the CAP^R mutants recovered exhibit better growth and the size of the colonies obtained following 12 days of selection in 25 μg/ml CAP increases significantly. At about 72 h the recovery of CAP^R per viable cell decreases and it is during this plateau interval from 24 to 72 h that there is progressive mitochondrial DNA loss and degradation. These observations are consistent with the theory that cell killing by MTX is the consequence of thymidylate pool depletion that in turn causes progressive mitochondrial mutagenesis and strand breakage leading to mitochondrial DNA degradation and mitochondrial death. In the absence of a functional mitcohondria, cell death will normally occur in mammalian cells, though not in yeast. Experiments to be reported elsewhere demonstrate that the viability of MTX treated cells can be significantly increased by culture in medium containing high levels of glucose. Such conditions permit anaerobic glycolysis for energy and decrease the demands for a functional mitochondria.

4.4. Summary

The cancer chemotherapeutic antifolate drug methotrexate or the thymidylate synthetase inhibit FUdR are capable of inducing mitochondrial specific mutagenesis and DNA degradation when sources of purines and amino acids are present. These findings demonstrating a mitochondrial and not nuclear genetic target for MTX in human cells suggest a possible explanation for the tumor cell preferential cytotoxicity exhibited by this drug. As tumor cells in general have both higher energy requirements and glycolytic rates [8, 32], and as a consequence are more dependent than normal cells on mitochondrial integrity, the differential cytotoxicity could be a result of the mitochondrial specificity of the genotoxicity of methotrexate.

ACKNOWLEDGMENTS

This work was supported by grants from the Alberta Heritage Foundation for Medical Research and the Medical Research Council of Canada (MA 6928). We wish to thank K. Neote who isolated the mitochondrial clone used in the hybridization studies and also M. McPherson for typing this manuscript.

REFERENCES

1. C. T. M. Anderson and E. C. Friedberg, The presence of nuclear and mitochondrial uracil-DNA-glycohydrolase in extracts of human KB cells, Nucleic Acid Res., 8:875-888 (1980).
2. J. R. Andrews, in: "Radiotherapy," pp. 395-416, University Park Press, Baltimore (1978).
3. B. J. Barclay and J. G. Little, Genetic damage during thymidylate starvation in Saccharomyces cerevisiae, Molec. Gen. Genet., 160:33-40 (1978).
4. H. Blanc, C. T. Wright, M. J. Bibb, D. C. Wallace, and D. A. Clayton, Mitochondrial DNA of chloramphenicol-resistant mouse cells contains a single nucleotide change in the region encoding the 3'-end of the large ribosomal RNA, Proc. Natl. Acad. Sci., U.S.A., 78:3789-3793 (1981).
5. D. Bogenhagen and D. A. Clayton, The number of mitochondrial deoxyribonucleic acid genomes in mouse L and human HeLa cells, J. Biol. Chem., 249:7991-7995 (1974).
6. J. Borsa and G. F. Whitmore, Cell killing studies on the mode of action of methotrexate on L-cells in vivo, Cancer Research, 29:737-744 (1969).
7. G. G. Carmichael and G. K. McMaster, The analysis of nucleic acids in gels using glyoxal and acridine orange, in: "Methods in Enzymology, Vol. 65, pp. 380-391, Academic Press, New York (1980).

8. R. Carroll, J. Ash, P. Vogt, and J. Singer, Reversion of transformed glycolysis to normal by inhibition of protein synthesis in rat kidney cells infected with temperature sensitive mutant of Rous sarcoma virus, Proc. Natl. Acad. Sci., U.S.A., 75: 5015-5019 (1978).
9. B. Demple and S. Linn, DNA N-glycosylases and UV repair, Nature (London), 287:203-208 (1980).
10. L. Dimnik, Mutagenesis of the mitochondrial genome by methotrexate, M.Sc. Thesis, University of Calgary (1982).
11. L. Dimnik, R. B. Church, and D. I. Hoar, Induction of mitochondrial mutants by methotrexate, submitted for publication (1983).
12. M. Goulian, B. Bleile, and B. Y. Tseng, Methotrexate-induced misincorporation of uracil into DNA, Proc. Natl. Acad. Sci., U.S.A., 77:1956-1960 (1980).
13. T. C. Hall and G. A. Gudouskas, Chemical pharmacology and biochemical interactions of currently useful anticancer drugs, Meth. Can. Res., 17:313-349 (1979).
14. K. R. Harrap, B. T. Hill, M. E. Furness, and L. I. Hart, Sites of action of amethopterin: Intrinsic and acquired drug resistance, Ann. N.Y. Acad. Sci., 186:312-324 (1971).
15. D. I. Hoar and P. Sargent, Chemical mutagen sensitivity in Ataxia telangiectasia, Nature (London), 261:590-592 (1976).
16. H. A. Ingraham, B. Y. Tseng, and M. Goulian, Mechanism for exclusion of 5-fluorouracil from DNA, Cancer Research, 80: 998-1001 (1980).
17. H. A. Ingraham, B. Y. Tseng, and M. Goulian, Nucleotide levels and incorporation of 5-fluorouracil and uracil into DNA of cells treated with 5-fluorodeoxyuridine, Molec. Pharmacol., 21:211-216 (1982).
18. S. Kearsey and I. W. Craig, Altered ribosomal RNA genes in mitochondria from mammalian cells with chloroamphenical resistance, Nature, 290:607-608 (1981).
19. S. Kit, D. R. Dubbs, and P. M. Frearson, HeLa cells resistant to bromodeoxyuridine and deficient in thymidine kinase activity, J. Cancer, 1:19-30 (1966).
20. S. Kit, W.-C. Leung, and L. A. Kaplan, Distinct molecular form of thymidine kinase in mitochondria of normal and BudR-resistant HeLa cells, Eur. J. Biochem., 39:43-48 (1973).
21. T. A. Kunkel and L. A. Loeb, Fidelity of mammalian DNA polymerases, Science, 211:765-767 (1981).
22. B. A. Kunz, B. J. Barclay, J. C. Game, J. G. Little, and R. H. Haynes, Induction of mitotic recombination in yeast by starvation for thymidine nucleotides, Proc. Natl. Acad. Sci., U.S.A., 77:6057-6061 (1980).
23. B. A. Kunz and R. H. Haynes, DNA repair and the genetic effects of thymidylate stress in yeast, Mutation Research, 93:353-375 (1982).
24. J. Littlefield, Selection of hybrids from matings of fibroblasts in vitro and their presumed recombinants, Since, 145: 709-710 (1964).

25. H. Mahler, Biogenetic autonomy of mitochondria, C.R.C. Crit. Rev. Biochem., 1:381-460 (1973).
26. V. McKusick, Medelian inheritance in Man, John Hopkins Press (1979).
27. J. A. Montgomery, Synthetic chemicals, in: "Methods in Cancer Research: Cancer Drug Development," Part A (V. T. DeVita, Jr., and H. Busch, eds.), pp. 3-25, Academic Press, New York (1979).
28. G. Pontecorvo, Production of mammalian somatic cell hybrids by means of polyethylene glycol treatment, Somatic Cell Genet., 1: 377-400 (1975).
29. J. W. Shay, Selection of reconstituted cells from karyoplasts fused to chloramphenicol-resistant cytoplasts, Proc. Natl. Acad. Sci., U.S.A., 74:2461-2464 (1977).
30. E. M. Southern, Detection of specific sequences among DNA fragments separated by gel electrophoresis, J. Mol. Biol., 89: 503-517 (1975).
31. C. M. Spolsky and J. M. Eisenstadt, Chloramphenicol-resistant mutants of human HeLa cells, FEBS Letters, 25:319-324 (1972).
32. E.-M. Suolinna, D. R. Long, and E. Racher, Quercetin, and artifical regulation of the high aerobic glycolysis of tumor cells, J. Natl. Cancer Inst., 53:1515-1519 (1974).
33. D. C. Wallace, C. L. Bunn, and J. M. Eisenstadt, Cytoplasmic transfer of chloramphenicol resistance in human tissue culture cells, J. Cell Biol., 67:174-199 (1975).
34. D. C. Wallace, N. A. Oliver, H. Blanc, and C. W. Adams, A system to study human mitochondrial genetics: Application to chloramphenicol resistance, in: "Mitochondrial Genetics" (P. Slonimski, et al., eds.), pp. 105-116, Cold Spring Harbor Laboratories, New York (1982).
35. R. Waters and E. Moustacchi, The fate of UV-induced pyrimidine dimers in the nuclear and mitochondrial DNAs of Saccharomyces cerevisiae on various postirradiation treatments and its influence on survival and cytoplasmic "petite" induction, in: "Molecular Mechanisms for DNA Repair" (P. C. Hanawalt and R. B. Setlow, eds.), pp. 556-565, Plenum Press, New York (1975).
36. A. Wiseman and G. Attardi, Reversible ten fold reduction in mitochondrial DNA content of human cells treated with ethidium bromide, Mol. Gen. Genet., 167:51-63 (1978).
37. A. Wiseman and G. Attardi, Cytoplasmically inherited mutations of human cell line resulting in deficient mitochondrial protein synthesis, Somatic Cell Genet., 5:241-262 (1980).
38. E. A. Wurtz, B. B. Sears, D. K. Rabert, H. S. Sheperd, N. W. Gillham, and J. E. Boynton, A specific increase in chloroplast gene mutations following growth of Chlamydomonas in 5-fluorodeoxyuridine, Molec. Gen. Genet., 170:235-242 (1979).

ADDENDUM

Experiments completed since the submission of this manuscript indicate that the CAP^R mutants reported here may not be true mutants. The general interpretations of these findings remain unchanged.

INDUCTION OF MUTATION AND CHROMOSOME DAMAGE

BY EXCESS BASES AND NUCLEOSIDES

Diana Anderson

The British Industrial Biological Research Association
Woodmamsterne Road
Carshalton, Surrey, SM5 4DS

SUMMARY

The studies reported in this communication were conducted to determine if naturally occurring cell constituents could themselves cause mutation. All the bases and their corresponding nucleosides have been shown in culture to produce chromosome damage in P388 mouse lymphoma and Chinese hamster ovary cells (CHO) and adenosine has produced an increased incidence of sister chromatid exchanges in CHO cells. In addition thymidine has produced an increase in V79 cells resistant to 8-azaguanine and ouabain, although no effects were seen in six strains of *Salmonella typhimurium*. Such damage probably arises as a result of imbalanced DNA-precursor pools. Thus mutagenic events may arise by mechanisms unrelated to direct alterations of DNA.

INTRODUCTION

Mutagenic effects can occur as a result of direct interaction of physical and chemical agents with DNA, from errors which arise during its self-replication, from interference with nuclear division and possibly from perturbations in DNA-precursor pools. The present communication will be concerned with this last possibility. Toxic and mutagenic effects resulting from thymidine imbalance in the DNA

Abbreviations: dCTP, deoxycytidine triphosphate; dTTP, deoxythymidine triphosphate; BUdR, 5-bromo-2-deoxyuridine; PBS, phosphate-buffered saline.

precursor pools have been reported in bacteriophage [5] and microbial systems [11, 13]. Thymidylate deprivation induced in yeast by imbalance is lethal and recombinogenic whereas excess thymidylate is a potent mutagen [4, 20]. Excess of thymidine produces an increase in the intracellular dTTP:dCTP ratio [8, 32, 33] and if DNA synthesis continues in conditions of this pyrimidine nucleotide imbalance, it would seem likely that deoxycytidine will be substituted by thymidine leading to an A:T transition mutation. This has been shown to be the case for bacteriophage T4, where thymine starvation has produced the reverse effect, A:T to G:C transition [34].

Thymidine can block the cell cycle in S phase [2, 36] rescue cells from the toxic effects of amethopterin [18], and exhibit selective lethality to tumor cells in vitro and in vivo [26, 27]. Such effects are probably mediated by thymidine blocking DNA synthesis so preventing the reduction of CDP to dCDP, thus depressing the pool of dCDP [8, 24, 18, 32]. Morris and Fischer [29] have shown that the toxic effect of thymidine can be reversed by the addition of dCTP. Recent results have shown that bromodeoxyuridine mutagenesis to ouabain resistance in HAB-2E cells in culture is stimulated by thymidine and suppressed by deoxycytidine [14] and purine deoxyribonucleosides stimulate bromodeoxyuridine mutagenesis in Syrian hamster melanoma cells [22]. Nitrosoguanidine mutagenesis is facilitated by thymidine and deoxycytidine for ouabain and azaguanine resistance in V79 Chinese hamster cells [31] and excess thymidine potentiates the cytotoxic and mutagenic effects of monofunctional alkylating agents in Chinese hamster ovary cells [30]. Thymidine itself has been reported to be mutagenic in various rodent lines for resistance to excess thymidine and 5-iodo-2-deoxyuridine [2], or purine analogs [9], whilst Brennand and Fox [10] suggest a possible mechanism to explain such observations involving preferential killing of wild-type cells.

Thymidine has also been shown to produce chromosome damage in Chinese hamster cells [37] as has deoxyadenosine in human leucocytes [23].

In the present study, the effect of thymidine on the induction of 8-azaguanine- and ouabain-resistant V79 cells has been investigated as has the effect of all the bases and their corresponding nucleosides on the chromosomes of both P388 mouse lymphoma cells and Chinese hamster ovary cells. The effect of adenosine on sister chromatid exchanges in CHO has also been studied. In addition, all the bases and nucleosides were examined in a spot test in 6 strains of Salmonella typhimurium and in a plate incorporation assay in strain TA100.

TABLE 1. Survival and Mutation Frequency in V79 Cells Treated with Thymidine Using 8-Azaguanine (A) and Ouabain (0) as Selective Agents

Dose		Expt.	Selective agent	Cell survival				Frequency of mutants per survivor
Expression time (h)		1	A		24	42	66	
		2	A		22	43	67	
		3.4	0		27	43	67	
Dose (M)	Treatment time (h)				Mean number of mutants per plate			
0, negative control		1	A	100	27	26	20	2.3×10^{-4}
		2	A	100	4	5.6	0	4.2×10^{-5}
		3.4	0	100	0	0	0	2.5×10^{-7}
Thymidine								
4×10^{-3}	18	1	A	100	68	60	68	5.3×10^{-4}
4×10^{-3}	18	2	A	87	1.3	4.6	11	9.4×10^{-5}
4×10^{-3}	18	3	0	87	1	1.6	5.6	4.8×10^{-5}
2×10^{-2}	4	1	A	92	27	43	19	4.6×10^{-4}
2×10^{-2}	4	2	A	68	2	14	7.3	$2\ \times10^{-4}$
4×10^{-3}	18	4	0	100	2.5	0	1.3	1.9×10^{-5}
Expression time (h)					24	46	56	
EMS	2	1	A	13	3.6	47	55	4.2×10^{-3}
8×10^{-5}	2	3	0	13	2.3	9.2	15	1.1×10^{-3}

Expts. 1 and 2 were selected with 8-azaguanine at 0.197 mM and Expts. 3 and 4 with ouabain at 1 mM Four mutation plates were used for each dose expression time. The frequency of mutation for both treated and spontaneous cultures was calculated from the number of resistant colonies at the optimum expression time. The frequency of ouabain resistance in the control series has been derived from 200 plates in 5 separate Expts. In the Expt. shown above no resistant colonies were observed.

MATERIALS AND METHODS

Chemicals

The bases and corresponding nucleosides were obtained from Sigma Chemical Co., St. Louis, Missouri (U.S.A.).

Cell Lines and Mutation Assays

The V79 line was obtained from Dr. C. F. Arlett (MRC, Cell Mutation Unit, University of Surrey, Brighton, Great Britain). The mutation assay in V79 cells and the way in which the cells were grown and maintained have been previously described [3]. Cells were treated for 4 or 18 h with thymidine and varying expression times were used (Table 1).

The origin of the P388 cell line and method of suspension in culture have also previously been described [16]. The CHO cells were obtained from Flow Laboratories, Irving (Great Britain) or from Dr. J. P. O'Neill (Oak Ridge, Tennessee, U.S.A.) and grown as a monolayer in a MEM (Flow Labs.) supplemented with 15% fetal calf serum (Flow Labs.) or F12 (Gibco) supplemented with 5% fetal calf serum (Sera Labs.) respectively.

Treatment and Processing of P388 and CHO Cells for Chromosome Studies

For both P388 and CHO cells the bases or nucleosides were in contact with the cells for approximately 2 doublings (32 and 24 h respectively) except that adenosine was in contact with the cells for 5 or 30 h in the SCE assay in the presence of 3 μg/ml BUdR. At the end of this period the medium was removed by centrifugation and washing in PBS for P388 cells and washing in PBS alone for CHO cells and both cell lines were supplemented with fresh growth medium containing colchicine or Colcemid (Sigma Chemical Co. (London) Ltd., Poole, Dorset, Great Britain) to give a final concentration of 0.05 μg/ml. Before harvesting the monolayer cells were trypsinized (0.25%) in saline and the trypsin neutralized by addition of serum contained in the culture medium. After harvest the cells were centrifuged, the supernatant removed and replaced with 0.075 M KCl and then left at room temperature (CHO for 4 or 15 min and P388 for 15 min). Cells were respun, the supernatant removed and replaced with fixative (1:3, glacial acetic acid:methanol). After 3 subsequent changes of fixative, slides were prepared by air-drying and stained in 4% Giemsa, Gurrs R66 (Hopkin and Williams, Chadwell Heath, Essex, Great Britain) or 2% Giemsa in 0.3 M phosphate buffer at pH 10.4.

Cells were scored for the incidence of chromatid and isochromatid gaps and breaks, fragments, rings, dicentrics, interchanges

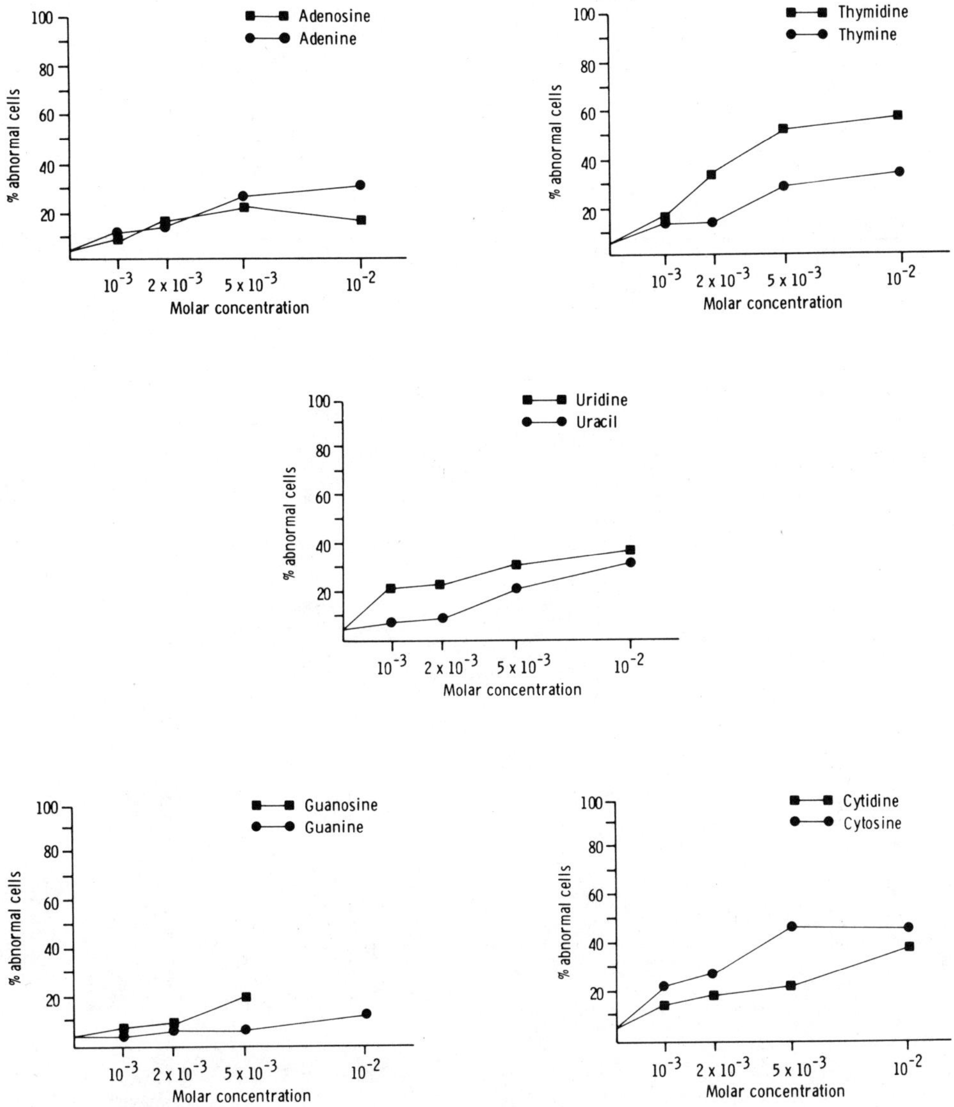

Fig. 1. Chromosome damage in P388 cells.

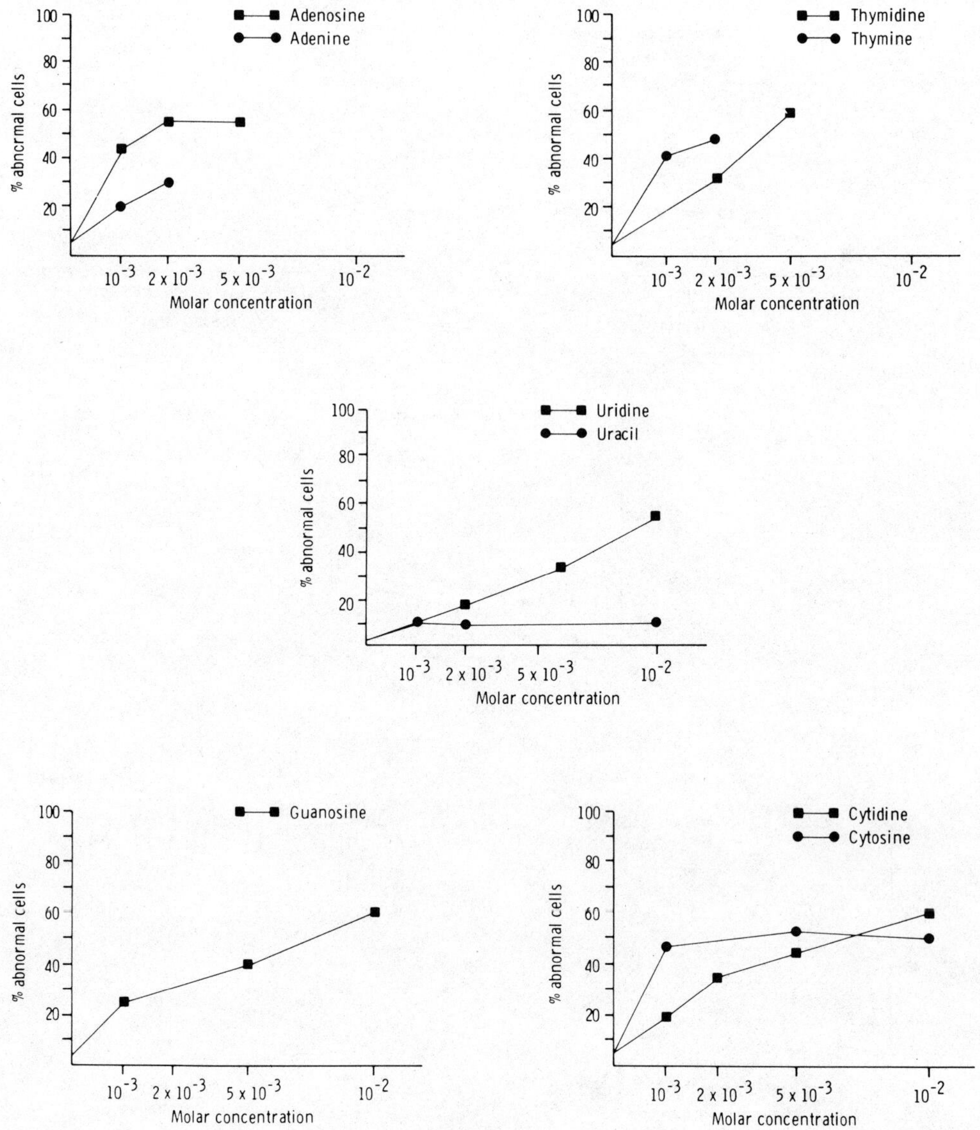

Fig. 2. Chromosome damage in CHO cells.

TABLE 2. Sister Chromatid Exchanges (SCE) per Cell (CHO) After Treatment with Adenosine

DOSE (5 HR TREATMENT)			DOSE (30 HR TREATMENT)		
UG/ML	MEAN SCE	%GROWTH	UG/ML	MEAN SCE	% GROWTH
0	10.0	100	0	9.96	100
	[2.27]			[1.93]	
125	11.8	96	12.5	10.24	106
	[1.64]			[1.98]	
250	9.04	90	25	10.92	101
	[1.74]			[2.40]	
500	8.16	93	50	10.56	97
	[1.57]			[2.57]	
1000	10.72	91	100	13.48***	83
	(2.91]			[2.65]	
2000	10.32	95	200	NM	55
	[2.54]				

FIGURES IN PARENTHESES = STANDARD DEVIATION

*** = $P<0.001$ STUDENT'S T TEST

NM = NO MITOSES STAINED DIFFERENTIALLY

and other more complex structural rearrangements. One hundred cells were scored per treatment for chromosome aberrations and 25 cells were examined for sister chromatid exchanges.

Salmonella typhimurium Assays

The strains TA1535, TA1537, TA1538, TA97, TA98, and TA100 of *Salmonella typhimurium* were obtained from Prof. B. N. Ames (University of California, Berkeley, U.S.A.).

The methods used for the plate incorporation assay were as described by Ames et al. [1] and 0.5 ml of 10% S-9 mix were added per plate. For the spot test S-9 mix was incorpoated into the top agar and a few crystals of the bases and nucleosides were placed in the center of each plate.

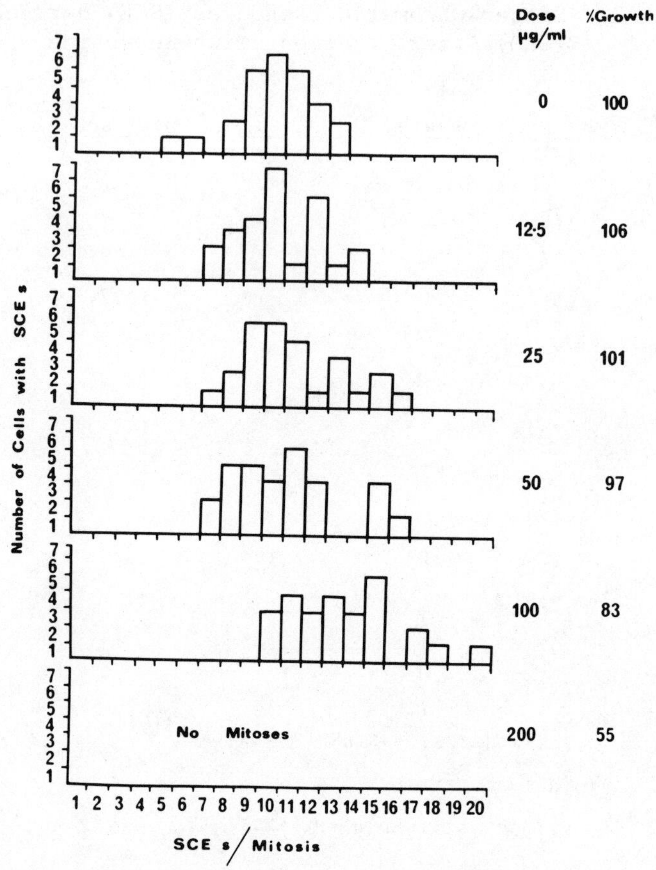

Fig. 3. The effects of adenosine on SCE frequency in CHO cells.

RESULTS

The results for survival and mutation frequency in V79 cells treated with thymidine using azaguanine and ouabain as selective agents are shown in Table 1. The doses chosen produced little kill with an absolute increase in mutant colony numbers above control values.

The results from the scoring of chromosome damage in P388 mouse lymphoma and Chinese hamster ovary cells are shown in Figs. 1 and 2 respectively. There was an increase in chromosome damage above control values with all agents (except guanine in CHO cells for which the preparations failed). Some of the bases and nucleosides tended to be insoluble, particularly guanosine and guanine, so that the concentrations quoted may be inaccurate. Generally all categories of chromosome damage were seen with pulverization induced at some of the highest doses.

TABLE 3. The Treatment of Salmonella typhimurium Strain TA100 with Bases and Nucleosides in the Plate Incorporation Assay

COMPOUND	DOSE/PLATE (μ MOLES)	MEAN REVERTANTS/PLATE	
		-S9 MIX	+S9 MIX
CONTROL	(100 μL DMSO)	118	102
ADENINE	0.5	89	83
ADENOSINE	5	89	99
CYTIDINE	10	103	97
CYTOSINE	0.5	108	92
GUANINE	0.5*	97	111
GUANOSINE	10	88	102
THYMINE	5	86	114
THYMIDINE	5	89	88
URACIL	2.5	100	91
URIDINE	10	102	92
BENZO(A)PYRENE	(5μG)	NT	1542

PIT = PLATE INCORPORATION TEST

NT = NOT TESTED

* = SOLUBILITY PROBLEM

Adenosine produced an increase in sister chromatid exchanges in CHO cells after a 30 h treatment but not after 5 h (Table 2) and an altered sister chromatid exchange distribution as shown in Fig. 3.

No increases in mutant yield above control values were seen for any of the six Salmonella strains in the spot test or in the plate incorporation assay in strain TA100 (Table 3).

DISCUSSION

Brennand and Fox [10] found a small increase in V79 cells resistant to thioguanine after treatment with thymidine and thought that the increase was due to selection of spontaneous mutants, since they demonstrated that thymidine selectively killed wild-type cells. The findings of the present study have shown that thymidine can produce an increase in V79 cells resistant to azaguanine and ouabain at doses producing high levels of survival or no kill and as a consequence there could be little or no selective killing of wild-type cells by comparison with resistant cells. For ouabain resistance there was over a 100-fold increase in mutation frequency. These findings would tend to confirm the observation of Anderson [2] that mutants can be induced as well as selected. The mechanisms by which thymidine can cause base transition were discussed earlier, and it is probable that similar mechanisms might operate for the other bases and nucleosides. The lack of effect of the bases and nucleosides in the Salmonella typhimurium strains, however, is not readily explainable. Perhaps a fluctuation assay may yield different results.

There was evidence of increased sister chromatid exchanges after the 30 h adenosine treatment at the highest dose level, but because of the reduction in survival observed at this dose level it cannot be ruled out that the observed increase may have been due to a selective sampling artefact, as postulated by Carrano et al. [12]. All the bases and their corresponding nucleosides were also shown to cause increased yields of chromosomal aberrations in 2 cell lines. The extent of the chromosome damage, however, is difficult to explain by transition alone. With such a high incidence of chromosome damage there may be a combination of events where repair processes are also involved causing mispairing. There is a known complexity of effects on intracellular pool sizes brought about by perturbation of the de novo and salvage pathways of purine and pyrimidine biosynthesis [8, 19, 25, 28, 32]. It has been shown [15, 21] that the cells of ataxia telangiectasia patients can replicate before repair processes are initiated, thus causing increased chromosome-breaking events, so suggesting another mechanism by which genetic damage can be mediated.

The fact that bases and nucleosides can cause mutagenic effects has implications for chemicals which may not directly attack DNA. In the in vitro situation it has been shown that nucleotide biosynthesis is markedly altered by a variety of chemical mutagens [6, 35] including chromium [7]. In the in vivo situation the balance of bases and nucleosides in the precursor pools would normally be maintained by feed-back mechanisms, but chemicals could disrupt such feed-back control causing pool imbalance and perhaps mutation in a manner unrelated to the direct attack of DNA by the chemical. The significance of such events for an unknown test compound is questionable and at present the means are not available to make risk assessments for potential mutagens acting through such a mechanism.

ACKNOWLEDGMENTS

Part of this report has been previously published (D. Anderson, C. R. Richardson, and P. J. Davies. The genotoxic potential of bases and nucleosides, Mutat. Res., 91:265-272 (1981)). The author would like to thank Dr. B. J. Phillips, A. Tee, and S. D. Blowers of this laboratory for their contribution.

REFERENCES

1. B. N. Ames, J. McCann, and E. Yamasaki, Methods for detecting carcinogens and mutagens with the Salmonella/mammalian microsome mutagenicity test, Mutation Res., 31:347-364 (1975).
2. D. Anderson, The selection and induction of 5-iodo-2-deoxyuridine and thymidine resistant variants of P388 mouse lymphoma cells with agents which are used for selection, Mutation Res., 33:399-406 (1975).
3. C. F. Arlett, Mutagenicity testing with V79 hamster cells, in: "Handbook of Mutagenicity Test Procedures" (B. J. Kilbey, M. Legator, W. Nichols, and C. Ramel, eds.), pp. 175-192, Elsevier/North-Holland, Amsterdam (1977).
4. B. J. Barclay and J. G. Little, Genetic damage during thymidylate starvation in Saccharomyces cerevisiae, Mol. Gen. Genet., 160:33-40 (1978).
5. C. Bernstein, M. Bernstein, S. Mufti, and B. Strom, Stimulation of mutation in phage T4 by lesions in gene 32 and by thymidine imbalance, Mutation Res., 16:113-119 (1972).
6. V. Bianchi, Nucleotide pool imbalance induced in cultured cells by treatments with different chemicals, Toxicol., 25:13-18 (1981).
7. V. Bianchi, P. Debetto, A. Zantedeschi, and A. G. Levis, Effects of hexavalent chromium on the adenylate pool of hamster fibroblasts, Toxicol., 25:19-30 (1982).
8. G. Bjursall and P. Reichard, Effects of thymidine on deoxyribonucleoside triphosphate pools and deoxyribonucleic acid synthesis in Chinese hamster ovary cells, J. Biol. Chem., 248: 3904-3909 (1973).
9. M. O. Bradley and N. A. Sharkey, Mutagenicity of thymidine to cultured Chinese hamster cells, Nature (London), 274:607-608 (1978).
10. J. Brennand and M. Fox, Excess thymidine is not mutagenic in Chinese hamster V79 fibroblasts, Cell Biol. Int. Rep., 4:923-932 (1980).
11. S. Bresler, M. Mosevitsky, and L. Vyacheslavov, Complete mutagenesis in a bacterial population induced by thymine starvation on solid media, Nature (London), 225:764-766 (1956).

12. A. V. Carrano, L. H. Thompson, P. A. Lindl, and J. L. Minkler, Sister Chromatid Exchanges as an Indicator of Mutagenesis, Nature (London), 271:551-553 (1978).
13. C. A. Coughlin and E. A. Adelberg, Bacterial mutation induced by thymine starvation, Nature (London), 178:531-532 (1956).
14. R. L. Davidson and E. R. Kaufman, Bromodeoxyuridine mutagenesis in mammalian cells is stimulated by thymidine and suppressed by deoxycytidine, Nature (London), 276:722-723 (1978).
15. M. J. Edwards and A. M. R. Taylor, Unusual levels of (ADP-ribose)$_n$ and DNA synthesis in ataxia telangiectasia cells following -irradiation, Nature (London), 287:745-747 (1980).
16. M. Fox, Repair synthesis and induction of thymidine revertant variants in mouse lymphoma cells of different radiosensitivity, Mutation Res., 23:129-145 (1974).
17. G. A. Gentry, P. A. Morse, D. H. Ives, R. Gerbert, and V. R. Potter, Pyrimidine metabolism in tissue culture cells derived from rat hepatomas, II. Thymidine uptake in suspension cultures derived from the novikoff hepatoma, Cancer Res., 25: 509-516 (1965).
18. G. A. Gentry, P. A. Morse, and V. R. Potter, Pyrimidine metabolism in tissue culture cells derived from rat hepatomas, III. Relationship of thymidine to the metabolism of other pyrimidine nucleosides in suspension cultures derived from the novikoff hepatoma, Cancer Res., 23:517-525 (1965).
19. L. J. Gudas, A. Cohen, B. Ullman, and D. W. Martin, Analysis of adenosine-mediated pyrimidine starvation using cultured wild type and mutant mouse T-lymphocyte cells, Somatic Cell Genet., 4:201-219 (1978).
20. R. H. Haynes, J. G. Little, B. A. Kunz, and B. J. Barclay, Non-DNA primary targets for the induction of genetic change, in: "Environmental mutagens and carcinogens" (T. Sugimura, S. Kondo, and K. Takebe, eds.), pp. 121-127, Alan R. Liss, Inc., New York (1981).
21. J. Houldsworth and M. F. Lavin, Effect of ionizing radiation on DNA synthesis in ataxia telangiectasia cells, Nucl. Acid Res., 8:3709-3720 (1980).
22. E. R. Kaufman and R. L. Davidson, Bromodeoxyuridine mutagenesis in mammalian cells is stimulated by purine deoxyribonucleosides, Somatic Cell Genet., 5:653-663 (1979).
23. B. A. Kihlman, W. W. Nichols, and A. Levan, Effects of deoxyadenosine and cytosine arabinoside on the chromosomes of human leucocytes in vitro, Hereditas, 50:139-143 (1963).
24. J. H. Kim, S. H. Kim, and M. L. Eidinoff, Cell Viability and Nucleic Acid Metabolism after Exposure of HeLa Cells to Excess Thymidine and Deoxyadenoside, Biochem. Pharmacol., 14:1821-1829 (1965).
25. D. Kuebbing and R. Werner, A model for compartmentation of de novo and salvage thymidine nucleotide pools in mammalian cells, Proc. Natl. Acad. Sci., U.S.A., 72:3333-3336 (1975).

26. S. S. Lee, B. G. Giovanella, and J. S. Stehlin, Selective lethal effects of thymidine on human and mouse tumor cells, J. Cell. Physiol., 92:401-406 (1977).
27. S. S. Lee, B. C. Giovanella, J. S. Stehlin, and J. C. Brunn, Regression of human tumors established in nude mice after continuous infusion of thymidine, Cancer Res., 39:2928-2933 (1979).
28. E. C. Moore and R. B. Hurlbert, Regulation of mammalian deoxyribonucleotide biosynthesis by nucleotides as activators and inhibitors, J. Biol. Chem., 241:4802-4809 (1966).
29. N. R. Morris and G. A. Fischer, Studies concerning the inhibition of cellular reproduction by deoxyribonucleosides, I. Inhibition of the synthesis of deoxycytidine by a phosphorylated derivative of thymidine, Biochim. Biophys. Acta, 68:84-92 (1963).
30. A. R. Peterson and H. Peterson, Facilitation by pyrimidine deoxyribonucleosides and hypoxanthine of mutagenic and cytoxic effects of monofunctional alkylating agents in Chinese hamster cells, Mutation Res., 61:319-331 (1979).
31. A. R. Peterson, J. R. Landolph, H. Peterson, and C. Heidelberger, Mutagenesis of Chinese hamster cells is facilitated by thymidine and deoxycytidine, Nature (London), 276:508-510 (1978).
32. E. C. Reynolds, A. W. Harris, and L. R. Finch, Deoxyribonucleoside triphosphate pools and differential thymidine sensivities of cultured mouse lymphoma and myeloma cells, Biochim. Biophys. Acta, 561:110-123 (1979).
33. L. Skoog and G. Bjursall, Nuclear and cytoplasmic pools of deoxyribonucleoside triphosphates in Chinese hamster ovary cells, J. Biol. Chem., 249:6434-6438 (1974).
34. M. D. Smith, R. R. Green, L. Ripley, and J. W. Drake, Thymineless mutagenesis in bacteriophage T4, Genetics, 74:393-403 (1973).
35. E. Volkin, M. E. Boling, W. M. Lee, and M. H. Jones, The Effect of Chemical Mutagens on Purine and Pyrimidine Nucleotide Biosynthesis, Biochim. Biophys. Acta, 755:271-274 (1983).
36. N. Xeros, Deoxyriboside control and synchronization of mitosis, Nature (London), 194:682-683 (1962).
37. S. J. Yang, G. M. Hahn, and M. A. Bagshaw Chromosome Aberrations induced by thymidine, Exp. Cell Res., 42:130-135 (1966).

THE GENETIC CONSEQUENCES OF THE THY$^-$ MUTATION TO CHO CELLS

Mark Meuth†, Otelinda Gonçalves,
and Marie Trudel†

Clinical Research Institute of Montreal
110 Pine Avenue West
Montreal, Quebec H2W 1R7
Canada

1. INTRODUCTION

The behavior and regulation of dNTP pools in somatic cells has long led to speculation that they may play a role in DNA synthesis besides that of simple precursors [2, 20]. While controversy has developed concerning the relationship of dNTP pools to DNA synthesis [18, 19], observations from several laboratories provide convincing evidence that pool variations in somatic cells strongly influence spontaneous mutational rates and sensitivity to DNA damaging agents. The objective of the work in our laboratory has been to elucidate the basis of these effects on the accuracy of DNA synthesis by using defined somatic cell mutations which alter dNTP synthesis and pools. In this communication we summarize the properties of one of these mutant strains of Chinese hamster ovary (CHO) cells, called Thy$^-$, and we speculate how the pool alterations produced by this metabolic defect might contribute to increased mutational rates and altered sensitivity to DNA damaging agents.

2. ISOLATION AND PROPERTIES OF THY$^-$ STRAINS

The Thy$^-$ strains were originally detected among arabinosyl cytosine-(araC) resistant mutants by their inability to grow in medium lacking thymidine [16]. Thy$^-$ strains are rare among araC^R strains representing only 10 to 20% of those screened. In addition to being resistant to low levels of araC, they are also cross resistant to the cytotoxic effects of thymidine and 5-fluorouracil

†Present address: Imperial Cancer Research Fund, Mill Hill Laboratories, London NW7 IAD, U.K.

TABLE 1. Effect of the Thy^- Mutation and Excess Thymidine on Mutation Rates in CHO Cells

Marker	*Altered Protein*	*Stimulation of mutational rate in Thy^- strains relative to control*[a]	
		Low Thymidine (4 μM)	*Excess Thymidine (2 mM)*
6-thioguanine resistance ($6\text{-}tg^r$)	Hypoxanthine guanine phosphoribosyl transferase (hprt)	13-35 fold	3-7 fold
Ouabain resistance (oua^R)	Na^+/K^+ dependent[1] ATPase	7-300 fold	3-7 fold
Emetine resistance (emt^r)	S20 ribosomal[2] protein	.4-2.4 fold[b]	2-3.5 fold
8-azaadenine resistance[c]	adenine phosphoribosyl transferase (aprt)	4.5-6.5 fold	6-14 fold

[a] *Stimulation of mutational rate is the mutational rate in Thy^- strains divided by the rate in the wild type strains, in the case of low thymidine. In excess thymidine, wild type cells are non viable, so stimulation of mutational rate is the rate at high thymidine divided by the mutational rate at thymidine concentrations giving minimal values (~ 0.1 mM). the values used in this table were calculated as described earlier.*[10] *The range of values obtained is from different Thy^- strains.*

[b] *The 2.4 fold stimulation is not significantly greater than the standard deviation of these measurements.*

[c] *The values for stimulation of mutational rates at the aprt locus are from Thy^- derivatives of strains hemizygous for the locus.*[17] *The diploid nature of the aprt gene in most cell strains precludes measurement of mutational rates to aprt deficiency.*

(5FU) [14]. The cross resistance of Thy⁻ mutants to araC and 5FU has enabled us to develop a selection system specific for Thy⁻ strains. When 5FU is included in the selection medium with araC, all strains isolated are also auxotrophic for thymidine [14].

Though the frequency of Thy⁻ mutants is very low as determined with our araC-5FU selection (spontaneous frequency, $<10^{-7}$; and $\sim 10^{-5}$ in EMS mutagenized cultures), the phenotype is unstable [16]. Thy⁺ revertants occur at a frequency of 10^{-3} to 10^{-4}. To determine whether this instability extended to other independent genetic markers, we examined mutational rates in Thy⁻ strains at four independent loci (Table 1). Rates of mutation at three of these loci were increased 5- to 300-fold depending upon the thymidine content of the medium and the Thy⁻ strain used [10-15]. Thus Thy⁻ is a mutator gene in CHO cells.

Somatic cell hybrids between Thy⁻ and wild type strains show resistance to the toxic effects of araC, thymidine and 5FU and retain increased mutational rates [15, 16]. Thus the resistance to drugs and the mutator gene are dominant or codominant.

3. BIOCHEMICAL CHARACTERIZATION OF THE THY⁻ MUTATOR PHENOTYPE

3.1. Altered dNTP Pool Balance

Intracellular dCTP content was significantly increased in all Thy⁻ strains examined - 5- to 10-fold higher than the level in wild type strains [10, 14-16]. None of the other dNTP pools were significantly altered when cells were grown in thymidine. Deprivation of thymidine led to a rapid and substantial decrease in the level of dTTP consistent with the cells being unable to synthesize dTTP. dGTP also decreased, since dTTP is an essential positive effector of dGTP synthesis [5]. Additionally, Robert de Saint Vincent, et al., observed that Thy⁻ strains have a 5-fold increase in CTP [22]. The pools are regulated in a normal pattern. Varying the thymidine in the medium of Thy⁻ strains modulates the dCTP and dGTP pools in a manner consistent with the pattern for the regulation of dNTP synthesis via regulation of ribonucleotide reductase, though at all concentrations of thymidine the dCTP level in Thy⁻ strains exceeds that in wild type [16]. CTP pools are not altered by variations in exogenous thymidine (our own observations).

3.2. Biochemical Lesion in Thy⁻ Strains

To determine the biochemical lesion responsible for the complex phenotype of Thy⁻ strains, we and others examined theenzymes responsible for dCTP and dTTP synthesis (Fig. 1). The key enzyme involved in the synthesis of both dCTP and dTTP, as well as the key regulatory point, is ribonucleotide reductase [20]. Therefore, our in-

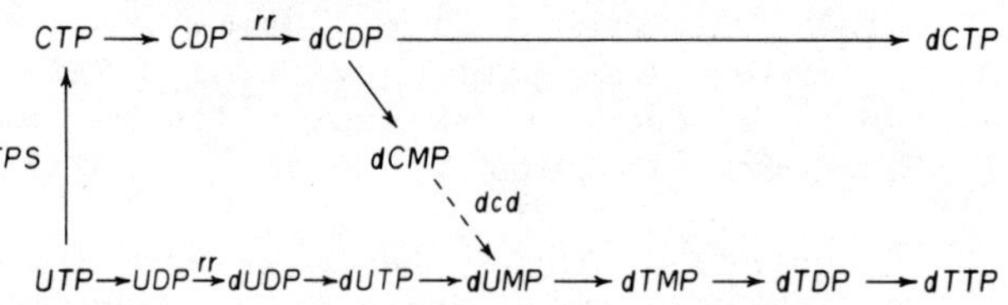

Fig. 1. Pathways of pyrimidine deoxynucleoside triphosphate synthesis in CHO cells. CTPs, CTP synthetase; dcd, deoxycytidylate deaminase; rdr, ribonucleotide reductase. Dashed line for dcd indicates that this activity cannot be detected in our CHO strains.

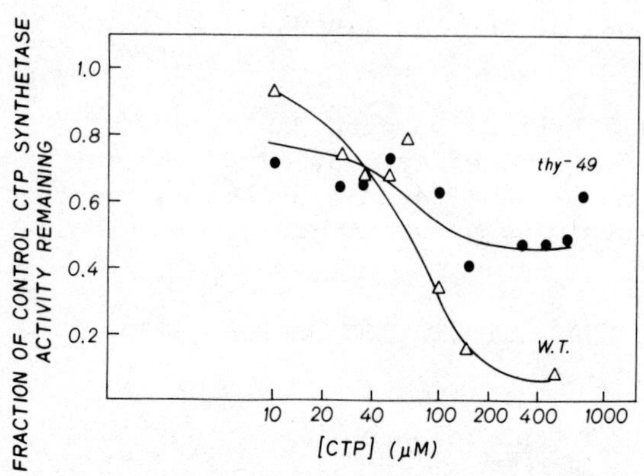

Fig. 2. Effect of CTP on CTP synthetase activity of wild type and thy⁻49. Cell-free extracts were purified of effector molecules by ammonium sulfate precipitation and dialyzed before assay. △) Wild type, pro⁻; ●) thy⁻49.

TABLE 2. CTP Synthetase Activity in Wild Type and Thy⁻ Strains

	Cell Strain	Specific Activity pmole/μg/h No addition	+ 500 μM CTP	Fraction of Control Activity Remaining in 500 μM CTP
Experiment 1	W.T.	5.3	0.089	0.016
	Thy⁻49	3.3	1.4	0.42
	$Thy^{+}_{2}49$-2	0.73	< 0.4	< 0.05
Experiment 2	W.T.	10.7	0.16	0.015
	$Thy^{-}_{3}49$-2-1	2.7	2.8	1.0
	$Thy^{-}_{3}49$-2-2	3.2	3.0	0.91

CELLULAR PHENOTYPE	W.T.	$\xrightarrow{10^{-5}}$	THY^{-}_{1}	$\xrightarrow{10^{-4}}$	THY^{+}_{2}	$\xrightarrow{10^{-5}}$	THY^{-}_{3}
CTP SYNTHETASE	100% CTP^{S}		50% CTP^{S} 50% CTP^{R}		100% CTP^{S}		100% CTP^{R}
PROPOSED GENOTYPE	$^{+}/_{+}$		$^{+}/_{MUT}$		$^{+}/_{-}$		$^{MUT}/_{-}$

Fig. 3. Proposed model for the generation of thy⁻ phenotypes. The frequencies of each event are indicated. CTP^{R}, CTP synthetase resistant to CTP. CTP^{S}, CTP synthetase sensitive to CTP. The figures above each step indicate the frequencies of each event - the first and last are in EMS mutagenized cultures while the second represents a spontaneous frequency.

vestigations began with analysis of this enzyme. Reduction of three substrates, CDP, UDP, and GDP, were examined in wild type and Thy⁻ cell-free extracts purified of effector molecules. It became clear from these analyses that neither the reduction of these substrates nor their feedback regulation by dNTP's was altered in Thy⁻ strains [23]. Therefore, enzymes involved in the interconversion of dCTP and dTTP were examined. Assay of dCMP deaminase in mutant and wild type strains showed that neither had detectable levels of this enzyme. CTP synthetase, on the other hand, was clearly altered [23]. The enzyme activity in mutant strains responded in a complex manner to inhibition by its end product, CTP, in that half of the activity appeared normally sensitive to CTP while half was resistant (Fig. 2, Table 2). Similar alterations were reported earlier by Robert de Saint Vincent and Buttin [21] for hamster strains having a phenotype similar to Thy⁻. Clearly, such an alteration of CTP synthetase could explain the behavior of pools in Thy⁻ strains. Loss of regulation of CTP synthesis could account for the increased CTP. This might well increase CDP levels leading to increased dCTP. The thymidine auxotrophy could be explained by the higher level of CDP competing out reduction of UDP.

To examine the biochemical basis of the high reversion frequency of Thy⁻ strains to the Thy⁺ state, CTP synthetase activity was examined in two independent revertants [23]. CTP resistant CTP synthetase activity was lost in both strains. But additionally, total activity was reduced (Table 2) and the revertant strains became hypersensitive to araC. These observations suggested that Thy⁻ strains might be reverting by loss or inactivation of the mutant allele. Consequently, Thy⁻ mutants were reisolated from the Thy⁺

revertants, the idea being that, if the mutant allele were lost or inactivated, CTP synthetase activity in the new Thy^- mutants (Thy^-3 or third step) should be completely resistant to inhibition by CTP (see Fig. 3). This was clearly found to be the case in four independent Thy^-3 mutants, isolated using our specific araC-5FU selection (Table 2). Thus the high frequency reversion event occurring in Thy^- strains appears to be due to the loss (deletion) or inactivation (via new point mutation, gene rearrangement or modification) of the CTP synthetase structural gene.

Fig. 4. Response of mutational rates in wild type and thy^- mutator strains to manipulations of exogenous thymidine (A) and endogenous dNTP content (B). In panel A, replica cultures of wild type (●), thy^-49 (○), or thy^-303 (△) were grown in medium containing defined concentrations of thymidine. At the appropriate density the entire culture was plated in medium containing ouabain (oua) to determine the fraction of cultures containing oua^R cells. The mutational rate to oua^R was calculated on the basis of the cultures with no mutants. In panel B, cultures of wild type (●), thy^-49 (○), and thy^-303 (△) growing in defined thymidine concentrations were analyzed for dNTP content. The mutational rates obtained in part A were then plotted as either a function of dCTP (thymidine concentration $<10^{-4}$ M); or dTTP (thymidine concentrations $>10^{-4}$ M). The figures are aligned so that the exogenous thymidine concentrations in panel A correspond roughly to the dCTP or dTTP level thy produce in panel B.

3.3. CTP Resistant CTP Synthetase is a Mutator Gene in CHO Cells

There are several lines of evidence providing convincing evidence that the CTP resistant CTP synthetase is the mutator gene in Thy^- strains: (1) Several independent Thy^- mutator strains have CTP resistant CTP synthetase [23]. (2) The frequency of Thy^- mutants in mutagenized cultures of CHO cells is consistent with mutation at a single locus being responsible for the phenotype [14]. Revertants of Thy^- strains to the wild type state (with respect to drug sensitivity, thymidine requirement, and mutational rate) lose mutant enzyme activity [23]. Finally (4) the mutator phenotype and altered CTP synthetase are cotransferred by DNA transfection into

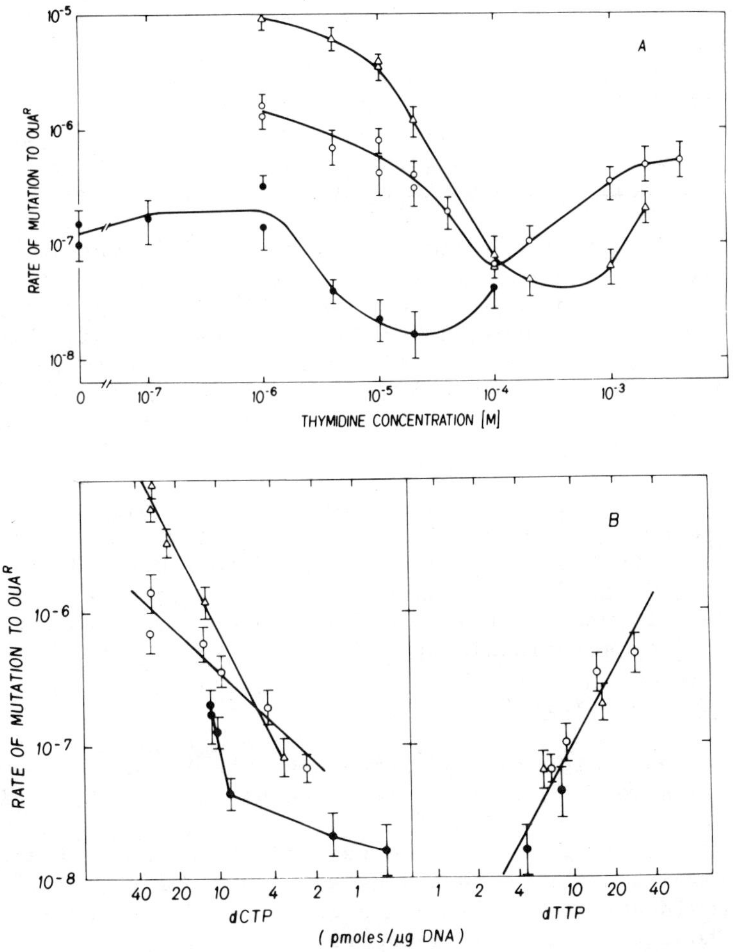

Fig. 4

wild type cells [23]. Left unexplained is the mechanism by which an enzyme far removed from DNA replication or repair so strongly influences mutational rates and sensitivity to DNA damaging agents.

4. THE GENETIC CONSEQUENCES OF THE THY$^-$ MUTATION

4.1. Increased Spontaneous Mutational Rates

The increased mutational rates observed in Thy$^-$ strains are conditional and locus specific. Exogenous thymidine strongly affects mutational rates in Thy$^-$ strains (Fig. 4) [10]. At low concentrations of thymidine mutational rates are at their maximum. They decrease with increasing thymidine until about 100 μM and then increase again with high concentrations of thymidine (>1 mM). It is also evident that mutational rates in Thy$^-$ strains, vary greatly, as much as 10-fold, eventhough there is no discernible difference in phenotype. Mutational rates in wild type cells are also conditional with exogenous thymidine, though the variations are not nearly as great as with the Thy$^-$ strains. Excess thymidine (>1 mM) is mutagenic as well as cytotoxic to wild type strains.

The dNTP content of wild type and Thy$^-$ cells was measured at these different concentrations of thymidine and when mutational rates were expressed as a function of dCTP in low thymidine (<100 mM) or dTTP in high thymidine (>100 mM), clear relationships emerged (Fig. 4). In the case of mutational rates to ouabain resistance (oua^R) in low thymidine, these functions were complex and unique for each Thy- strain. In the case of high dTTP a single relationship emerged for wild type and two Thy$^-$ strains. The difference in response to variations in the two pools may well reflect differences in the intracellular distributions of the two pools [18]. Thus, while these experiments provide good evidence that the conditionality of the Thy$^-$ mutator gene is related to fluctuations of the dCTP level (CTP levels are unaffected by variations of thymidine in the medium), very little light is shed on the mechanism by which the pools alter mutational rates.

4.2. Locus Specificity

Mutational rates in Thy$^-$ strains can be increased as much as 300-fold in some Thy$^-$ strains. But the effects are not uniform, as the mutational rates at other loci are unaffected (Table 1) [10]. This locus specificity cannot be readily explained by the type of mutations which can be detected at a locus. For example, both ouabain resistant (oua^R) [1] and emetine resistant (emt^r) mutants [3] are thought to be strictly point mutants, while thioguanine resistant (tg^r, HPRT$^-$) mutants can include point mutants as well as deletions and insertions since the target is a nonessential gene. The thy$^-$ mutator gene increases the rate of mutation 20- to 300-fold

TABLE 3. Stability of Thy^- Induced $HPRT^-$ Mutants

$HPRT^-$ Strain	Isolation[a] μM Thymidine)	Rate of Reversion to HAT^+ in[b]		Frequency of HAT^+ revertants[c] after EMS treatment	
		4 μM Thymidine	2 mM Thymidine	Control	1.6 mM
Thy^-49 tg^r 6.13-1	1 μM	$< 4 \times 10^{-9}$	1.6×10^{-8}	8×10^{-8}	2×10^{-7}
Thy^-49 tg^r 6.16-1	1 μM	9.2×10^{-9}	9.9×10^{-9}	$< 10^{-7}$	$< 10^{-7}$
Thy^-49 tg^r 6.26-1	1 μM	9.2×10^{-9}	2.4×10^{-8}		
Thy^-49 tg^r 6.27-1	1 μM	$< 3.3 \times 10^{-9}$	2.1×10^{-8}	$< 10^{-7}$	1.4×10^{-6}
Thy^-404 tg^r 6.4-2	4 μM	$< 9.7 \times 10^{-9}$	1.7×10^{-8}	$< 10^{-7}$	2.9×10^{-6}

[a] Forward mutants were isolated by fluctuation test in 1 μM or 4 μM thymidine as indicated. Rates of mutation in Thy^- strains to 6-tg^r under these conditions is about 1.0×10^{-6} mutants/cell/generation.

[b] Values given are reversion rates as determined by Luria Delbrük fluctuation test plated in HAT medium as described previously.[10] Cells were grown to desired densities in the indicated thymidine concentrations before plating in HAT.

[c] Cultures of the indicated mutants were treated with EMS (1.6 mM) as described previously.[11] Control cultures were grown in the same medium, but were not treated with EMS. The frequency of revertants represents the colonies forming in HAT medium/cells plated x the fraction forming colonies in control medium).

TABLE 4. Sensitivity of Thy⁻ and Wild Type Strains to Various DNA Damaging Agents

	D_{10}[a] in the presence of					
	4 μM Thymidine			100 μM Thymidine		
Cell Strain	EMS (mM)	U.V. (sec)	X-rays (rads)	EMS (mM)	U.V. (sec)	X-rays (rads)
W.T.	2.9	13	550	1.0	12	610
Thy^-49	3.4	12	460	1.7	12	400
Thy^-303	3.8	NT	550	2.5	NT	530

NT = not tested.

[a] D_{10} is the dose of the indicated agent reducing the plating efficiency of treated cultures to 10% of that of unexposed cultures. In these experiments indicated cell strains were grown in medium in the indicated thymidine concentrations 24 h before and 7 days after treatment (to allow formation of surviving colonies).

for oua^R mutants, 20-50-fold for $6\text{-}tg^r$ but does not alter the rate at the emt locus.

It is also possible that the site specificity relates to the site of the gene on the chromosome (position effect) or the ploidy level. The emt locus maps in CHO cells at a region on chromosome 2 which is deleted on the homologous chromosome [24]. Thus emt is hemizygous in CHO cells. The Thy^- mutator does not increase mutational rates at the emt locus and works weakly at the adenine phosphoribosyl transferase (aprt) locus in a strain hemizygous for that gene. Thus one could argue that somatic recombination might be involved in the mechanism as two copies of the gene are necessary. On the other hand, mutational rates are increased at the X-linked hprt locus. Mutagenesis by excess thymidine in Thy^- strains does not show the same locus specificity (Table 1).

4.3. Thy^- Induced $HPRT^-$ Mutants are Stable

As the Thy^- phenotype is unstable, it was interesting to examine the stability of Thy^- induced mutants. The rate of reversion of Thy^- induced $HPRT^-$ mutants to $HPRT^+$ (as determined by plating in HAT medium) was 10^{-9} revertants per cell per generation or lower (Table 3), which is 1000-fold below the forward rate (10^{-6} mutants/cell/generation). Additionally many of the "revertants" isolated from HAT appeared to be second site revertants in that they remained tg^r. Treatment of $HPRT^-$ strains with a DNA alkylating agent (EMS) or 2 mM thymidine increased reversion frequencies slightly, though revertants were still rare. Thus mutants induced by the Thy^- mutator gene are stable.

4.4. Altered Sensitivity to DNA Damaging Agents

As there may be components in common between cellular mechanisms involved in maintaining the accuracy of DNA replication and those responsible for repair of lesions induced by DNA damaging agents, it seemed possible that pool imbalances could influence the cytotoxicity and/or mutagenicity. Thus, the sensitivity of Thy^- and wild type strains to DNA alkylating agents, UV radiation and X-rays were examined. Thy^- strains were slightly resistant to EMS, though they were normally sensitive to the cytotoxic effects of UV and X-rays (Table 4). Due to the high background of spontaneous mutations in Thy^- strains, it is more difficult to judge how the pool imbalance alters the mutagenicity of these agents. In wild type cells, mutagenicity of DNA alkylating agents is decreased by high levels of dCTP [11]. Presumably, high dCTP reduces the likelihood of the suspected promutagenic lesion produced by alkylating agents, O^6 alkylguanine [7], mispairing with thymine to produce a GC → AT transition mutation [4]. Consistent with this explanation is the effect of a mutation in CHO cells reducing dCTP levels. Deoxycytidine kinase deficient mutants have lower intracellular levels of dCTP and become hypersensitive to the cytotoxic and mutagenic effects of ethylmethane sulfonate [12].

5. THE MOLECULAR BASIS OF MUTATIONS INDUCED BY THE THY^- MUTATOR GENE

DNA precursor imbalances produce misincorporation of dNTPs by in vitro replication systems [2, 6]. However, the complex responses elicited in vivo by thymine deprivation in prokaryotes and lower eukaryotes, combined with the unusual properties of the Thy^- mutator gene (instability and locus specificity), provides an impetus for analyzing mutations induced by the Thy^- mutator gene. This analysis has recently become possible as a result of the cloning of a 3.8 kb hamster genomic fragment containing the structural gene for aprt by Lowy et al. [9]. Using this molecular probe for a selectable genetic locus, mutations induced by Thy^- can be examined by Southern blotting [13, 17]. Basically DNAs from $aprt^-$ mutants are purified, digested with restriction endonucleases, and fractionated on agarose gels. These DNA digests are then transferred to nitrocellulose and aprt containing sequences are probed using a ^{32}P labeled probe. Gross rearrangements (deletions or insertions) of mutant aprt genes are immediately detectable by the loss or gain of aprt containing sequences - to a limit of resolution on agarose gels of about 50 nucleotide base pairs [17]. Point mutants have a normal pattern of

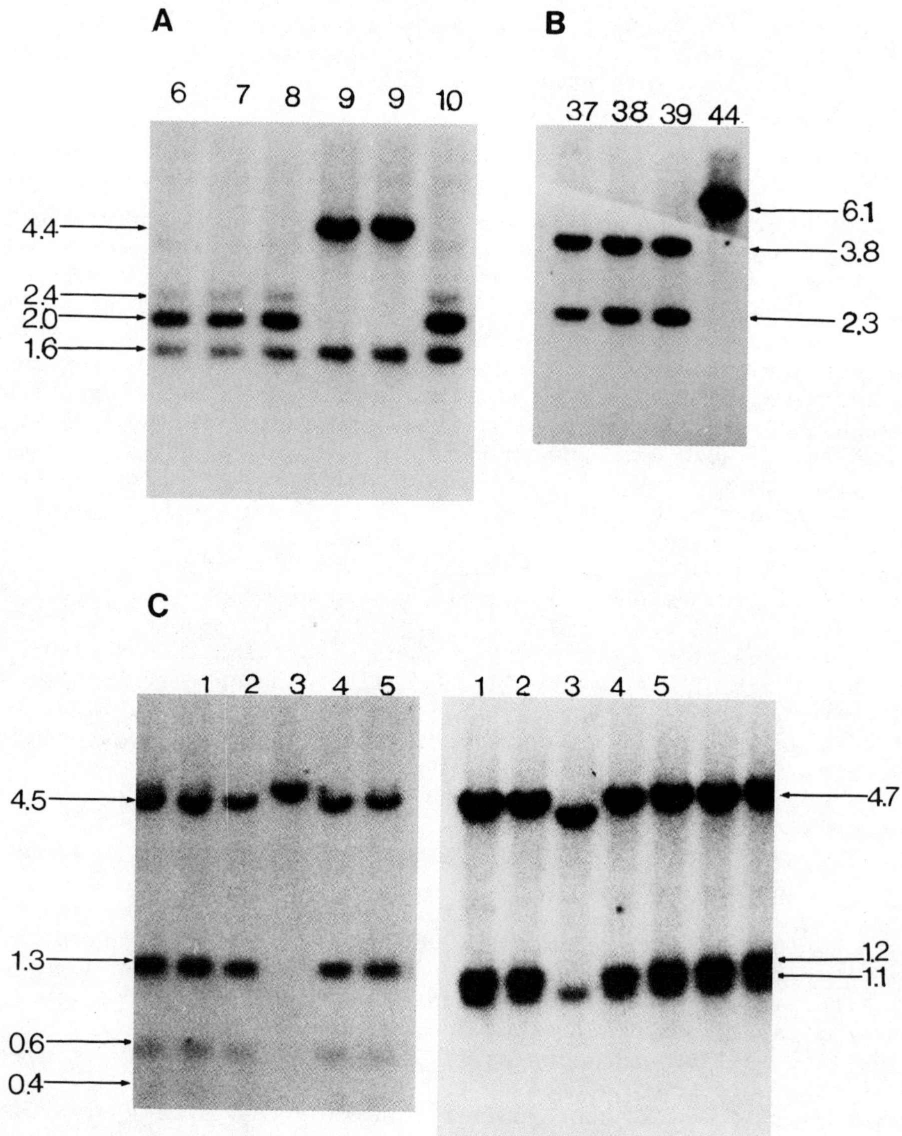

fragments unless they occur within a restriction enzyme recognition sequence leading to the loss of the site and the generation of a new larger fragment.

One limitation of this approach is that aprt is an autosomal gene and is therefore present in two copies in most strains making

←

Fig. 5. Southern blot analysis of aprt⁻ mutants induced by the Thy⁻ mutator gene or excess thymidine. A) Analysis of Thy induced aprt⁻ mutants from strain D422 Thy⁻2-1. Mutant 9 clearly shows an altered pattern consistent with a modification of the Pvu 11 recognition sequence. The two aprt containing fragments normally produced by cleavage of the aprt gene by Pvu 11, have been lost and a novel fragment equal in size to these two fragments is generated. No other differences were detected in aprt containing sequences from these mutant strains after digestion with nine other restriction endonucleases. B) Pattern of aprt containing sequences in mutants of D422 induced by excess thymidine (10 mM for 24 h). The pattern of one mutant dT44 is altered in such a way as to suggest modification of the Pst 1 site in the aprt gene of this mutant. C) Southern blot analysis of DNA's of aprt⁻ mutants isolated from D422 Thy⁻1-1 in high concentrations of thymidine (2 mM). Patterns of aprt containing sequences in all restriction endonucleases digestions of DNA from one mutant, D422 Thy⁻1-1 dT3, were altered in that some aprt containing fragments disappeared while others were altered in size. Such alterations indicate a deletion in the aprt gene of this mutant. Shown here are Taql (left) and Mspl (right) digests of DNA's from the deletion mutant and other aprt⁻ mutants with apparently normal patterns of aprt sequences. The numbers on the side of each figure indicate the size (in kilobases, b) of the indicated aprt fragment.

it difficult to obtain Thy⁻ induced aprt⁻ mutants. However, recently we found that a CHO strain, D422, is hemizygous for the aprt gene [17] making it feasible to analyze spontaneous and Thy⁻ induced aprt⁻ mutants.

The patterns of aprt containing sequences of some of these Thy⁻ induced aprt⁻ mutants from D422 can be seen in Fig. 5. Basically, patterns were unchanged in most of the mutants suggesting that they occurred either as a result of single base pair change or small rearrangements below our limit of resolution (about 30 base pairs). In one mutant an alteration was observed which appeared to be the result of the loss of a Pvull site. Mutants induced by excess thymidine (10 mM) were also examined. Most patterns were again unchanged, but two deletions were observed in our collection of 60 mutants and several lost or gained restriction enzyme sites - Taql, Mspl, or Pstl. Loss of these sites has also been observed in spontaneous and EMS induced mutants. The restriction fragment length polymorphisms observed in these mutants allow localization of damage induced by Thy⁻ or excess thymidine, facilitating subsequent cloning and sequencing for specific base pair changes. Though induction of large rearrangements of gene structure by these agents can be ruled out, sequencing data will be essential to precisely define their mode of action.

6. CONCLUSIONS

The dCTP imbalance induced by the Thy⁻ mutation has significant genetic consequences for somatic cells. (1) Spontaneous mutational rates are increased up to 300-fold, depending on exogenous thymidine and the locus involved, and (2) sensitivity to DNA alkylating agents is altered. Undoubtedly, misincorporation, perhaps aided by altered proofreading [6, 8], can explain some of the effects. However, more complex responses, such as induction of error prone repair, are certainly not outside the realm of possibility. We think that defining the specific base pair changes induced by Thy⁻ at the aprt locus will go a long way to elucidating the effects of pool fluctuations on mutation.

ACKNOWLEDGMENTS

The work done in this laboratory which is described here was supported by grants from the Medical Research Council and the National Cancer Institute of Canada.

REFERENCES

1. R. M. Baker, Nature and use of ouabain-resistant mutants. in: "Banbury Report 2. Mammalian Cell Mutagenesis: The Maturation of Test Systems" (A. W. Hsie, J. P. O'Neill, V. K. McElheney, eds.), Cold Spring Harbor Laboratory, pp. 237-247 (1979).
2. G. Bjursell and P. Reichard, Effects of thymidine on deoxyribonucleoside triphosphate pools and deoxyribonucleic acid synthesis in Chinese hamster ovary cells, J. Biol. Chem., 248: 3904-3909 (1973).
3. D. Boersma, S. M. McGill, J. W. Mollenkamp, and D. J. Roufa, Emetine resistance in Chinese hamster cells in linked genetically with an altered 40S ribosomal subunit protein, S20, Proc. Natl. Acad. Sci., U.S.A., 76:415-419 (1979).
4. C. Coulondre and J. H. Miller, Genetic studies of the lac repressor. IV. Mutagenic specificity in the lac 1 gene of Escherichia coli, J. Mol. Biol., 117:557-606 (1977).
5. S. Ericksson, L. Thelander, and M. Åkerman, Allosteric regulation of calf thymus ribonucleoside diphosphate reductase, Biochemistry, 18:2948-2952 (1979).
6. A. R. Fersht, Fidelity of replication of phage ϕX174 DNA by DNA polymerase III holoenzyme: spontaneous mutation by misincorporation, Proc. Natl. Acad. Sci., U.S.A., 76:4946-4950 (1979).
7. P. D. Lawley and C. J. Thatcher, Methylation of deoxyribonucleic acid-cultured, mammalian cells by N-methyl-N-nitro-N-nitrosoguanidine. The influence of cellular thiol concentrations on the extent of methylation and the 6-oxygen atom of guanine as a site of methylation, Biochem. J., 116:693-707 (1979).
8. L. Loeb and T. A. Kunkel, Fidelity of DNA synthesis, Ann. Rev. Biochem., 52:429-457 (1982).
9. I. Lowy, A. Pellicer, J. F. Jackson, G. K. Sim, S. Silverstein, and R. Axel, Isolation of transforming DNA: cloning the hamster aprt gene, Cell, 22:817-823 (1980).
10. M. Meuth, Sensitivity of a mutator gene in Chinese hamster ovary cells to deoxynucleoside triphosphate pool alterations, Molec. Cell. Biol., 1:652-660 (1981).
11. M. Meuth, Role of deoxynucleoside triphosphate pools in the cytotoxic and mutagenic effects of DNA alkylating agents, Somatic Cell Genet., 7:89-102 (1981).
12. M. Meuth, Deoxycytidine kinase-deficient mutants of Chinese hamster ovary cells are hypersensitive to DNA alkylating agents, Mut. Res., 110:383-391 (1983).
13. M. Meuth and J. E. Arrand, Alterations of gene structure in ethyl methane sulfonate-induced mutants of mammalian cells, Molec. Cell. Biol., 2:1459-1462 (1982).
14. M. Meuth, O. Gonçalves, and P. Thom, A selection system specific for the Thy mutator phenotype, Som. Cell. Genet., 8: 423-432 (1981).

15. M. Meuth, N. L'Heureux-Huard, and M. Trudel, Characterization of a mutator gene in Chinese hamster ovary cells, Proc. Natl. Acad. Sci., U.S.A., 76:6505-6509 (1979).
16. M. Meuth, M. Trudel, and L. Siminovitch, Selection of Chinese hamster cells auxotrophic for thymidine by 1-β-D-arabinofuranosyl cytosine, Somatic Cell Genet., 5:303-318 (1979).
17. J. Malbantoglu, O. Gonçalves, and M. Meuther, Structure of mutant alleles at the aprt Locus of Chinese hamster ovary cells, J. Mol. Biol., 167:575-594 (1983).
18. B. Nicander and P. Reichard, Dynamics of pyrimidine deoxynucleoside triphosphate pools in relationship to DNA synthesis in 3T6 mouse fibroblasts, Proc. Natl. Acad. Sci., U.S.A., 80:1347-1351 (1983).
19. G. P. V. Reddy and A. B. Pardee, Multienzyme complex for metabolic channeling in mammalian DNA replication, Proc. Natl. Acad. Sci., U.S.A., 77:3312-3316 (1980).
20. P. Reichard, From deoxynucleotides to DNA synthesis, Fed. Proc., 37:9-14 (1978).
21. B. Robert de Saint-Vincent and G. Buttin, Studies on 1-β-D-arabinofuranosylcytosine-resistance mutants of Chinese hamster fibroblasts, Biochim. Biophys. Acta, 610:352-359 (1980).
22. B. Robert de Saint-Vincent, M. deChamps, and G. Buttin, The Modulation of the Thymidine Triphosphate Pool of Chinese Hamster Cells by dCMP Deaminase and UDP Reductase, J. Biol. Chem., 255:162-167 (1980).
23. M. Trudel, T . Van Genechten, and M. Meuth, Biochemical characterization of the hamster Thy mutator gene and its revertants, Submitted for publication.
24. R. C. Worton, C. Duff, and C. E. Campbell, Marker Segregation without chromosome loss at the emt locus in Chinese hamster cell hybrids, Somat. Cell. Genet., 6:199-213 (1980).

DEOXYRIBONUCLEOSIDE-INDUCED SELECTIVE MODULATION OF CYTOTOXICITY AND MUTAGENESIS

A. R. Peterson,* P. V. Danenberg,†
L. L. V. Ibric, and H. Peterson

*The Cancer Research Laboratory
University of Southern California Cancer Center and
The Institute for Toxicology
School of Pharmacy
1303 N. Mission Road
Los Angeles, California 90033

†Department of Biochemistry
School of Medicine and the Cancer Center
Univerisity of Southern California
Los Angeles, California

1. INTRODUCTION

Treatment of Chinese hamster cells with deoxythymidine (dThd)‡ and deoxycytidine (dCyd) leads to phosphorylation of these nucleosides and imbalances of the cells' deoxyribonucleotide (dNTP) pools [17, 36]. Treatment of the cells with alkylating agents leads to alkylation of bases in DNA, and the alkylated bases may mispair during DNA replication [39]. Imbalances of pyrimidine dNTP pools or alkylation of DNA can induce cytotoxicity and mutagenesis [17, 39]. Synergistic effects of alkylating agents and deoxyribonucleosides on mutagenesis and cytotoxicity have been demonstrated by us [30-32] and more recently by other groups [5, 22, 26]. In the present article we have collected the results of our studies in this area. We also report data pertaining to the mechanisms of the above synergistic effects, and we discuss some possible practical applications of our findings to the control of the development of drug resistance in cells treated with chemicals used in cancer chemotherapy.

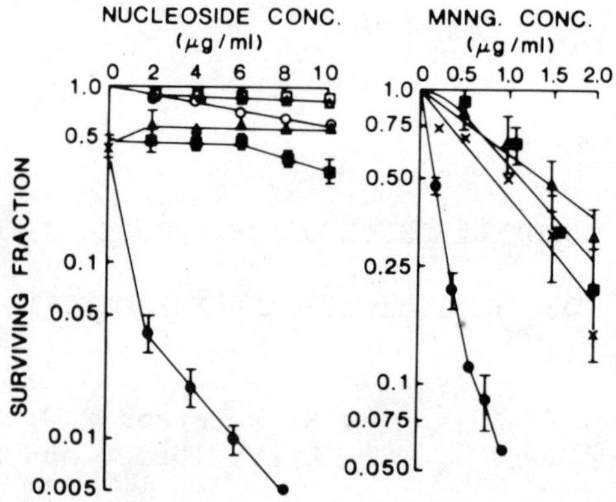

Fig. 1.
The cytotoxicity of MNNG in medium containing dThd (●), or dCyd (▲), equal concentrations of dThd plus dCyd (■), or in medium supplemented with dialyzed serum only (×). The cytotoxicity of the deoxyribonucleosides in the absence of MNNG is shown by the open symbols. The MNNG concentration was 6.8 μM (in the left panel), and the concentrations of dThd and of dCyd were 10 μM (in the right panel). (From Ref. 32.)

1.1. Effects of Thymidine and Deoxycytidine on Mutagenesis and Cytotoxicity Induced by N-Methyl-N'-nitro-n-nitrosoguanidine

The origin of the Chinese hamster lung fibroblasts (V79 cells) used in our experiments and methods for their culture in Dulbecco's medium supplemented with 10% dialyzed fetal calf serum are described in our publications [29, 34, 35]. For measurements of mutagenesis and cytotoxicity, at 4-6 h after plating in 55 mm or 100 mm plastic petri dishes the cells were treated with mutagen for 2 h in complete medium, washed once and covered with fresh medium. Cytotoxicity was measured by counting colonies of 50 or more cells that grew from single cells that survived treatment with mutagen and nucleosides. Mutagenesis to 8-azaguanine (AG) resistance was assayed with our extensively characterized techniques [29, 34, 35], with 3-10 days of exponential growth of mutagen-treated cells and replating ($\geq 10^5$ cells/100 mm dish) in medium to allow time (the expression time) for phenotypic lag prior to mutant selection in medium containing 40 μg AG/ml.

The effects of dThd and dCyd on the cytotoxicity of MNNG are shown in Fig. 1. The cytotoxicity of MNNG increased as an exponential function of the concentration of dThd and was increased more than 10-fold by 2 μg/ml ($\sim 10^{-5}$ M) dThd, which by itself reduced the surviving fraction by only 10%. By contrast, the cytotoxicity of MNNG was slightly (~10%) decreased by dCyd, which also abolished the increase induced by dThd.

The effects of dThd and dCyd on the induction of AG^r colonies by MNNG are shown in Fig. 2. The frequencies of AG^r colonies induced by MNNG increased as functions of the concentrations of dThd, dCyd, and dThd plus dCyd. In medium lacking these nucleosides, the

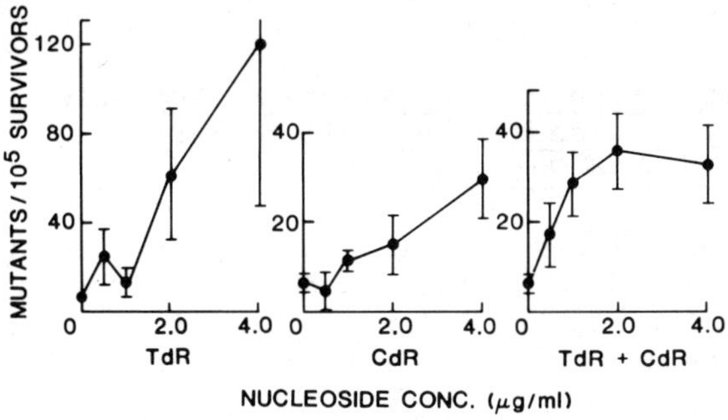

Fig. 2. The mutagenicity of MMNG (6.8 μM) in medium containing dThd, dCyd or dThd plus dCyd. Cells were replated in fresh medium containing the nucleosides 48 h after treatment with MNNG, and AG (40 μg/ml) was added 24 h later. (From Ref. 32.)

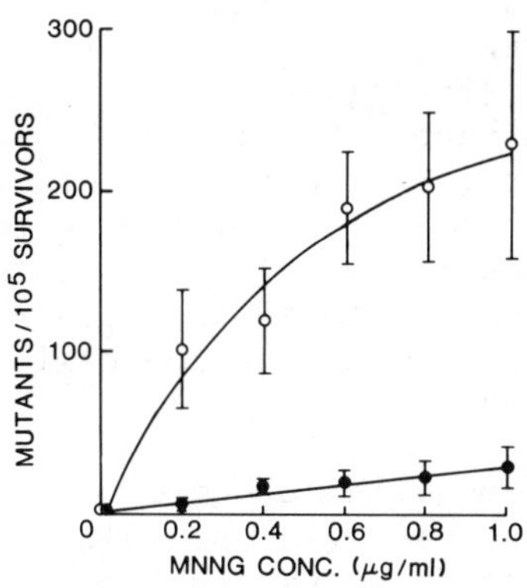

Fig. 3. Dose-response for MNNG-induced mutagenesis in V79 cells grown in the presence (○ 4 expts.) or in the absence (● 3 expts.) of dThd (10 μM) plus dCyd (10 μM). The expression time was 6 days, and the nucleosides were present throughout. (From Ref. 32.)

AG^r-colony frequency increased as a linear function of the dose of MNNG (Fig. 3). This linearity is consistent with previous data [25, 34] and with dose-responses for mutagenesis predicted by theory [20]. In medium containing dThd plus dCyd, the AG^r-colony frequencies induced by MNNG were increased 3-15-fold in 4 experiments (Fig. 3). Note that dCyd and dCyd plus dThd, which did not enhance the cytotoxicity of MNNG (Fig. 1), did increase the frequency of MNNG-induced AG^r colonies (Figs. 2 and 3), indicating that MNNG-induced mutagenesis and cytotoxicity in V79 cells could be selectively modulated.

2. MUTAGENESIS

2.1. Characterization of the Effects of Thymidine and Deoxycytidine as the Facilitation of MNNG-Induced Mutagenesis

The following lines of evidence show that the AG^r colonies induced by MNNG in medium containing dThd and dCyd do not merely reflect differential sensitivities, to AG, of cells treated with the deoxyribonucleosides and cells not treated with the deoxyribonucleosides [5]:

(i) Neither in the presence nor in the absence of dCyd plus dThd were the doubling times of AG^r clones different from those of the wild-type V79 cells. Therefore, the growth rate of AG^r colonies was not selectively increased in the presence of dThd plus dCyd.

(ii) Incubations with dCyd plus dThd increased the MNNG-induced AG^r-colony frequency but did not increase MNNG-induced cytotoxicity (Fig. 1), and a 2 h treatment with 10 μM dThd maximally enhanced MNNG-induced cytotoxicity but did not significantly increase the MNNG-induced AG^r-colony frequency [31]. These data indicate that the increased AG^r-colony frequencies were not due to alterations in the selecting efficiency of AG that can result from alteration in the growth kinetics of the cells exposed to dThd or dCyd plus dThd [5].

(iii) Maximal increases in AG^r-colony frequencies induced by MNNG were observed even when dThd or dThd plus dCyd were removed 5 days prior to treatment of the cells with AG. Moreover, exposure of MNNG-treated cells to dThd for 3 days prior to selection or only for the period of mutant selection produced, respectively, less than maximal or no increases in AG^r-colony frequencies [31, 32]. These data show that the increased AG^r-colony frequencies were not produced by effects of the deoxyribonucleosides on the selecting efficiency of the AG.

The following lines of evidence show that dThd and dCyd facilitate MNNG-induced mutagenesis at two genetic loci in V79 cells:

(i) Twenty AG^r colonies induced by MNNG in medium containing dThd plus dCyd were isolated and grown to mass cultures, and the phenotypes of these cultures were characterized [32]. None were AG^s, hypoxanthine-aminopterin-thymidine resistant (HAT^r), which is the phenotype of the wild-type cells; 70-80% were AG^r, HAT^s, which is the phenotype of hypoxanthine:guanine-phosphoribosyl-transferase deficient

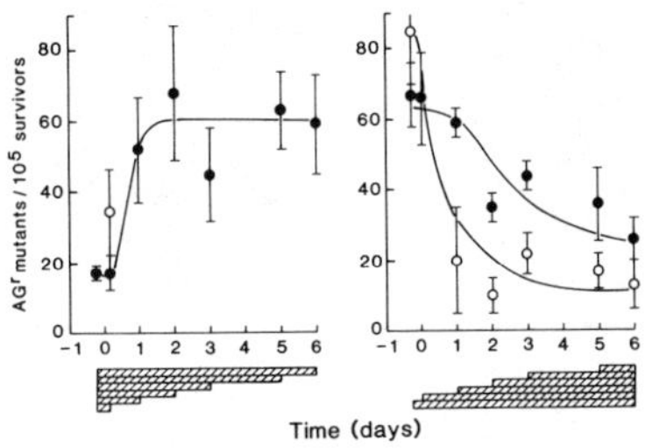

Fig. 4. Temporal aspects of the facilitation by 10 μM dThd (○) and 10 μM dThd plus 10 μM dCyd (●) of induction of AG^r mutants by 3.4 μM MNNG. MNNG was added to the culture medium at 0 h and removed by changing the medium at 2 h. The hatched horizontal bars show the periods when the cells were incubated in medium containing deoxyribonucleosides. Each data point shows the mean and SEM from two to five experiments. The spontaneous mutant frequencies were 0.3-1.4 mutants per 10^5 survivors. (From Ref. 31.)

(HGPRT$^-$) mutants [4, 29, 35], and 20-30% were AG^r, HAT^r, or AG^s, HAT^s, which are indicative of mutants containing HGPRT with altered substrate affinities [3, 7]. These observations show that dCyd plus dThd increased the frequencies of MNNG-induced AG^r mutants.

(ii) The MNNG-induced mutagenesis was maximally increased when dThd plus dCyd were present in the medium throughout the first 24 h after treatment of cells with MNNG, but further, significant increases did not occur on maintaining the MNNG-treated cells in medium containing dCyd plus dThd for longer than 24 h (Fig. 4), which is the doubling time of the MNNG-treated cells. Significant increases in the MNNG-induced mutagenesis did not occur in medium that contained the deoxyribonucleosides only until the end of the MNNG-treatment period of 2 h (Fig. 4), which is an order of magnitude less than the doubling time of the cells. Thus, one doubling of the MNNG-treated cells, which entails one round of replication of their DNA, in medium containing the deoxyribonucleosides was necessary and also sufficient for maximal enhancement of MNNG-induced mutagenesis.

(iii) Significant enhancement of mutagenesis did not occur in medium that contained dThd for only the 5 days immediately prior to mutant selection, and was significantly reduced in medium that contained dThd plus dCyd for only 4 days immediately prior to mutant selection (Fig. 4). During these periods 3.3 and 4.0 population doublings occurred, respectively. Thus, 5 population doublings in medium containing dThd prior to exposure to AG were necessary for maximal production of AG^r mutants. This number of dou-

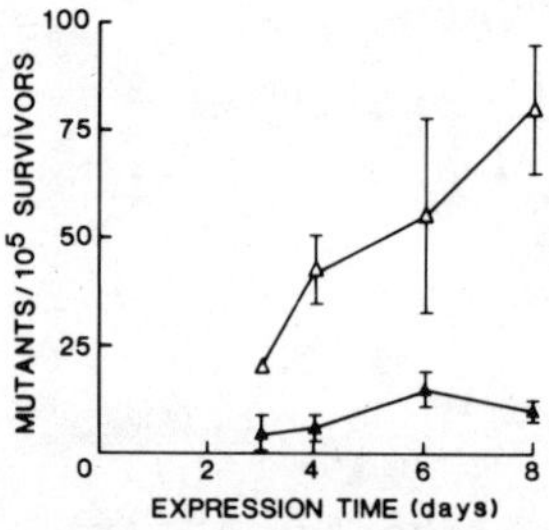

Fig. 5. Dependence of mutation frequency on expression time (the period between the end of treatment of the cells with mutagen and exposure to AG) in V79 cells treated with 6.8 μM MNNG and grown in medium supplemented only with dialyzed serum (▲) or, throughout the experiment, with dThd (10 μM) plus dCyd (10 μM) (△). (From Ref. 32.)

blings is consistent with the result in Fig. 5, which shows that with or without deoxyribonucleosides the duration of the phenotypic lag of MNNG-induced AG^r mutants is 4-6 doubling times. Methylating agents, such as MNNG, that contain N-methyl-N-nitroso groups can methylate all possible oxygen and nitrogen atoms in the bases of DNA [38]. Several of these methylated bases violate the Watson-Crick rules of base pairing in DNA replication and transcription in vitro [24, 39]. Moreover, treatment of Chinese hamster cells with dThd or dCyd results in expansion of the dTTP or dCTP pools of the cells [6, 36], and unbalanced dCTP and dTTP pools facilitate mispairing during replication of methylated DNA in vitro [1, 24]. Our data (Figs. 2-5) indicate that unbalanced dCTP and dTTP pools facilitate mispairing during replication of DNA that is methylated by MNNG in living V79 cells. Figures 4 and 5 suggest the following chronology for the process of mutagenesis by MNNG and dThd plus dCyd at the cells' HGPRT loci. In the course of the first day after treatment of the cells with MNNG and dCyd plus dThd, the surviving cells divide once entailing a single round of DNA replication that presumably suffices to produce most of the mispairing that results in AG^r mutants. During the next 5 days of the period allowed for phenotypic lag, the pre-existing HGPRT levels in the newly mutated cells presumably are depleted [4], and the resultant HGPRT-deficient mutants subsequently grow to form colonies during mutant selection with AG.

(iv) We have also recently shown that MNNG-induced mutagenesis to resistance to 10 μg/ml of 6-thioguanine (TG) was reproducibly enhanced more than 2-fold by dCyd plus dThd, providing further evidence of facilitation of MNNG-induced

TABLE 1. Effects of Deoxyribonucleosides on Background- and MNNG-Induced Mutagenesis. Expression Times were 6 Days for AG^r and TG^r Mutants and 2 Days for Oua^r Mutants. Means and Standard Errors of Results of 2 or More Experiments are Shown

	Relative Mutant Frequencies		
	$AG^{r\,e}$	$TG^{r\,e}$	$Oua^{r\,e}$
Background Mutations			
No Nucleosides	1 ± 0.2 (0.5)[a]	1 ± 0.5 (1.1)	1 ± 0.3 (1.9)
10^{-5}M dThd	2.7 ± 0.8		
2×10^{-5}M dCyd	2.2 ± 0.5		
10^{-5}M dThd + 10^{-5}M dCyd	2.2 ± 0.6[b]	6.2 ± 5.4	1.3 ± 0.6
5×10^{-6}M dGuo	3.2 ± 0.8		
10^{-5}M dAdo	2.9 ± 0.1		
Mutation Induced by 3.4 μM MNNG			
No Nucleosides	1 ± 0.2 (18.6)[a]	1 ± 0.2 (22)	1 ± 0.1 (4.9)
10^{-5}M dThd	11.0 ± 3.1		
2×10^{-5}M dCyd	4.5 ± 1.41[c]		
10^{-5}M dThd + 10^{-5}M dCyd	3.6 ± 0.5[d]	2.2 ± 0.04	2.5 ± 0.5
5×10^{-6}M dGuo	1.5 ± 0.4		
10^{-5}M dAdo	1.5 ± 0.1		

a : Mutants/10^{-5} survivors in brackets b : ν=44, $p > 0.05$ c : ν= 2, $p < 0.05$ d : ν=44, $p < 0.01$ (Student's t test, ν = degrees of freedom, p = probability that the difference from 1.0 is due to chance). e : Selection media did not contain deoxyribonucleosides.

mutagenesis at the HGPRT locus (Table 1). Moreover, Landolph and Heidelberger demonstrated facilitation of MNNG-induced mutagenesis to ouabain (Oua) resistance, which results from mutations at the sodium/potassium ATPase locus [30]. They used expression times of 3 days, with replating after 2 days and selection in 3 mM ouabain. However, even using more conventional protocols, namely, expression time of 2 days, no replating, and selection with 1 mM ouabain [4], we found that MNNG-induced mutagenesis was facilitated by dCyd and dThd (Table 1).

2.2. Background Mutagenesis and Effects of Purine Deoxyribonucleosides

Table 1 shows that dThd and dCyd did not significantly increase the background mutagenesis at the two genetic loci. Thus, the dNTP-pool imbalances that are presumably produced by 10^{-5} M dThd and 10^{-5} M dCyd were not sufficient by themselves to significantly induce mutagenesis.

The purine deoxyribonucleosides, deoxyadenosine (dAdo) and deoxyguanosine (dGuo), were incorporated into the DNA of the cells and, at concentrations greater than 2×10^{-5} M, increased the cytotoxicity of MNNG by more than 2-fold (Peterson, unpublished data). These findings indicate that dAdo and dGuo are phosphorylated in V79 cells and suggest that they affect the balance of the dNTP pools. However, at concentrations that did not significantly enhance MNNG-induced cytotoxicity, neither dAdo nor dGuo significantly affected background- nor MNNG-induced mutagenesis (Table 1), suggesting that mispairing of methylated bases with G or A were not major causes of MNNG-induced mutagenesis in the V79 cells. Moreover, since the purine deoxyribonucleosides were present while the cells were being treated with MNNG, the lack of any effects on mutagenesis suggests that the purine-dNTP pools of V79 cells are not as significant targets for MNNG as are the dNTP pools of C3H/10T1/2 cells for MNU [40].

2.3. Effects of Pyrimidine Deoxyribonucleosides

The mutant frequencies produced by treating cells with dThd plus dCyd that had been incubated with MNNG were not different from the background mutant frequency, showing that the facilitation of MNNG-induced mutagenesis did not result from any action of MNNG on the nucleosides (Peterson and Peterson, unpublished data). On exposure of MNNG-treated cells to the deoxyribonucleosides, dThd produced consistently higher AG^r mutant frequencies than did dThd plus dCyd (Figs. 2 and 4 and Table 1). This observation is consistent with the finding that elevated dTTP levels facilitate mispairing of T with O^6-methyl guanine (m^6G) [1, 24], which is produced by methylation of DNA with MNNG [19]. The elevated dTTP levels produced by treatment of Chinese hamster cells with dThd decreases the dCTP:dTTP ratio in the dNTP pools [6], but dCyd, which is phosphorylated to dCTP [6, 36] and should restore this ratio, did not abolish the facilitation of mutagenesis with dThd and, indeed, by itself facilitated MNNG-induced mutagenesis (Figs. 2, 3, and Table 1). These findings suggest either that dCyd is converted into dTTP, a process which is well known to occur in mammalian cells [36], or that increased dCTP:dTTP ratios, which are produced by phosphorylation of dCyd [6], facilitate mutagenesis. The latter suggestion is consistent with the finding that constitutively elevated dCTP pools in mutant Chinese hamster cells are associated with increased rates of background mutagenesis in these cells [23, 41]. Further evi-

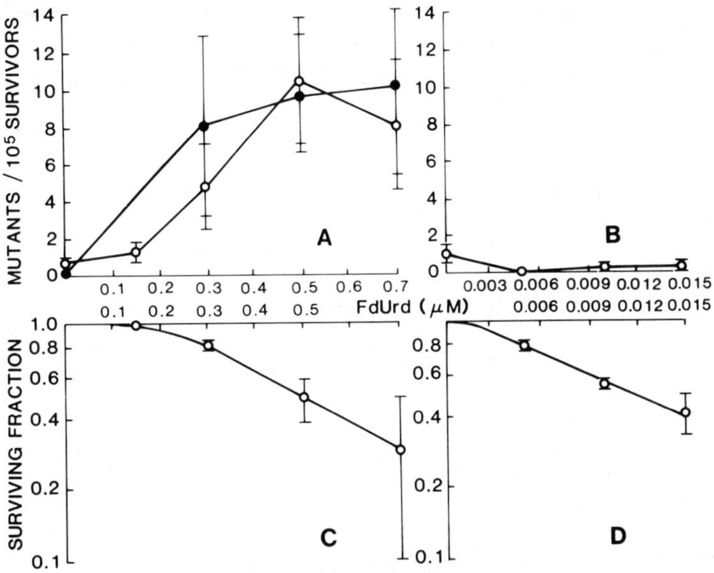

Fig. 6. AG^r-(○) and TG^r-(●) mutant frequencies (A and B) and cytotoxicities (C and D) induced by 2 h treatments with FdUrd in V79 cells in medium containing (A and C) or lacking (B and D) 10 μM dCyd. Each data point is the mean ± standard error of 2-5 independent experiments. (From Ref. 33.)

dence that increased dCTP:dTTP ratios in V79 cells facilitate mutagenesis was obtained in our experiments with 5-fluorodeoxyuridine (FdUrd).

Treatment of mammalian cells with FdUrd results in a powerful inhibition of thymidylate synthetase and, hence, a severe depletion of dTTP in dNTP pools that are used for DNA synthesis [9]. However, although FdUrd produces chromosome aberrations, cytotoxicity and oncogenic transformation in mammalian cells (Ref. 17 and Peterson, Heidelberger, and Benedict in this volume), it was reported to be only weakly mutagenic in C3H/10T1/2 cells [18] and in synchronized V79 cells [2] and, at concentrations less than 1 μM, not mutagenic to nuclear genes in asynchronous V79 cells (Refs. 2, 4, 17, 18, and Grosovsky and Little in this volume). We reasoned that in the presence of imbalanced precursor pools, mammalian DNA polymerases can continue to catalyze DNA synthesis, producing replication errors in the process [15, 16]. Therefore, treating V79 cells with FdUrd in medium containing dCyd, it should be possible to provide the cells' DNA polymerases with sufficient pyrimidine deoxyribonucleotides to catalyze DNA synthesis, while the dCTP:dTTP ratio in the cells is greatly increased because of the inhibition of thymidylate synthetase. Under these conditions, DNA replication errors and mutations should occur.

In medium containing 10^{-5} M dCyd, a two hour treatment with FdUrd produced dose dependent increases in AG^r- and TG^r-mutant frequencies up to 10-fold and 90-fold greater, respectively, than background mutant frequencies (Fig. 6). Moreover, with an expression time of 2 days, FdUrd in medium containing dCyd produced Oua^r mutant frequencies an average of 6-fold greater than background Oua^r mutant frequencies [33]. However, equitoxic concentrations of FdUrd in medium that did not contain dCyd, induced no mutations (Fig. 6 and Ref. 33). These results are consistent with the rationale for our experiments, but further work is required to fully elucidate the mechanism of FdUrd-induced mutagenesis in medium containing dCyd.

2.4. Self-Potentiated Drug Resistance Induced by FdUrd and dCyd

Our findings, of mutant frequencies induced by FdUrd and dCyd that are an order of magnitude greater than spontaneous frequencies, suggested that self potentiated induction of $FdUrd^r$ mutants could occur in V79 cells exposed to FdUrd and concentrations of dCyd $\sim 10^{-5}$ M. We have now found that three consecutive 2 h treatments with FdUrd plus 10^{-5} M dCyd induced ⩾8 fold more colonies of $FdUrd^r$ cells than did equitoxic treatments with FdUrd without dCyd. Seven colonies of $FdUrd^r$ cells have been isolated and cultured for 8-21 passages in medium without FdUrd. The $FdUrd^r$ cells grow to confluence in 2×10^{-7} M FdUrd, which is more than 30 times onies induced by FdUrd plus dCyd contained more than 10-fold elevated levels of thymidylate synthetase (Danenberg and Peterson, manuscript in preparation).

These observations indicate the FdUrd, in medium containing 10 μM dCyd, does potentiate the induction of stable $FdUrd^r$ phenotypes. Moreover, 10^{-5} M concentrations of dCyd have been found in rat plasma and the ascites fluid of mice bearing L1210 ascites tumors, and up to 4×10^{-6} M dCyd has been found in human plasma [10, 11], indicating that self potentiated induction of $FdUrd^r$ mutants could occur *in vivo*. These considerations suggest a possible practical application of our observations with FdUrd, namely, that treatments in which FdUrd is administered in combination with inhibitors of dCyd phosphorylation may effectively inhibit the emergence of FdUrd-resistant cells.

2.5. The Role of DNA Modification in Mutagenesis

Table 2 shows that mutagenesis by BCNU, which induces cross-links in DNA [12], is not facilitated by dCyd plus dThd. Mutagenesis produced by treatment of V79 cells with rat-liver post mitochondrial supernatant plus benzo(a)pyrene or plus aflatoxin B1, which induce frame shift mutations [21], is only slightly (less than

TABLE 2. Effects of 10^{-5} M dCyd + 10^{-5} M dThd on AG^r Mutant Frequencies Induced by Chemical Mutagens. a) 10^{-5} M dThd Only and Metabolic Activation with Liver Post-Mitochondrial Supernatant. Means and Standard Errors of Results of 2 or More Experiments Are Shown

Compound	(Molar)	Induced Mutants/10^5 Survivors	
		With Nucleosides	Without Nucleosides
MNNG	3.4×10^{-6}	67.3 ± 9.2	18.6 ± 4.2
MNU	9.7×10^{-4}	143 ± 39	71 ± 29
ENNG	1.9×10^{-5}	114 ± 34	31.3 ± 18
EMS	9.7×10^{-3}	44 ± 34	9.5 ± 2.5
MMS	2.7×10^{-4}	4.7 ± 1.3	2.7 ± 1.3
BCNU	4.7×10^{-6}	2.9 ± 0.2	4.5 ± 0.5
Aflatoxin B1[a]	1.6×10^{-7}	4.6	2.5
Benzo(a)Pyrene[a]	1.2×10^{-5}	3.6	2.3

2-fold) enhanced by dThd. These findings are consistent with reports that mutagenesis by UV radiation and nitrogen mustard is not facilited by deoxyribonucleosides [5, 22]. By contrast, mutagenesis by the monofunctional alkylating agents MNNG, MNU, and ENNG, were consistently facilitated by dCyd plus dThd (Table 2) or by dThd (Table 1). Facilitation of EMS-induced mutagenesis was less reproducible (Table 2 and Ref. 32), and MMS-induced mutagenesis was not significantly facilitated by dThd plus dCyd nor by dThd (Tables 1 and 2 and Ref. 22).

In medium with or without dCyd plus dThd, V79 cells were treated with equitoxic concentrations of [^{14}C]-MNNG and [^{14}C]-MMS. The extents of DNA methylation and quantities of the major products of oxygen and nitrogen methylation were determined using established techniques (see Ref. 31), and trwated cells were incubated with or without the deoxyribonucleosides for measurement of cytotoxicity and AG^r mutant frequencies. The extent of DNA methylation and the quantities of the major products of oxygen and nitrogen methylation were not affected by dThd and dCyd (Table 3). However, MNNG produced 10 times more m^6G than did MMS, and only the MNNG-induced mutagenesis was significantly facilitated by the deoxyribonucleosides (Table 3). These findings indicate that O-alkylation of DNA was necessary for facilitation of mutagenesis. However, treatment of DNA polymerases with MNNG and MMS may diffentially affect the fidelity of these enzymes in DNA synthesis [24], and modulation by the deoxyribonucleosides of mechanisms that repair methylated DNA may occur.

TABLE 3. Methylation of DNA, Cytotoxicity, and Mutagenesis by Equitoxic Concentrations of Methylating Agents in Medium With or Without dThd and dCyd (from Ref. 31)

Measurement	MNNG Without nucleosides	MNNG With nucleosides	MMS Without nucleosides	MMS With nucleosides
Methylation, μM methyl/mol DNA phosphate				
Total DNA methylations	16 ± 6	15 ± 6	ND	ND
m^7G	12 ± 4	11 ± 4	20 ± 5	28 ± 2
m^3A	0.7 ± 0.2	0.5 ± 0.2	1.6	3.2
m^6G	1 ± 0.4	0.8 ± 0.4	0.07 ± 0.07	0.06 ± 0.06
m^6G/m^7G	0.08 ± 0.04	0.07 ± 0.04	0.003 ± 0.003	0.002 ± 0.002
m^3A/m^7G	0.06 ± 0.03	0.05 ± 0.03	0.08 ± 0.02	0.11 ± 0.02
Cytotoxicity				
Surviving fraction	0.52 ± 0.02	0.52 ± 0.0	0.63 ± 0.03	0.58 ± 0.04
AG^r Mutant frequency				
Mutants/10^5 survivors	21 ± 3[a]	49 ± 3[a]	5.5 ± 2[b]	10.5 ± 2[b]

[a]$P < 0.03$ (Student's test) that the difference between these two values could have arisen by chance.
[b]$P > 0.2$ (Student's test) that the difference between these two values could have arisen by chance.

3. CYTOTOXICITY

3.1. Deoxyribonucleoside-Induced Potentiation of the Cytotoxicity of Methylating Agents

The cytotoxicity of monofunctional alkylating agents in Chinese hamster cells was potentiated by concentrations of dThd that were themselves nontoxic or only marginally cytotoxic [5, 22, 30-32]. In our experiments, concentrations of dCyd that were less than one twentieth of the concentrations of dThd that produced the potentiation abolished it (Figure 7). This effect is similar to that of dCyd on the cytotoxicity of dThd (Fig. 7 and Ref. 36) that is attributed to decreases in the dCTP:dTTP ratio in dNTP pools [5, 36]. Decreases in this ratio were associated with both the facilitation of mutagenesis and the potentiation of the cytotoxicity of EMS in Chinese hamster ovary cells exposed to dThd [22]. Moreover, the enhancement by dThd of MNNG-induced cytotoxicity was greater than that of MMS-induced cytotoxicity, and the latter enhancement was abolished by a lower concentration of dCyd than was the former (Fig. 7). These observations indicate that both mutagenesis and cytotoxicity induced by alkylating agents are associated with mispairing of T with O^6-alkyl guanines in DNA.

On the other hand, dCyd abolishes the dThd-induced potentiation of cytotoxicity (Fig. 1) but facilitates mutagenesis (Figs. 2 and 3). This suggests either that dCyd does not facilitate mutagenesis of genes whose products are essential for survival, or that the mechanisms by which alkylating agents produce most of their cytotoxicity are fundamentally different from the mechanisms by which they produce mutations.

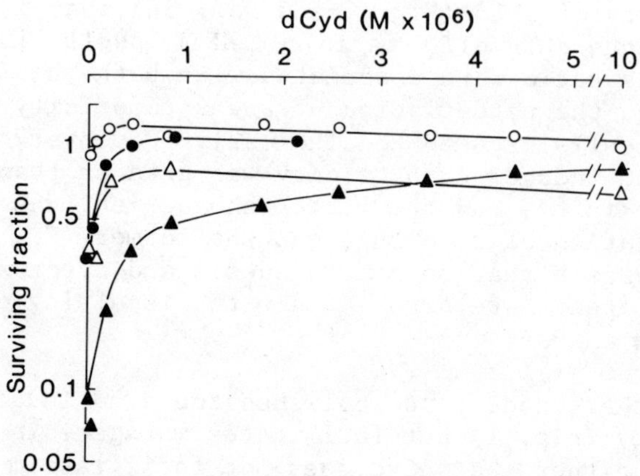

Fig. 7. Effects of dCyd on the cytotoxicities of 10 μM dThd (○), 80 μM dThd (●), 182 μM MMS plus 10 M dThd (△), and 1.7 μM MNNG plus 10 μM dThd (▲) in V79 cells. The nucleosides were introduced when the cells were plated, and the medium was not changed (●) or was changed after a 2 h treatment with MMS, MNNG, or 0.5% acetone (○, △, ▲). The surviving fractions of the cells treated with 182 μM MMS and 1.7 μM MNNG in medium that did not contain added nucleosides were 0.66 ± 0.11 and 0.61 ± 0.05 (mean ± SEM), respectively. Each data point shows the mean of two to six experiments. The root mean square of the standard errors of the surviving fractions was 0.08. (From Ref. 31.)

TABLE 4. Temporal Aspects of the Cytotoxicity of MNNG plus dThd (from Ref. 31)

Nucleoside treatment		Surviving fraction with MNNG at 0 to 2 hr		
Nucleosides	Time	0 μM	3.4 μM	6.8 μM
None		1.00 ± 0.06 (4)	0.50 ± 0.05 (7)	0.17 ± 0.03 (4)
10 μM dThd	-6 hr to 2 hr	0.98 ± 0.01 (4)	0.06 ± 0.03 (6)	0.004 ± 0.002 (4)
10 μM dThd	0 hr to 2 hr	1.06 ± 0.06 (2)	0.11 ± 0.04 (6)	ND
10 μM dThd	2 hr to day 8	0.96 ± 0.03 (2)	0.09 ± 0.01 (4)	ND
10 μM dThd	-6 hr to day 8	0.92 ± 0.10 (4)	0.05 ± 0.02 (7)	ND
10 μM dThd	-6 hr to day 8			
+10 μM dCyd	2 hr to day 8	1.07 ± 0.01 (2)	ND	0.16 ± 0 (4)

V79 cells were plated at -6 hr and treated with MNNG from 0 hr to 2 hr. The times of the nucleoside treatments, with dThd or dThd plus dCyd, are shown. Data are mean ± SEM of surviving fractions, and the numbers of independent experiments are given in parentheses. The plating efficiency of V79 cells in medium without introduced nucleosides was 72% (four experiments). ND, not done.

3.2. Differences in Temporal Aspects of Mutagenesis and Cytotoxicity

We reasoned that by treating V79 cells with MNNG and incubating the treated cells for various times in medium containing dThd, it should be possible to determine whether the replication of DNA containing m^6G is necessary for both mutagenesis and cytotoxicity. We found maximal cytotoxicity on exposure of the cells to dThd during treatment of the cells with MNNG for 2 h (Table 4), which is an order of magnitude less than the doubling time of the MNNG-treated cells [31]. Thus, not even one round of replication of the DNA of the cells in medium containing dThd was necessary for maximal potentiation of the cytotoxicity of MNNG in V79 cells.

We have also examined the temporal aspects of cytotoxicity in murine leukemia L1210 cells. The L1210 cells grow in suspension culture. As a measure of cytotoxicity, we determined IC_{50}, which is the drug concentration required to inhibit by 50% the growth of the cells over 48-72 h [14]. Using this method with these cells, we avoid possible effects of cell-cell interactions in colonies and of dThd on cell attachment that may occur in monolayer cultures, such as those of V79 cells. The IC_{50} of MNNG was reduced 3.0-fold and 2.7-fold by incubation with 50 μM dThd for 48 h and 2 h, respectively, (Peterson and Danenberg, manuscript in preparation). Thus, the exposure to dThd for 2 h during treatment with MNNG produced maximal growth inhibition. Since the doubling time of the L1210 cells is 16 h, this finding shows that not even 1 round of DNA replication in medium containing dThd was necessary for potentiation of the cytotoxicity of the MNNG in L1210 cells. Thus, these studies of the temporal aspects of mutagenesis and cytotoxicity in V79 cells and of cytotoxicity in L1210 cells show that MNNG produces cytotoxicity by mechanisms that are fundamentally different from those by which it produces mutagenesis.

3.3. Studies on Possible Mechanisms of Cytotoxicity and Mutagenesis

Alkylating agents inhibit DNA synthesis. Our previously published data suggested that this inhibition is uniformly associated with cytotoxicity but is not uniformly associated with mutagenesis [28]. Monofunctional alkylating agents also induce poly (ADP-ribose) synthesis in mammalian cells [8, 37]. Treatment of the cells with poly (ADP-ribose) synthesis inhibitors, one of which is dThd, enhances the cytotoxicity of monofunctional alkylating agents [37]. Moreover, it has been suggested that the induction of poly (ADP-ribose) synthesis leads to DNA-synthesis inhibition, which allows time for repair of DNA lesions that would otherwise cause cytotoxicity and mutations [8, 27]. It has also been suggested that poly (ADP-ribose) synthesis is necessary for the later stages of excistion repair of DNA damaged by alkylating agents [37].

We measured the effects of 2 h and 24 h treatments with dThd on the incorporation of [^{3}H]-dThd into the DNA and on NAD depletion in V79 cells treated with MMS and MNNG. In mammalian cells treated with alkylating agents, the incorporation of [^{3}H]-dThd has been extensively characterized as measuring DNA synthesis [28], and NAD depletion has been characterized as measuring poly (ADP-ribose) synthesis [37]. We found that equitoxic concentrations of MMS and MNNG, which induce different amounts of mutagenesis (Table 2 and Refs. 25 and 34), induced equal inhibition of DNA synthesis [28] and equal depletion of NAD. Moreover, inhibition of DNA syn; thesis and depletion of NAD were not significantly affected by exposure of the cells to dThd that potentiated cytotoxicity and facilitated mutagenesis (Peterson and Ibric, manuscript in preparation). These findings are consistent with the suggested association between poly (ADP-ribose) synghesis and inhibition of DNA synthesis in cells treated with alkylating agents [8]. They also show that neither poly (ADP-ribose) synthesis nor inhibiton of DNA synthesis are uniformly associated with mutagenesis and cytotoxicity induced by methylating agents and dThd in V79 cells. Our interpretation of these findings is that poly (ADP-ribose) synthesis and inhibition of DNA synthesis do not prevent the expression of mutagenic and cytotoxic damage induced by methylating agents and low concentrations of dThd in V79 cells, and that the inhibition of the mutagenicity and cytotoxicity that is associated with T:m^{6}G mispairing in V79 cells treated with MNNG and dThd.

4. PROSPECTS FOR SELECTIVE MODULATION OF MUTAGENESIS AND CYTOTOXICITY

Four lines of evidence that mechanisms of mutagenesis are different from those of cytotoxicity in V79 cells treated with monofunctional alkylating agents and FdUrd are:

(i) Equitoxic concentrations of methylating agents produce different mutant frequencies [25, 34].

(ii) Lines of V79 cells have been isolated that have susceptibilities to mutagenesis different from their susceptibilities to cytotoxicity produced by alkylating agents [13].

(iii) Whereas dThd facilitated both the mutagenicity and cytotoxicity of MNNG, dCyd abolished the facilitation of cytotoxicity but itself facilitated the mutagenesis (Table 1). Furthermore, dCyd reduced the cytotoxicity of FdUrd but facilitated mutagenesis by FdUrd (Fig. 6).

(iv) Maximal facilitation of mutagenesis occurred when dThd and dCyd were present throughout at least one round of DNA replication after treatment of V79 cells with MNNG; but for maximal cytotoxicity, in both V79 and L1210 cells, it was necessary for dThd to be present only for a brief period, during which most of the DNA cannot have been replicated (Table 4 and Peterson and Danenberg, manuscript in preparation).

These lines of evidence create a rationale for seeking combinations of alkylating agents and modulators of nucleotide metabolism that will produce cytotoxicity without mutagenesis. Our data further lead us to seek compounds that specifically inhibit dCTP synthesis, because such compounds should selectively potentiate the cytotoxicity of monofunctional alkylating agents and of FdUrd, inhibit the self potentiated induction of FdUrd^r resistant mutants and also the inhibit induction of oncogenic transformation by FdUrd (Peterson, Heidelberger, and Benedict in this volume).

The prospects for selective modulation of cytotoxicity and mutagenesis will undoubtedly be improved by a deeper understanding of mechanisms. This will entail careful measurements of dNTP pools. In the area of mutagenesis, the influence of dNTP metabolism on mutagenesis in DNA repair deficient mutants should be explored (S. Shall, personal communication). In the area of cytotoxicity, the possibility that cytotoxic treatments selectively activate apoptosis [42] should be investigated.

5. SUMMARY

Treatment of Chinese hamster V79 cells with dThd, dCyd, or dThd plus dCyd increased MNNG-induced AG^r-, TG^r-, and Oua^r-mutant frequencies but did not significantly increase background mutant frequencies. All the AG^r colonies that were isolated possessed phenotypes characteristic of HGPRT-deficient mutants, and the deoxyribonucleosides did not selectively affect the growth of the mutants, nor the selecting efficiency of AG, and did not significantly enhance background mutagenesis. These data show that both dThd and dCyd facilitated MNNG-induced mutageneis. This facilitation was maximal when cells were exposed to the deoxyribonucleosides throughout the first doubling time (24 h) after treatment with MNNG and for 4 more doubling times prior to mutant selection with AG. This indicates that one round of DNA replication was sufficient for mispairing of methylated bases in the DNA with the C and T provided by the deoxyribonucleosides, and that 4-6 doublings prior to mutant selection with AG were necessary to deplete pre-existing hypoxanthine: guanine phosphoribosyl transferase in newly mutated cells. The dCyd facilitated mutagenesis by FdUrd, which was not mutagenic without dCyd, indicating that increased dCTP:dTTP ratios were mutagenic. Treatment with FdUrd plus dCyd also induced $FdUrd^r$ cells, suggesting that inhibition of dCyd utilization may prevent the development of FdUrd-resistance in cancer chemotherapy. Although dCyd and dThd facilitated mutagenesis in cells treated with monofunctional alkylating agents that methylate DNA oxygens, facilitation of mutagenesis did not occur in cells treated with BCNU, which cross links DNA, nor with benzo(a)pyrene and aflatoxin B1, which are frame shift mutagens, nor with MMS, which produces barely detectable levels of O-methylation in DNA.

Virtually non toxic concentrations of dThd potentiated the cytotoxicity of MNNG more than 10-fold but that of MMS was potentiated only about 2-fold showing that O-alkylation of DNA was associated not only with the facilitation of mutagenesis but also with the potentiation of cytotoxicity. The potentiation of MNNG-induced cytotoxicity was maximal in V79 and L1210 cells after only 2 h treatment with dThd, showing that not even one round of DNA replication was necessary for this potentiation. Moreover, dCyd abolished the potentiation, and, at equitoxic concentrations, MNNG induced higher mutant frequencies than did MMS. These data show that the mechanisms by which methylating agents plus dThd induce mutagenesis are fundamentally different from their mechanisms of cytotoxicity. The DNA-synthesis inhibition and NAD depletion induced by MMS and MNNG with dThd were not uniformly associated with mutagenesis nor with cytottoxicity, indicating that these repair mechanisms play no more than a minor role in the genetic effects of MMS and MNNG with dThd in V79 cells. Our data show that the mutagenesis and cytotoxicity induced by alkylating agents and FdUrd can be selectively modulated by manipulating dNTP metabolism. These findings create a rationale for seeking treatments that will selectively modulate cytotoxicity, mutagenesis, the development of drug resistance and oncogenic transformation in cancer chemotherapy.

ACKNOWLEDGMENTS

We gratefully acknowledge the support, encouragement, and wisdom of Charles Heidelberger, to whose memory we dedicate this work. The experimental work was supported by grants from R. J. Reynolds Industries, Inc., the American Cancer Society, and Grant CA09320 from the National Cancer Institute.

REFERENCES

1. P. J. Abbott and R. Saffhill, DNA Synthesis with Methylated Poly (dC-dG) Templates, Evidence for a Competitive Nature to Miscoding by O^6-Methylguanine, Biochim. Biophys. Acta, 562: 51-61 (1979).
2. P. M. Aebersold, Mutation Induction by 5-Fluorodeoxyuridine in Synchronous Chinese Hamster Cells, Cancer Res., 39:808-810 (1979).
3. P. J. Benke, N. Herrick, and A. Herbert, Hypoxanthine Guanine Phosphoribosyl-transferase Variant Associated With Accelerated Purine Synthesis, J. Clin. Invest., 52:2234-2244 (1973).
4. M. D. Bradley, B. Bhuyan, M. C. Francis, R. Langenbach, A. R. Peterson, and E. Huberman, Mutagenesis by Chemical Agents in V-79 Chinese Hamster Cells: A Review and Analysis of the Literature, Mutation Res., 87:81-142 (1981).
5. J. Brennand, S. McMillan, and M. Fox, Effects of Metabolic Inhibitors on Cell Lethality and Mutation Induction in Chinese Hamster Cells, II. The Effect of Post-treatment with Non-Toxic Concentrations of Thymidine, Chem.-Biol. Interactions, 36:89-106 (1981).
6. G. Bjursell and P. Reichard, Effects of Thymidine on Deoxyribonucleoside Pools and Deoxyribonucleic Acid Synthesis in Chinese Hamster Ovary Cells, J. Biol. Chem., 248:3904-3909 (1973).
7. C. T. Caskey and G. D. Kruh, The HPRT Locus, Cell, 16:1-9 (1979).
8. J. E. Cleaver, W. J. Bodell, D. C. Gruenert, L. N. Kapp, W. K. Kaufmann, S. D. Park, and B. Zelle, Repair and Replication Abnormalities in Various Human Hypersensitive Diseases, in: Mechanisms of Chemical Carcinogenesis (C. C. Harris and P. A. Cerutti, eds.), pp. 409-418, Liss, New York (1982).

9. P. V. Danenberg, Thymidylate Synthetase, A Target Enzyme for Cancer Chemotherapy, Biochim. Biophys. Acta, 473:73-85 (1977).
10. L. L. Danhauser and Y. M. Rustum, Effect of Thymidine on the Toxicity, Antitumor Activity and Metabolism of 1-β-D-Arabinofuranosylcytosine in Rats Bearing a Chemically Induced Colon Carginoma, Cancer Res., 40:1274-1280 (1980).
11. E. M. Egan, L. Sargent, A. Rosowsky, and D. W. Kufe, Rescue of Thymidine Cytotoxicity in L1210 Ascites by Elevated Endogenous Levels of Deoxycytidine, Cancer Treatment Reports, 65: 853-859 (1981).
12. R. A. Ewig and K. W. Kohn, DNA-Protein Crosslinking and DNA Interstrand Crosslinking by Haloethylnitrosoureas in L1210 Cells, Cancer Res., 38:3197-3203 (1978).
13. J. Friedman and E. Huberman, Post-replication Repair and the Susceptibility of Chinese Hamster Cells to Cytotoxicity and Mutagenic Effects of Alkylating Agents, Proc. Natl. Acad. Sci., U.S.A., 77:6072-6076 (1980).
14. G. B. Grindey and C. A. Nichol, Interaction of Drugs Inhibiting Different Steps in the Synthesis of DNA, Cancer Res., 32:527-531 (1972).
15. T. A. Kunkel, J. R. Silber, and L. A. Loeb, The Mutageneic Effect of Deoxynucleotide Substrate Imbalances During DNA Synthesis with Mammalian DNA Polymerases, Mutation Res., 94: 413-419 (1982).
16. T. A. Kunkel and L. A. Loeb, Fidelity of Mammalian DNA Polymerases, Science, 213:765-767 (1981).
17. B. A. Kunz, Genetic Effects of Deoxyribonucleotide Pool Imbalances, Environ. Mutagen., 4:695-725 (1982).
18. J. R. Landolph and P. A. Jones, Mutagenicity of 5-Azacytidine and Related Nucleosides, Cancer Res., 42:817-823 (1982).
19. P. D. Lawley and C. J. Thatcher, Methylation of Deoxyribonucleic Acid in Mammalian Cells by N-Methyl-N'-nitro-N-nitrosoguanidine, Biochem. J., 116:693-707 (1970).
20. P. D. Lawley, Some Chemical Aspects of Dose-Response Relationships in Alkylation Mutagenesis, Mutation Res., 23:283-295 (1974).
21. J. McCann, E. Choi, E. Yamasaki, and B. N. Ames, Detection of Carinogens as Mutagens in the Salmonella/Microsome Test: Assay of 300 Chemicals, Proc. Natl. Acad. Sci., U.S.A., 72:5135-5139 (1975).
22. M. Meuth, Role of Deoxynucleoside Triphosphate Pools in the Cytotoxic and Mutagenic Effects of DNA Alkylating Agents, Somatic Cell Genet., 7:89-102 (1981).
23. M. Meuth, N. L. Hereux-Huard, and M. Trudel, Characterization of a Mutator Gene in Chinese Hamster Ovary Cells, Proc. Natl. Acad. Sci., U.S.A., 76:6505-6507 (1979).
24. M. Miyaki, K. Suzuki, M. Aihara, and T. Ono, Misincorporation in DNA Synthesis After Modification of Template for Polymerase by MNNG, MMS, and UV Radiation, Mutation Res., 107:203-218 (1983).

25. F. F. Newbold, W. Warren, A. S. C. Medcalf, and J. Amos, Mutagenicity of Carcinogenic Methylating Agents is Associated with a Specific DNA Modification, Nature (London), 283:596-599 (1980).
26. A. Onfelt and D. Jenssen, Enhanced Mutagenic Response of MNU by Post-Treatment with Methyl Mercury, Caffeine, or Thymidine in V79 Chinese Hamster Cells, Mutation Res., 106:297-303 (1982).
27. R. B. Painter, Effect of Caffeine on DNA Synthesis in Irradiated and Unirradiated Mammalian Cells, J. Mol. Biol., 143: 289-301 (1980).
28. A. R. Peterson, DNA Synthesis, Mutagenesis, Cytotoxicity, and DNA Damage in Cultured Mammalian Cells, Cancer Res., 40:682-688 (1980).
29. A. R. Peterson, D. F. Krahn, H. Peterson, C. Heidelberger, B. K. Bhuyan, and L. H. Li, The Influence of Serum Components on the Growth of Chinese Hamster Cells in Medium Containing 8-Azaguanine, Mutation Res., 36:345-356 (1976).
30. A. R. Peterson, J. R. Landolph, H. Peterson, and C. Heidelberger, Mutagenesis of Chinese Hamster Cells is Facilitated by Thymidine and Deoxycytidine, Nature (London), 276:508-510 (1978).
31. A. R. Peterson and H. Peterson, Differences in Temporal Aspects of Mutagenesis and Cytotoxicity in Chinese Hamster Cells Treated with Methylating Agents and Thymidine, Proc. Natl. Acad. Sci., U.S.A., 79:1643-1647 (1982).
32. A. R. Peterson and H. Peterson, Facilitation by Pyrimidine Deoxyribonucleosides and Hypoxanthine of Mutagenic and Cytotoxic Effects of Monofunctional Alkylating Agents in Chinese Hamster Cells, Mutation Res., 61:319-331 (1979).
33. A. R. Peterson, H. Peterson, and P. V. Danenberg, Induction of Mutations by 5-Fluorodeoxyuridine: A Mechanism of Self-Potentiated Drug Resistance? Biochem. Biophys. Res. Commun., 110:573-577 (1983).
34. A. R. Peterson, H. Peterson, and C. Heidelberger, Oncogenesis, Mutagenesis, DNA Damage and Cytotoxicity in Cultured Mammalian Cells, Cancer Res., 39:131-138 (1979).
35. A. R. Peterson, H. Peterson, and C. Heidelberger, The Influence of Serum Components on the Growth and Mutation of Chinese Hamster Cells in Medium Containing Aminopterin, Mutation Res., 24:25-33 (1974).
36. P. Reichard, From Deoxyribonucleosides to DNA Synthesis, Fed. Proc., 37:9-14 (1978).
37. S. Shall, B. W. Durkacz, D. A. Gray, J. Irwin, P. J. Lewis, M. Perera, and M. Tavassoli, (ADP-Ribose)$_n$ Participates in DNA Excision Repair, in: "Mechanisms of Chemical Carcinogenesis" (C. C. Harris and P. A. Cerutti, eds.), pp. 389-407, Liss, New York (1982).
38. B. Singer, All Oxygens in Nucleic Acids React with Carcinogenic Ethylating Agents, Nature (London), 264:333-339 (1976).

39. B. Singer, The Effect of Base Modification on Fidelity in Transcription, in: "Carcinogenesis: Fundamental Mechanisms and Environmental Effects" (B. Pullman, P. O. P. Ts'o, and H. Gelboin, eds.), pp. 91-102, Reidel, Boston (1980).
40. M. D. Topal and M. S. Baker, DNA Precursor Pool: A Significant Target for N-Methyl-N-nitrosurea in C3H/10T1/2 Clone 8 Cells, Proc. Natl. Acad. Sci., U.S.A., 79:2211-2215 (1982).
41. G. Weinberg, B. Ullman, and D. W. Martin, Jr., Mutator Phenotypes in Mammalian Cell Mutants with Distinct Biochemical Defects and Abnormal Deoxyribonucleotide Pools, Proc. Natl. Acad. Sci., U.S.A., 78:2447-2451 (1981).
42. A. H. Wyllie, J. F. R. Kerr, and A. R. Currie, Cell Death: The Significance of Apoptosis, Intern. Rev. Cytology, 68:251-305 (1980).

---

‡Abbreviations: AG, 8-azaguanine; BCNU, 1,3-bis(2-chloroethyl)-1-nitrosourea; dAdo, deoxyadenosine; dCyd, deoxycytidine; dGuo, deoxyguanosine; dNTP, deoxyribonucleoside triphosphate; dThd, deoxythymidine; EMS, ethyl methane sulfonate; ENNG, N-ethyl-N'-nitrosoguanidine; FdUrd, 5-fluorodeoxyuridine; HAT, medium containing 10^{-5} M hypoxanthine, 10^{-6} M aminopterin and 5×10^{-6} M thymidine; HGPRT, hypoxanthine: guanine phosphoribosyl transferase; m^3A, 3-methyladenine; m^6G, O^6-methylguanine; m^7G, 7-methylguanine; MMS, methyl methane sulfonate; MNNG, N-methyl-N'-nitro-N-nitrosoguanidine; MNU, methyl nitrosourea; oua, ouabain; r, resistant; s, sensitive; TG, 6-thioguanine.

VIEWING THE COMPLEXITIES OF MUTAGENESIS FROM AN EVOLUTIONARY PERSPECTIVE

John W. Drake

Laboratory of Genetics
National Institute of Environmental Health Sciences
Research Triangle Park, North Carolina 27709

We have been treated to several examples of organismally distinctive nucleotide pool anomalies at this workshop. This morning, as we turn from quantitative to qualitative aspects of nucleotide pools and consider base analogue mutagenesis, we shall learn of additional complexities, notably the induction of transversions and deletions by base analogs that might classically have been expected to produce only transitions. Such phenomena need not discourage us from seeking generalities, however, since even apparent inconsistencies must eventually be accommodated within a general theory, particularly once we know more about the evolution of mechanisms for generating and avoiding mutations. I will therefore briefly describe several distinct phylogenetic trends that may assist our thinking about such matters.

The main balancing forces that might determine an equilibrium mutation rate for a population are the adaptive advantage of genetic variability and the disadvantages of producing suboptimal genotypes. When new mutations arise, they do so more often in organisms with higher-than-average mutation rates. Therefore, when a new mutation proves advantageous, its selective advantage indirectly selects for an increased mutation rate. This can be readily demonstrated experimentally in a chemostat, where a mutator gene provides a powerful competitive advantage to an adapting population [2]. Conversely, because far more mutations are deleterious than are advantageous [11], the offspring of rapidly mutating organisms will display reduced average fitness. (These couplings between mutation rates and more directly selected traits are much tighter in organisms with little sexual reproduction than in those with obligate sexual reproduction.)

There is little basis to anticipate what might be a specific optimal mutation rate determined by these and other balancing forces. Different organisms certainly display very different mutation rates. However, when average mutation rates are surveyed, a remarkable relationship can be observed. The best way to obtain average mutation rates is to measure forward rates at one or several loci that are sufficiently well characterized to permit extrapolations from locus to genome or to base pair. When this was done for the bacteriophages λ and T4, the bacteria *Escherichia coli* and *Salmonella typhimurium* and the fungus *Neurospora crassa*, the average forward mutation rate per genome per asexual generation turned out always to be in the neighborhood of 0.001 [3]. The original *N. crassa* value was probably a few-fold low, and values for two yeasts that fall into the same range have since become available. When guesstimates are made to correct for the efficiency of mutation detection, the result is about 0.005 mutations per genome per replication, with only few-fold variations for at least seven microbial species (excluding mutator variants). It is interesting that organisms of such diverse ecology and evolutionary history should display such similar mutation rates. This constancy is even more remarkable when we note that, since their genome sizes vary by 1000-fold, their mutation rates per base pair also vary by 1000-fold. However, spontaneous mutation rates measured in cultured mammalian somatic cells seem not to have decreased from those of microbial eukaryotes on a per-locus basis, and it is unclear what fraction of a mammalian genome should be taken as comparable to that of a fungal genome in the sense of being a target for spontaneous mutation at a rate selectable during the course of evolution. On the other hand, mutation rates in the small RNA viruses are at least as large as predicted by extrapolation from the general relationship [8].

If mutation rates decrease by over 10,000-fold as genomes increase by a similar factor, then powerful determinants must be operating at the molecular level. Even before the mechanistic details are sorted out, however, it is possible to anticipate some basic principles. Any genetic system will be subject to a considerable variety of qualitatively different kinds of error and damage. Consider these ordered in monotonically decreasing importance as threats to genetic integrity. Then, as genomes expand and selection acts to decrease average mutation rates per base pair, the organism can initially profit by evolving a mechanism to combat but a single, predominant source of mutations. In order to achieve further decrements of mutation rate, however, the organism must next combat two causes of mutations, and eventually many causes. Therefore, as average mutation rates decrease, more and more diverse mechanisms of mutation can be expected to contribute significantly to the total rate. Note, however, that this argument refers to the sources of mutations, not to the mechanisms used to avoid mutations: a complex enzyme such as a DNA polymerase might well encompass multiple avoidance mechanisms [6].

My comments to this point have dealt mainly with microbial systems. There exists another striking trend, however, in the responses of eukaryotes to mutagens. This trend was first formulated as the "ABCW" relationship [1], which stated that the rate of induced mutation per locus per rad of ionizing radiation was directly proportional to total genome size over a 1000-fold range. A similar relationship was later proposed for ethyl methanesulfonate (EMS) [7]. The ABCW relationship was examined in critical detail, wherein many of the data were challenged [10]. Nevertheless, while strict proportionality seems not to hold, there remains a definite and strongly upwards trend in specific-locus mutability per unit dose of mutagen with increasing genome size. The molecular basis of the trend remains vague. It is likely to involve disruption of polypeptide-encoding sequences by introns, the presence of large amounts of DNA of unknown function in higher eukaryotes, and perhaps other yet to be discovered factors as well. It also probably reflects the kinds of mutations induced, in particular the frequencies of deletions whose ends may be targeted well outside of the functional domains of genes.

The final trend I wish to mention today may help to link the first few. It concerns the relative contribution of active cellular processes to the production, rather than the avoidance, of mutation [4, 5]. It appears that, as mutation avoidance has improved with increasing genome size, the relative role of direct mispairing has decreased while the role of error-prone or mutagenic repair has increased. This may be seen most simply in the complete independence of EMS mutagenesis from viral _WXY_-mediated mutagenesis in bacteriophage T4, in the partial requirement for _recA_-mediated mutagenic repair for EMS mutagenesis in _E. coli_, and in the stringent requirement for _rad6_-mediated mutagenic repair for EMS mutagenesis in _Saccharomyces cerevisiae_. Indeed, it is difficult to demonstrate mutagenesis with any chemical in yeast defective in error-prone repair. Perhaps, therefore, a large fraction of mutagenesis also requires mutagenic repair in higher eukaryotes, a condition that would raise interesting possibilities for antimutagenic intervention.

We are left, however, with a puzzle: if mutation avoidance is so effective in eukaryotes that mutation requires an active cellular response to damage, then what is the selective advantage of such a process? It has been possible to speculate that the rather small increment of survival associated with error-prone repair of ultraviolet damage, compared to the large increments of survival available through error-proof mechanisms such as excision and recombination repair, implies that the usefulness of error-prone repair lies in the very mutations it generates. Until recently, the only counter-argument seemed to be error-prone repair in mammalian somatic cells, where genetic diversity has not been implicated in anything except decreased fitness for the whole organisms. Recently, however, it has been shown that the _E. coli_ _umuC_ system (the portion of the recA system specifically responsible for mutagenesis) confers a relatively large survival increment upon cells photosensitized with isopsoralen [9]. It thus seems likely that mutagenic repair serves at least two distinct purposes: increasing genetic diversity in the face of environmental challenge, and increasing survival as well.

These several phylogenetic and evolutionary trends may provide some insights into the differences among mutation processes in different organisms, such as we have already encountered in our conference on nucleotide pool dynamics. Whether they will be sufficient to rationalize all of the diversity remains an open question.

REFERENCES

1. S. Abrahamson, M. A. Bender, A. D. Conger, and S. Wolff, Uniformity of radiation-induced mutation rates among different species, Nature, 245:460-462 (1973).
2. E. C. Cox, Bacterial mutator genes and the control of spontaneous mutation, Ann. Rev. Genet., 10:135-156 (1976).
3. J. W. Drake, Comparative rates of spontaneous mutation, Nature, 221:1132 (1969).
4. J. W. Drake, Fundamental mutagenic mechanisms and their significance for environmental mutagenesis, in: "Progress in Genetic Toxicology" (D. Scott, B. A. Bridges, and F. H. Sobels, eds.), pp. 43-55, Elsevier/North Holland, Amsterdam (1977).
5. J. W. Drake, Perspectives in molecular mutagenesis, in: "Molecular and Cellular Mechanisms of Mutagenesis" (J. F. Lemontt and W. M. Generoso, eds.), pp. 361-378, Plenum Press, New York (1982).
6. J. W. Drake, B. W. Glickman, and L. S. Ripley, Updating the theory of mutation, Am. Sci., 71:621-630 (1984).
7. J. A. Heddle and K. Athanasiou, Mutation rate, genome size and their relation to the rec concept, Nature, 258:359-361 (1975).
8. J. Holland, K. Spinkler, F. Horodyski, E. Grabau, S. Nichol, and S. VandePol, Rapid evolution of RNA genomes, Science, 215:1577-1585 (1982).
9. S. Miller and E. Eisenstadt, Lethal effects of isopsoralen on a umuC mutant of *Escherichia coli*, J. Cell. Biochem., Suppl. 7B:211 (1983).
10. A. P. Schalet and K. Sankaranarayanan, Evaluation and reevaluation of genetic radiation hazards in man. I. Interspecific comparison of estimates of mutation rates, Mutat. Res., 35:341-370 (1976).
11. M. J. Simmons and J. F. Crow, Mutations affecting fitness in Dorosphila populations, Ann. Rev. Genet., 11:49-78 (1977).

MUTAGENESIS BY INCORPORATION OF ALKYLATED NUCLEOTIDES

Michael D. Topal

Departments of Pathology and the Cancer Research Center
University of North Carolina Medical School
Chapel Hill, North Carolina 27514

SUMMARY

The cellular DNA precursor pool was shown to be a target for N-methyl-N-nitrosourea, a potent mutagen and carcinogen. O^6medGTP, a product of this interaction, was chemically synthesized and shown to be incorporated into DNA *in vitro* by Klenow *E. coli* pol I and phage T4 DNA polymerases. O^6medGTP incorporated predominantly opposite T template residues and to a lower extent opposite C. At some loci incorporation of O^6medGTP caused DNA synthesis arrest. The significance of the behavior of O^6medGTP for mutagenesis *in vivo* is discussed.

INTRODUCTION

Chemical mutagenesis is generally thought to occur through modification of the double-strand DNA duplex *in vivo*. However, many potential DNA sites for chemical modification are shielded by participation in hydrogen bonds that stabilize base-pairing and by nearest-neighbor effects. In addition, double-strand DNA may be protected from chemical modification *in vivo* by interaction with histone and nonhistone binding proteins and by *in vivo* packaging [1, 11, 14, 18, 20]. Consideration of nucleic acid target sites free of such Interactions led us to study modification of the DNA precursor pool as a basis for mutagenesis by chemical mutagens [24-27]. Nucleosides and -tides are more susceptible to alkylation *in vitro* when compared to DNA polymers (reviewed by Singer [22]), but the situation *in vivo* was unknown. Thus, the precursor pool was expected to be relatively free of the interactions that mask potential sites of modification in chromosomal DNA. For the purposes

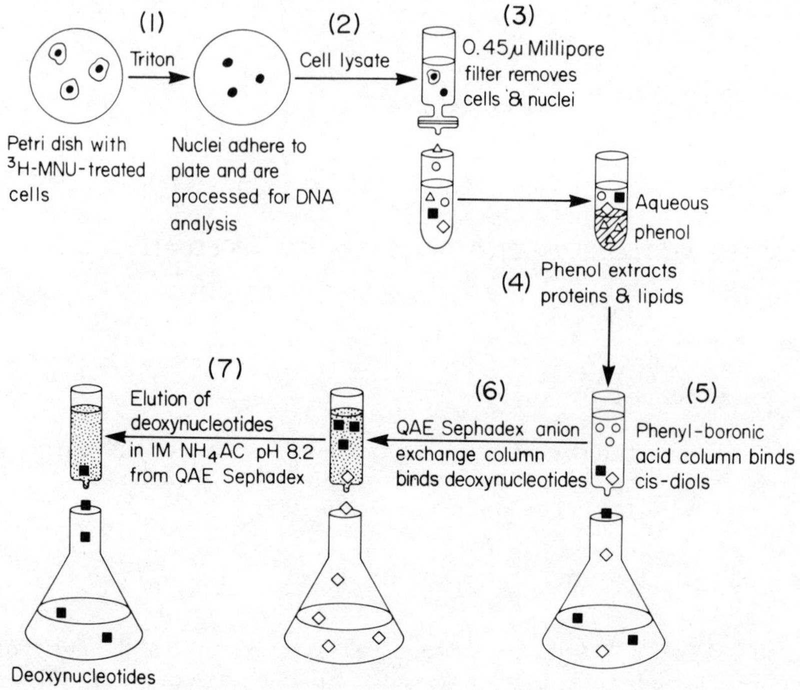

Fig. 1. Purification of cellular deoxyribonucleotides precursors. △) Protein and lipid; ○) molecule with cis-diol groups; ■), deoxynucleotide; ◊) deoxynucleosides and amino acids.

of this article I used the term "precursor pool" in its broadest sense since earlier nucleic acid intermediates may be chemically modified by mutagens and then converted to the deoxyribonucleoside triphosphate (dNTP) through normal pathways.

That modification of DNA precursors might produce mutagenic intermediates was not surprising in light of the behavior of base-analog mutagens such as 2-aminopurine [5] and N-4-hydroxycytidine [15]. In fact, production of mutagenic base analogs through alkylation of the DNA precursor pool may explain some of the characteristics of chemical mutagenesis and carcinogenesis. For example, chemical mutagens such as N-methyl-N-nitrosourea (MNU) and N-ethyl-N-nitrosourea act at the chromosomal replicating point in procaryotes [4, 8] and are specific to the S phase in mammalian cells [6, 23]. This behavior is consistent with the rapid turnover of DNA precursors and their variation in concentration with phases of the cell cycle (reviewed by Hauschka [7]). Modified precursors must be produced during the cell's S phase in order to be present for incorporation into replicating DNA.

Production of Modified DNA Precursors by Alkylation of Cells in Culture with MNU

For alkylated precursors to have a role in chemical mutagenesis they must be (a) produced when cells are subjected to a chemical mutagen and (b) incorporated into the DNA during DNA synthetic processes. Therefore, to ascertain whether alkylated precursors are indeed a product of carcinogen treatment of cells, we treated mouse embryo fibroblast cells in culture with MNU [25]. Cells that had been synchronized with respect to growth cycle by post-confluence inhibition of cell division were treated for one-half hour during their S phase with [^{3}H]MNU (0.04 mM). This dose of MNU killed less than 20% of the cells. The cells were then lysed on the dish with detergent and the DNA isolated by repeated banding in cesium chloride after treatment with proteinase K and RNAase. The cell lysate was treated to remove debris, lipids, proteins, RNA, and ribonucleotides as shown in Fig. 1. The deoxynucleotides were isolated by ion exchange chromatography. Determination of the radioactivity lost from the extract after each step in the procedure measured the amount of methylation of the cell components removed by each of those steps. The results show large amounts of methylation of protein/lipid and RNA and relatively small amounts of modification of DNA and the DNA precursor pool (Table 1). Most important to these studies, the ratio of pool methylation to DNA methylation varied from 1 to 8 for three independent determinations. These results confirmed our expectation of significant pool-alkylation by MNU. However, the results had not yet taken into account the significant differences in nucleic acid residue content per cell between the pool and the DNA. Determined the nucleotide-residue content of the DNA to be approximately 1000-fold greater than that in the pool [25]. Combining this result with the 1- to 8-fold preferential pool methylation over DNA methylation, indicated that the average susceptibility of a nucleic acid residue in the nucleic acid precursor pool was 1000 to 8000 times that in the DNA.

Acid hydrolysis of the DNA precursor pool and the DNA and resolution of the products by HPLC enabled us to compare susceptibilities of the individual potential methylation sites of the pool and DNA. The results (Table 2) indicated a strong dependence of methylation on the site being modified. For example, the N-1 position of Ade was 13,000 times more susceptible to methylation by MNU when in the pool than when in the DNA. This was consistent with the involvement of this position in hydrogen bonding of A·T base pairs in DNA. The N-7 position of guanine, however, was only 190-fold more susceptible to modification in the pool than in the DNA helix. This was consistent with its position in the major groove of the DNA helix free from many of the interactions that involve sites such as the N-1 position of adenine. Thus, the DNA precursor pool is a significant target for the chemical mutagen MNU and perhaps chemical mutagens in general. Mutagenic and toxic effects of pool alkyla-

TABLE 1. Distribution of Radioactivity during Isolation of dNTP from [^{3}H]MNU-Treated 10T^1/$_2$ Cells

Exp.	Radioactivity, cpm × $10^{-3}/10^6$ cells remaining after step						
				Precursor pool QAE-Sephadex elution			
	Enucleated cell lysate (step 3)	Phenol extraction (step 4)	Affi-Gel (step 5)	0.1 M NH_4OAc (step 6)	1.0 M NH_4OAc (step 7)	DNA	Pool/DNA
1	386	149	40	25	7.6 (4.7)	1.0 (0.63)	7.6
2	298	78	28	20	6.7 (4.2)	1.4 (0.88)	4.8
3	580	86	29	25	1.5 (0.91)	1.2 (0.75)	1.2

All data were normalized for differences in cell number between experiments. Results in parentheses represent pmol of methyl group per 10^6 cells and were calculated by dividing cpm × 10^{-3} by specific activity of the [^{3}H]MNU.

TABLE 2. Site Susceptibility of Nucleotide Residues in Precursor Pool and DNA to Methylation by MNU

	Methyl groups, pmol per 10^6 cells			Relative pool
Modification	Pools	DNA*	Pool/DNA	susceptibility†
1MeAde	0.069	0.011 (0.005-0.017)	6.3	13,000
3MeAde	0.028	0.056 (0.039-0.071)	0.50	1,000
7MeGua	0.069	0.502 (0.457-0.585)	0.13	190
O^6MeGua	0.013	0.047 (0.026-0.058)	0.28	400
Unidentified‡	0.644	0.146 (0.112-0.184)	4.4	5,000

*Data for extent of methylation of DNA at these sites was obtained from references 14 to 17 listed in reference 25 and are mean (range). The mean values were used to determine the pool/DNA ratio and the relative pool susceptibility. The reported values agree well with each other and with our preliminary measurements.

†Determined from the pool/DNA methylation ratio (this table), for a particular site, normalized for the large difference in nucleotide content between pool and DNA. Normalization was achieved by multiplying the pool/DNA methylation ratio by the appropriate ratio of DNA nucleotide content/pool nucleotide content per cell. The latter ratios were determined to be 1,432 for guanine, 2,095 for adenine, and 1,130 for total nucleotide assuming a DNA G/C content of 40% for 10T½ cells and using the measured nucleotide content per cell values of Table 2 in reference 25.

‡Determined to be mainly phosphate methylation by treatment of the deoxynucleotides with acid to hydrolyze purines and with alkaline phosphatase to remove unmethylated phosphates. With both treatments, the majority of radioactivity cochromatographed on reverse-phase HPLC with a propyl phosphate standard provided by Richard Wolfendon but not with standards indicating base methylation.

tion may come directly from incorporation of modified precursors during DNA biosynthesis or indirectly from interaction of modified precursors with nucleotide-binding proteins. The latter could, for example, alter pool sizes by affecting metabolic processes and alter DNA repair by interacting with DNA repair proteins.

Incorporation of Methylated Precursors into DNA

To determine whether alkylated precursors are incorporated into DNA by DNA polymerase a method was required that was not only sensitive but also would allow us to study many possible sites of incorporation simultaneously. Sensitivity was required since small effects can have great importance for mutagenesis. The ability to study many sites simultaneously was required since the same template residue may display very different incorporation properties depending on the DNA sequence context in which it lies. Such a method was developed [26] through the application of the plus and minus DNA sequencing technique [21]. The method involved starting with a mix of primed DNA templates. The initial primer was chosen to be just before the sequence of interest. The primer was extended for different lengths of time using ^{32}P-radiolabelled triphosphates to give a mix of primers terminating at early, middle, and late regions within the DNA sequence of interest. These extended primer-templates were pooled, purified away from the radiolabelled precursors, and used in plus or minus reactions. Incubation of the purified template-primer with one dNTP and phage T4 DNA polymerase resulted in chewing back of the primers by the 3'-5'-exonuclease activity of the polymerase until a template site was reached where the dNTP could incorporate. Repeated incorporation and removal of the triphosphate at this site idled the DNA polymerase. Electrophoresis of the reaction products on a sequencing gel capable of separating oligonucleotides differing in length by only one residue results in a sequencing ladder. The use of a modified dNTP in place of the normal precursor in such reactions enabled elucidation of the incorporation properties of that modified precursor by comparison of bands produced in the sequencing gel.

A similar sequencing ladder is generated by incubating the purified template-primer with three normal triphosphates plus DNA polymerase. In this case the sequencing ladder is generated by blocks to DNA replication caused by the missing triphosphate. Addition to the reactions of the modified precursor in place of the missing triphosphate allows analysis of the incorporation properties of the modified precursor. If the modified precursor can incorporate for the missing normal precursor it will remove blocks to DNA replication. Such "read through" can be ascertained by the attenuation of bands on the sequencing gel. The facility of the method was demonstrated by using N-4-hydroxydCTP. N-4-hydroxydCTP was known to incorporate ambiguously as C and as T from site-directed mutagenesis studies [15]. The results from plus and minus reactions were clearly able to show that hydroxydCTP incorporated as T and C (Fig. 2). It was also apparent from these sequencing gels that hydroxydCTP incorporated at many more sites during minus reactions than during plus reactions. The basis for this effect lies in the heavy dependence of plus reactions on the 3'-5' proof-reading activity of the phage T4 DNA polymerase. An incorporated modified precursor is continuously scrutinized by the proof-reading activity and thus must be able to incorporate quite well in order to idle the polymerase. Minus reactions on the other hand are mainly dependent on the polymerizing activity of the DNA polymerase. The proof-reading function

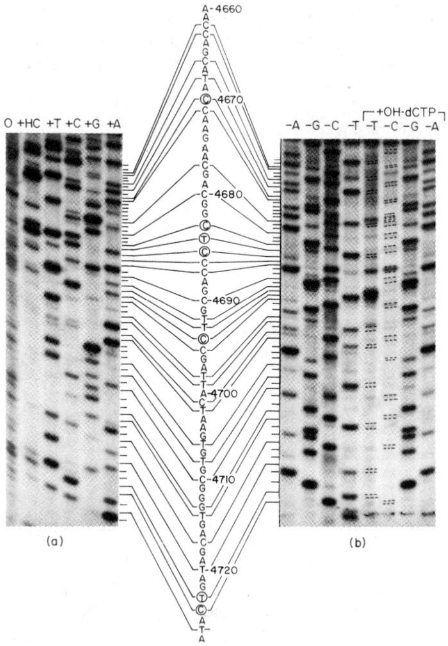

Fig. 2. Incorporation of OH-dCTP (HC) during; a, plus and b, minus sequencing reactions. Plus strand ϕX174 am3 cs70 DNA annealed with the Hae III restriction fragment Z9 was incubated at 37°C for 20 min with 0.4 μg of bacteriophage T4 DNA polymerase in the presence of 0.2 mM dNTP for plus reactions, or 10 μM dNTP for minus reactions. OH-dCTP (500 μM) was used where indicated. a) Products of normal plus reaction and products of a plus reaction in the presence of OH-dCTP are analyzed on this gel. The sequence of the minus strand (growing strand) is shown. Sites at which OH-CTP incorporated during plus reactions are indicated by circled bases in the sequence. Only those sites where incorporation was unambiguously assigned are indicated. Caution is required analyzing this gel because the bands slant downwards as you proceed from left to right across the gel. b) Products of normal minus reactions and products of four minus reactions in the presence of OH-dCTP. Bands that have been diminished in intensity by the presence of OH-dCTP during minus reactions are indicated by pairs of dotted lines. The control slot, labelled O, contains the protein-free products of the asynchronous extension of template-primer (by Klenow pol I) in the absence of phage T4 DNA polymerase. All sequence sites are represented by radio-labelled primer extensions. These extended primers are either hydrolyzed or elongated by DNA polymerase during plus and minus reactions, respectively. Reprinted by permission from Nature, Vol. 298, No. 5877, p. 863, Copywright (c) 1982 McMillan, Journals Limited.

has only a short time to scrutinize the incorporated modified precursor since incorporation of the next normal triphosphate present in the reaction protects the incorporated modified precursor from removal. Thus, minus reactions are a sensitive test for misincorporation of a modified precursor.

Analysis of the Incorporation Properties of O^6medGTP

Armed with what appeared to be a powerful technique, we set out to analyze the incorporation properties of O^6medGTP. O^6meG lesions in DNA have been the source of intense study ever since it was suggested by Loveless [12] that this lesion would exhibit unusual pairing characteristics. Since then the presence of O^6meG lesions in DNA has been correlated with the occurrence of mutations and cancer [16]. In addition, O^6medGTP was a modified DNA precursor produced when cells were treated with MNU in culture [25].

O^6medGTP was synthesized as described elsewhere [24] and purified to greater than 99.99% freedom from contaminating dNTPs by HPLC. Analysis of O^6medGTP in minus reactions indicated that it incorporated as dATP and dGTP but not as dCTP on dTTP [24]. In addition, comparison of the incorporation of O^6medGTP and dATP in minus-A reactions suggested that dATP has only about a ten-fold advantage over O^6medGTP for incorporation opposite T template residues. The concentrations of O^6medGTP required to give equivalent incorporation opposite T and C template residues indicated a 20- to 50-fold preference for pairing opposite T. Unmodified dGTP under the same experimental conditions showed a 10,000-fold preference for pairing with C. The basis for the preferential pairing of O^6medGTP with T template residues can be seen from the possible pairing schemes for O^6meG·C base pairs and O^6meG·T base pairs (Fig. 3). Complementary pairing schemes are possible for both base pairs. O^6meG can make a complementary base pair with T using two hydrogen bonds and normal isomers of both bases. O^6meG can also make a complementary base pair with C stabilized by three hydrogen bonds by using the imino tautomer of O^6meG. The significant level of occurrence of O^6meG·C base pairs, only 20- to 50-fold below that of O^6meG·T base pairs, suggested that the imino tautomer of the O^6meG occurs with significant frequency. The ability of O^6meG to readily tautomerize is consistent with its structural similarities to 2-amino- and 2,6-diamino-purine. The amino-purines are base-analog mutagens whose mutagenicity appears to reflect their ability to tautomerize relatively frequently [5].

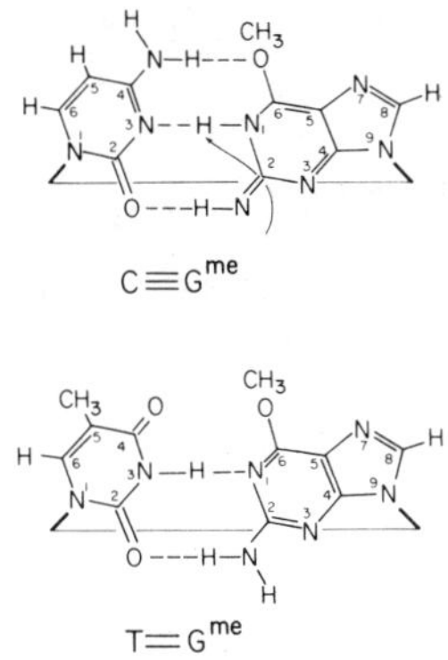

Fig. 3. O^6medGdr·pyrimidine base pairs. O^6medG can form a base pair with Tdr using two hydrogen bonds and the most prevalent isomers of both bases. O^6medG can form a base pair with Cdr with three hydrogen bonds by using an imino tautomer of O^6meGdr. The tautomerization is indicated by an arrow. Note the similarity between the electronic structure of O^6meGdr and 2-aminopurine, which also appears to form the imino tautomer with significant frequency [5].

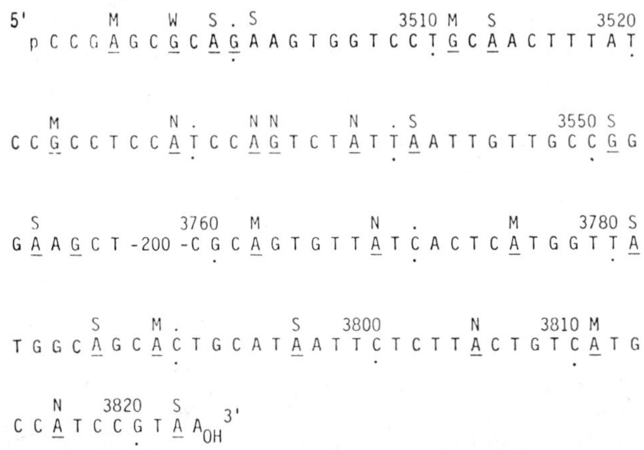

Fig. 4. Growing-strand sites of O^6medGTP incorporation. Incorporation was determined from minus reactions using two f1-R208 sequence regions 200 bases apart as template and with "Klenow" E. coli pol I as the polymerase. A and G growing strand sites where O^6medGTP incorporated are underlined. G incorporation sites were not determined in the second region. Above these sites are an indication of the relative strength of DNA synthesis arrest by O^6medGTP incorporation. Strong arrest (S) - 50 to 100% of intensity lost at minus A or G band held immediately above. Moderate arrest (M) - 25 to 50% held above incorporation site; weak to no arrest (N) - less than 25% of lost intensity held above incorporation site.

O^6methyldGTP Incorporation Arrested DNA Synthesis at Non-Pyrimidine-Rich Sequences

Besides incorporation opposite T and C template residues, O^6medGTP also arrested DNA synthesis upon incorporation at specific sites [24]. This was shown in minus reactions by bands being shifted to and held at the next higher molecular weight positions. Precursors that did not arrest DNA synthesis shifted bands to the higher molecular weight regions of the gel. A summary of those sites where O^6medGTP incorporated, and a semi-quantitative indication of the amount of DNA synthesis arrested is shown in Fig. 4. Analysis of DNA sequences where DNA synthesis arrest occurred indicated first, that two O^6meG·pyrimidine base pairs cannot incorporate sequentially. This is possibly due to steric considerations and the melting of the DNA growing point due to the presence of two weak base pairs. Second, O^6medGTP was found to arrest DNA synthesis at all growing-strand A sites where it incorporated except those surrounded by pyrimidine-rich sequences. The basis for the DNA synthesis arrest activity of O^6medGTP would appear to derive from the nonplanarity of the tetrahedral methyl group at the O^6 position. This bulky methyl group must be accommodated by the geometry of the DNA helix in order for the triphosphate to correctly position itself to be covalently incorporated by the DNA-polymerase enzyme. The lack of DNA-synthesis arrest when incorporating at pyrimidine-rich sequences may be due to the greater ability of the methyl group to be accommodated by a pyrimidine-rich growing-strand sequence. The thermodynamic cost of accommodation may be less at pyrimidine-rich sequences because of the lower stacking interactions involved. In addition, the smaller size of pyrimidine residues versus purine residues may play a role.

Significance of Mutagenesis

The behavior of O^6medGTP in minus reactions suggests at least three mechanisms for mutagenesis. First, the ambiguity exhibited by O^6medGTP in its pairing with T and C template residues suggests the induction of a low level of substitution mutations due to O^6meG template residues directing the incorporation of a low level of precursors other than dTTP at each round of DNA replication [24]. Second, repair of O^6meG lesions in the DNA to normal G by removal of the methyl group will convert O^6meG·C base pairs to normal G·C but will convert O^6meG·T base pairs to the mutagenic G·T intermediate. Therefore, methyl-transferase repair [2, 13, 17, 18] of O^6meG lesions should fix a high level of A·T to G·C transition mutations by production of G·T mutagenic intermediates. This level should be 20- to 50-fold higher than that caused by inherent ambiguity of O^6meG base pairing. Thus, methyl-transferase repair of O^6meG·T base pairs represents a possible major pathway for mutagenesis by O^6meG lesions in DNA.

Finally, DNA synthesis arrest by incorporation of O^6medGTP into replicating DNA may mimic the behavior of the drugs ara A and ara C. These drugs also arrest DNA synthesis and have been shown to cause chromosomal abnormalities [3], mutation [4], neoplastic transformation [10], and death [10] in cells in culture.

CONCLUSIONS

The mutagenic mechanisms suggested by the incorporation of O^6medGTP during DNA replication indicate that alkylation of the precursor pool may be an important area for investigation. Moreover, the methods introduced present a sensitive way to determine the incorporation and base-pairing properties of modified nucleic acid residues. DNA synthesis arrest by O^6medGTP incorporation plus the possible production of mutagenic intermediates by "repair" of incorporated O^6medG·T base pairs suggests that other interesting mutagenic mechan-sms may yet be uncovered by studying the incorporation of modified DNA precursors during DNA replication.

ACKNOWLEDGMENTS

This work was supported by grants from the U.S. Publich Health Service (GM24798, CA28632, and CA16086). M.D.T. is a Leukemia Society of America Scholar.

REFERENCES

1. G. S. Bailey, J. E. Nixon, J. D. Hendricks, R. O. Sinnhuber, and K. E. van Holde, Carcinogen Aflatoxin B_1 is located preferentially in internucleosomal deoxyribonucleic acid following exposure in vivo in Rainbow Trout, Biochemistry, 19:5836-5842 (1980).
2. J. M. Bogden, A. Eastman, and E. Bresnick, A system in mouse liver for the repair of O^6-methylguanine lesions in methylated DNA, Nucleic Acids Res., 9:3089-3103 (1981).
3. J. G. Brewer and N. T. Christie, Studies on the induction of chromosomal aberrations in human leukocytes by cytosine arabinoside, Exptl. Cell Res., 46:276-291 (1967).
4. E. Cerda-Olmeda, P. C. Hanawalt, and N. Guerola, Mutagenesis of the replication point by nitrosoguanidine: map and pattern of replication of the Escherichia coli chromosome, J. Mol. Biol., 33:705-719 (1968).
5. E. Freese, The specific mutagenic effect of base analogs on phage T4, J. Mol. Biol., 1:87-105 (1959).
6. J. W. Grisham, D. S. Greenberg, D. G. Kaufman, and G. Smith, Cycle-related toxicity and transformation in 10T $^1/_2$ cells treated with N-methyl-N'-nitro-N-nitrosoguanidine, Proc. Natl. Acad. Sci., U.S.A., 77:4813-4817 (1980).
7. P. V. Hauschka, Analysis of nucleotide pools in animal cells, in: "Methods in Cell Biology" (D. Prescott, ed.), Vol. 8, pp. 361-462, Academic Press, New York (1973).
8. T. A. Hince and S. Neale, A comparison of the mutagenic action of the methyl and ethyl derivatives of nitrosamides and nitrosamidines on Escherichia coli, Mutat. Res., 24:383-387 (1974).
9. E. Huberman and C. Heidelberger, The mutagenicity to mammalian cells of pyrimidine nucleoside analogs, Mutat. Res., 14:130-132 (1972).
10. P. A. Jones, M. S. Baker, J. S. Bertram, and W. F. Benedict, Cell cycle-specific oncogenic transformation of C3H/10T $^1/_2$ clone 8 mouse embryo cells by 1-β-D-arabinofuranosylcytosine, Cancer Res., 37:2214-2217 (1977).
11. A. M. Kolchinsky, A. D. Mirzabekov, W. Gilbert, and L. Li, Preferential protection of the minor groove of non-operator DNA by lac repressor against methylation by dimethyl sulfate, Nucleic Acids Res., 3:11-18 (1976).
12. A. Loveless, Possible relevance of O^6 alkylation of deoxyguanosine to the mutagenicity and carcinogenicity of nitrosamines and nitrosamides, Nature, 223:206-207 (o969).
13. J. R. Mehta, D. B. Ludlum, A. Renard, and W. G. Verly, Repair of O^6 ethylguanine in DNA by a chromatin fraction from rat liver: transfer of the ethyl group to an acceptor protein, Proc. Natl. Acad. Sci., U.S.A., 78:6766-6770 (1981).
14. H. L. Moses, R. A. Webster, G. D. Martin, and T. C. Spelsberg, Binding of polycyclic aromatic hydrocarbons to transcriptionally active nuclear subfractions of AKR mouse embryo cells, Cancer Res., 36:2905-2910 (1976).

15. W. Muller, H. Weber, F. Meyer, and C. Weissman, Site-directed mutagenesis in DNA: generation of point mutations in cloned β globin complementary DNA at the positions corresponding to amino acids 121 to 123, J. Molec. Biol., 124:343-358 (1978).
16. R. F. Newbold, W. Warren, A. S. C. Medcalf, and J. Amos, Mutagenicity of carcinogenic methylating agents is associated with a specific DNA modification, Nature, 283:596-599 (1980).
17. M. Olsson and T. Lindalhl, Repair of alkylated DNA in Escherichia coli methyl group transfer from O^6-methylguanine to a protein cysteine residue, J. Biol. Chem., 255:10569-10571 (1980).
18. A. E. Pegg, M. Roberfroid, C. von Bahr, R. S. Foots, S. Mitra, H. Bresil, A. Likhachev, and R. Montesano, Removal of O^6-methylguanine from DNA by human liver fractions, Proc. Natl. Acad. Sci., U.S.A., 79:5162-5165 (1982).
19. S. Rajalakshmi, P. M. Rao, and D. S. R. Sarma, Studies on carcinogen chromatin-DNA interaction: inhibition of N-methyl-N-nitrosourea-induced methylation of chromatin-DNA by spermine and Distamycin-A, Biochemistry, 17:4515-4518 (1978).
20. R. Ramanathon, S. Rajalakshmi, D. S. R. Sarma, and E. Farber, Nonrandom nature of in vivo methylation by dimethylnitrosamine and the subsequent removal of methylated products from rat liver chromatin DNA, Cancer Res., 36:2073-2079 (1976).
21. F. Sanger and A. R. Coulson, A rapid method for determining sequences in DNA by primed synthesis with DNA polymerase, J. Mol. Biol., 94:441-448 (1975).
22. B. Singer, The chemical effects of nucleic acid alkylation and their relation to mutagenesis and carcinogenesis, in: "Prog. Nucl. Acid Res. and Molec. Biol." (J. N. Davidson and W. B. Cohn, eds.), Vol. 15, pp. 219-284, Academic Press, New York (1975).
23. C. Tong, M. Fazio, and G. M. Williams, Cell cycle-specific mutagenesis at the hypoxanthine phosphoribosyltransferase locus in adult rat liver epithelial cells, Proc. Natl. Acad. Sci., U.S.A., 77:7377-7379 (1980).
24. D. Toorchen and M. D. Topal, Mechanisms of chemical mutagenesis and carcinogenesis: effects on DNA replication of methylation at the O^6-guanine position of dGTP, Carcinogenesis, 4:1591-1597 (1983).
25. M. D. Topal and M. S. Baker, DNA precursor pool: a significant target for N-methyl-N-nitrosourea in C3H/10T $^1/_2$ clone 8 cells, Proc. Natl. Acad. Sci., U.S.A., 79:2211-2215 (1982).
26. M. D. Topal, C. A. Hutchison, III, and M. S. Baker, DNA precursors in chemical mutagenesis: a novel application of DNA sequencing, Nature, 298:863-865 (1982).
27. M. D. Topal, C. A. Hutchison, III, M. S. Baker, and C. Harris, Studies of the effect of chemical mutagens on the fidelity of DNA replication, in: "Mechanistic Studies of DNA Replication and Genetic Recombination, ICN-UCLA Symposia on Molecular and Cellular Biology" (B. Alberts and C. F. Fox, eds.), Vol. 19, pp. 725-734, Academic Press, New York (1980).

2-AMINOPURINE MUTAGENESIS IN ESCHERICHIA COLI

Barry W. Glickman

National Institute of Environmental Health Sciences
Research Triangle Park
North Carolina 27709

INTRODUCTION

Base analog mutagens are structurally modified bases which sufficiently mimic the natural bases to be incorporated into DNA but whose specificity of incorporation and subsequent templating properties lead to mutation [17, 19]. Two base analogs are quite predominant in mutational literature, the pyrimidine analog 5-bromouracil (5BU) and the purine analog 2-aminopurine (2AP). While this study does draw from our knowledge of 5BU-induced mutagenesis, our experimental efforts have concentrated on 2-aminopurine and some related purine analogs. The chemical and biological properties of 2AP have recently been reviewed by Ronen (see Ref. 49).

It is important to appreciate early on that mutagenesis by base analogs is not simply the result of mispairing during DNA replication. Rather, base analogs appear capable of exerting their mutagenic effects through either of two distinct pathways. These are, the inherent mispairing directed by the analogs themselves [17] and the perturbations in nucleotide triphosphate pools (e.g., Ref. 27). For example, under certain conditions mutagenesis by 5BU is not correlated with the amount of the base analog incorporated into the DNA, but rather with the relative concentration of the base analog on the growth medium [1, 16, 33, 34]. In addition, mutagenesis by base analogs is complicated by the properties of the DNA polymerases; their selectivity, proof-reading capabilities as well as the existence of repair mechanisms such as post-replication mismatch correction that may function to reduce mutation induced by base analogs. A variety of ways in which the properties of DNA polymerase influence mutagenesis as well as how 2AP provides insights into the workings of DNA polymerase is presented elsewhere in this volume (see M. F. Goodman, this volume).

METABOLISM OF 2AP

2AP is taken up by E. coli from growth medium [26, 62] and can, for a short while at least, satisfy the purine requirement of a purine auxotroph. The incorporation of 2AP into nucleic acids has been well studied [48]. The free base is probably converted to a nucleoside by nucleoside deoxyribosyl transferase to yield 2APR. E. coli extracts supplemented with purine and pyrimidine deoxyribosides can catalyze this reaction in vitro. The conversion of the nucleoside to the monophosphate occurs efficiently in vivo but has proven difficult to detect in vitro. The conversion of the monophosphate to the diphosphate appears to be most efficiently carried out by guanine monophosphate (GMP) kinase, at least in vitro. In the case of the kinase reaction, the 2-aminopurine derivative appears to the enzyme more like a G than an A residue! The conversion of the d2APDP to d2APTP by nucleoside diphosphate kinase occurs readily in vitro with crude extracts of E. coli.

The incorporation of 2AP into the DNA of a number of organisms has been demonstrated (see Ronen, Ref. 49). Wacker [62] found that after growing an E. coli B strain in the presence of 3H-labelled 2AP, most of the radioactivity was recovered as A and G residues, not 2AP. Moreover, of the label remaining in 2AP, only 10% was in the DNA, the rest was recovered in RNA. However, by using an adenine requiring strain and growing in the presence of exogenous A and G, more than 90% of the label was recovered as 2APdR [26]. Under these conditions about 0.4% of the adenine was replaced by 2AP, though the nature of the distribution of the 2AP in the DNA remains unknown. Rogan and Bessman [48] reported similar results in another strain of E. coli but Rudner [54, 55], using a purine requiring strain of S. typhimurium found as much as 5% of the adenine in DNA could be substituted for by 2AP. Whether the high level of incorporation in the Salmonella strain is due to the strain or treatment conditions is unknown.

2AP is also incorporated into bacteriophage DNA, though apparently with less efficiency than seen in bacteria. Gottschling and Freese [26] found 1 2AP residue incorporated for every 3364 adenine in the DNA of bacteriophage T4 grown in E. coli B. Using much higher exogenous 2AP concentrations Goodman et al. [25] found 1 2AP residue to be incorporated for every 1700 adenine residues.

In vitro synthesis systems allow a much higher ratio of 2AP to be incorporated depending upon the condition utilized. With equimolar concentrations of each of the four normal triphosphates and d2APTP, Rogan and Bessman [48] found that DNA synthesized on a calf thymus template by DNA polymerase I had about 10% of its adenine residues replaced by 2AP. Thus, consistent with its suspected properties as a mutagenic base analog, 2AP may pair less effectively with T than the natural base. Alternatively, the polymerase itself may not utilize the d2APTP as effectively. Using the same template and T4 polymerase, Bessman et al. [4] found about 1 in 12 adenines replaced by 2AP. However, for every three 2AP residues incorporated, two d2APMP molecules were recovered free in the reaction buffer. These molecules are presumably the product of the turn-over reaction in which triphosphates are incorporated by the polymerase but then removed by the 3'-5'-exonucleolytic proof-reading activity associated with the T4 polymerase. Indeed, different polymerases having different degrees of proof-reading capability appear to incorporate 2AP to different degrees [10, 47, 48]. The ability of polymerases to discriminate between the natural bases and 2AP, both as substrates and templates will be addressed later when the influence of DNA polymerase on the mutagenic consequences of 2AP is discussed.

THE MUTAGENIC CONSEQUENCES OF 2AP

2-Aminopurine whose mutagenic properties were first determined in bacteriophage T4 by Freese [19] has since been found to be mutagenic in several bacteria including E. coli, S. thyphimurium, B. subtilus, and their phages. However, the bacteria Haemophilus influenzae which is resistant to mutation by a number of generally mutagenic treatments, e.g., ultra-violet light, was found by Kimball and Perdue [35] not to be mutated by 2AP. The lack of mutation induction by 2AP is not due to the inability of these cells to utilize 2AP, as it has been shown that 2AP is incorporated into the cellular DNA. These results indicate that mechanisms may exist to avoid mutagenesis by 2AP. Among the possibilities to be considered: a high selectivity of DNA polymerase resulting in effective discrimination against the incorporation of the base analog; an extremely efficient proof-reading capacity which effectively removes all the base analog from the DNA; or the operation of some extremely efficient post-replicational error-avoidance system, such as mismatch repair, that prevents the fixation of replicational errors directed by the base analog. The degree with which repair affects mutagenesis by 2AP will be discussed later.

2AP has been reported to be mutagenic in fungi [6], though the effects were quantitatively less dramatic than seen in bacteria. Increased mutagenesis in the presence of 2AP has also been reported in Saccharomyces cerevisiae [59, 60], though a purine requiring strain was required to obtain mutagenesis. Other purine analogs may be more effective mutagens in yeast, e.g., for N^6-hydroxyaminopurine was found to be highly mutagenic ([60], Glickman and Kunz, unpublished observations). In plants mutagenic data is still rather sparse. However, the induction of 2AP of chlorophyl mutations has been reported in a number of plants such as sweet clover [36] and rice [29]. In insects, 2AP has not been found to be mutagenic, even at relatively toxic concentrations (see Ronen, Ref. 49 for a review).

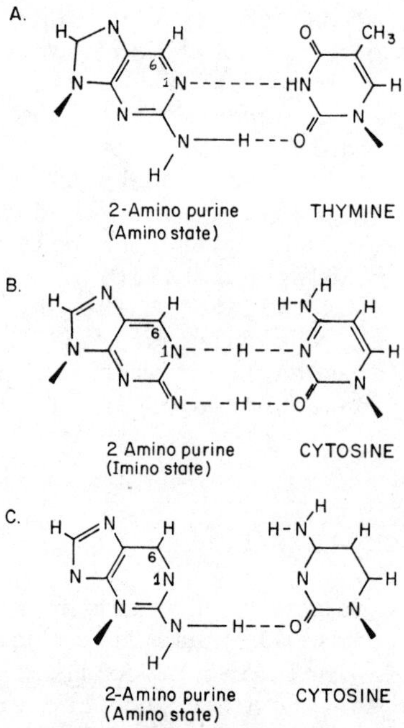

Fig. 1. Hydrogen bonding properties of 2AP and base pairing. A) "Normal" pairing of 2AP with thymine by means of two hydrogen bonds. This mode of pairing results in the replacement of adenine by thymine during DNA replication. B) The rare imino form of 2AP can pair with cytosine by means of two hydrogen-bonds. The second bond is made possible by the tautomeric shift of a hydrogen from the N2 to the N1 position. C) 2AP can also pair with cytosine by means of a single hydrogen bond. Although the strength of such a pairing is low, this pairing results in relatively little distortion of the DNA structure. (From Watenabe and Goodman, 1982.)

In early studies of mammalian cells 2AP was not found to be mutagenic, rather toxic and radiomimetic affects were reported. However, later experiments revealed 2AP to be an effective mutagen in a number of mammalian mutational systems [2, 7, 11].

THE MECHANISM OF MUTAGENESIS BY 2AP: LIFE AS A TAUTOMER

Early recognition of the potential properties of base analogs led Freese [19] to propose that 2AP, in its normal amino form, can pair with thymine creating 2 hydrogen bonds and thereby replacing adenine in the DNA (see Fig. 1a). Like adenine, in its rare imino form, 2AP can create 2 hydrogen bonds with cytosine (Fig. 1B). However, unlike adenine, with little distortion of the double-helical symmetry the amino tautomer can form one hydrogen bond with cytosine (Fig. 1C). The ability of 2AP to pair with either T or C led Freese [19] to suggest that the mutation induced by 2AP was the result of frequent pairing errors of the analog.

An examination of the possible role of the imino tautomer of 2AP in mutagenesis suggested that 2AP was less likely to take on the imino form than adenine [15, 45]. This would seem to negate the possibility that AP_{imino}·C replication intermediates are responsible for 2AP-induced mutagenesis. On the other hand, 2AP·C intermediates share only a single hydrogen bond and are therefore very weak and hence equally unlikely to be reponsible for the observed mutations. A solution this dilemma was put forward by Freese [20] who suggested that once 2AP·C intermediates have formed with a single hydrogen bond, a subsequent tautomeric shift of either the 2AP or C residue permits a second hydrogen bond to form and stabilize the intermediate. It has been suggested by Danilov et al. [15] that the formation of the single hydrogen bond in the initial $2AP_{amino}$·Camino base pair might provide a "trigger mechanism" that forces the tautomeric shift.

THE MECHANISM OF MUTATION: 2AP·C VERSUS C·2AP INTERMEDIATES

The hypothesized mechanism of 2AP induced mutagenesis involves mispaired intermediates and predicts that 2AP will result in the induction of transition mutations. Whether the specificity of the mutational event involves a G:G → A:T or A:T → G:C transition, the mutational intermediate must involve the "incorrect" pairing of the base analog with a C. In the case of transitions involving A:T sites, 2AP pairs initially with T and once residing in the DNA, has the potential of "mispairing" with a C at each replication. Since mutagenesis no longer requires exogenous 2AP, mutation by "misreplication" is often referred to as "clean mutagenesis." In contrast, mutation at a G:C site involves an "error of incorporation" during which 2AP is inserted opposite a C in the template. Since "misincorporation" errors must occur in the presence of the base analog this is often referred to as "dirty" mutagenesis.

The ultimate in clean growth experiments was carried out by Ripley [47]. In her experiments, T4 DNA replicated in the presence of 2AP was packaged, isolated and reintroduced into cells that had never been exposed to 2AP. Among the progeny phase A:T → G:C revertants were recovered. This work confirms that in vivo, mispairing by 2AP can lead to the A:T → G:C transition predicted by the model for 2AP mispairing.

Notice that while mutagenesis at both A:T and G:C sites involves a 2AP:C intermediate, there is a distinct difference. Mutation at an A:T site is driven by 2AP in the template, while mutation at a G:C site involves the misincorporation of 2AP, now a substrate. A definite asymmetry in the accuracy of DNA replication has been observed when 2AP occupies a template versus substrate role in mutagenesis [47]. This is a curious observation in light of models for DNA replication in which fidelity is based upon hydrogen bonding energies, i.e., there is no reason to expect any difference of hydrogen-bonding energy between the two orientations. (For a more detailed discussion of this point the reader is referred to the contribution of M. F. Goodman, this volume.)

THE MECHANISM OF MUTAGENESIS BY 2AP: LIFE VIEWED FROM A CELL

Based upon the current models of mutagenesis by 2AP, a difference in the time of expression (and fixation) can be predicted for A:T and G:C base pairs in the target DNA. Simply stated, mutation at G:C sites requires only a single round of replication following the incorporation of 2AP before the transition event is fixed (i.e., G:G → 2AP:C → 2AP:T) while mutation at an A:T site involves incorporation of 2AP followed by a number of replications during which the 2AP may pair with a C and after which the 2AP:C intermediate still requires an additional round of replication to yield the G:C base pair. Thus, mutation at G:C sites can be expected to be expressed rapidly after the 2AP incorporation event while mutation at A:T sites may be expected to occur later. This is, of course, in addition to questions of which strand is the "sense" strand since this may affect the expression of an altered phenotype by a single round of replication. This phenomenon is known as "immediate" versus "delayed" segregation and is related to the question described above as "clean" versus "dirty" mutagenesis.

In most studies, the pattern of mutant segregation has generally conformed with the predictions of these models [53-55] and hence the likelihood that 2AP normally pairs as an A which once in the DNA "misreplicates" and is rarely "misincorporated" as a G. However, when it is incorporated opposite a C, it leads to mutation fixation the next round of replication. Indeed, many <u>in vitro</u> studies with base analogs as either templates or substrates support this model for 2AP mutagenesis. However, an interesting potential effect of the repair of mismatched lesions is the production of pure mutant clones which have been recovered following base analog mutagenesis (e.g., Witkin and Sicurrela, Ref. 65). This observation represents another indication that DNA repair may modify the mutational consequences of base analog mutagenesis.

UNEASY EQUILIBRIUMS: MUTATION AND POOL IMBALANCES

The reader is by now well acquainted with the potential costs to genetic integrity posed by nucleotide pool imbalances. A number of factors indicate that alterations of nucleotide pools may also contribute to mutagenesis by 2AP. Hopkins and Goodman [27] have pointed out that the 2AP metabolite, 2AP-deoxyribose is an effective inhibitor of adenosine deaminase, ADAase (see Bessman et al., Ref. 4). ADAase is a catabolic enzyme that functions in the salvage pathway for purine (deoxyribo)nucleotide synthesis. It catalyzes the conversion of adenosine and deoxyadenosine to inosine and deoxyinosine respectively. Normally, the capacity of this enzyme is sufficient to prevent the phosphorylation of deoxyadenosine. However, when the ADAase is inhibited, deoxyadenosine is rapidly phosphorylated. The high concentrations of dATP are toxic, strongly inhibiting ribonucleotide reductase, and as a consequence dramatically reducing the intracellular pools of TTP, dGTP, and dCTP [9, 42].

The above reasoning led Hopkins and Goodman [27] to predict that during 2AP mutagenesis, 2-aminopurinedeoxyribose inhibition of ADAase would result in high levels of dATP and consequently reduced pools of TTP, dGTP, and dCTP. This imbalance is then likely to enhance the mutagenic potential of the base analog. Indeed, experimental evidence exists that strongly supports this notion. Caras et al. [7] studied 2AP induced mutagenesis in mouse T-lymphoma cell line, AraC-6-1. This cell line has an abnormally high dCTP pool and a very low TTMP pool. Such an imbalance would be expected to lead to increased levels of 2AP mutagenesis since the chance of a template 2AP mispairing with a C would be increased relative to expectations at normal pool ratios. Indeed, the AraC-6-1 cell line was found to be 3 to 5-fold more mutable than the parent cell line. In addition, supplementing the medium with thymine lowered their high dCTP pools (2 to 3-fold) and raised their low TTP pools (also by 2 to 3-fold) and simultaneously suppressed the cell line's hypermutability to 2AP (8 to 10-fold).

A now classical observation supporting the mispairing hypothesis of 2AP induced mutagenesis might also be reinterpreted in light of the potential influence of pool imbalance effects. Howard and Tessman [28] showed that 2AP induced all potential transitions in the single-stranded phage S13. However, the C → T and G → A transitions but not the T → or A → G transitions were suppressed by the addition of deoxyguanine (GdR). One possible interpretation is that the addition of GdR increases the intracellular pools of dGTP which pairs substantially better with C than does 2AP so that 2AP is essentially excluded from the DNA. However, it was also observed that the addition of deoxyadenosine (AdR) did not affect mutagenesis at A or T sites as would be expected if the same argument was valid, i.e., that A pairs better with T than does 2AP as has in fact been shown to be the case [10]. Thus a reasonable reinterpretation of these observations might be that pool imbalances created by the addition of 2AP are restored to some degree by the addition of GdR but not AdR. To date, however, no reports of actual pool sizes of the dCTPs have been reported under conditions of 2AP mutagenesis.

As discussed by B. Kunz (this volume), thymine deprivation severely perturbs the nucleotide pools. 2AP treatment in conjunction with a treatment leading to thymine deprivation is synergistic in both E. coli [43] and bacteriophage T4 [58], thus indicating again the potential influence of pool effects on base analog mutagenesis. Mutagenesis by 2AP need not be related to pool imbalances. The in vitro coding properties of 2AP reveal its potential to produce transitions as do the 'clean' growth experiments described above.

ERROR-AVOIDANCE: REPAIR OF 2AP LESIONS?

Besides the potential of proof-reading functions associated (or unassociated) with DNA polymerase to mediate the amount of 2AP incorporated into DNA (see Ref. 49 and M. F. Goodman, this volume), evidence has accumulated that cellular metabolism can modify the mutagenic consequences of base analog mutagenesis. An example already sited is the immutability of H. influenzae by 2AP [35]. These authors found no 2AP induced mutagenesis even at dose levels sufficient to inactivate the cells. The cells were not mutable by BUdR either, even when more than 20% of the thymine in the DNA had been substituted for by BU. This bacteria lacks an inducible recA-like "error-prone" repair system, but in other microorganisms 2AP mutagenesis is independent of recA function. The authors suggest that possibly the DNA replication complex can effectively discriminate against 2AP:C mispairings. This possibility makes it attractive to consider measuring the replicational fidelity of H. influenzae's DNA polymerase in vitro in a system such as described by Weymouth and Loeb [64]. An interesting alternative is that the polymerase is not endowed with extreme accuracy but that the cell has evolved a very active and very accurate method of mismatch repair (for a review of the potentials of mismatch repair see Glickman, Ref. 22).

In the opposite direction, mutants are known which increase mutation induced by 2AP. Among this class of mutants in E. coli are a number of mutator mutants defective in mismatch correction. These include uvrD and uvrE (or mutU) [56, 57] which are hypermutable by 2AP and the dam mutant, defective in the methylation of adenine residues at the sequence 5'-GATC-3', which are greatly sensitized to killing by 2AP [23]. The relationship between DNA methylation and mismatch repair as well as an hypothesis to account for the sensitivity of dam mutants to the lethal effects of 2AP have been described previously by Glickman and Radman [24] and Glickman [22]. The possible influence of excision repair on base analog mutagenesis has been discussed by Ronen [49]. In addition, the base analog benomyl has been found to be only mutagenic in the absence of excision repair; it is apparently recognized and removed by the excision repair pathway of E. coli.

PITFALLS IN THE INTERPRETATION OF MUTATIONAL DIRECTIONALITY

A number of studies have attempted to examine the mechanism of in mutagenesis by 2AP by determining the preferential directionality of 2AP. Many of these studies have been summarized by Ronen [49] and the reader is referred there. However, a word of caution is required before attempting to interpret these results. The immense site to site variation in mutation frequencies (often more than two orders of magnitude) and the clear neighboring base pair effects make these results difficult to interpret [8, 12, 13, 19, 50]. In the extreme case a hotspot may be predominant in one gene and be absent in another. Moreover, it must be realized that the mutational system employed to analyze mutational specificity will also influence the outcome. In addition, it is important to be aware of how in each case the mutational directionality is assigned. The basis for such assignments ranges form the security of DNA sequence knowledge to the reversion profile with several or even a single agent whose mutational specificity is though to be known. A stunning example of the power of the analysis of mutation at the DNA sequence level is the study of 5BU mutagenesis by Skopek and Hutchinson [57]. Finally, it need also be recognized that studies involving the examination of directionality by the comparison of foward and reverse frequencies at the same site suffer the unavoidable objection of not studying the same DNA sequence as a mutational target.

MUTATIONAL SPECIFICITY IN _E. COLI_

Weinberg and Boyer [63] showed using reversion analysis by hydroxylamine and 5BU that 2AP induced are mutants of _E. coli_ were the consequence of both G:C → A:T and A:T → G:C transitions. In a system not relying on predicted specificites of mutagenic treatments but rather information obtained from protein sequences, Yanofsky et al. [66] analyzed the reversion data for several sites within the _trpA_ gene of _E. coli_. They showed the induction of A:T → G:C transitions by 2AP but unfortunately are non-informative concerning the induction of G:C → A:T transitions. In a more extensive study of reversion of mutants in the same gene Pershing et al. [44] found that both A:T → G:C and G:C → A:T transitions were induced by 2AP, though the former appared to be preferred. In addition, they found that 2AP enhanced transversions at three sites and frameshifts at another. These latter events were _recA_-independent in origin and though generally lower in frequency than observed at most transition sites, were clearly induced to levels significantly above background. While the transition events are predicted by the mispairing properties of the base analog, the source of the transversion and frameshift mutations is as yet unclear.

Mutational studies of 2AP have also been undertaken in the _lacI_ gene of _E. coli_ by Miller and his colleagues. In this system the induction of single base substitutions leading to the production of nonsense mutations can be monitored at more than 65 sites within the gene [12, 13]. All four transversions and the G:T → A:T transition can be monitored. The A:T → G:C transition can not be monitored in this system as this transition cannot give rise to de novo nonsense mutations. In their study of the mutational specificity of 2AP, Coulondre and Miller [13] confirmed that 2AP could potentiate the G:C → A:T transition. Moreover, mutation did not occur equally at all sites. Specifically, hotspots were observed among the amber mutants isolated. In fact, more than 90% of the amber mutants recovered mapped to three locations, each having the common sequence 5'-CCAGG-3' (see Ref. 14). We have questioned whether the hotspot at this site is related to the DNA sequence [for examples of nearest neighbor or local sequence effects see Koch [37], Ronen et al. [51]] or to the fact that the cytosine residue at this sequence in E. coli K12 is modified to a 5-methyl cytosine? In addition, since these hotspots are also spontaneous hotspots due to deamination, we have asked whether mutagenesis by 2AP might be influenced by the presence of an ung mutation which results in the loss of uracil-N-glycosylase and therefore the cell's ability to remove uracil residues from the DNA.

To probe the nature of the mutational profile obtained with 2AP we have also examined the mutation spectra obtained with another purine analog, 2-amino-N^6-hydroxyadenine [(OH)N^6-2AP], which has been shown to be mutagenic in bacteria [31, 32].

MATERIAL AND METHODS

The E. coli LacI system allows the monitoring of base substituion leading to 72 nonsense mutations at 65 sites within the lacI gene. These mutations can be identified by a combination of genetic mapping and an analysis of their suppression properties [12, 13] and as the DNA sequence of the lacI gene is known, each mutation can be correlated with a specific base substitution [41]. The experimental rationale, procedures, strains, and selection media have been described previously [12, 13, 21, 61]. The strains used in this study are all isogenic derivatives of the wild type strain KMBL 3835 and only differ from each other in the trait being compared.

RESULTS

The Induction of Mutation by 2AP in E. coli: Effect of Dcm and Ung Mutations

Growth overnight in 600 μg/ml 2AP resulted in a large increase of mutation in all four strains examined. These included the wild type, Dcm^-, Ung^-, and Dcm^--Ung^- double mutant. In the case of the wild

TABLE 1. Induction of lacI Mutations in the Wild Type Strain by 2AP

MUTATIONS	SPONTANEOUS FREQUENCY	2AP (600μg/ml) FREQUENCY	FOLD-INCREASE OVER CONTROL
$lacI^-$	1.5	103	69
ambers	0.013 (1%)	13.5 (13%)	1040
ochres	0.006 (0.4%)	10.4 (9.9%)	1700

Mutation frequencies are x 10^{-6}.

TABLE 2. The Effect of dcm and ung Mutations on Mutation Induction of $lacI^-$ by 2AP

STRAIN	SPONTANEOUS FREQUENCY		2AP (600μg/ml) FREQUENCY	
	TOTAL	AMBER+OCHRE	TOTAL	AMBER+OCHRE
Wild type	1.5	0.019	103	24
Ung^-	4.7	0.27	136	24
Dcm^-	1.9	0.061	88	16
Dcm^--Ung^-	4.4	0.24	177	27

Mutation frequencies are x 10^{-6}.

type strain the increase in lacI mutations over the spontaneous background was about 70-fold. When this is viewed in terms of the induction of nonsense mutations, i.e., as the increase in base substitusion events, the increase was over 3-orders of magnitude (see Table 1). The induction of mutation in the other strains was very similar and is summarized in Table 2.

The nature of the mutations induced by 2AP was also analyzed. As described above, 2AP is known to cause both G:T → A:T and A:T → G:C transitions in both bacterial and bacteriophage systems. How-

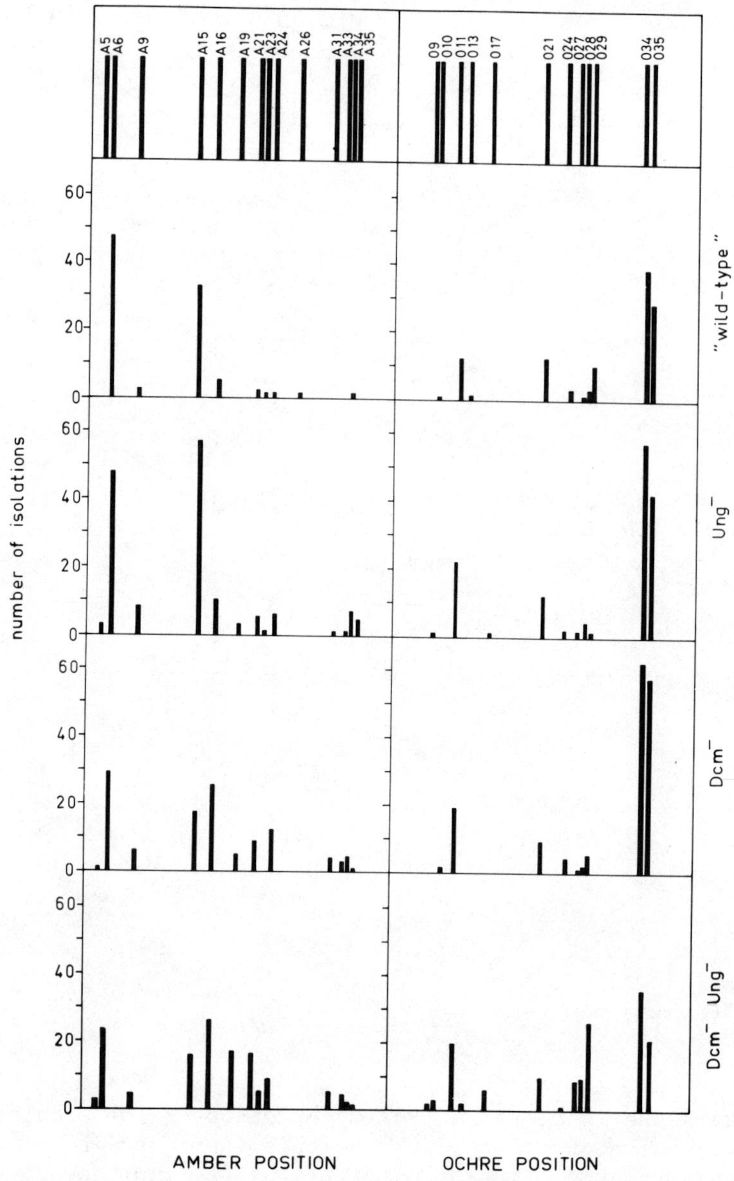

Fig. 2. The distribution of 2AP induced lacI⁻ amber and ochre mutations in the wild type, Ung⁻, Dcm⁻, and Dcm⁻-Ung⁻ double mutant. In the case of the amber mutations, 94, 157, 111, and 133 mutants were analyzed in the wild type, Ung⁻, Dcm⁻, and Dcm⁻-Ung⁻ double mutant respectively. In the case of the ochre mutants, 121, 169, 166, and 149 mutants were analyzed.

TABLE 3. The Induction of str^R Mutants by 2AP

STRAIN	SPONTANEOUS FREQUENCY	2AP (600μg/ml) FREQUENCY
Wild type	2.4 (1.2 - 2.8)	3800 (2400-4700)
Ung$^-$	2.7 (1.7 - 2.9)	2000 (1600-2800)
Dcm$^-$	2.4 (1.2 - 2.6)	2800 (2200-4100)
Dcm$^-$-Ung$^-$	2.2 (1.7 - 2.6)	2600 (2000-2700)

Frequency x 10^{-10}. Numbers in parentheses denote the range for 48-60 independent cultures in 6 independent experiments.

ever, in the lacI system only the G:C → A:T transitions can be monitored. An analysis of nonsense mutations induced by 2AP in the lacI gene reveals that they are exclusively the result of G:C → A:T transitions, i.e., transversions were induced. The mutational spectra resulting from 2AP in the wild type, Dcm$^-$, Ung$^-$, and Dcm$^-$-Ung$^-$ strains are shown in Fig. 2. Although the lacI system does not permit the detection of A:T → G:C transitions, 2AP did however effectively induce mutaants resistant to streptomycin (see Table 3), a mutational step reflecting A:T → G:C transition events [21].

As described earlier, not all sites respond equally to the 2AP treatment and hotspots for mutation can be identified. In the case of the lacI gene, mutatinal hotspots have been correlated with the sequence 5'-CCAGG-3' within which the G:C = >A:T transition creates an amber mutation [14]. As seen in Fig. 2, the wild type strain shows hotspots at amber site A6 and A15 which have that sequence. In fact, in the experiment presented in Fig. 2, 83% (78/94) of the amber mutants occur at these two sites. In all experiments with the wild type strain 79% (606/769) of the amber mutants characterized mapped at these two sites (see Fig. 3).

The questions arises as to why the amber mutations are induced preferentially at these sites. Normally, in E. coli K12, this sequence is methylated so that the residue at the mutational site is not a C but rather a 5-methylcytosine (5MeC). These sites are also spontaneous hotspots [13, 14]. In the case of spontaneous mutage-

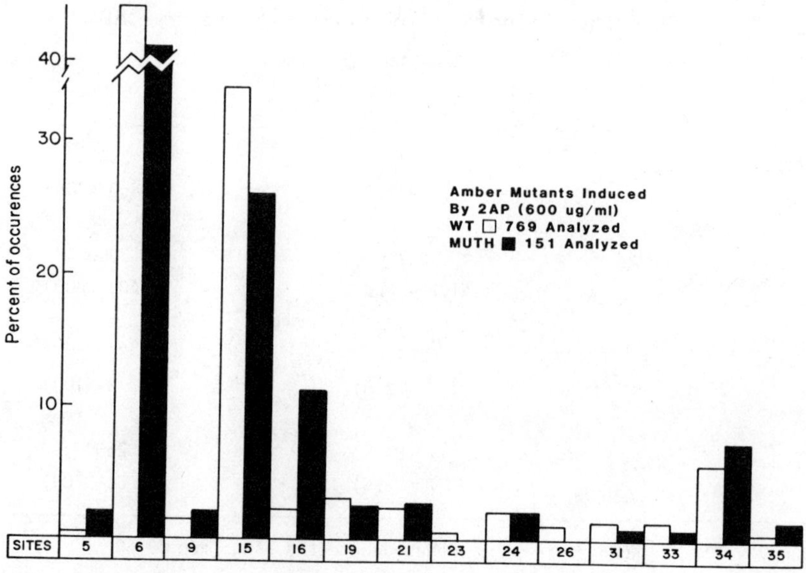

Fig. 3. The mutational spectra for 2AP induced lacI amber mutants in the MutH⁻ and wild type strains. 151 and 769 amber mutants were analyzed respectively.

nesis, mutations at these sites are thought to arise as a consequence of deamination [14]. Normally the deamination of C results in a U in the DNA which is efficiently removed by the action of DNA-uracil-glycosylase. In the case of the deamination of 5MeC, however, the product is 5MeU or thymine! This product remains in the DNA and upon replication r esults in a G:C → A:T transition.

The possibility has been put forward that the hotspots resulting from 2AP treatment may be in some way related to this phenomenon [32]. She suggested that since in 2AP:C pairs, the cytosine C4 amino group is unable to be hydrogen-bonded, this results in an instability promoting the deamination event. Thus as in the case of spontaneous deamination, repair by DNA-uracil-glycosylase would reverse mutations at all but the 5Me sites. According to the current view of the coding properties of 2AP however, it would be predicted that every time a 2AP:C pair formed at what was a G:C position in the sequence, the next round of replication would result in the incorporation of a T opposite the 2AP and hence a G:C → A:T transition. Nontheless, a possible contribution of deamination to the hotspots can readily be assessed by determining the mutational specificity of 2AP in an Ung⁻ strain defective in DNA-uracil-glycosylase. If deamination did contribute significantly to the production of mutation by 2AP, then it would be expected that in the presence of

TABLE 4. Effect of Genotype on 2AP-Induced Mutational Hotspot Frequencies

	SITES					
STRAIN	A6	A15	A34	O34	O35	SITE AVERAGE*
Wild type	630	470	68	300	290	32
Ung$^-$	380	450	55	370	240	43
Dcm$^-$	130	75	17	440	320	31
Dcm$^-$-Ung$^-$	170	110	14	490	250	80

Frequencies X 10^{-8}. A indicates amber sites; O ochre sites. *The notation SITE AVERAGE indicates the mutation frequency averaged at all transition sites excluding the hotspots sites A6, A15, O34 and O35.

the ung mutation, mutational events a non-5MeC sites would be relatively enhanced, provided of course that the presence of the 5-methyl group was not required for the promotion of the deamination event. The presence of the ung mutation did not alter the mutational specificity of 2AP (Fig. 2), though it did have the expected effect of increasing the spontaneous transition frequency at G:C sites [18]. No increase was observed among the spontaneous Str mutations (Table 3) as would be expected for A:T sites. It is also possible that 5MeC is responsible for the mutational hotspots because it alters the templating properties of the DNA. The relationship between the methylated base and the mutational hotspot can be probed by determining the mutational specificity of 2AP in the absence of cytosine methylation. We therefore have determined the spectrum for 2AP mutagenesis in a strain carrying a dcm mutation and thus lacking 5MeC in its DNA [40]. As can be seen in Fig. 2, while mutants are certainly recovered at these two amber sites, they no longer predominate the amber spectrum. As shown in Table 4, mutations at these sites are reduced approximately 5-fold. However, 2AP-induced mutation is still 2 to 4-fold greater at these sites than the site average calculated for all possible transition sites excluding the hotspots. In Table 4 it should be noted that mutation frequencies at the ochre hotspots are not affected by the dcm mutation.

The double mutant Dcm$^-$-Ung$^-$ was also examined for its effect on the specificity of 2AP induced mutagenesis. This strain behaved

TABLE 5. Comparison of 2AP Induced Mutagenesis in Wild Type and MutH Strains

CONTROL	MUTH	WILD TYPE
FREQ LACI$^-$	2.0×10^{-4}	7×10^{-6}
% AMBERS	0.65%	1.2%
% OCHRES	1.7%	0.7%
# TESTED	9034	9000
A+O FREQ	4.7×10^{-6}	1.3×10^{-7}
FREQ lacI$^-$	3.7×10^{-4}	7.0×10^{-4}
% AMBERS	6.0%	6.9%
% OCHRES	7.3%	5.7%
# TESTED	2767	1082
A+O FREQ	4.9×10^{-5}	8.8×10^{-5}
FOLD INCREASE		
TOTAL lacI$^-$	1.9	100
A + O	10.4	680

Cells were grown overnight in LB medium with or without the addition of 2AP (600μg/ml) as described in materials and methods.

as the Dcm$^-$ strain in showing the reduced induction of mutation by 2AP at A6 and A15 (Fig. 2, Table 4). The effect of the ung mutation on the spontaneous mutation frequency was also apparent (Table 2).

Influence of Mismatch Repair on the 2AP Hotspots

Mismatch repair involves the correction of bases incorrectly incorporated during DNA replication. The sensitivity of dam$^-$ mutants (defective in strand-discrimination during mismatch repair) to 2AP [23] suggested that mismatch repair may take an active role in avoiding mutation by base analogs. Thus, the possibility exists that the mutational hotspots at A6 and A15 following growth in 2AP reflect the lack of mismatch correction at these sites. If that were the case, then mutants defective in mismatch repair would be expected to show an altered mutational spectrum in which mutation at other

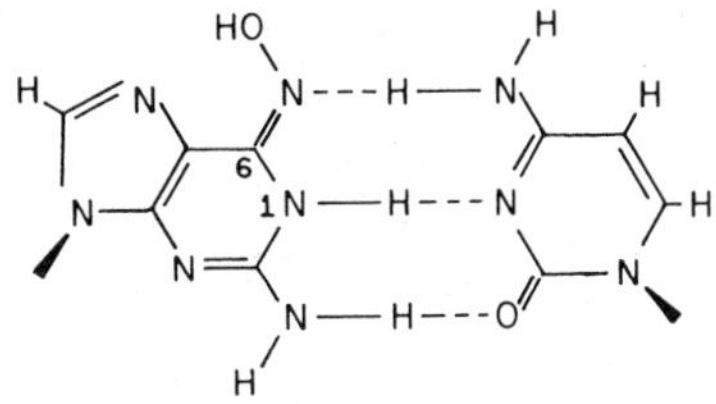

Fig. 4. Proposed pairing configuration for N^6-hydroxy-2,6-diamino-purine [(OH)N^6-2AP] and cytosine.

(usually efficiently corrected) sites would increase relative to the hotspot sites and mask their presence. We therefore have examined the mutational consequence of 2AP in a MutH strain.

The mutational response to 2AP of the MutH strain is compared to that of the wild-type strain in Table 5. As can be seen, at the dose of 2AP examined, the MutH strain shows a level of mutation induction that is very similar to that seen in the wild type strain. Possibly at lower 2AP concentrations the mismatch defective strain may show an increased sensitivity to mutation induction by 2AP, however, the mutator property of the MutH strain makes this difficult to determine with any degree of certainty (data not shown). Independent amber mutants were collected after overnight growth in 600 μg/ml 2AP and the responsible base substitutions determined. As can be seen in Fig. 3 the hotspots at A6 and A15 remain. It would thus appear that mismatch repair is not related to their origin.

Mutation Induction by 2-Amino-N^6-hydroxyadenine

The reason for the mutational hotspots at A6 and A15 (and indeed the ochre sites 034 and 035) remain obscure. Indeed, we have no way of knowing whether the hotspots reflect increased levels of incorporation of the base analog or reduced levels of its removal by either DNA polymerase or some alternative repair pathway. One possibility for the hotspots at the 5MeC sites is that 5MeC residues have slightly different base pairing properties with 2AP than do C residues. In that case, another base analog closely resembling 2AP but having the potential to form hydrogen bonds with the C4 amino group of cytosine might pair as well with C as 5MeC. Perhaps then, the mutational hotspots seen with 2AP at 5MeC sites might be absent from the mutational spectrum. The base analog N^6-hydroxy-2,6-diaminopurine [(OH)N^6-2AP] resembles 2AP greatly, yet has the ability to form more normally hydrogen-bonded base pairing with cytosine (see Fig. 4). The mutagenic properties of this base analog has been previously described [30-32, 60].

TABLE 6. Comparison of the Mutagenic Potentials of 2AP and [(OH)N^6-2AP]

DOSE	2AP (200 μg/ml)	(OH)N6-2AP (200μg/ml)
FREQ LACI$^-$	26, 30, 42 X=33	26, 26, 30 X=27
NUMBER TESTED	2855	6000
NUMBER OF AMBERS	215 (7.5%)	403 (6.7%)
NUMBER OF OCHRES	292 (10.2%)	342 (5.7%)
FREQ OF NONSENSE MUTATIONS*	5.8	3.3
FOLD INCREASE OVER SPONTANEOUS	200	175

*Frequencies are per 10^6. The control data is contained in Table I.

The wild type strain was grown in the presence of [(OH)N^6-2AP] under conditions similar to that used for the 2AP experiments. [(OH)N^6-2AP] was clearly mutagenic. A comparison of mutation induction at various concentrations of 2AP and [(OH)N^6-2AP] revealed them to be of similar potencies; mutational data for the dose of 200 μg/ml is given in Table 6. The mutational spectrum for [(OH)N^6-2AP] in lacI is given in Fig. 5. All the mutants analyzed were the result of G:C → A:T transitions, though A:T → G:C transitions are induced by [(OH)N^6-2AP] as demonstrated by the induction of Str mutants (data not shown). An analysis of the mutational spectrum produced by [(OH)N^6-2AP] shows it to be quite different from that recovered following growth in the presence of 2AP. The hotspots at A6 and A15 are clearly absent and as a result, mutation at other transition sites is evident. Intriguingly, the ochre mutational hotspots at sites 034 and 035 are also greatly reduced (relative to other sites). The mutational spectrum of [(OH)N^6-2AP] reveals a very different profile than obtained with 2AP. This reduces the likelihood that a position effect during e.g., mismatch repair is responsible for the differences in their mutational consequences.

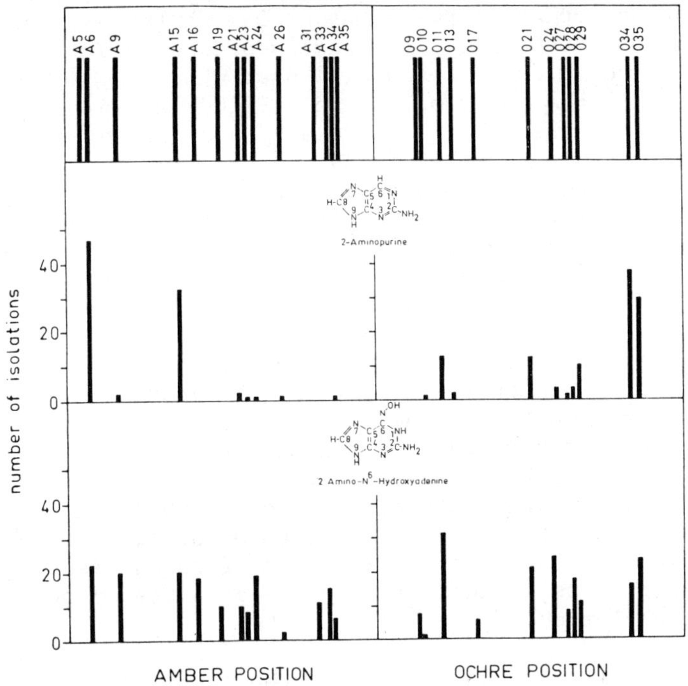

Fig. 5. Spectra of lacI amber and ochre mutants induced by 2AP (upper) and [(OH)N^6-2AP] (lower) in the wild type strain. 94 amber and 121 ochre mutations were analyzed following growth in 2AP; 166 ambers and 162 ochres were analyzed after growth in [(OH)N^6-2AP]. The concentration of 2AP was 600 μg/ml and that of [(OH)N^6-2AP], 200 μg/ml.

DISCUCUSSION

As can be seen in Table 1, the induction of lacI mutations in all four strains is rather similar, both in the frequency of lacI mutants and the frequency of nonsense mutants among them. Mutations were induced effectively in both the G:C → A:T and A:T → G:C direction, though no quantitative comparison between the efficiencies of these transitions can be made considering the low number of sites available for the induction of Str (only 3) and that the surrounding DNA sequences remain unknown.

The Origin of the Amber Hotspots

In considering the spectra presented in Fig. 2, a number of points can be made. In the first place, the appearance in the wild type of hotspots at amber sites 6 and 15 is confirmed [13, 14].

These mutations occur at positions of 5-methylcytosine residues at the sequence 5'-CCAGG-3'. This same sequence occurs an additional time in the lacI gene at the position at which A34 mutations occur. The paucity of this mutation in the 2AP induced spectrum in the wild type strain (see also Fig. 3 where a greater number of mutants have been analyzed) suggests that if DNA methylation is responsible for the hotspots, perhaps this sequence is less efficiently methylated than the other two. This would also be consistent with the observation that spontaneous mutations at this site which are likely to be dependent upon the degree of methylation is also lower than seen at the other two sites [13, 38].

The question whether the amber hotspots found after 2AP treatment are due to the methylation of the DNA is addressed in the analysis of the mutational spectrum in a strain carrying a dcm mutation and thus lacking methylation at these sequencies [40]. As seen in Fig. 2, the mutational response at these sites is greatly diminished. However, the level of mutation remaining at the A(amber) 6 and 15 sites is still greater than the average mutational frequency of all sites (Table 4). Note that the mutation frequencies observed at the ochre hotspot sites is unaltered by the presence or absence of cytosine methylation.

The enhancing effect of 5MeC at the mutational hotspots is clearly demonstrated by the reduction in the importance of these hotspots in the Dcm strain. However, the mechanism of this enhancement remains unclear. The lack of influence of the ung mutation on these hotspots rules out any role that deamination may have in some way been responsible. We cannot be certain whether the presence of the 5MeC residue leads to an increased incorporation of 2AP at those sites, or whether proof-reading is hampered, be it by the goodness of fit of 2AP opposite the 5MeC, or immunity to mismatch correction somehow resulting from the modified base. Alternatively, the hotspots could be related to a lack of mismatch repair at these sites, again either because of the goodness of fit of the 2AP:5MeC intermediate, or an immunity to mismatch repair provided by the sequence. The spectrum of amber mutants obtained in the MutH strain, however, suggests that mismatch repair has little effect of the mutagenesis at these modified sites. The hotspots at A6 and A15 continue to dominate the spectra. That does not necessarily mean that mismatch repair has no influence on base analog mutagenesis, only that these sites are not abnormal with respect to the mismatch repair process.

Nearest Neighbors and the Influence of Mismatch Repair

Clearly a multiplicity of observations indicate a major determinant of 2AP mutation frequencies is related to the DNA context of the mutation, i.e., the local sequence; hence the coining of what has come to be known as nearest neighbor effects. This problem was recognized with the original observation of Freese [19] and Benzer

TABLE 7. Sites of Highest Reversion; Influence of Nearby Sequences

OCHRE SITES*	SEQUENCE (5'→3')
034	G-G-C-A-A
035	T-G-C-A-A
029	T-G-C-A-A
011	C-G-C-A-A
021	A-G-C-A-A
017	C-G-C-A-A
AMBER SITES**	**SEQUENCE (5'→ 3')**
A6	A-C-C-A-G
A15	A-C-C-A-G
A16	A-C-C-A-G
A21***	G-C-C-A-G
A24***	T-C-C-A-G

*These sites account for 96% and 93% of the mutants recovered at the 13 ochre G:C → A:T transition sites in the Dcm$^-$ and wild type strains respectively.

**These sites account for 89% of the 2AP-induced amber mutants in the Dcm$^-$ strain.

***With the exception of sites Am21 and 24, all the transitions involve a C to T change at a CAA or CAG codon. Am21 and 24 are involve G:C → A:T transitions at TGG codons; the "none sense" strand is shown.

[3] that 2AP-induced mutations in the rII cistrons of bacteriphage T4 were non-randomly distributed. Since such a distribution cannot be explained solely by the existence of target sites, i.e., A:T or G:C base pairs in the DNA, Benzer [3] suggested that other factors, e.g., neighboring nucleotides must influence mutation rates. Indeed, Koch [37] and others (see review by Ronen, Ref. 49) have amply illustrated the existence of neighboring base pair effects. Benzer [3], in the absence of any sequencing data suggested that 2AP mutation would occur in A:T rich regions of the DNA because the weaker hydrogen bonding would facilitate mismatches. Others, notably Bessman and Reha-Kranz [5] hold the opposite view, that G:C neighbors have greater hydrogen-bonding strengths and would therefore contribute more greatly to the stability of the mismatched region during its replication. This argument may only apply to the incorpora-

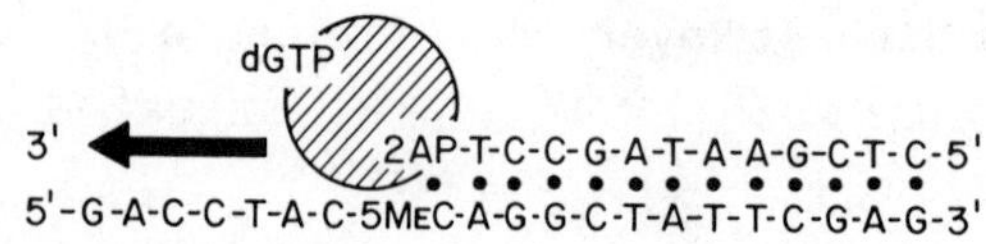

Fig. 6. The hottest sites for 2AP induced G:C A:T transitions have the sequence 5'-C-C-A-C-3' (amber sites) or 5'-G-C-A-A-3' (ochre sites) and thus as demonstrated above, a G:C base pair follows immediately upon the 2AP insertion site. While it has been suggested (e.g., Bessman and Reha-Kranz, 1977) that the greater hydrogen bonding strength of the G:C base pair helps stabilize the 2AP:C mismatch, whether this involves some property ot he insertion or proof-reading step remains unknown. The choice of the 5MeC residue in the template for 2AP insertion is for illustrative purposes only.

tion of 2AP in place of A is unlikely to be viewed as a mismatch. In the lacI gene, however, we are restricted to G:C → A:T transitions and as can be seen in Table 7, in either the wild type of Dcm$^-$ strain, the six 'hottest' ochre sites are proceeded by a G:C base pair. Similarly, the 5 most mutable amber sites (Table 7) are preceded by 2 G:C base pairs. There is thus clearly an influence of the nearest neighbor on the mutation frequency. Among the amber mutation sites, none are preceded by a G, likewise, among the available ochre mutation sites, none are preceded by a C. Thus, the orientation of the G:C base pair adjacent to the site of the mutation cannot be compared under similar circumstances. Nevertheless, the association of G:C base pairs with the hotspots is remarkable. In terms of the mutational mechanism involved, the base immediately following the incorporation of the 2AP opposite a template C, is part of a G:C base pair (see Fig. 6) and thus in agreement with the predictions of Bessman and Reha-Kranz [5].

As discussed earlier, the amber mutational spectrum obtained in the mutH$^-$ strain is indistinguishable from that obtained in the wild type strain. From this we conclude that mismatch repair dependent upon mutH is not responsible for either the mutational hotspots or the nearby neighbor effects (i.e., the origin of these sites is not that they are poorly repaired). Thus both the 5MeC associated hotspots and the sequence enhanced hotspots must be due to an increased level of base analog insertion at those sites.

Mutagenic Consequences of N^6-Hydroxy-2,6-diaminopurine

An important aspect of mutagenesis N^6-hydroxy-2,6-diaminopurine [(OH)N^6-2AP] is the difference in the mutational spectra produced by this analog compared with that of 2AP. The hotspots so characteristic of the 2AP spectra are absent following growth in [(OH)N^6-2AP]. This observation further reduces the likelihood that mismatch repair is responsible for the mutational hotspots seen in 2AP. Rather, it seems likely that the base pairing properties of these two bases are sufficiently different to lead to a totally different pattern of distribution in the DNA. The strong preference demonstrated by 2AP for sites of methylation makes it an interesting probe, however, one to be used with caution with the discovery of an alternative base analog that displays less sequence specificity.

The base analog N^6-hydroxy-2,6-diaminopurine showed a mutagenic potency very similar to that of 2AP (Table 6). This does not mean that the miscoding potentials of the two analogs are in any way equivalent! The metabolism of these purine analogs need not be similar. The two bases could be incorporated at widely different positions and as a result of very different coding properties, still display similar mutation frequencies.

In fact, it still needs to be shown that N^6-hydroxy-2,6-diaminopurine exerts its mutagenic action as a base analog. No information is yet available on the effect of this purine analog on the nucleotide pools, etc. However, the spectra produced by this agent is very specific, exclusively transitions. This is very much in keeping with that would be expected from a base analog. From its structure, it appears that this analog would be able to form reasonable base pairings with both thymine and cytosine. The duality of the pairing properties is essential to its ability to behave as a base analog. Should a purine analog behave all of the time, precisely like a G or an A, then it would not be mutagenic.

REFERENCES

1. P. M. Aebersold, Mutagenic Mechanism of 5-Bromodeoxyuridine in Chinese Hamster Cells, Mutat. Res., 36:357-362 (1976).
2. J. C. Barret, Induction of Gene Mutation in and Cell Transformation of Mammalian Cells by Modified Purines: 2-Aminopurine and 6-N-hydroxylaminopurine, Proc. Natl. Acad. Sci., U.S.A., 78:5685-5689 (1981).
3. S. Benzer, On the topography of the genetic fine structure, Proc. Natl. Acad. Sci., U.S.A., 47:403-415 (1961).
4. M. J. Bessman, N. Muzyczka, M. F. Goodman, and R. L. Schnaar, Studies on the Biochemical Basis of Spontaneous Mutation. II. The Incorporation of a Base and Its Analogue into DNA by Wild-Type, Mutator, and Antimutator DNA Polymerase, J. Mol. Biol., 88:409-421 (1974).
5. M. J. Bessman and L. J. Reha-Krantz, Studies on the Biochemical Basis of Spontaneous Mutation. V. Effect of Temperature on Mutation Frequency, J. Mol. Biol., 116:115-123 (1977).

6. H. E. Brockman and F. J. de Serres, Induction of ad-3 mutants of Neurospora crassa by 2-Aminopurine, Genetics, 48:597-604 (1963).
7. I. W. Caras, M. A. MacInnes, D. H. Pershing, P. Coffino, and A. W. Martin, Mechanisms of 2-Aminopurine Mutagenesis in Mouse T-Lymphosarcoma Cells, Mol. Cell. Biol., 2:1096-1103 (1982).
8. S. P. Champe and S. Benzer, Reversal of Mutant Phenotypes by 5-Fluorouracil: An Approach to Nucleotide Sequences in Messenger-RNA, Proc. Natl. Acad. Sci., U.S.A., 48:532-546 (1962).
9. T. Chan, Deoxyguanosine Toxicity on Lymphoid Cells as a Cause for Immunosuppression in Purine Nucleoside Phosphorylase Deficiency, Cell, 14:523-530 (1978).
10. L. K. Clayton, M. F. Goodman, E. Q. Branscomb, and D. J. Galas, Error Induction and Correction by Mutant and Wild Type T4 DNA Polymerases, J. Biol. Chem., 254:1902-1912 (1979).
11. D. Clive, K. O. Johnson, J. F. S. Spector, A. G. Bateson, and M. H. Brown, Validation and Characterization of the L5178Y/TK+/-mouse Lymphoma Mutagen Assay System, Mutat. Res., 59:61-108 (1979).
12. C. Coulondre and J. H. Miller, Genetic Studies of the lac Repressor. III. Additional Correlation of Mutational Sites with Specific Amino Acid Residues, J. Mol. Biol., 117:525-575 (1977a).
13. C. Coulondre and J. H. Miller, Genetic Studies of the lac Repressor. IV. Mutagenic Specificity in the lacI Gene of Escherichia coli, J. Mol. Biol., 117:577-606 (1977b).
14. C. Coulondre, J. H. Miller, P. J. Farabaugh, and W. Gilbert, Molecular Basis of Base Substitution Hotspots in Escherichia coli, Nature, 274:775-780 (1978).
15. V. I. Danilov, Yu. A. Kruglyak, V. A. Kupriyevich, and O. V. Shramko, Mechanism of the Mutagenic Effect of 2-Aminopurine, Biophysics (USSR), 12:840-844 (1967).
16. R. L. Davidson and E. R. Kaufman, Bromodeoxyuridine Mutagenesis in Mammalian Cells is Stimulated by Thymidine and Suppressed by Deoxycytidine, Nature, 276:722-723 (1978).
17. J. W. Drake, The Molecular Basis of Mutaton, Holden-Day, San Francisco (1970).
18. B. K. Duncan and J. H. Miller, Mutagenic Deamination of Cytosine Residues in DNA, Nature, 287:560-561 (1980).
19. E. Freese, The Specific Mutagenic Effect of Base Analogs on Phate T4, J. Mol. Biol., 1:87-105 (1959).
20. E. Freese, The Molecular Mechanisms of Mutations, Proc. 5th Int. Congr. Biochem., Moscow (1961).
21. B. W. Glickman, Spontaneous Mutagenesis in E. coli Strains Lacking 6-Methyladenine Residues in Their DNA: An Altered Mutational Spectrum in dam^- Mutants, Mutation Res., 62:227-237 (1979).
22. B. W. Glickman, Methylation-Instructed Mismatch Repair as a Post-Replication Error Avoidance Mechanisms in Escherichia

coli, in: "Molecular and Cellular Mechanisms of Mutagenesis" J. F. Lemontt and W. M. Generoso, eds.), pp. 65-87, Plenum Press, New York (1982).
23. B. W. Glickman, P. van der Elsen, and M. Radman, Induced Mutagenesis in dam mutants of Escherichia coli: A Role for 6-Methyladenine in Error-Avoidance, Molec. Gen. Genet., 163: 307-312 (1978).
24. B. W. Glickman and M. Radman, The Isolation and Characterization of Mutants of Escherichia coli K12 Defective in Mismatch Correction, Proc. Natl. Acad. Sci., U.S.A., 77:1063-1067 (1980).
25. M. F. Goodman, R. Hopkins, and W. C. Gore, 2-Aminopurine-Induced Mutagenesis in T4 Bacteriophage: A Model Relating Mutation Frequency to 2-Aminopurine Incorporation in DNA, Proc. Natl. Acad. Sci., U.S.A., 74:4806-4810 (1977).
26. H. Gottschling and E. Freese, Incorporation of 2-Aminopurine into the Deoxyribonucleic Acid of Bacteria and Bacteriophages, Z. Naturforsch., 16B:515-519 (1961).
27. R. I. Hopkins and M. F. Goodman, Deoxyribonucleotide Pools, Base Pairing, and Sequence Configuration Affecting Bromodeoxyuridine- and 2-Aminopurine-Induced Mutagenesis, Proc. Natl. Acad. Sci., U.S.A., 77:1801-1805 (1980).
28. B. D. Howard and I. Tessman, Identification of the Altered Bases in Mutated Single-Stranded DNA. II. In vivo Mutagenesis by 5-Bromodeoxyuridine and 2-Aminopurine, J. Mol. Biol., 9: 364-371 (1964).
29. M. Inou, S. Mori, and S. Hori, Mutagenic Effect of Base Analogs Incorporated into Rice, Japn. J. Breed., 26:110-120 (1976).
30. C. Janion, The Synthesis and Properties of N^6-Substituted 2-Aminopurine Derivatives, Acta Biochim. Pol., 23:57-68 (1976).
31. C. Janion, Hydroxy-N^6-2-aminopurine, N^4-hydroxycytidine, and 5-Methyl-N^4-hydroxy-deoxycytidine: Potent Mutagens of the Base Analog Type, Mutat. Res., 41:37-38 (1977).
32. C. Janion, The Efficiency and Extent of Mutagenic Activity of Some New Mutagens of Base-Analog Type, Mutat. Res., 56:225-234 (1978).
33. E. R. Kaufman and R. L. Davidson, Bromodeoxyuridine Mutagenesis in Mammalian Cells, Mutagenesis is Independent of the Amount of Bromouracil in the DNA, Proc. Natl. Acad. Sci., U.S.A., 75: 4982-4986 (1978).
34. E. R. Kaufman and R. L. Davidson, Bromodeoxyuridine Mutagenesis in Mammalian Cells is Stimulated by Purine Deoxyribonucleotides, Somatic Cell Genetics, 5:653-663 (1979).
35. R. F. Kimball and S. W. Perdue, Attempts to Induce Mutations in Haemophilus influenzae with the Base Analogs 5-Bromodeoxyuridine and 2-Aminopurine, Mutat. Res., 44:197-206 (1977).
36. A. Kleinhofs, H. J. Gorz, and F. A. Haskins, Mutation Induction in Melilotus alba annua by Chemical Mutagens, Crop. Sci., 8: 631-632 (1968).
37. R. E. Koch, The Influence of Neighboring Base Pairs upon Base-pair Substitution Mutation Rates, Proc. Natl. Acad. Sci., U.S.A., 68:773-776 (1971).

38. B. A. Kunz and B. W. Glickman, Infidelity of Conjugal DNA Transfer, Genetics, 105:489-500 (1983).
39. G. Maenhaut-Michael and P. Caillet-Fauqet, 2-Aminopurine Induced DNA Repair in E. coli, Molec. Gen. Genet., 188:143-148 (1982).
40. M. G. Marinus, Location of DNA Methylase Genes on the Escherichia coli Map, Molec. Gen. Genet., 127:47-55 (1973).
41. J. H. Miller, C. Coulondre, and P. J. Farabaugh, Correlation of Nonsense Sites in lacI Gene with Specific Codons in the Nucleotide Sequence, Nature, 274:770-775 (1978).
42. E. C. Moore and R. B. Hurlbert, Regulation of Mammalian Deoxyribonucleotide Biosynthesis by Nucleotides as Activators and Inhibitors, J. Biol. Chem., 241:4802-4809 (1966).
43. C. Pauling, The Specificity of Thymineless Mutagenesis, in: "Structural Chemistry and Molecular Biology" (A. Rich and N. Davidson, eds.), pp. 383-398, Freeman, New York (1968).
44. D. H. Pershing, L. McGinty, C. W. Adams, and R. G. Fowler, Mutational Specificity of the Base Analog, 2-Aminopurine in Escherichia coli, Mutat. Res., 83:25-37 (1981).
45. B. Pullman and A. Pullman, La tautomerie des bases puriques et pyrimidiques et la theorie des mutations, Biochim. Biophys. Acta, 64:403-405 (1962).
46. L. J. Reha-Krantz and M. J. Bessman, Studies on the Biochemical Basis of Mutation. IV. Effect of Amino Acid Substitution on the Enzymatic and Biological Properties of Bacteriophage T4 DNA Polymerase, J. Mol. Biol., 116:99-113 (1977).
47. L. S. Ripley, Influence of Diverse Gene 43 DNA Polymerases on the Incorporation and Replication at A:T Base Pairs in Bacteriophate T4, J. Mol. Biol., 150:197-216 (1981).
48. E. G. Rogan and M. J. Bessman, Studies on the Pathway of Incorporation of 2-Aminopurine into the Deoxyribonucleic Acid of Escherichia coli, J. Bacteriol., 103:622-633 (1970).
49. A. Ronen, 2-Aminopurine, Mutat. Res., 75:1-47 (1980).
50. A. Ronen and A. Rahat, Mutagen Specificity and Position Effects on Mutation in T4rII Nonsense Sites, Mutat. Res., 34:21-34 (1976).
51. A. Ronen, A. Rahat, and C. Halevy, Marker Effects on Reversion of T4rII Mutants, Genetics, 84:423-436 (1976).
52. E. Rosenberg, C. Mora, and D. L. Edwards, Selection of Extranuclear Mutants of Neurospora crassa, Genetics, 83:11-24 (1976).
53. R. Rudner, Mutation as an Error in Base Pairing, Biochem. Biophys. Res. Commun., 3:275-280 (1960).
54. R. Rudner, Mutation as an Error in Base Pairing. I. Mutagenicity of Base Analogs and their Incorporation into the DNA of Salmonella typhimurium, Z. Vererbusl., 92:336-360 (1961).
55. R. Rudner, Mutation as an Error in Base Pairing. II. Kinetics of 5-Bromodeoxyuridine and 2-Aminopurine-Induced Mutagenesis, Z. Vererbungsl., 92:361-379 (1961).
56. B. Rydberg, Bromouracil Mutagenesis in Escherichia coli. Evidence for Involvement of Mismatch Repair, Mol. Gen. Genet., 152: 19-28 (1977).

57. T. R. Skopek and F. Hutchinson, DNA Base Changes Induced by Bromouracil Mutagenesis of Lambda Phage, J. Mol. Biol., 159: 19-33 (1982).
58. M. D. Smith, R. R. Green, L. S. Ripley, and J. W. Drake, Thymineless Mutagenesis in Bacteriophage T4, Genetics, 74: 393-403 (1973).
59. S. Sora, L. Panzeri, and G. E. Magni, Molecular Specificity of 2-Aminopurine in Saccharomyces cerevisiae, Mutat. Res., 20: 207-213 (1973).
60. W. G. Sorenson, J. P. Simpson, and Tong-Man Ong, Comparison of Mutagenic and Recombinogenic Effects of Some Adenine Analogs in Saccharomyces cerevisiae D7, Mutat. Res., 82:95-100 (1981).
61. P. A. Todd and B. W. Glickman, The Mutational Specificity of UV Light in Excision Repair Proficient and Excision Repair Deficient Strains of Escherichia coli: Implications for a Role for Secondary Structure in Determining Mutational Hotspots, Proc. Natl. Acad. Sci., U.S.A., 79:4123-4127 (1982).
62. A. Wacker, S. Kirschfeld, and S. Trager, Uber den Einbau Purinanaloger Verbindungen in die Bakterien-Nukleinsaure, J. Mol. Biol., 2:241-242 (1960).
63. R. Weinberg and H. W. Boyer, Base Analog Induced Arabinose-Negative Mutants of Escherichia coli, Genetics, 51:545-553 (1964).
64. L. A. Weymouth and L. A. Loeb, Mutagenesis during in vitro DNA Synthesis, Proc. Natl. Acad. Sci., U.S.A., 75:1924-1928 (1978).
65. E. M. Witkin and N. A. Sicurella, Pure Clones of Lactose-Negative Mutants Obtained in Escherichia coli after Treatment with 5-Bromouracil, J. Mol. Biol., 8:610-613 (1964).
66. C. Yanofski, J. Ito, and V. Horn, Amino Acid Replacements and the Genetic Code, Cold Spring Harbor Symp. Quant. Biol., 31: 151-162 (1966).
67. D. A. Youngs, E. Van der Schueren, and K. C. Smith, Involvement of uvrD, exrA, and recB Genes in the Control of the Postreplicational Repair Process, in: "Molecular Mechanisms for Repair of DNA" (P. C. Hanawalt and R. B. Setlow, eds.), pp. 331-333, Plenum Press, New York (1975).

MUTAGENICITY OF 2-AMINOPURINE, 6-N-HYDROXYLAMINOPURINE,

AND 2-AMINO-N^6-HYDROXYADENINE IN NEUROSPORA CRASSA

Frederick J. de Serres,* Herman E. Brockman,†
Ching-Yuan Hung,† and Laurie K. Overton†

*Office of the Director
National Institute of Environmental Health Sciences
Research Triangle Park
North Carolina 27709

†Genetics Group
Department of Biological Sciences
Illinois State University
Normal, Illinois 61761

INTRODUCTION

We have been interested in using base analogs as probes to study the process of mutation in a eukaryotic organism. It was twenty years ago when we first tried to evaluate the mutagenicity of an especially promising purine analog in haploid wild-type strains of Neurospora crassa [1]. 2-Aminopurine (2AP) had been shown to induce a high frequency of mutants in phage [8] and bacteria [3], but no quantitative data existed for its mutagenic effect on eukaryotic organisms. The high mutagenic activity of 2AP in phage and bacteria was attributed by Ernst Freese [9] to the fact that it would pair with either thymine or cytosine and give rise to GC → AT or AT → GC base-pair transitions. In our initial studies with 2AP in Neurospora, 2AP was an extremely weak mutagen, causing only a 6.5-fold increase over the spontaneous incidence of ad-3 mutants. Furthermore, the genetic characterization of the 2AP-induced mutants showed that they were most probably a mixture of induced mutants and those of spontaneous origin [1]. This approach of using base analogs to study the process of mutation-induction at specific loci in Neurospora was abandoned by us for other mutagens where mutagenic potency was not a problem and where, therefore, the mutant yields were well in excess of those of spontaneous origin.

More recently, however, data on two other purine analogs suggested that they might fulfill our original expectations for 2AP. The purine analog 6-N-hydroxylaminopurine (HAP) was shown to be mutagenic in phage T4 [10], but more importantly was shown to induce specific-locus mutations in yeast [15]. Another analog, 2-amino-N^6-hydroxyadenine (AHA) was shown to be an extremely potent mutagen in Salmonella [11]. As a result of this work, we decided to test both analogs in Neurospora in a new line of work.

In the present experiments, we have used a two-component heterokaryon instead of a haploid wild-type strain so that we could better evaluate the types of genetic lesions produced by these purine analogs in a eukaryotic organism. By assaying for the induction of specific-locus mutations in the ad-3 region, it is possible to recover a wide range of genetic damage at the ad-3A and ad-3B loci. This damage includes point mutations at each locus as well as multilocus deletions that cover one or both loci simultaneously [16, 6, 4].

These studies have shown, in contrast to 2AP, that HAP and AHA are effective mutagens in Neurospora. Furthermore, in contrast to ad-3 mutants recovered after treatment with 2AP, which include point mutations and multilocus deletions, AHA induces exclusively point mutations that may result from AT → GC base-pair transitions. Brief reports of the data from these experiments have been published [2, 14].

CHEMICAL STRUCTURES OF PURINE ANALOGS UNDER STUDY

The chemical structures of each of the 3 purine analogs used in the present study are given in Fig. 1. They enable us to compare the genetic effect of changing the amino group position from C^6 to C^2 and of substituting a hydroxylamino group for the amino group at the C^6 position. We have assumed that these analogs will induce mutations as a result of base mispairing as originally proposed by Ernst Freese in the late 1950's [8].

INDUCTION AND GENETIC CHARACTERIZATION OF ad-3 MUTANTS

The ad-3 forward-mutation test provides a two-step mechanism for the evaluation of mutagenic activity: (a) the recovery of ad-3 mutants (as reddish-purple colonies in a background of white ($ad\text{-}3^+$) colonies) provides a direct method to determine whether the analogs are mutagenic, as well as a measure of mutagenic potency [6], and (b) the characterization of ad-3 mutants by a series of genetic tests [4] provides a mechanism for distinguishing point mutations from multilocus deletions, as well as a presumptive identification of the genetic lesions in the ad-3B mutants at the molecular level.

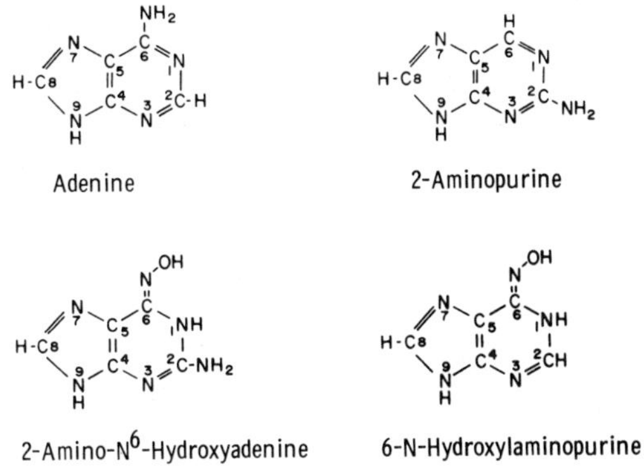

Fig. 1. Adenine and its structural analogs used in the present study.

Ad-3 mutants result from mutation at two closely linked loci, *ad-3A* and *ad-3B*, that control sequential steps in purine biosynthesis. The precursor is not accumulated as such but is converted into a reddish-purple pigment that is stored in the vacuoles of the mycelium. *Ad-3A* and *ad-3B* mutants can be isolated directly on the basis of this pigment accumulation rather than their requirement for adenine.

These loci are located on the right arm of linkage group I. To mimic a diploid eukaryotic organism, we use a two-component heterokaryon, which is heterokaryotic for the *ad-3A* and *ad-3B* loci (Table 1). Heterokaryotic conidia, which are heterokaryotic for both loci in the *ad-3* region form white colonies, whereas heterokaryotic conidia that are homokaryotic for one or both *ad-3* loci due to mutation form reddish-purple colonies. In the *ad-3* forward-mutation test, relatively simple techniques are used to recover and characterize the mutants.

In the first step, recovery of specific-locus mutations in the *ad-3* region, conidia from untreated or analog-treated growth cultures are incubated in a soft agar medium in 12-liter Florence flasks [6]. When these flasks are incubated with aeration for 1 week in the dark at 30°C, individual conidia form colonies about 1-2 mm in diameter. The contents of each flask are harvested by placing aliquots of the suspension in large white photographic developing trays, and the reddish-purple colonies are collected. The number of *ad-3* mutants is determined directly during isolation of the reddish-purple colonies. The total number of background colonies is determined by direct count of the number of colonies in 10-ml aliquots. This step enables us to determine whether treatment with any of the purine analogs results in a significant increase over the spontaneous frequency obtained from untreated cultures. Comparison of the

TABLE 1. Genetic Composition of a Two-Component Heterokaryon Used to Recover Specific Locus Mutants in the *ad-3* Region (Heterokaryon 12).

Com-ponent	Strain No.	Linkage Group									
		I						III	IV	V	VI
I	74-OR60-29A	*A*	*his-2*	*ad-3A*	*ad-3B*	*nic-2*	+	*ad-2*	+	*inos*	+
II	74-OR31-16A	*A*	+	+	+	+	*al-2*	+	*cot-1*	+	*pan-2*

frequency of ad-3 mutants in control and treated series provides a precise measure of mutagenic potency.

In the next step, the ad-3 mutants are subcultured and then subjected to a series of genetic tests to determine their genotype. Because the two-component heterokaryon used in these experiments is heterokaryotic for the ad-3A and ad-3B loci, ad-3 mutants can result from point mutations at either locus ($ad\text{-}3^R$) or from multilocus deletions ($ad\text{-}3^{IR}$) that include one or both loci. Point mutations in the ad-3 region result from intragenic alterations at the ad-3A or ad-3B locus. Multilocus deletions result from deletion of the ad-3A locus or the ad-3B locus, or both loci simultaneously, as well as other loci in the immediately adjacent regions. Point mutations are of genotype $ad\text{-}3A^R$ or $ad\text{-}3B^R$, whereas multilocus deletions are $ad\text{-}3A^{IR}$, $ad\text{-}3B^{IR}$, $(ad\text{-}3A\ ad\text{-}3B)^{IR}$, $(ad\text{-}3B\ nic\text{-}2)^{IR}$, or $(ad\text{-}3A\ ad\text{-}3B\ nic\text{-}2)^{IR}$.

Point mutations at the ad-3B locus may show allelic complementation (e.g., some pairwise combinations of mutants grow on minimal medium, whereas neither mutant alone grows on minimal medium). The interaction matrix from such tests on large numbers of ad-3B mutants can be described in terms of a linear complementation map where mutants are complementing with nonpolarized or polarized complementation patterns or are noncomplementing [7]. In previous studies [12], we have shown a correlation between complementation pattern and genetic alteration at the molecular level. Mutants with nonpolarized patterns result mainly from base-pair transitions of AT → GC or GC → AT. However, the higher the frequency of mutants with nonpolarized patterns, the higher the frequency of mutants that result from AT → GC transitions [13]. Mutants with polarized patterns and noncomplementing mutants result in the main part from frameshift mutations. The genetic characterization of ad-3B mutants in this second step in the analysis provides a mechanism for the presumptive identification of the genetic lesions produced by a mutagen at the molecular level.

INDUCTION OF ad-3 MUTANTS BY 2AP, HAP, AND AHA

Data from forward-mutation experiments with each of the 3 analogs show that each gives a significant increase over the average spontaneous frequency of 0.9×10^{-6} survivors in these experiments.

The experiments with 2AP confirm our previous data with haploid wild-type strains, showing less than a 10-fold (Fig. 2) increase over the spontaneous frequency with the highest concentration tested. In addition, no ad-3 mutants were recovered from either 0.1 or 1.0 μg/ml exposures where numerous mutants were recovered with AHA and HAP. The data from experiments with HAP show an increase in ad-3 mutants with increasing concentrations at low concentrations, with a peak increase of about 400-fold between 3 and 10 μg/ml. Higher concentra-

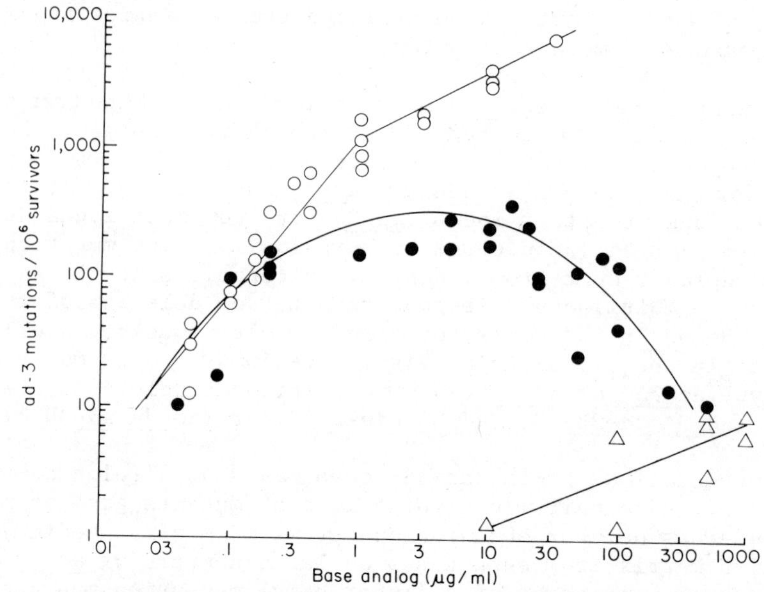

Fig. 2. Dose-response curves for three base analogs in the ad-3 forward-mutation test. ○) 2-amino-N^6-hydroxyadenine; ●) 6-N-hydroxylaminopurine; △) 2-aminopurine.

tions of HAP result in lower forward-mutation frequencies. The most striking increase in mutagenic activity was shown with AHA. With this compound, there was an increased recovery of ad-3 mutants with increasing concentrations over the entire range of exposures tested. The frequency of mutants appears to increase slightly greater than linearly from 0.05 to 1.0 μg/ml and then at less than the 0.5 power of the dose at doses up to 30 μg/ml. At the highest concentration tested, the mutant yield is 7 per 1000 background colonies. This is the highest yield of ad-3 mutants that we have obtained with this assay for any mutagen tested.

The forward-mutation dose-response curves enable us to classify these 3 mutagens with regard to mutagenic potency. The data indicate that 2AP is a weak mutagen (maximum yield in the range of 1 to 10 ad-3 mutants/10^6 survivors), HAP is a strong mutagen (maximum yield in the range of 100 to 1000 ad-3 mutants/10^6 survivors), and AHA is a potent mutagen (maximum yield in the range of 1000 to 10,000 ad-3 mutants per 10^6 survivors) [5].

The next step in the analysis is to characterize the ad-3 mutants by a series of genetic tests to determine their genotypes. At this time, data are available from the genetic analysis of mutants induced by 2AP and AHA, but not by HAP.

CHARACTERIZATION OF 2AP- AND AHA-INDUCED *ad-3* MUTANTS BY GENETIC TESTS

In the genetic analysis of *ad-3* mutants recovered after treatment with 2AP, a total of 232 *ad-3* mutants (induced by 500 μg/ml) were tested. A total of 204 were point mutations (*ad-3*^{R}), and 28 were multilocus deletions (*ad-3*^{IR}). The point mutations were distributed among the two loci as follows: 66 *ad-3A*^{R} and 138 *ad-3B*^{R}. The multilocus deletions were almost exclusively small deletions, covering only the *ad-3A* locus or the *ad-3B* locus with a distribution as follows: 16 *ad-3A*^{IR}, 11 *ad-3B*^{IR}, and 1 (*ad-3A* *ad-3B*)^{IR}, with none in the other 2 classes, which include the largest deletions that can occur in this region. The occurrence of deletions is totally unexpected after treatment with 2AP because deletions are not expected to arise from AT → GC or GC → AT transitions. One possibility is that because the mutant recovery is about 6.6-fold over the spontaneous frequency these mutants are a mixture of spontaneous and induced mutants. Multilocus deletions are known to occur spontaneously. Another possibility is that the high concentrations of 2AP used in the experiments have resulted in some nucleotide pool imbalance that causes chromosome breakage and, therefore, multilocus deletions.

The genetic analysis of the AHA-induced *ad-3* mutants gives a much different picture than that observed with 2AP. A total of 546 mutants (induced by 0.05 or 0.15 μg/ml) were tested, and all of these resulted from point mutations (*ad-3*^{R}), with a total of 248 *ad-3A* mutants and 298 *ad-3B* mutants. None of the AHA-induced *ad-3* mutants resulted from multilocus deletion.

CHARACTERIZATION OF 2AP- AND AHA-INDUCED *ad-3* MUTANTS BY MEANS OF TESTS FOR ALLELIC COMPLEMENTATION

The next step in the genetic characterization of the *ad-3* mutants is to test the *ad-3B*^{R} mutants for allelic complementation. The *ad-3B* mutants define a linear complementation map, with 17 complementation subgroups [7]. Complementing mutants have nonpolarized or polarized complementation patterns. All other mutants are noncomplementing. By using a series of *ad-3B* mutants with nonpolarized complementation patterns as testers, we determined the complementation patterns of the *ad-3B* mutants recovered after treatment with 2AP or AHA. The data from these experiments show that the frequency of 2AP-induced *ad-3B* mutants with nonpolarized patterns is 54.3%, in contrast to 83.6% for AHA-induced mutants. In general, the higher the frequency of *ad-3B* mutants with nonpolarized patterns the higher the frequency that result from base-pair transitions of the type AT → GC. Prior to these experiments, the highest frequency of *ad-3B* mutants with nonpolarized patterns (79.9%) was found among N-methyl-N'-nitro-N-nitrosoguanidine induced *ad-3B* mutants. We have shown from an analysis of nitrous acid-induced ad-3B mutants that mutants with a high frequency of GC base pairs at the mutant site leads to a high frequency of mutants with nonpolarized complementation patterns [13]. From the present data, we can conclude that AHA-induced mutants result predominantly from AT → GC base-pair transitions, but a final proof requires experiments at the DNA sequence level.

SUMMARY AND GENERAL CONCLUSIONS

These data from our experiments with 3 purine analogs reveal striking differences in mutagenic potency. It seems highly likely that these analogs substitute readily for adenine and that they cause mutations in the main part, and in the case of AHA perhaps predominantly, by mispairing with cytosine. The most potent mutagens are those with the hydroxylamino group at the C^6 position (AHA and HAP). Of these, the most potent is the analog with an amino group in the C^2 position (AHA).

The most interesting aspect of the present studies is their implications for other eukaryotic organisms. We have determined that AHA, which was shown to be a potent mutagen in bacteria [11], is an extremely potent mutagen in a eukaryotic organism. AHA is active at relatively low concentrations, and it gives rise to point mutations that appear to arise predominantly by AT → GC base-pair transitions. AHA should be an extremely useful genetic probe for studies on higher eukaryotic organisms. Its potency and specificity make it an unusual mutagen that can be expected to produce specific-locus mutants at high frequency with the genetic damage confined to the boundaries of the gene. These characteristics should make it useful not only for studies of specific-locus mutations and sex-linked recessive lethal mutations in Drosophila but also for specific-locus studies in mammalian cells in culture and in the whole animal. In these latter systems, it is extremely time consuming and sometimes impossible to distinguish between point mutations and multilocus deletions. The use of AHA as a mutagen in these systems should provide a useful new approach to genetic fine structure analysis.

REFERENCES

1. H. E. Brockman and F. J. de Serres, Induction of ad-3 mutants of *Neurospora crassa* by 2-aminopurine, Genetics, 48:597-604 (1963).
2. H. E. Brockman, C. Y. Hung, F. J. de Serres, and T. M. Ong, Potent mutagenicity of 6-N-hydroxylaminopurine and 2-amino-N^6-hydroxyadenine at the ad-3 region of *Neurospora crassa* (Abst.) Environ. Mut., 1:133 (1979).
3. M. Demerec, Frequency of deletions among spontaneous and induced mutations in *Salmonella*, Proc. Natl. Acad. Sci., U.S.A., 46: 1075-1079 (1960).

4. F. J. de Serres, Induction and genetic characterization of specific locus mutations in the ad-3 region in two-component heterokaryons of Neurospora crassa, in: "Short-Term Tests for Chemical Carcinogens" (H. F. Stich and R. H. C. San, eds.), Springer-Verlag, New York, pp. 175-186 (1981).
5. F. J. de Serres, The use of Neurospora in the evaluation of the mutagenic activity of environmental chemicals, Environmental Mutagenesis, 5:341-351 (1983).
6. F. J. de Serres and H. V. Malling, Measurement of recessive lethal damage over the entire genome and at two specific loci in the ad-3 region of a two-component heterokaryon of Neurospora crassa, in: "Chemical Mutagens: Principles and Methods for Their Detection," Vol. II (A. Hollaender, ed.), Plenum Press, New York, pp. 311-342 (1971).
7. F. J. de Serres, H. E. Brockman, W. E. Barnett, and H. G. Kølmark, Allelic complementation among nitrous acid induced ad-3B mutants of Neurospora crassa, Mutation Res., 4:415-424 (1967).
8. E. Freese, The specific mutagenic effect of base analogs on phage T4, J. Mol. Biol., 1:87-105 (1959).
9. E. Freese, The difference between spontaneous and base-analog induced mutations of phage T4, Proc. Natl. Acad. Sci., U.S.A., 45:622-633 (1959).
10. E. B. Freese, The mutagenic effect of hydroxylaminopurine derivatives on phage T4, Mutation Res., 5:299-301 (1968).
11. C. Janion and K. Myszkowska, Mutagenicity and inhibitory properties of some new purine analogs on Salmonella typhimurium TA 1530, Mutation Res., 91:193-197 (1981).
12. H. V. Malling and F. J. de Serres, Relation between complementation patterns and genetic alterations in nitrous acid-induced ad-3B mutants of Neurospora crassa, Mutation Res., 4:425-440 (1967).
13. H. V. Malling and F. J. de Serres, Correlation between type of base-pair transition and complementation patterns in nitrous acid-induced ad-3B mutants of Neurospora crassa, Mutation Res., 5:359-371 (1968).
14. L. K. Overton, C. Y. Hung, H. E. Brockman, and F. J. de Serres, Mutagenic potency and specificity of 2-amino-N^6-hydroxyadenine in the ad-3 forward-mutation test in nucleotide excision repair-deficient and proficient dikaryons of Neurospora crassa, (Abst.), Environmental Mut., 5:501-502 (1983).
15. A. S. H. Shawky, Studies on 6-hydroxyaminopurine (6-HAP) induced mutations in Saccharomyces cerevisiae, (Abst.), Egypt. J. Genet. Cytol., 5:485 (1976).
16. B. B. Webber and F. J. de Serres, Induction kinetics and genetic analysis of X-ray induced mutations in the ad-3 region of Neurospora crassa, Proc. Natl. Acad. Sci., U.S.A., 53:430-437 (1965).

BROMODEOXYURIDINE MUTAGENESIS AND DEOXYRIBONUCLEOTIDE POOL IMBALANCE IN MAMMALIAN CELLS

Charles R. Ashman, Elliot R. Kaufman,
and Richard L. Davidson

Center for Genetics
University of Illinois College of Medicine
808 South Wood Street
Chicago, Illinois 60612

BrdU Effects and DNA

The thymidine (dT) analog 5-bromodeoxyuridine (BrdU) produces a variety of biological effects in mammalian cells, such as the induction of latent viruses [10], inhibition or induction of differentiated functions [20], and mutagenesis [11]. Because BrdU is such an excellent analog of dT, it is readily phosphorylated and incorporated into mammalian DNA in place of dT. Numerous studies have demonstrated a correlation between the incorporation of BrdU into DNA and the occurrence of its biological effects. As a result, it has been assumed that all or most of the biological effects of BrdU result from the presence of the analog in DNA.

The interest of our laboratory in BrdU initially stemmed from attempts to isolate new mutant mammalian cell lines resistant to BrdU toxicity. At that time, it was thought that BrdU toxicity was due to the presence of bromouracil (BU) residues in DNA, and that BrdU resistance generally resulted from a deficiency of thymidine kinase, the enzyme required for phosphorylation of the analog, and thus, the failure to incorporate BrdU into DNA. However, we isolated from the Syrian hamster melanoma cell line 3460 [16] a number of BrdU-resistant mutants which retained both thymidine kinase activity and the ability to incorporate BrdU into DNA [3, 5, 12]. Among the more interesting mutants isolated was a cell line which we termed "BrdU-dependent" since it grew better in the presence of BrdU than in its absence [5]. We also isolated a number of cell lines which could grow indefinitely with all (at least 99.8%) of their thymine (T) residues in DNA replaced by BU [3]. These mutants were called HAB cells because they were isolated in medium containing hypoxanthine, aminopterin, and BrdU. The presence of aminopterin, an inhibitor of de novo purine and thymidine nucleotide biosynthesis, forced the cells to utilize BrdU in place of dT nucleotides for DNA synthesis. The ability of these mutant cell lines to grow with all of their T residues replaced by BU suggested to us that BrdU might be producing at least some of its biological effects at sites outside of DNA.

One biological effect of BrdU which was thought to be clearly due to its incorporation into DNA was mutagenesis. A model proposed by Freese in 1958 has gained wide acceptance [9]. This model proposed, based on the greater electronegativity of the Br atom of BU relative to the CH_3 group of T, that BU assumes the rare enol tautomeric form more often than does T. Since this tautomer can base pair with guanine (G), it follows that BU will mispair with G more frequently than will T. Freese proposed that mutations could be induced by BrdU by two mechanisms. If BU is incorporated into DNA opposite a G residue, a GC to AT transition could result. Transition mutations produced by this mechanism have been termed errors of incorporation. Alternatively, BU residues already correctly incorporated into DNA (in place of T) could mispair with G during subsequent replication, resulting in an AT to GC transition. Such mutations have been termed errors of replication.

BrdU Mutagenesis: Concentrations vs Substitution Effects

Freese's model proposes that BU residues in DNA will be mutagenic. If the mispairing process occurs at a measurable frequency, one would expect that mutagenesis will increase as the BrdU substitution (the percentage of T residues in DNA replaced by BU) increases. However, the existence of the HAB cell line capable of indefinite growth with a level of BrdU substitution of 100% suggested that neither errors of incorporation nor errors of replication were occurring at a high frequency in these cell lines. In order to examine this possibility, we carried out a study of the relationship between BrdU substitution and mutagenesis in one HAB line and the parental 3460 cell line [13].

Initial experiments demonstrated that BrdU can induce mutations in HAB cells, but revealed a very nonlinear relationship between BrdU substitution and mutagenesis. In the experiment shown in Fig. 1, the BrdU substitution was varied in HAB cells either by increasing the BrdU concentration in the medium (E-B medium) or by changing the ratio of BrdU to dT in medium containing hypoxanthine, aminopterin BrdU, and dT (HABT medium), with a total combined concentration of BrdU and dT of 10 μM. The results from this experiment suggested that the level of BrdU substitution does not play a major role in determining the frequency of ouabain-resistant (Oua^r) mutants. Conditions could be created in which low mutant frequencies were ob-

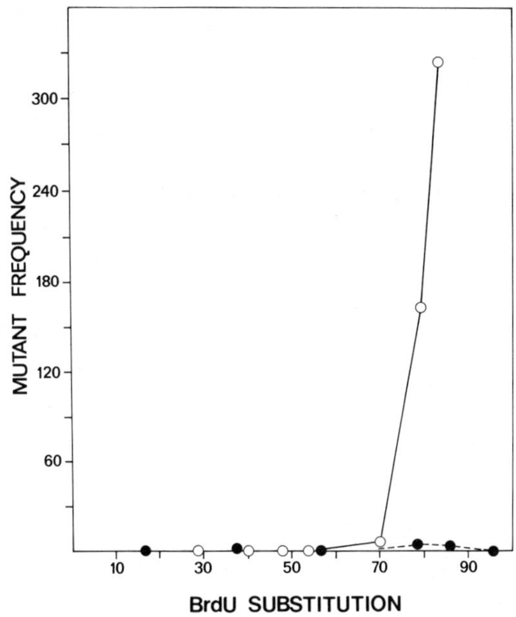

Fig. 1. Mutant frequency (Oua resistance) as a function of BrdU substitution. ○) HAB cells grown in E-B medium with BrdU concentrations of 3, 5, 7, 10, 20, 30, and 40 μM, respectively; ●) HAB cells grown in HABT medium with ratios of BrdU/dT of 2:8, 4:6, 6:4, 8:2, 9:1, and 10:0, respectively; with the concentration of BrdU plus dT fixed at 10 μM. The mutant frequencies are expressed as the number of Oua^r colonies per 1×10^6 cells, corrected for plating efficiency in the absence of drug. Reprinted, with permission from Kaufman and Davidson (1978) [13].

tained at much lower BrdU substitution levels in E-B medium. These data also suggested that the level of mutagenesis is determined by the concentration of BrdU to which the cells are exposed. When the cells were grown in E-B medium a significant increase in mutation frequency was observed only when the BrdU concentration was increased to greater than 20 μM. In these initial experiments the cells which were grown in HABT medium and which did not show an increase in mutant frequency were never exposed to BrdU concentrations above 10 μM. Thus, these data suggested that the extent of BrdU mutagenesis in these mammalian cells is determined by BrdU concentration and is independent of BrdU substitution.

This hypothesis was tested in a group of mutagenesis experiments in which BrdU concentration and substitution were varied independently of one another. This was accomplished by treating the

TABLE 1. BrdU Concentration and Mutagenesis[a]

Additives (uM)[b]			%BrdU	Mutant
Apt	BrdU	dT	Substitution	Frequency[c]
0	0	30	-	<1
0	30	0	73	321
0.4	10	0	100	3
0.4	30	0	100	119
0.4	8	2	77	5
0.4	30	7.5	80	368
0.4	50	12.5	80	3100
0.4	10	20	31	<1
0.4	20	40	31	148
0.4	30	60	29	235

a) Levels of BrdU substitution and mutant frequencies were determined as described in (13).

b) Cells were grown in HABT medium containing 0.1 mM hypoxanthine, 0.4uM aminopterin (Apt) and the indicated concentrations of BrdU and dT.

c) The frequencies of Oua^r mutants are expressed as the number of Oua^r colonies per 10^6 cells after correction for plating efficiencies.

These data were taken from Kaufman and Davidson, 1978 (13).

cells with BrdU in the presence of an inhibitor of de novo thymidine biosynthesis (aminopterin or fluorodeoxyuridine). Under these conditions, the ratio of BU to T residues in newly synthesized DNA is approximately equal to the ratio of BrdU to dT in the medium. Thus, the BrdU substitution can be varied at a constant BrdU concentration by varying the dT concentration, while the BrdU concentration can be varied at a constant BrdU substitution by maintaining a constant BrdU to dT ratio in the medium. The results of experiments of this type utilizing HAB cells are summarized in Table 1. As can be seen, the mutant frequency increased with increasing BrdU concentration when the BrdU substitution was held constant at either 100, 80, or 30%. In fact, much higher mutant frequencies were observed at 30% BrdU substitution when the cells were exposed to 30 μM BrdU than at 100% BrdU substitution when the cells were grown in 10 μM BrdU. Clearly, it was the concentration of BrdU in the medium which determined the level of BrdU mutagenesis in these experiments. A similar relationship between mutagenesis and BrdU concentration has also been observed in the parental Syrian hamster 3460 cell line [13] and in the Chinese hamster cell lines CHO [8] and V79 [C. Ashman and E. Kaufman, unpublished observations].

Suppression of BrdU Mutagenesis by dC

The finding that levels of mutagenesis were determined by the concentration of BrdU to which the cells were exposed rather than the level of BrdU substitution suggested that BrdU might be producing its mutagenic effect by acting at a site outside of DNA. We had shown previously that HAB cells with all their T residues in DNA replaced by BU were still sensitive to high concentrations of BrdU [3]. Clearly, such a result involved effects unrelated to the amount of BU in DNA. The work of Meuth and Green [15] showed that one non-DNA effect of BrdU was to reduce the supply of deoxycytidine (dC) nucleotides by inhibiting the ribonucleotide-reductase catalyzed reduction of CDP to dCDP. Meuth and Green showed that the toxicity of BrdU could be greatly reduced by the addition of dC to the growth medium.

We have attempted to determine if a reduction in the level of dC nucleotides plays a role in BrdU mutagenesis. The results of an experiment on the effect of dC on BrdU mutagenesis can be seen in Fig. 2. It was found that the addition of dC to the medium had a strong suppressive effect on BrdU mutagenesis. In this particular experiment 200 μM dC was added, but in subsequent experiments it was shown that as little as 10 μM dC produces a similar effect [2]. The effect of dC on BrdU mutagenesis was observed in the presence of inhibitors (aminopterin and fluorodeoxyuridine) which block the conversion of dC to dT nucleotides. Thus, as can be seen in Fig. 2, dC suppressed mutagenesis without reducing the incorporation of BrdU into DNA in place of dT. The suppressive effect of even 200 μM dC can be overcome, however, by further increasing the concentration

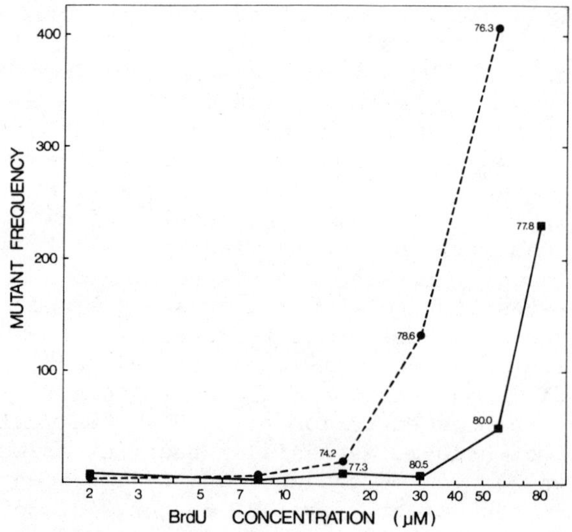

Fig. 2. Effect of BrdU concentration on the suppression of mutagenesis by dC. HAB cells were grown in HABT medium with BrdU and dT in a 4:1 ratio. The cells were grown either in the presence (■) or absence (●) of 200 μM dC. The frequency of Oua^r mutants was expressed as the number of mutants per 10^6 cells, corrected for plating efficiency. The numbers next to the data points indicate the levels of BrdU substitution for those points. Mutagenesis conditions and determinations of mutant frequencies and BrdU substitution were as described in [6]. Reprinted, with permission from Davidson and Kaufman (1978) [6].

of BrdU in the medium. As can be seen in Fig. 2, dC appears to shift the mutagenesis curve to higher BrdU concentrations by a factor of approximately 2.5. These results suggest that the perturbation of dC nucleotide metabolism plays a key role in BrdU mutagenesis. However, since mutagenesis can occur even in the presence of high levels of dC, simple dC nucleotide starvation cannot account for the high levels of mutagenesis caused by BrdU.

BrdU Induction of Sister Chromatid Exchange

In addition to producing mutations, BrdU also increases the frequency of sister chromatid exchange (SCE) in mammalian cells. For example, when Chinese hamster ovary (CHO) cells were grown in medium containing BrdU at concentrations ranging from 12 to 300 μM, the frequency of SCEs increased by over 4-fold [8]. Following our results with BrdU mutagenesis, we evaluated the roles of BrdU concentration and substitution in SCE induction. CHO cells were grown in

TABLE 2. BrdU Induction of SCE in CHO Cells[a]

Additive (uM)			%BrdU	
BrdU	dT	FdU	Substitution	SCE/Cell
12	8	10	65	10
120	80	10	64	18
180	120	10	59	23
300	200	10	61	33
12	24	10	34	7
12	12	10	55	8
12	8	10	65	9
12	1	10	84	12
12	0	10	95	12

[a] Determinations of SCE/cell and levels of BrdU substitution were carried out as described in (8). These data are taken from Davidson et al., 1980 (8).

FBT medium containing 5-fluorodeoxyuridine (FDU) plus BrdU and dT at varying concentrations but at a fixed ratio. When the BrdU concentration in FBT medium was increased from 12 to 300 μM, the frequency of SCEs increased by over 3-fold (Table 2) despite the fact that the level of BrdU substitution remained fixed at approximately 60%. In similar experiments, it was found that increasing the BrdU substitution from 30% to 95% while maintaining a constant BrdU concentration had a smaller effect on the frequency of SCEs. These results suggest that, as in BrdU mutagenesis, the major factor in determining the frequency of induction of SCEs is the concentration of BrdU in the medium rather than the amount of BU present in the DNA.

As a result of this apparent similarity, we decided to determine the effect of dC on BrdU-induced SCEs. As with the mutagenesis experiments involving dC, these experiments were carried out in the presence of FdU, to prevent the conversion of dC to dT nucleotides. It was found that the addition of 200 μM dC largely suppressed the increase in SCEs due to increasing the BrdU concentration from 12 to 300 μM. However, dC was unable to suppress the smaller component of SCE induction of SCEs caused by increasing the level of BrdU substitution while maintaining a fixed concentration of BrdU.

TABLE 3. Stimulation of Mutagenesis by Deoxynucleosides[a]

Additives (uM)				%BrdU	Mutant
BrdU	dT	dG	dA	Substitution	Frequency [b]
30				66	238
10				49	12
10	100			9	220
10		100		52	333
10			100	47	75

[a] Mutagenesis conditions and determinations of BrdU substitution and mutant frequencies were as described in (14).

[b] See footnote c, Table 1

These data were taken from Kaufman and Davidson, 1979 (14).

These results suggest that, as in BrdU mutagenesis, BrdU may be inducing SCEs as a result of its action at a site outside DNA. They raise the possibility that perturbation of dC metabolism due to the inhibition of CDP reduction by BrdUTP plays a key role in the induction of both SCEs and mutations by high concentrations of BrdU. The fact that dC did not suppress the small increase in SCEs caused by increasing the level of BrdU substitution suggests that the separable effects of BrdU substitution and concentration on SCE induction might involve different mechanisms.

Stimulation of BrdU Mutagenesis by Deoxynucleosides

The mechanism that we have proposed for BrdU mutagenesis, involving the perturbation of dC nucleotide metabolism due to the inhibition of ribonucleotide reductase activity, predicts that other agents that inhibit CDP reduction should stimulate BrdU mutagenesis. Since the triphosphates of dT, deoxyguanosine (dG), and deoxyadenosine (dA) have been shown to inhibit this activity [17], the effects of these deoxynucleosides on BrdU mutagenesis were tested [6, 14]. HAB cells were exposed to a concentration of BrdU that was only weakly mutagenic and the effects of adding dT, dG, and dA were determined (Table 3). The frequency of Oua^r mutants and the levels of BrdU substitution in nuclear DNA were measured. It was found that the addition of 100 μM dT or dG to medium containing 10 μM BrdU resulted in a 20-30-fold increase over the frequency of mutants produced by 10 μM BrdU alone. Addition of 100 μM dA produced a small (6-fold) increase in mutant frequency. The addition of dG or dA had no effect on the level of BrdU substitution. Most dramatically, however, dT was found to stimulate BrdU mutagenesis while actually decreasing BrdU substitution by approximately 5-fold. The stimulation of BrdU mutagenesis by dT, dG, and dA was suppressed by the addition of dC.

The mutagenicity of dG and dT alone were also determined in this study. At very high concentrations (1 mM or greater) mutagenesis by dT alone was observed, but to a much smaller extent than caused by BrdU. The mutagenicity of dT at these high concentrations was enhanced by dG and suppressed by dC. Concentrations of dG alone as high as 1 mM were found to be nonmutagenic. The fact that dT and dG are not effective mutagens relative to BrdU suggests that perturbation of dC metabolism alone is not sufficient for mutagenesis. Apparently, some BrdU is required for high levels of mutagenesis to occur, but a relatively small amount of BrdU appears sufficient.

BrdU Mutagenesis in Mutants of Nucleotide Metabolism

The relationship between ribonucleotide reductase activity and BrdU mutagenesis hypothesized above was examined further by the study of mutants with alterations in ribonucleotide reductase activity. A series of mutants resistant to hydroxyurea (HU), an inhibitor of ribonucleotide reductase activity, was selected from the Syrian hamster 3460 cell line. A number of mutants resistant to 0.3 mM HU were isolated, and, from some of these, second step mutants resistant to 1.0 mM HU were isolated [7].

The ribonucleotide reductase activity of the HU^r mutants was studied using an in vitro assay of CDP reductase activity [2]. Seven of the nine mutants were found to have increased specific activities of CDP reductase (Table 4). In those lines with increased activity, the activities ranged from 1.5-4 times as great as the parental line. All of the second step mutants had activities greater than the single step mutants from which they were derived. Enzyme assays were carried out to see if there were qualitative as well as quantitative changes in the CDP reductase in some of the mutants. The inhibitory effects of BrdUTP, and in some cases HU, TTP, and dGTP, were tested against the reductase activity of several of the mutant lines. In no case was the activity from a mutant cell extract more resistant to the effect of an inhibitor than was the wild-type enzyme. The HU resistance in these mutants, therefore, appears associated with the presence of increased levels of the wild-type enzyme.

In most of the HU^r lines, resistance to BrdU mutagenesis was associated with increased levels of ribonucleotide reductase activ-

TABLE 4. Ribonucleotide Reductase Activity and BrdU Mutagenesis in HU^r Lines[a]

Cell Line	CDP Reductase Activity[b]	Mutant Frequency[c] 18uM BrdU	30uM BrdU
3460	1.0	308	
HU-4	0.7	14	114
HU-8	1.0		
HU-14	1.7		
HU-12	1.8		
HU-1	1.8	15	161
HU-6	2.0		
HU-4A1	1.5	10	175
HU-1B1	4.0	NT	16
HU-6B1	4.0	3	24

[a] Mutagenesis conditions and determinations of CDP reductase activity and mutant frequencies were as described in (2).

[b] CDP reductase activity relative to the specific activity of the parental 3460 cells. The value obtained for the 3460 line was 0.6 nanomoles of CDP reduced/mg of protein/hour.

[c] See footnote c, Table 1.

These data were taken from Ashman and Davidson, 1981 (2).

ity. The induction of Oua^r mutants by BrdU was examined in five of the mutants (Table 4). All of the lines tested were relatively resitant to a concentration of BrdU (18 μM) which was highly mutagenic in the parental 3460 line. Three of the lines showed high levels of mutagenesis when the BrdU concentration was increased to 30 μM BrdU, while the mutants with the highest levels of CDP reductase activity were resistant to even this concentration of BrdU. Thus, it appears that among the HU^r cell lines, those with the highest levels of reductase activity are the most resistant to BrdU mutagenesis.

Recently, we have isolated other mutants with alterations in nucleotide metabolism and have attempted to determine if these mutants have altered sensitivities to BrdU mutagenesis. Three such mutants, two resistant to the drug cytosine arabinoside (ara-C) and one resistant to the drug 5-fluorouracil (FU) have been isolated in the Chinese hamster V79 cell line. Work from other laboratories has demonstrated that resistance to ara-C usually arises by one of two basic mechanisms [4]. One class of mutant has a deficiency in deoxycytidine kinase activity and is unable to phosphorylate the analog to its toxic form. The other class of mutant is resistant to ara-C as a result of an expanded dCTP pool which can arise by one of several mechanisms. We have isolated ara-C^r mutants which apparently represent both of these two classes. One mutant, AC^rC9-1, is cross-resistant to high dT, a property characteristic of mutants with expanded dCTP pools. The second mutant, AC^rB1-1, is sensitive to high dT but, unlike the parental cells, this sensitivity is not suppressed by the addition of dC. In addition, mutant AC^rB1-1 is considerably more resistant to ara-C than is AC^rC9-1. These findings are consistent with AC^rB1-1 being a dC kinase deficient mutant. The FU^r mutant appears to have a complex phenotype which has not yet been fully analyzed. One aspect of this phenotype is that its dCTP pool is 3-5 times larger than that of its parental line.

Measurements of BrdU mutagenesis in these cell lines have shown that mutant AC^rC9-1 and the FU^r mutant are both extremely resistant to BrdU mutagenesis (relative to their parental cell lines). For example, exposure to 90 μM BrdU resulted in frequencies of thioguanine-resistant (TG^r) mutants of 1.6×10^{-3} for the parental cells but only 7×10^{-5} for AC^rC9-1. Similar experiments showed that 50 μM BrdU was completely nonmutagenic in the FU^r cells while this concentration of BrdU produced a 500-600-fold increase above background in the frequency of TG^r mutants in the parental cells. In contrast, the mutant AC^rB1-1 has been found to be at least as sensitive and possibly more sensitive to BrdU mutagenesis than its parental cells. Thus, the mutants which apparently have expanded dCTP pools are resistant to BrdU mutagenesis while the ara-C^r mutant which is presumed to be dC kinase deficient is at least as sensitive to BrdU mutagenesis as the parental cells. Again, these results are consistant with the hypothesis that a perturbation in dC nucleotide metabolism plays a key role in BrdU mutagenesis.

TABLE 5. BrdU Mutagenesis and dNTP Pools in 3460 Cells[a]

Additives (uM)			Mutant	dNTP Pools (pmol/10^6 cells)				
FdU	BrdU	dT	Frequency[b]	dGTP	dATP	TTP	BrdUTP	dCTP
0	0	0	NT	19	41	67	0	30
10	18	12	308	52	49	72	107	10

[a] Mutagenesis conditions and dNTP pool measurements were as described in (2).

[b] See Footnote c, Table 1.

These data were taken from Ashman and Davidson, 1981 (2).

BrdU Effects on Deoxyribonucleotide Pools

The relationship between changes in nucleotide pool levels and BrdU mutagenesis was examined more directly by measuring the effect of BrdU on deoxynucleoside triphosphate (dNTP) pools in Syrian hamster 3460 cells [1, 2]. It was found that the addition of a highly mutagenic concentration of BrdU (18 μM) caused a rapid expansion of the BrdUTP pool and a rapid drop in the dCTP pool, with maximal alterations occurring within 6 to 12 h. This concentration of BrdU also produced a 2-3-fold increase in the dGTP pool and about a 20% increase in the dATP pool (Table 5).

The analysis of nucleotide pool levels under a variety of conditions demonstrated that BrdU mutagenesis is normally associated with both the formation of a large BrdUTP pool and a drop in the dCTP pool, but that neither by itself is sufficient for mutagenesis to occur. Weakly mutagenic BrdU concentrations were found to produce a greater than 50% reduction in the size of the dCTP pool, while completely nonmutagenic concentrations of dT reduced the dCTP pool as much as highly mutagenic concentrations of BrdU [1]. The fact that a large BrdUTP pool is not, by itself, sufficient for mutagenesis was demonstrated by the finding that 10 μM dC, which almost completely suppresses the mutagenic effect of 18 μM BrdU and completely prevents the drop in dCTP levels, did not prevent the formation of as a large a BrdUTP pool as is usually produced by 18 μM BrdU alone [2]. Thus, under most conditions, high levels of mutagenesis were found to be associated with both the formation of a large BrdUTP pool and a drop in the dCTP level. We have proposed that a critical ratio of BrdUTP: dCTP may be necessary for mutagenesis to occur.

TABLE 6. Deoxyguanosine Stimulation of BrdU Mutagenesis[a]

Additives (uM)					Mutant Frequency	Ratio
Fdu	BrdU	dT	dG	dC	($TG^r/10^5$ cells)	BrdUTP/dCTP[b]
30	-	30	-	-	0.2	(6.1)[c]
30	-	30	300	-	0.9	(98)[c]
30	15	15	-	-	0.5	2.1
30	15	15	300	-	16	26
30	15	15	300	100	0.2	0.9

[a] Chinese hamster fibroblast, V79.5 cells were grown for 24 hours in the presence of the various additives. After 7 days expression time the cells were exposed to 18uM TG and the number of TG^r mutants determined. The frequency is corrected for plating efficiency in the absence of selective agent.

[b] The dNTP pools were measured in perchloric acid extracts of duplicate cultures of the mutagenesis protocol. Extracts were made after 24 hours growth in the presence of the various additives. dNTPs were separated and quantitated using high performance liquid chromatography on an anion exchange column, after degradation of the ribonucleotides with periodate.

[c] The numbers in parenthesis indicate the dTTP/dCTP ratio.

Measurements of dNTP pool levels in the HU^r mutants supported the idea that the relative levels of the BrdUTP and dCTP pools are important in BrdU mutagenesis [2]. Pool levels were measured in two HU^r mutants which are resistant to BrdU mutagenesis, and it was found that the dNTP pool perturbations produced in the mutants by 18 μM BrdU (a relatively non-mutagenic concentration for these cells) were considerably smaller than those produced by 18 μM BrdU in the parental 3460 cells (which are strongly mutagenized by this concentration). Significant pool perturbations were produced in the mutants at higher BrdU concentrations, however, and it was found that BrdU concentrations which were highly mutagenic in the mutant lines produced ratios of BrdUTP/dCTP which were similar in magnitude to those produced by 18 μM BrdU in 3460.

Recently, we have obtained additional evidence that high levels of BrdU mutagenesis are associated with high BrdUTP/dCTP ratios. In experiments using the V79 Chinese hamster cell line, 300 μM dG was added to medium containing a weakly mutagenic concentration (15 μM) of BrdU (Table 6). The addition of dG resulted in a greater than 30-fold increase in the frequency of TG^r mutants. As with the HAB cells described above, the dG stimulation was suppressed by the addition of dC. Measurements of dNTP pools showed that dG stimulation of mutagenesis was accompanied by a large (13-fold) increase in the BrdUTP/dCTP ratio. This increase in ratio was due completely to decreased dCTP levels as the BrdUTP levels were unchanged. The addition of dC to the medium sharply dropped the BrdUTP/dCTP ratio. Control cultures showed that dG did not stimulate mutagenesis by dT despite the fact that a very high ratio of TTP/dCTP was produced. This is in agreement with our earlier finding that high TTP/dCTP ratios were not associated with high levels of mutagenesis in the 3460 cell line [1].

We have observed one exception to the association between high BrdUTP/dCTP ratios and mutagenesis [2]. This occurred in 3460 cells when the suppression of BrdU mutagenesis by dC was overcome by increasing the concentration of BrdU. In this situation a very large BrdUTP pool was produced but the total intracellular dCTP pool did not drop, and a large BrdUTP/dCTP ratio was not established (Table 7). The absence of a drop in dCTP levels is not unexpected under these conditions, since much of the dCTP would be produced by the phosphorylation of exogenous dC rather than by the reduction of CDP. The production of dCTP by this pathway should be insensitive to the level of BrdUTP. The occurrence of mutagenesis in the absence of the high BrdUTP/dCTP ratio that we have observed in all other cases might indicate that dC nucleotides produced by the reduction of CDP rather than by the phosphorylation of dC are preferentially utilized for DNA synthesis. Evidence in support of this suggestion has been obtained in several studies [2, 18, 19]. Such preferential utilization of nucleotides could occur by channeling through a multi-enzyme complex containing ribonucleotide reductase and DNA polymerase, as has been proposed for mammalian cells [19]. If dNTPs produced by the reduction of ribonucleotides are preferentially utilized for DNA synthesis in our cells, as studies with permeabilized cells indicate [2], then the size of the total dCTP pool may not accurately reflect the concentration of dCTP actually available for DNA synthesis. Under such conditions, the effective BrdUTP/dCTP ratio for DNA synthesis could be much higher than that actually measured. Obviously, a more complete understanding of the multi-enzyme complex and nucleotide channeling is required before the role which nucleotide pool alterations play in mutagenesis can be understood.

Mechanisms of BrdU Mutagenesis in Mammalian Cells

The mechanism by which a high BrdUTP/dCTP ratio results in mutation remains to be elucidated. The simplest model consistent with

TABLE 7. Effect of dC on BrdU Mutagenesis and dNTP Pools[a]

Additives (uM)				Mutant	dNTP Pools (pmol/10^6 cells)			BrdUTP/
Fdu	BrdU	dT	dC	Frequency[b]	BrdUTP	TTP	dCTP	dCTP[c]
0	0	0	0	NT	0	52	24	2.2
10	0	30	0	2	NT	NT	NT	NT
10	12	8	0	210	75	50	10	7.6
10	18	12	0	250	94	62	8	11.8
10	18	12	10	15	91	61	41	2.2
10	30	20	10	73	109	73	52	2.1
10	60	40	10	192	168	113	63	2.7
10	90	60	10	290	202	134	66	3.1

[a] Mutagenesis conditions and determinations of mutant frequencies and dNTP pool levels were as described in (2).

[b] The frequencies of TG^r mutants are expressed as the number of TG^r colonies per $2X10^5$ cells after corrections for plating efficiencies.

[c] In cases where the cells were not exposed to medium containing BrdU, the ratio given is the ratio of TTP/dCTP.

These data were taken from Ashman and Davidson, 1981 (2).

our data is that the high BrdUTP/dCTP ratio produced by the inhibition of CDP reductase by BrdUTP drives the misincorporation of BU in place of C residues in DNA, leading to the production of errors of incorporation. Such a model is consistent with our findings that: 1) the BrdU concentration determines the level of mutagenesis rather than the degree of replacement of T residues by BU; 2) BrdU inhibits CDP reductase activity, as do the deoxynucleosides (dT, dG, and dA) which enhance BrdU mutagenesis, while the addition of dC suppresses mutagenesis; 3) mutagenesis is reduced in mutants which overproduce ribonucleotide reductase and in mutants which have increased dCTP levels; and 4) highly mutagenic BrdU concentrations result in the formation of large BrdUTP pools and reduced dCTP levels.

Our finding that the level of BrdU substitution does not influence the level of BrdU mutagenesis suggests that errors of replication are not an important mechanism of BrdU mutagenesis in mammalian cells. Recently, however, we have obtained evidence that errors of replication may be induced in V79 cells under certain special conditions of nucleotide pool perturbation (E. Kaufman, unpublished observations). In these experiments mutations were induced under conditions of "clean growth" mutagenesis; that is in cells containing BU-substituted DNA but in the absence of BrdU in the growth medium. In order for significant numbers of mutations to be produced under clean growth conditions, however, it was necessary to grow the BU-substituted cells in the presence of high concentrations (1 mM) of dT. Under these conditions, the frequency of TG^r mutants produced was proportional to the level of BrdU substitution pre-existing in the DNA. Pool measurements demonstrated that large alterations in dNTP levels occurred during the clean growth period in high dT. The TTP pool increased by approximately 40-fold and the dGTP pool increased by about 9-fold in size, undoubtedly as a result of the stimulation of GDP reduction by TTP. The dATP pool, however, increased in size by only about 50%. These results are consistent with the proposal (but certainly do not prove) that the increase in the ratio of dGTP/dATP may drive the misincorporation of G opposite a BU residue already in the DNA. Further experimenta are currently in progress to detemine the mechanism by which these mutations are generated. While errors of replication may not normally be an important mechanism of BrdU mutagenesis, it appears that clean growth mutagenesis can be induced under conditions of nucleotide pool imbalance, as we had previously observed for mutagenesis occurring during BrdU incorporation into DNA.

The actual mechanism of BrdU mutagenesis may be considerably more complex than we have proposed. For example, in addition to its effects on the BrdUTP and dCTP pools mutagenic concentrations of BrdU produce a large increase in the dGTP pool. This pool imbalance may also influence the frequency at which mutants are generated. Also, we have observed that the induction of both mutations and SCE by BrdU appears to occur by similar mechanisms. It is difficult to imagine how merely the substitution of BU for C residues could lead to the production of an SCEs. In bacteria, it has been shown that a mismatch repair system modifies levels of BU mutagenesis [21]. Little is known about DNA repair mechanisms which may act on BrdU-induced mispairs in mammalian cells. The elucidation of the mechanism of BrdU mutagenesis thus may depend upon the understanding of the process of DNA repair in mammalian cells. An important step in elucidating the mechanism of BrdU mutagenesis will be a determination of the actual base sequence changes occurring in BrdU induced mutants of mammalian cells. Such experiments are currently in progress in our laboratory.

REFERENCES

1. C. R. Ashman and R. L. Davidson, Bromodeoxyuridine Mutagenesis in Mammalian Cells is Related to Deoxyribonucleotide Pool Imbalance, Mol. Cell Biol., 1:254-260 (1981).
2. C. R. Ashman, G. P. V. Reddy, and R. L. Davidson, Bromodeoxyuridine Mutagenesis, Ribonucleotide Reductase Activity, and Deoxyribonucleotide Pools in Hydroxyurea-Resistant Mutants, Somat. Cell Genet., 7:751-768 (1981).
3. M. D. Bick and R. L. Davidson, Total Substitution of Bromodeoxyuridine for Thymidine in the DNA of a Bromodeoxyuridine-Dependent Cell Line, Proc. Natl. Acad. Sci., U.S.A., 71:2082-2086 (1974).
4. B. R. de Saint Vincent and G. Buttin, Studies on 1-β-D-Arabinofuranosyl Cytosine-Resistant Mutants of Chinese Hamster Fibroblasts: III. Joint Resistance to Arabinofuranosyl Cytosine and to Excess Thymidine - A Semidominant Manifestation of Deoxycytidine Triphosphate Pool Expansion, Somat. Cell Genet., 5:67-82 (1979).
5. R. L. Davidson and M. D. Bick, Bromodeoxyuridine Dependence - A New Mutation in Mammalian Cells, Proc. Natl. Acad. Sci., U.S.A., 70:138-142 (1973).
6. R. L. Davidson and E. R. Kaufman, Bromodeoxyuridine Mutagenesis in Mammalian Cells is Stimulated by Thymidine and Suppressed by Deoxycytidine, Nature, 276:722-723 (1978).
7. R. L. Davidson and E. R. Kaufman, Resistance to Bromodeoxyuridine Mutagenesis and Toxicity in Mammalian Cells Selected for Resistance to Hydroxyurea, Somat. Cell Genet., 5:873-885 (1979).
8. R. L. Davidson, E. R. Kaufman, C. P. Dougherty, A. M. Ouellete, C. M. DiFolco, and S. A. Latt, Induction of Sister Chromatid Exchanges by BUdR is Largely Independent of the BUdR content of DNA, Nature, 284:741-76 (1980).
9. E. Freese, The Specific Mutagenic Effect of Base Analogs on Phage T4, J. Mol. Biol., 1:87-105 (1959).
10. B. Hampar, J. G. Derge, L. M. Martos, and J. L. Walker, Persistence of a Repressed Epstein-Barr Virus Genome in Burkitt Lymphoma Cells Made Resistant to 5-Bromodeoxyuridine, Proc. Natl. Acad. Sci., U.S.A., 68:3185-3189 (1971).
11. E. Huberman and C. Heidelberger, The Mutagenicity to Mammalian Cells of Pyrimidine Nucleoside Analogs, Mutat. Res., 14:130-132 (1972).
12. E. R. Kaufman and R. L. Davidson, Novel Phenotypes Arising from Selection of Hamster Melanoma Cells for Resistance to BUdR, Exp. Cell Res., 107:15-24 (1977).
13 E. R. Kaufman and R. L. Davidson, Bromodeoxyuridine Mutagenesis in Mammalian Cells: Mutagenesis is Independent of the Amount of Bromouracil in DNA, Proc. Natl. Acad. Sci., U.S.A., 75:4982-4986 (1978).
14. E. R. Kaufman and R. L. Davidson, Bromodeoxyuridine Mutagenesis in Mammalian Cells is Stimulated by Purine Deoxyribonucleosides, Somat. Cell Genet., 5:653-663 (1979).

15. M. Meuth and H. Green, Induction of a Deoxycytidineless State in Cultured Mammalian Cells by Bromodeoxyuridine, Cell, 2:109-112 (1974).
16. G. E. Moore, In vitro Cultures of a Pigmented Hamster Melanoma Cell Line, Exp. Cell Res., 36:422-423 (1964).
17. E. C. Moore and R. B. Hurlburt, Regulation of Mammalian Deoxyribonucleotide Biosynthesis by Nucleotides as Activators and Inhibitors, J. Biol. Chem., 241:4802-4809 (1966).
18. B. Nicander and P. Reichard, Dynamics of Pyrimidine Deoxynucleoside Triphosphate Pools in Relationship to DNA Synthesis in 3T6 Mouse Fibroblasts, Proc. Natl. Acad. Sci., U.S.A., 80:1347-1351 (1983).
19. G. P. V. Reddy and A. B. Pardee, Multienzyme Complex for Metabolic Channeling in Mammalian DNA Replication, Proc. Natl. Acad. Sci., U.S.A., 77:3312-3316 (1980).
20. W. J. Rutter, R. L. Pictet, and P. W. Morris, Toward Molecular Mechanisms of Developmental Processes, Ann. Rev. Biochem., 26:601-646 (1973).
21. B. Rydberg, Bromouracil Mutagenesis in Escherichia coli: Evidence for Involvement of Mismatch Repair, Molec. Gen. Genet., 152:19-28 (1977).

THE BIOCHEMICAL BASIS of 5-BROMOURACIL-
AND 2-AMINOPURINE-INDUCED MUTAGENESIS

Myron F. Goodman, Randi L. Hopkins,
Roger Lasken, and Dhanajaya N. Mhaskar

Department of Biological Sciences
Molecular Biology Section
University of Southern California
Los Angeles, California 90089-1481

ABSTRACT

We describe in vitro measurements of heteroduplex base mispaired intermediates involving 5-bromouracil and 2-aminopurine in A•T → G•C and G•C → A•T transition mutation pathways. For the case of 2-aminopurine, 2-aminopurine•cytosine mispairs are formed at a much higher frequency than adenine•cytosine mispairs in either transition pathway. For the case of 5-bromouracil, at least a 40-fold increase in 5-bromouracil•guanine mispairs are observed over thymine•guanine mispairs but only in the G•C → A•T pathway. In the A•T → G•C pathway, mispairs involving 5-bromouracil are formed 2.5-fold more frequently to those involving thymine suggesting perhaps that 5-bromouracil may exhibit substantially different base-pairing behavior depending on whether it is present as a template base or as a deoxyribonucleosides triphosphate substrate. The effect of the base analogs on dNTP pool size perturbations is discussed. A measurement of dNTP pools in 2-aminopurine mutagenized bacteriophage T4-infected cells is presented. An approximate eight-fold expansion in common dNTP pools is observed in a ts L141 antimutator genetic background compared to wild type T4 43^{+} and ts L56 mutator backgrounds. The effects of distorted dNTP pools on mutagenesis will be considered.

INTRODUCTION

When DNA is being replicated, competition for insertion by DNA polymerase occurs at each template site between a deoxyribonucleotide satisfying Watson-Crick base-pairing rules and all other noncomplimentary deoxyribonucleoside triphosphates (dNTPs). The presence of Watson-Crick base pairs is generally sufficient to exert a powerful bias favoring insertion of a correct nucleotide into DNA. However, errors in DNA replication still occur, and it has been shown recently in several different cell-free DNA synthesizing systems that the frequency of inserting a mispaired deoxynucleotide is directly proportional to the ratio of mismatched to correctly matched dNTP substrates [7, 11, 12]. Thus, an increased frequency of base substitution mutations is perhaps the most obvious consequence of intracellular perturbations which cause significant dNTP pool imbalances.

In certain instances, there is sufficient biochemical information available to predict the effects on dNTP (and NTP) pools following addition of exogenous metabolites to growing cells. An excellent example is provided by the base analog bromodeoxyuridine (BrdUrd) and to a lesser extent by 2-aminopurine (AmPur) - see, e.g., Ref. 19. Both analogs are strong mutagens in *Escherichia coli* and bacteriophage T4 infected *E. coli* [2, 13, 29, 31, 32]; BrdUrd is also mutagenic in animal cells [21].

It is straightforward to predict the effects of exogenous BrdUrd on dNTP pools and base substitution errors [19]. The key components in the biochemical pathway for utilization of BrdUrd are the salvage pathway enzyme thymidine kinase and ribonucleoside diphosphate reductase. It is known that *E. coli* and animal cell ribonucleotide reductases are feedback inhibited by dTTP and presumably BrdUTP while the T4 enzyme is not [33]. Feedback inhibition by BrdUTP is believed to act selectively to reduce the reduction of CDP to dCDP [24]. The subsequent reduction in dCTP pools should result in an enhanced frequency of BrdUrd-induced G·C → A·T transitions in *E. coli* and animal cells. In T4 infected cells, no strong bias is expected favoring G·C → A·T over A·T → G·C transitions; both transitions should be induced by BrdUrd.

A fascinating aspect of dNTP pool effects on mutagenesis is that the simple predictions alluded to above do not appear to be uniformly adhered to by the various cells. Skopek and Hutchinson [32] have recently shown that BrdUrd-induced mutations in the Cl region of prophage lambda are overwhelmingly in the A·T → G·C direction. Earlier data from Champe and Benzer [5] and Ronen and Rahat [30] indicates that G·C → A·T transitions are preferentially induced by BrdUrd in T4 infected cells.

In this paper, we will consider two fundamental aspects of BrdUrd- and AmPur-induced mutagenesis. The first involves the types of heteroduplex base mispairs formed in the analog-induced mutagenic pathways. The second involves what is currently known about the analogs' effects on dNTP pools. We will present data on the formation of BrdUrd·G and AmPur·C mispaired intermediates in both A·T → G·C and G·C → A·T directions obtained *in vitro* using purified DNA polymerases. We will also offer some preliminary data on an interesting distortion in dNTP pools during AmPur mutagenesis in a T4 L141 antimutator DNA polymerase background.

MATERIALS AND METHODS

Enzymes

The T4 DNA polymerase was purified by the method as in [26]. Terminal deoxynucleotidyl transferase was purified from calf thymus essentially as in [6].

Synthesis of Copolymer and Homopolymer Templates

pd(A,T)·oligo d(T), pd(A,BrU)·oligo d(T), pd(T,G)·olig d(A), pd(T)·oligo d(A), pd(A)·oligo d(T), and pd(A,C)·oligo d(T) were synthesized with terminal deoxynucleotidyl transferase as previously described [34].

Fidelity Assays

T4 DNA polymerase reactions were carried out essentially as described [7]. The specific activity of [^{3}H]-BrdUTP ranged from 1×10^7 cpm/nmol to 1.8×10^9 cpm/nmol. That of [^{3}H]-dAmPurTP was around 2×10^7 cpm/nmol.

Digestion of ^{32}P Labelled Polynucleotides

Incorporation of [^{32}P] BrdUTP was carried out as described above. Products in the reaction mixture were extracted twice with 24:1 chloroform:isoamyl alcohol followed by extensive dialysis against 0.01 M NaPPi and distilled water. The polynucleotides were then digested to mononucleotides as previously described [34] and chromatographed by high performance liquid chromatogrphy on a Waters Associates Radial-Pak μBondapak C_{18} cartridge in a mobile phase of 0.2 M NH_4PO_4 pH 5.5.

Labelled Analog dNTP's

[^{32}P] dCTP purchased from New England Nuclear was brominated as previously described [3], purified with 20% Norite, deaminated in nitric acid [20] and purified by column chromatography with Sephadex A-25.

2-Amino-6-chloropurine (6 chloroguanine), purchased from Sigma Chemical Co., St. Louis, Mo. was converted to [6-^{3}H]-2-aminopurine through catalytic reduction by ICN Pharmaceuticals, Inc., Irvine, Ca., 2-aminopurine deoxynucleoside triphosphate (dAmPurTP) was synthesized from the free base as described in [7].

dNTP Pools

AmPur mutagenesis is performed as described in [18]. Infected cells are harvested 20 min post infection on a 15 cm diam. 0.45 μ millipore filter (Millipore, Mass.). Cells are PCA extracted, and ribonucleotides are $NaIO_4$ oxidized according to [23]. The products of oxidation are removed from the crude extract by forming a Schiff's base with lysine, and ion exchange chromatography using a disposable NH_2 HPLC cartridge (J. T. Baker) and a 0.01 M to 0.1 M step gradient of NH_4PO_4 pH 9. The cut containing dNTP's are analyzed using a Waters C_{18} μ-Bondapak radial compression cartridge (Waters, Milford, Mass.) with a mobile phase of 0.2 M NH_4PO_4 pH 5.5. Effluent is monitored simultaneously at 254 and 313 nm.

RESULTS

Base analog-induced mutagenesis is believed to involve two central components. First, there is a direct participation of the analogs themsleves in the formation of heteroduplex base mispaired intermediates in A·T → G·C and G·C → A·T transition pathways. Second, one might have possible base analog-induced perturbations in dNTP pools. Concerning the question of base-mispaired intermediates, it is essential to establish whether the postulated intermediates, AmPur·C and BrU·G, are observed in vitro and in vivo. It is also necessary to determine whether the analogs can form base mispairs at frequencies above those involving the normal bases, A·C and T·G.

1. Base Mispaired Intermediates Involving 2-Aminopurine

1.1. AmPur·T Base Pairs

There are two intermediates involving AmPur in the A·T → G·C direction, AmPur·T and AmPur·C. An AmPur·T base pair is formed whenever 2-aminopurine deoxynucleoside triphosphate (dAmPurTP) is inserted into DNA opposite a template Thy site in competition with dATP (Fig. 1). Measurements in vitro for a variety of DNA polymerases using natural DNA templates and equimolar concentrations of dAmPurTP and dATP show that A·T base pairs are formed preferentially to AmPur·T base pairs by a ratio of about 7:1 [4, 7]. Three independent criteria can be used to show that AmPur·T base pairs are less stable than A·T base pairs: (1) the K_M for the insertion of dAmPurTP opposite a template Thy is 7-fold larger than for the insertion of dATP opposite Thy [7]; (2) using DNA polymerase having 3'-exonuclease proofreading activity, the rate of removal of dAmPurMP opposite Thy is about 3-fold larger than the removal of dAMP opposite Thy [4, 7]; (3) a significant reduction in melting temperature is observed when AmPur replaces Ade in synthetic DNA copolymers annealed to pd(T) [17].

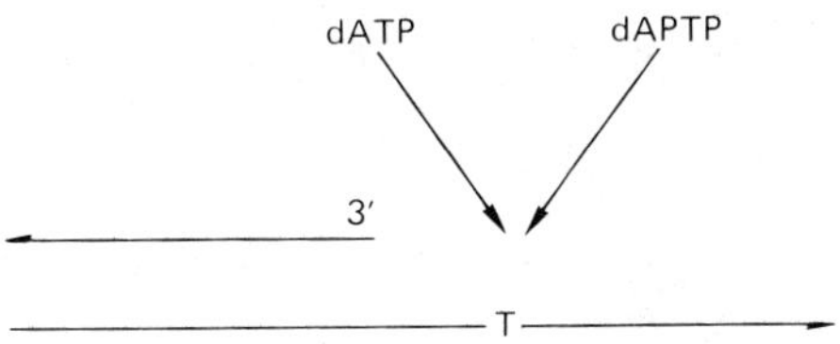

Fig. 1. Competition between A and AP for insertion opposite T. The next base to be attached to the 3'OH terminus of the primer strand (top) must pair with a T on the template strand (bottom). dATP and dAPTP are shown competing for association at this site.

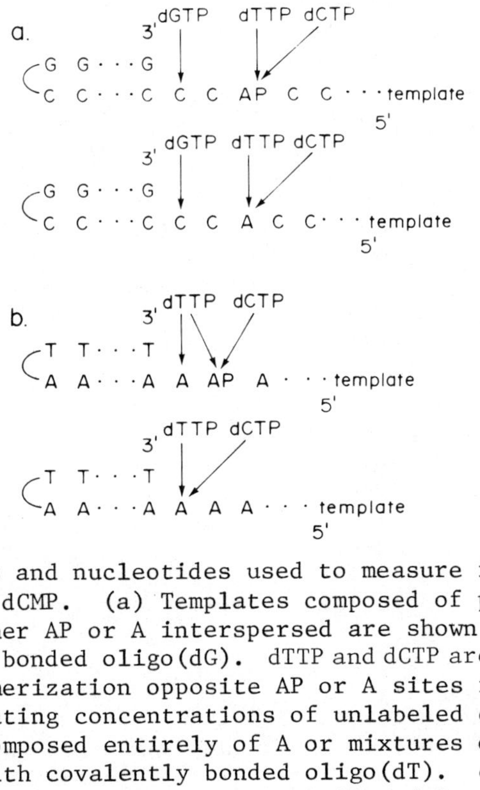

Fig. 2. Templates and nucleotides used to measure incorporation of dTMP and dCMP. (a) Templates composed of primarily dCMP with either AP or A interspersed are shown primed with covalently bonded oligo(dG). dTTP and dCTP are shown competing for polymerization opposite AP or A sites in the presence of saturating concentrations of unlabeled dGTP. (b) Templates composed entirely of A or mixtures of AP and A are primed with covalently bonded oligo(dT). dTTP and dCTP compete for sites opposite both AP and A, although, as indicated by the arrows, dCTP is able to compete far more successfully opposite AP.

The formation of AmPur·T base pairs has also been measured in vivo [16] using T4 bacteriophage DNA polymerase (gene 43) alleles. Here it is observed that the frequency of forming AmPur·T base pairs is correlated with spontaneous and AmPur-induced A·T → G·C mutation frequencies for the various alleles. The AmPur·T/A·T base pair ratio is about 1/1500, 1/1600, and 1/9000 in DNA isolated from tsL56 mutator, 43^{+}, and tsL141 antimutator phage, respectively. These ratios depend not only on differencesin the proofreading abilities of the mutant and wild type polymerases but also on the metabolic conversion of AmPur to dAmPurTP and the analog's effect on the pools of common dNTPs [19]. Data on dNTP pools including dAmPurTP will be discussed below.

1.2. AmPur·C Base Pairs

The AmPur·C base pair has been postulated to be the critical mispaired intermediate [13] common both to A·T → G·C and G·C → A·T pathways induced by 2-aminopurine. The mutagenic properties of AmPur might be reasonably understood if it could be shown that AmPur·C mispairs were to form at a much higher frequency than analogs A·C mispairs. We begin, therefore, by comparing the frequencies of forming AmPur·C and A·C base pairs in vitro [35].

In the A·T → G·C direction, AmPur·C base pairs can be formed when AmPur is present on the DNA template and dCMP replaces dTMP opposite AmPur. Polydeoxynucleotide primer-templates were synthesized with AmPur and Ade having different nearest neighbor bases (Fig. 2). With these primer-templates it is important to determine whether dCTP is inserted more frequently on templates containing AmPur in place of Ade and if misinsertion for each nucleotide is dependent on base context. The results obtained are in agreement with the postulated enhanced mispairing properties of AmPur. It was found, using DNA polymerase α, that AmPur·C base pairs are formed at a frequency of about 5% relative to AmPur·T base pairs [35]. Thus, when dCTP and dTTP are allowed to compete at equimolar concentrations for insertion opposite a template AmPur, the deoxycytidine nucleotide is inserted 5% of the time. This AmPur·C base mispairing frequency represents an increase of at least 230-fold over A·C base mispair formation in the presence of 5' nearest neighbor A·T base pairs (Fig. 2b). For the case of nearest neighbor G·C base pairs (Fig. 2a), the ratio of AmPur·C/A·C base mispairs is 35-fold. The in vitro data strongly supports the idea that base-mispairing frequencies are amplified substantially when Ade is replaced by AmPur in DNA.

There is in vivo data on the frequency of forming AmPur·C heteroduplex-heterozygotes in T4 43^{+} [18] and tsL56 mutator [28] backgrounds. Here it was found that AmPur·C hets were formed at a frequency of about 5%, a value which is in good agreement with in vitro DNA polymerase fidelity measurements [35]. It is important to note that methylation directed mismatch repair is believed to be absent

Fig. 3. Different competing substrate nucleotides for forming AP·C mispairs in the A·T → G·C direction (left hand side) and in the G·C → A·T direction (right hand side).

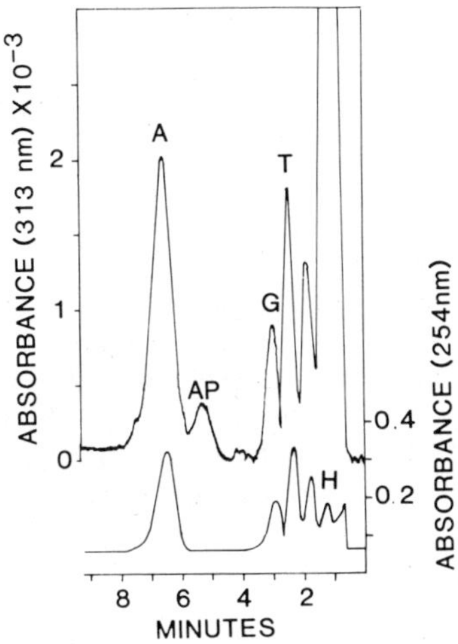

Fig. 4. 500 μl of cell extract representing about 10^{10} 2-aminopurine mutagenized T4 rII UV199-ts(43) L141-infected E. coli CR63 is injected into the reverse-phase HPLC described in Materials and Methods. Effluent, flowing at a rate of 1.5 ml/min, is monitored for absorbance at 313 and 254 nm simultaneously. Peaks are identified from right to left as follows: H,HMdCTP; T,TTP; G,dGTP; AP,dAmPurTP; A,dATP.

in T4, unlike E. coli. Hence, an average value of 5% is likely to represent a fairly reasonable estimate of the net frequency at which AmPur·C base mispairs are made by the T4 replication complex.

In the G·C → A·T direction, AmPur·C base pairs are formed with dAmPurTP competing against dGTP for insertion opposite a template Cyt (Fig. 3). In contrast to the A·T → G·C direction, one expects to find a much lower probability of forming AmPur·C hets in the G·C → A·T direction. The reason is that formation of the AmPur·C het must occur in competition with highly stable G·C base pairs in the G·C → A·T direction while in the A·T → G·C direction the competing AmPur·T base pair is itself relatively unstable. Measurements using wild type T4 DNA polymerase indicate that AmPur·C base mispair formation in the G·C → A·T direction is about 1/300

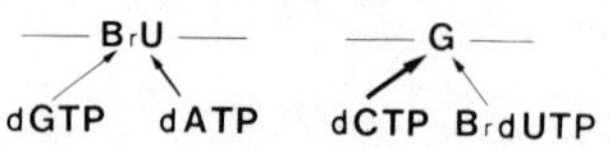

Fig. 5. Different competing substrate nucleotides for forming BrU·G mispairs in the A·T → G·C direction (left hand side) and in the G·C → A·T direction (right hand side).

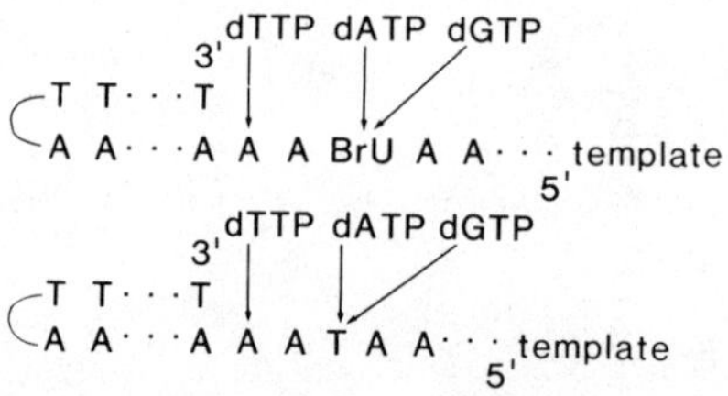

Fig. 6. Templates and nucleotides used to measure incorporation of dGMP and dAMP. Templates composed of primarily dAMP interspersed with either BrdUMP or TMP are shown primed with covalently bonded oligo (dT). dGTP and dATP are shown competing for polymerization opposite BrUra or Thy sites in the presence of saturating concentrations of unlabeled dTTP.

when AmPur is incorporated next to Thy on the primer strand and about 1/30 when the base analogue is incorporated next to Gua [25].

1.3. dNTP Pools during AmPur Mutagenesis

Preliminary data suggest that during AP mutagenesis, normal dNTP pools are 2.5- to 8-fold expanded in an antimutator (ts L141 (43) - r_{II} UV199) phage-infected cell as compared to dNTP pools in its gene 43 alleles (43^+ and mutator ts (43) L56). The pool dAmPurTP is similar in these phages. Thus the ratio of dAmPurTP/dATP is about 5% in 43^+ and mutator, but only 0.4 to 1% in antimutator. A representative chromatogram is shown in (Fig. 4).

2.2. Base Mispaired Intermediates Involving 5-Bromouracil

5-Bromouracil behaves as a close analog of thymine and is incorporated equally well into DNA opposite adenine. A suggested explanation for bromouracil's mutagenicity is that base mispairs involving BrU·G are able to form more readily than those involving T·G [13]. In this section we will consider BrU·G heteroduplexes formed in the A·T → G·C and G·C → A·T transition mutation pathways (Fig. 5). *In vitro* data using T4 DNA polymerase will be presented to examine the pathway dependency of BrU mutagenesis and specifically to address the question of whether BrU·G mispairs occur more frequently than T·G mispairs.

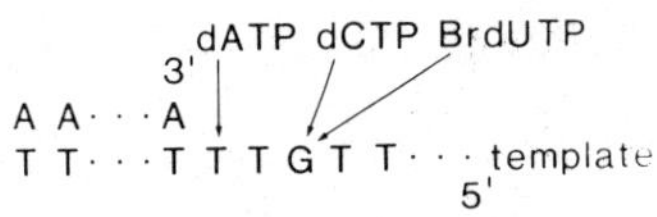

Fig. 7. Template and nucleotides used to measure the incorporation of BrdUMP and dCMP. A template composed of primarily Thy (94%) with Gua (6%) interspersed is shown primed with covalently bonded oligo(dA). BrdUTP and dCTP are shown competing for polymerization opposite a Gua site in the presence of a saturating concentration of unlabled dATP.

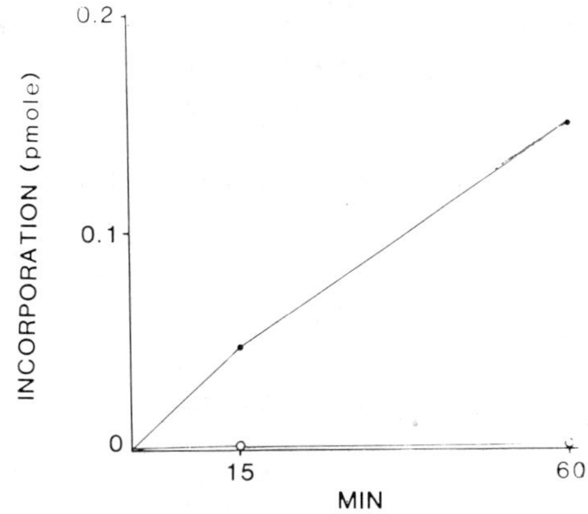

Fig. 8. Time course of [^{32}P]BrdUMP (●) and [^{3}H]dTMP (○) incorporation into pd(T_{94},G_6)·oligo(dA) template (Fig. 7).

2.1. BrU·G Base Pairs

In the A·T → G·C direction, a BrU·G mispair is formed whenever a substrate guanine deoxynucleotide is stably incorporated opposite a template BrU site (Fig. 5). Templates with short primer hooks were constructed using terminal deoxynucleotidyl transferase and misincorporation of dGMP was measured opposite template Thy or BrUra sites (Fig. 6). We observed a significant increase in dGMP misincorporation when BrU replaces Thy on the DNA template [22] in agreement with an earlier study using an alternating copolymer of BrUra and Ade [34].

Similar results were observed in the G·T → A·T transition pathway where Gua is situated on the template and BrdUTP (Fig. 7) and dTTP are present as substrates. A time course showing the misincorporation of BrdUMP and dTMP opposite Gua for the synthetic hook copolymers pd(T_{94},G_6)·oligo d(A) (Fig. 7) is given in Fig. 8. The formation of T·G mispairs are not significantly above background. From the data in Fig. 8 we observe that base mispairs involving BrU·G occur at a frequency of at least 40-fold higher than those involving T·G [22].

The [α-^{32}P]BrdUTP used in the assay was synthesized by bromination followed by deamination of [α-^{32}P]dCTP - see Materials and Methods. It was, therefore, important to show that the labelled nucleotide incorporated into DNA was indeed BrdUMP and not BrdCMP. This control is especially important since the presence of only a small contaminating level of BrdCTP in the BrdUTP, on the order of 1 part per hundred, could fully account for the data in Fig. 8.

We have verified that it is BrdUTP which is incorporated opposite Gua on the pd(T_{94},G_6)·oligo d(A) primer-templates shown in Fig. 7. The controls were as follows. First, following the DNA polymerase reaction, the primer-template DNA was digested to mononucleotides and chromatographed by high performance liquid chromatography. Greater than 90% of the radioactive label migrated with marker BrdUMP; no detectable cpm above background were found migrating either with BrdCMP or dCMP. Second, the addition of unlabelled dCTP to the polymerase assay reduced the incorporation of BrdUMP to levels which were not detectable above background. Third, BrdUMP was not incorporated opposite template strands in which Gua was absent, e.g., pd(T)· oligo d(A). Fourth, we have verified using HPLC that Ade is absent from the template strand (in the absence of the oligo d(A) hook) to less than 1 part in ten thousand; thus, the incorporation of BrdUMP (Fig. 8) is not caused by contamination of the template with Ade. Our failure to observe dTMP incorporation (Fig. 8) is further strong evidence that Ade is not present as a template contaminant.

In a double label assay with [^{32}P]BrdUTP and [^{3}H]dCTP we have allowed the two deoxynucleotides to compete for incorporation opposite a template Gua, as illustrated schematically in Fig. 5. Based on the relative incorporation of the two deoxynucleotide substrates we deduce that the BrU·G/C·G ratio is about 1/1000. The T4 DNA polymerase has an exceptionally active proofreading 3'-exonuclease activity. In preliminary measurements of DNA-dependent deoxynucleotide turnover, the conversion of dNTP $\rightarrow$ dNMP [27], we found that about 75% of the inserted bromouridine deoxynucleotide is excised from the DNA [22]. Thus, the insertion of BrUra opposite Gua in competition with Cyt might possibly be as a high as one in several hundred. Additional turnover data will be required to verify this point.

DISCUSSION

The base analogs 2-aminopurine and 5-bromouracil are strong mutagens which can, in principle, induce base substitutions bidirectionally, A·T $\rightarrow$ G·C and G·C $\rightarrow$ A·T. Two primary factors are believed to be responsible for their mutagenicity: (1) the ability of AmPur and BrUra to form heteroduplex base mispairs much more frequently than their normal counterparts; (2) the possibility that the analogs or their nucleotide derivatives cause significant perturbations in normal nucleotide pools.

In both the A·T → G·C and G·C → A·T directions, AmPur·C hets seem to occur at a much higher frequency than A·C hets. AmPur substitution for Ade results in a stimulation of erroneous insertion of dCTP in vitro by 35- to 240-fold depending on 5'-nearest neighbor base stacking partners (Fig. 2, Ref.

When dAmPurTP is present as a substrate for DNA polymerase (Fig. 3, Ref. 25), AmPur·C hets occur at easily measurable frequencies in vitro. Their precise elevation over A·C hets is currently unknown because the frequency of dATP mispairing with template Cyt in Pd(A_{80},C_{20})·oligo d(T) is too small to measure accurately. Our best estimate for ratio of dAmPurMP/dAMP incorporation is 1500/1 using pd(C)·oligo(G) as a template-primer [25].

In an analogous series of experiments, BrU·G base mispairs have been measured in vitro when 5'-bromouracil is located on the DNA template [22]. The replacement of Thy by BrUra (Fig. 6) appears to result in a large increase in the misincorporation of dGTP, i.e., the frequency of BrU·G vs. T·G het formation is significantly different when BrUra resides on the template [22]. Similarly, in the G·C → A·T transition pathway (Fig. 5), we find that BrdUTP mispairs with template Gua far more readily than does TTP. The rules governing the formation of AmPur·C and BrU·G heteroduplex base mispairs by the replication enzymes obviously represent one important facet of base analog-induced mutagenesis. However, the frequency of heteroduplex formation is also governed by the relative concentration of the base analog or one of its phosphorylated intermediates acts to distort the pools of common dNTP's, then one can envision an amplification in the rate of base mispair formation in which the mispairing partners might not include the base analog.

In an attempt to understand the large differences in mutation frequencies between T4(43) mutator and antimutator alleles, we made a comparison of dNTP pools in ts (43) L56 mutator, 43^+, and ts (43) L141 antimutator genetic backgrounds during AmPur mutagenesis. A reasonable explanation for the quantitative differences in AmPur-induced mutagenesis is not readily apparent from just a simple comparison of the mutant and wild type DNA polymerases in vitro [15].

The very low AmPur-induced A·T → G·C marker reversion frequency of T4 antimutator may partly result from a 2- to 8-fold expansion of normal dNTP pools. Expansion of the pool of dATP relative to dAmPurTP would, of course, be expected to diminish the frequency of AmPur substitution for Ade in DNA. This in turn leads to a decrease in dCTP misincorporation (A·T → G·C direction). Similarly, an increase in dGTP pools would be expected to diminish dAPTP misincorporation opposite template Cyt (G·C → A·T direction). These ideas assume that the observed pool ratios reflect the situation at the site of replication.

In marked contrast to the paucity of knowledge concerning the effects of AmPur on dNTP pools, one can, in the case of BrdUrd, construct a simple biochemical model to predict the dNTP pool perturbations induced by the analog [19]. BrdUTP is a potent feedback inhibitor of ribonucleoside diphosphate reductase in animal cells [24] and presumably in E. coli but not in T4. BrdUTP selectively inhibits the reduction of CDP. This should result in a lower than normal pool of dCTP. One is immediately led to speculate that a reduction of dCTP pools might allow more favorable competition by BrdUTP for insertion opposite a template Gua site (Fig. 5). Hence one might expect that G·C → A·T transitions would be more easily driven in E. coli and animal cells than in T4. These BrUrd-induced mutations might also be expected to be reduced in the presence of dCyt assuming that dCTP pools can be restored to normal by the action of kinases and that conversion of Cyt → Ura → Thy compounds does not complicate countermutagenicity; see Ref. 19 for a more detailed discussion of these points.

In an interesting series of studies carried out in R. L. Davidson's laboratory [1, 21], BrUrd mutagenesis in a BrdUrd-dependent Syrian hamster melanoma cell line seems to behave reasonably in accordance with the arguments presented above. For BrdUrd concentrations which are mutagenic in the Syrian hamster melanoma cells, there is about a 5-fold reduction in dCTP pools [1]. These data seem consistent with the appearance of BrUrd-induced G·C → A·T transitions in these animal cells but cannot, of course, verify the point directly.

An ideal system to study BrdUrd mutagenesis is E. coli and T4 because of the predicted difference in BrdUTP-mediated depletion of dCTP. It is to be expected that low concentrations of BrdUrd will not deplete the pool of dCTP while high concentrations will. Low BrdUTP concentrations should therefore induce mainly A·T → G·C transitions while high BrdUTP concentration favors the reverse transition direction in E. coli. The transition direction in T4, however, should be less easily mediated. The design of this experiment should provide that the dNTP pools for various BrdUrd concentrations be measured. Genetic studies performed subsequently should exploit possible pool differences.

ACKNOWLEDGMENTS

We thank Noah Bornstein and Sarah Wright for illustrations and preparation of the manuscript. This work was supported by National Institutes of Health Grants GM 21422 and CA 17358.

REFERENCES

1. C. R. Ashman and R. L. Davison, Bromodeoxyuridine Mutagenesis in Mammalian Cells is Related to Deoxynucleotide Pool Imbalance, Mol. Cell. Biol., 1:254-260 (1981).
2. S. Benzer, On the Topography of the Genetic Fine Structure, Proc. Natl. Acad. Sci., U.S.A., 45:1607-1620 (1959).
3. M. J. Bessman, I. R. Lehman, J. A. Adler, S. B. Zimmerman, E. S. Simms, and A. Kornberg, Enzymatic Synthesis of Deoxynucleic Acid. III. The Incorporation of Pyrimidine and Purine Analogs into Deoxynucleic Acid, Proc. Natl. Acad. Sci., 44: 633-640 (1958).
4. M. J. Bessman, N. Muzyczka, M. F. Goodman, and R. L. Schnaar, Studies on the Biochemical Basis of Spontaneous Mutation. II. The Incorporation of a Base and Its Analog into DNA by Wild-Type, Mutator, and Antimutator DNA Polymerases, J. Mol. Biol., 88:409-421 (1974).
5. S. P. Champe and S. Benzer, Reversal of Mutant Phenotypes by Fluorouracil: An Approach to Nucleotide Sequences in Messenger-RNA, Proc. Natl. Acad. Sci., U.S.A., 48:532-546 (1962).
6. L. M. S. Chang and F. J. Bollum, Deoxynucleotide-Polymerizing Enzymes of Calf Thymus Gland, J. Biol. Chem., 246:909-916 (1971).
7. L. K. Clayton, M. F. Goodman, E. W. Branscomb, and D. J. Galas, Error Induction and Correction by Mutant and Wild Type T4 DNA Polymerase, J. Biol. Chem., 254:1902-1912 (1979).
8. C. Coulondre, J. H. Miller, P. J. Farabaugh, and W. Gilbert, Molecular Basis of Base Substitution Hotspots, In *Escherichia coli*, Nature, 274:775-780 (1979).
9. W. Drake, Comparative Rates of Spontaneous Mutation, Nature, 221:1132 (1969).
10. J. W. Drake, The Molecular Basis of Mutation, Holden-Day, San Francisco (1970).
11. A. R. Fersht, Fidelity of Replication of Phage ϕX174 DNA by DNA Polymerase III Holoenzyme: Spontaneous Mutation by Misincorporation, PNAS U.S.A., 76:4946-4950 (1979).
12. A. Fersht, J. W. Knill-Jones, and W. C. Tsui, Kinetic Basis of Spontaneous Mutations Misertion Frequencies, Proofreading Specificities and Cost of Proofreading by DNA Polymerases of *E. coli*, J. Mol. Biol., 156:37-51 (1982).
13. E. Freese, The Specific Mutagenic Effect of Base Analogs on Phage T4, J. Mol. Biol., 1:87-105 (1959).
14. C. Garret and D. V. Santi, A Rapid and Sensitive High Pressure Liquid Chromatography Assay for Deoxyribonucleoside Triphosphates in Cell Extracts, Analyt. Biochem., 99:268-273 (1979).
15. M. F. Goodman and E. W. Branscomb, DNA Replication Fidelity and Base Mispairing Mutagenesis: Relating Experiments to a K_m Discrimination DNA Polymerase Model, in: "Accuracy in Molecular Biology" (D. J. Galas, ed.), Marcel Dekker, to be published (1984).

16. M. F. Goodman, R. Hopkins, and W. C. Gore, 2-Aminopurine-induced Mutagenesis in T4 Bacteriophage: A Model Relating Mutation Frequency to 2-Aminopurine Incorporation in DNA, Proc. Natl. Acad. Sci., U.S.A., 74:4806-4810 (1977).
17. M. F. Goodman and R. L. Ratliff, Evidence of 2-Aminopurine·cytosine Base Pairing Involving 2 Hydrogen Bonds: A 2-Aminopurine Tautomer? Submitted for publication (1983).
18. R. Hopkins and M. F. Goodman, Assymmetry in Forming 2-Aminopurine·hydroxymethylcytosine Heteroduplexes; A Model Giving Misincorporation Frequencies and Rounds of DNA Replication from Base-Pair Populations in vivo, J. Mol. Biol., 135:1-22 (1979).
19. R. Hopkins and M. F. Goodman, Deoxyribonucleotide Pools, Base Pairing and Sequence Configuration Affecting Bromodeoxyuridine- and 2-Aminopurine-Induced Mutagenesis, Proc. Natl. Acad. Sci., U.S.A., 77:1801-1805 (1980).
20. R. B. Innman and R. L. Baldwin, Helix-Random Coil Transitions in DNA Homopolymer Pairs, J. Biol. Chem., 8:452-469 (1964).
21. E. R. Kaufman and R. L. Davidson, Bromodeoxyuridine Mutagenesis in Mammalian Cells: Mutagenesis is Independent of the Amount of Bromouracil in DNA, Proc. Natl. Acad. Sci., 75:4982-4986 (1978).
22. R. L. Lasken and M. F. Goodman, The Biochemical Basis of 5-Bromouracil-induced Mutagenesis, J. Biol. Chem., in press (1984).
23. C. K. Mathews, Biochemistry of Deoxyribonucleic Acid-Defective Amber Mutants of Bacteriphage T4. (III Nucleotide Pools), The J. of Biol. Chem., 247:7430-7438 (1972).
24. M. Meuth and H. Green, Alterations Leading to Increased Ribonucleotide Reductase in Cells Selected for Resistance to Deoxynucleosides, Cell, 3:367-374 (1974).
25. D. N. Mhaskar and M. F. Goodman, Frequency of Forming 2-Aminopurine-Cytosine Base Mispairs in the G·C → A·T Mutational Pathway by T4 DNA Polymerase in vitro, J. Biol. Chem., in press (1984).
26. C. F. Morris, H. Hamma-Inaba, D. Mare, N. K. Sinha, and B. Alberts, Purification of the Gene 43, 44, 45, and 62 Proteins of the Bacteriophage T4 DNA Replication Apparatus, J. Biol. Chem., 254:6787-6796 (1979).
27. N. Muzyczka, R. L. Poland, and M. J. Bessman, Studies on the Biochemical Basis of Spontaneous Mutation. I. A Comparison of the Deoxyribonucleic Acid Polymerases of Mutator, Antimutator, and Wild Type Strains of Bacteriophage T4, J. Biol. Chem., 247:7116-7122 (1972).
28. L. S. Ripley, The Influence of Diverse Gene 43 DNA Polymerases on the Insertion and Replication in vivo of 2-Aminopurine at A·T Base Pairs in Bacteriophage T4, J. Mol. Biol., 150:197-216 (1981).
29. A. Ronen, 2-Aminopurine, Mutat. Res., 69:1-47 (1979).
30. A. Ronen and A. Rahat, Mutagen Specificity and Position Effects on Mutation T4 r_{II} Nonsense Sites, Mutation Rev., 34:21-34 (1976).

31. B. Rydberg, Bromouracil Mutagenesis and Mismatch Repair in Mutator Strains of Escherichia coli, Mutat. Res., 52:11-24 (1978).
32. T. R. Skopek and F. Hutchinson, DNA Base Sequence Changes Induced by Bromouracil Mutagenesis of Lambda Phage, J. Mol. Biol., 159:19-33 (1982).
33. L. Thelander and P. Reichard, Reduction of Ribonucleotides, Ann. Rev. Biochem., 48:133-158 (1979).
34. T. A. Trautner, M. N. Swartz, and A. Kornberg, Enzymatic Synthesis of Deoxyribonucleic Acid, X. Influence of Bromouracil Substitutions on Replication, Proc. Natl. Acad. Sci. U.S.A., 48:449-455 (1962).
35. S. M. Watanabe and M. F. Goodman, On the Molecular Basis of Transition Mutations: The Frequencies of Forming and 2-Aminopurine·cytosine Adene·cytosine Base Mispairs in vitro, Proc. Natl. Acad. Sci., U.S.A., 78:2864-2868 (1981).

MUTATION INDUCTION BY EXCESS DEOXYRIBONUCLEOTIDES

IN *SACCHAROMYCES CEREVISIAE*

Martin Brendel

Institut für Mikrobiologie
J. W. Goethe-Universität
Theodor-Stern-Kai 7
Haus 75
D-6000 Frankfurt/Main

SUMMARY

Excess dTMP is toxic and mutagenic with exponentially growing dTMP efficient uptaking yeast strains 831 rho^+ and 833 rho. The respiratory deficient strain 833 exhibits a tenfold sensitivity to the genotoxicity of excess dTMP. Mutant yield in the forward mutation system CAN1 → can1 after dTMP excess is comparable to that found after irradiation with UV_{254}nm. Excess dTMP is a poor mutagen in stationary phase cells of both strains. Mutagenicity of excess dTMP is not found in an ochre mutant allele (ade2-1). Exposure of exponentially growing cells to other deoxyribonucleotides (dCMP, dAMP, and dGMP) reveal these nucleotides to have mutagenic potential as well.

1. INTRODUCTION

While searching for an economic method of DNA specific labeling in the yeast *Saccharomyces cerevisiae* we were able to isolate efficient dTMP uptaking mutants which will utilize up to 25% of exogenously offered dTMP for DNA synthesis when endogenous thymidylate synthesis is blocked either chemically by antifolates or genetically by mutation in the gene coding for thymidylate synthase [2, 10]. The determination of the optimal dTMP supplement for such thymidylate low-requiring strains (*tlr*-mutants) revealed that either too low or too high supplementation with exogenous dTMP had cytotoxic effects [3, 4, 13]. Furthermore, *tlr*-mutants with normal thymidylate biosynthesis were not spared the cytotoxic effects of excess dTMP as evident in rapid killing of colony forming ability, a phenomenon called thymidylate excess death (TED) [13, 15]. Subsequently we could show that an excess offer of dTMP to the tlr-strains resulted in the uptake of large amounts of thymidylate (normally not present at any time during the growth cycle) into the cells, followed by its conversion to the corresponding di- and triphosphates [15].

Mutation induction by exogenous dTMP could be demonstrated in dTMP-uptaking mutants [1, 6] and a model was proposed in which efficient uptake of thymidylate by the cells may cause pyrimidine nucleotide pool imbalances which in turn may lead to irregularities in cellular DNA synthesis. With the present communication we wish to add further information on the mutagenicity of excess dTMP and also demonstrate that other deoxyribonucleoside monophosphates, when offerred in concentrations between 10^{-4} to 10^{-2} mole/liter, are not only cytotoxic [4], but may likewise act mutagenically on tlr-strains of *Saccharomyces cerevisiae*.

2. MATERIAL AND METHODS

2.1. Yeast Strains and Media

The two yeast strains used throughout this study are isogenic for nuclear markers and differ in their respiratory competence only. Strain A30-833, lacking mtDNA (rho) was selected from the corresponding rho^+ strain A30-831 by incubating the latter in YEPD medium supplemented with 10 μg/ml ethidium bromide. Loss of mtDNA was verified by analysis of yeast DNA in CsCl density gradients. The complete genotype of the strains is:

A30-831 MATa RAD ade2-1 ilv2 tlr_6 rho^+, named strain 831
A30-833 MATa RAD ade2-1 ilv2 tlr_6 rho, named strain 833

The marker tlr_6 describes the ability of these strains to regain (nearly) normal growth characteristics in a medium supplemented with 6 μg/ml dTMP when their endogenous synthesis of thymidylate is blocked [7].

Both media used in this study have a high contents of inorganic phosphate (∿7 mmole/liter) [2]. Medium N consists of 6.7 g yeast nitrogen base w/o aminoacids (Difco), 20 g glucose, 2 g casamino acids (Difco) per l of demineralized water. N + Ade is supplemented with 20 mg adenine/liter. Inhibition of endogenous thymidylate synthesis was achieved by adding 20 mg aminopterin and 6 g sulfanilamide/liter to medium N [10]. Forward mutation to canavanine resistance was measured on SC-Can medium, consisting of 6.7 g yeast nitrogen base w/o amino acids (Difco), 20 g glucose and a supplement of each 20 mg of adenine, uracil, histidine, lysine, and methionine/liter plus 20 mg canavanine. Solid medium was obtained by adding 10 g of agar (Oxoid No. 1).

2.2. Growth of, and Mutation Induction in, Yeast Strains

Both strains were routinely grown in medium N + Ade from single colony inocula. Except where especially noted all experiments were performed with exponentially growing cells at titers not exceeding 3×10^7 cells/ml.

Mutagenicity of dTMP or other deoxyribonucleoside 5'-monophosphates was tested by adding the appropriate concentration of nucleotide dissolved in medium N + Ade to exponentially growing cultures in medium N + Ade and incubation of these cultures for 3 hours (approx. one generation time) in a waterbath shaker at 30°C. Cells from controls were incubated with added medium only. Survival or growth of cells was determined after dilution in ice-cold phosphate buffer and plating onto N + Ade agar while mutation to canavanine resistance (CAN1 → can1) was assayed on SC-Can agar and reversion to adenine prototrophy (ade2-1 → ADE2) on medium N agar. Washing of nucleotide treated cells prior to plating was omitted since the plating of 0.2 ml aliquots of undiluted suspension onto 30 ml agar plates was shown to have no significant effect on either mutation induction or survival of mutants. For some experiments, the SC-Can agar overlay technique was applied as described previously [14].

Survival and mutation induction were assayed after 5 days of incubation at 30°C. Sensitivity of spontaneous can1 mutants towards dTMP was tested in the agar well assay: after streaking four suspensions of can1 isolates from the center outwards, a center hole was cut into the N + Ade agar and filled with 0.2 ml of 10^{-2} mole/liter or 10^{-1} mole/liter dTMP solution when testing for dTMP sensitivity of strain 833 or 831, respectively. After two days of incubation at 30°C sensitivity to dTMP lead to non-growth of the streaked cells for at least one cm distance from the rim of the center well; a resistant control culture would grow to the very edge of the well.

UV irradiation of cells occurred under constant stirring at room temperature at a density of 3 to 6×10^7 cells/ml and an exposure dose of 1.4 Joule $m^{-2}sec^{-1}$.

3. RESULTS

3.1. Genotoxic Activity of Excess dTMP

The main characteristics of exogenous dTMP interaction with tlr_6 (thymidylate low requiring) strains of *Saccharomyces cerevisiae* are shown in Fig. 1: From Fig. 1a we can obtain information on the optimal dTMP supplementation of the two yeast strains 831 and 833 when their thymidylate biosynthesis is blocked by aminopterin and sulfanilamide. Growth is restored best at dTMP exposure concentra-

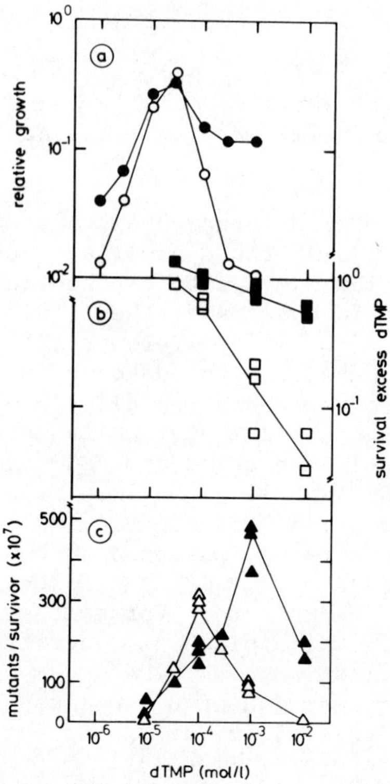

Fig. 1. Genotoxic activity of dTMP in two tlr₆ mutant strains of Saccharomyces cerevisiae. a) Restoration of growth by exogenous dTMP in cultures blocked in endogenous thymidylate biosynthesis by aminopterin and sulfanilamide in relation to that of unihibited controls. b) Cytotoxic action of excess dTMP on tlr₆ mutant strains growing exponentially in medium N + Ade (thymidylate synthesis is not blocked). Survival is expressed as fraction of culture able to form colonies. c) Mutagenic action of excess dTMP. Mutant frequency/survival in the forward mutation system CAN1 → can1. Open symbols: strain 833 rho; closed symbols: strain 831 rho$^+$.

tions between 6 and 9 μg/ml. At exposure concentrations either below or above, we find growth retardation due to thymidylate less death (dTMP too low) or thymidylate excess death (TED; dTMP too high). TED is a rapid process as can be seen in Fig. 1b which reveals strain 833 rho to be sensitive to dTMP at tenfold lower exposure concentrations than strain 831 rho$^+$. Since both strains are growing in normal medium N + Ade and are not blocked in biosynthesis of thymidylate, exposure concentrations of dTMP lower than 10 μg/ml do not have any adverse effects on both cultures. On the con-

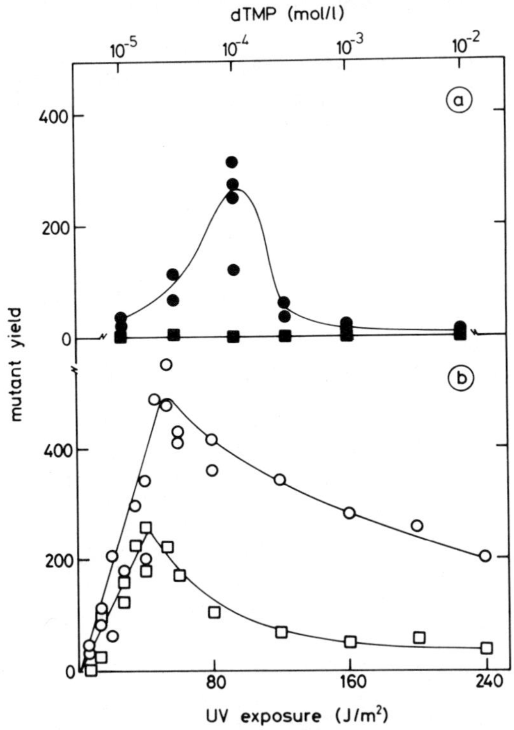

Fig. 2. Comparison of the mutagenic action of dTMP (a) and UV_{254}nm (b) in exponentially growing cells of strain 833 rho. Mutant yield versus exposure concentration or dose is given for the forward mutation system CAN1 → can1 (open and closed circles) and for the reversion test ade2-1 → ADE2 (open and closed squares).

trary, strains will continue growth during exposure to low concentrations of dTMP (Fig. 1b).

The cytotoxic effect of excess dTMP is coupled with genetical activity in these strains. Figure 1c shows that a three hour exposure to excess dTMP is mutagenic in both strains. Mutation induction of canavanine resistance (CAN1 to can1) occurs in both strains at similar frequencies. Strain 833 rho exhibits the genetical activity of excess dTMP already at one tenth the exposure concentration needed for strain 831 (Fig. 1c). Preferred selection of spontaneous can1 mutants from a subpopulation of dTMP resistant cells did not occur: 405 spontaneous can1 mutants from both strains were tested for dTMP sensitivity and none was found resistant to dTMP.

TABLE 1. Growth Phase Dependence of Genotoxicity of Excess dTMP

dTMP exposure[a] (mol/l)	survival (%) expon.	survival (%) stationary	yield of can1 mutants/10^7 cells expon.	yield of can1 mutants/10^7 cells stationary
none	100	100	45	6
10^{-2}	96	96	182	14
10^{-3}	100	98	530	9
10^{-4}	100	91	195	7

[a] strain 831 rho^+

TABLE 2. Induction and Expression of canl Mutants After dTMP Excess as a Function of Post-Treatment Time Prior to Selection by Canavanine

dTMP exposure[a]	post-treatment time(h)	strain 831	strain 833
none	0	55	37
	3.5	50	50
10^{-3}mol/l	0	385	140
	2.5	439	145
	3.5	447	118
10^{-4}mol/l	0	210	252
	2.5	237	284
	3.5	188	260

[a] Three hours exposure at 30°C.

While mutation induction in the forward system CAN1 → canl is clearly visible, both on a mutant per survivor or mutant per cell treated (mutant yield) basis, reversion of the ade2-1 allele to ADE2 could not be induced by excess dTMP in the same experiments (Fig. 2a). Reversion of the ochre triplet mutant ade2-1, however, did occur after exposure to UV_{254}nm (Fig. 2b) so that the absence of mutagenesis after excess dTMP is not due to non-mutability of the ade2-1 allele. Forward mutation to canavanine resistance, however, occurs with a higher probability than reversion to adenine prototrophy (Fig. 2).

TABLE 3. Mutagenicity of Different Deoxyribonucleotides

Nucleotide[a]	exposure (mol/l)	can1 mutant yield/ 10^7 cells strain 831	strain 833
	none	35	51
dTMP	10^{-2}	168	64
	10^{-3}	417	142
	10^{-4}	200	290
dCMP	10^{-2}	44	433
	10^{-3}	10	1,296
	10^{-4}	8	17
	none	10	16
dAMP	10^{-2}	144	11
	10^{-3}	72	9
	10^{-4}	18	110
dGMP	10^{-2}	44	61
	10^{-3}	18	310
	10^{-4}	10	60

[a] Three hours exposure at 30°C

Mutagenicity of excess dTMP is nearly abolished in both $\overline{tlr_6}$ strains when employing cells in stationary growth phase (Table 1). This and another type of experiment (cf. Table 2) indicate that full metabolic activity is a prerequisite for mutagenicity of excess dTMP. Since the expression of the canl phenotype may depend on the growth conditions of the mutated yeast cells [11] we controlled whether a post-treatment incubation prior to selection for canavanine resistance had any influence on canl mutant frequency. The results of such a delayed canavanine resistance test are summarized in Table 2. There is no significant change in mutant yield when adding canavanine immediately or up to 3.5 h later to dTMP mutagenized exponentially growing cells of either strain.

3.2. Mutagenic Activity of Other Deoxyribonucleotides

The three other deoxyribonucleoside 5'-monophosphates also act mutagenic upon both exponentially growing tlr_6 strains (Table 3). The second pyrimidine nucleotide dCMP is clearly more mutagenic under the conditions of the test than the two purine analogs dAMP and dGMP. However, it should be pointed out that these deoxyribonucleotides were offered to tlr_6 strains that, by definition, are only known to be excellent uptakers of dTMP and that other selection procedures might well yield yeast strains with a more sensitive response to purine nucleotides.

4. DISCUSSION

The ability of tlr yeast mutants to efficiently take up and utilize exogenous dTMP for their DNA synthesis by directly converting dTMP to the corresponding di- and triphosphates [2] has led us to study the influence of exogenously offered dTMP onto internal thymidylate pool sizes. Although dTMP uptaking yeast cells can also export dTMP [8] the balance of dTMP import to export seems in favor of dTMP uptake since Toper et al. [15] could clearly show that excess dTMP leads to increased pools of thymidylate nucleotides and that breakdown to thymidine only occurs after hours of massive uptake of dTMP. High internal concentration of thymidylate is likely to influence the pool sizes of other deoxyribonucleotides by allosteric inhibition of ribonucleotide reductase (cf. review by Kunz [12]); it also may inhibit DNA synthesis [13, 15] more directly by interfering with one or more functions associated with DNA replication [1].

Fidelity of DNA replication seems closely geared to strict control of DNA precursor pools and nucleotide pool imbalances have been shown to produce genetic change in diverse organisms [12]. Excess dTMP acts mutagenic on exponentially growing dTMP uptaking mutants [1, 6] and forward mutation to can1 in the CAN1 gene is readily detectable (Figs. 1c and 2a). Reversion to ADE2 of the ochre mutant allele ade2-1 could not be shown (Fig. 2a). This could be a general phenomenon for dTMP induced mutagenesis since other ochre mutant alleles also were not found to revert under excess dTMP (lys2-1, Eckardt, pers. comm.; his5-2 and lys1-1, Brendel, unpublished data). However, this should not lead to premature speculations as to putative mechanisms of mutation induction by excess dTMP since we have to bear in mind that excess of other deoxyribonucleoside 5'-monophosphates also may act mutagenic in tlr strains of yeast (Table 3). A quantitative statement as to the mutagenic potency of the different nucleotides should not be attempted at present because we do not know pool sizes of their corresponding di- and triphosphates nor do we have detailed knowledge of uptake and possible salvage pathways. We do know, however, that, proper precautions provided, dAMP will be nearly exclusively utilized for DNA synthesis in yeast tlr strains [4, 9]. Selection of efficient uptaking strains for dAMP and dGMP will facilitate the study of interrelation of purine nucleotide pool size and genetic change in yeast.

The poor mutagenicity of excess dTMP in stationary tlr cells (Table 1) demonstrates that full metabolic activity is needed in order to have induction of mutations within the three hour exposure period used in our experiments. Delayed selection of canavanine resistant mutants by applying the Can-agar overlay technique [14] showed that mutation to canl is perfected within the three hour exposure period to excess dTMP in exponentially growing cells (Table 2); this suggests that no further DNA replication is required for induction of genetic change.

Mutagenicity of excess dTMP in exponentially growing cells is highest when survival is around 60% (corresponding to 1/2 lethal hit). Calculation of mutant yield [5] clearly demonstrates that mutagenicity of excess dTMP is not due to selection of dTMP resistant spontaneous canl mutants (Fig. 2a). It also reveals excess dTMP to be a potent mutagen in the CAN1 gene as the mutant yield is comparable with that found after irradiation with UV_{254}nm (Fig. 2b).

ACKNOWLEDGMENTS

I wish to thank Dr. W. W. Fäth for the construction of the yeast strain A30 and F. Wilborn for selecting the rho derivative. The technical assistance of Ms. M. Marcovic, Ms. E. Schneider, and Ms. H. Weisser is greatly appreciated. This research was funded in part by a grant of the Deutsche Forschungsgemeinschaft.

REFERENCES

1. B. J. Barclay and J. G. Little Mutation induction in yeast by deoxythymidine monophosphate: A model, Mol. Gen. Genet., 181: 279-281 (1981).
2. M. Brendel and R. H. Haynes, Exogenous thymidine 5'-monophosphate as a precursor for DNA synthesis in yeast, Mol. Gen. Genet., 126:337-348 (1973).
3. M. Brendel and U. G. Langjahr, Thymineless death in a strain of Saccharomyces cerevisiae auxotrophic for deoxythymidine 5'-monophosphate, Mol. Gen. Genet., 131:351-358 (1974).
4. M. Brendel, W. W. Fäth, and R. Toper, in: "Radiation and Cellular Control Processes" (J. Kiefer, ed.), pp. 35-40, Springer-Verlag, Berlin-Heidelberg-New York (1976).
5. F. Eckardt and R. H. Haynes, Quantitative measures of mutagenicity and mutability based on mutant yield data, Mutat. Res., 74:439-558 (1980).

6. F. Eckardt, B. A. Kunz, and R. H. Haynes, Variation of mutation and recombination frequencies over a range of thymidylate concentrations in a diploid thymidylate auxotroph, Curr. Genet., 7:399-402 (1983).
7. W. W. Fäth, and M. Brendel, Isolation and properties of yeast mutants with highly efficient thymidylate utilization, Z. Naturforsch., 31c:468-478 (1976).
8. W. W. Fäth, and M. Brendel, Nucleic acid metabolism in yeast. V. Excretion of thymidylate, Mol. Gen. Genet., 188:115-120 (1982).
9. W. W. Fäth and M. Brendel, Nucleic acid metabolism in yeast. VI. Utilization of exogenous dAMP, Mol. Gen. Genet., 188: 121-127 (1982).
10. W. W. Fäth, M. Brendel, W. Laskowski, and E. Lehmann-Brauns, Economizing DNA-specific labelling by exogenous deoxythymidine 5'-monophosphate in Saccharomyces cerevisiae, Mol. Gen. Genet., 132:335-345 (1974).
11. E. Gocke and T. R. Manney, Expression of radiation-induced mutations at the arginine permease (CAN1) locus in Saccharomyces cerevisiae, Genetics, 91:53-66 (1979).
12. B. A. Kunz, Genetic effects of deoxyribonucleotide pool imbalances, Environmental Mutagenesis, 4:695-725 (1982).
13. U. G. Langjahr, E. M. Hartman, and M. Brendel, Nucleic acid metabolism in yeast. I. Inhibition of RNA and DNA biosynthesis by high concentrations of exogenous deoxythymidine 5'-monophosphate in 5'-dTMP low requiring strains, Mol. Gen. Genet., 143:113-118 (1975).
14. J. F. Lemontt and S. V. Lair, Plate assay for chemical- and radiation-induced mutagenesis of CAN1 in yeast as a function of post-treatment DNA replication: The effect of rad6-1, Mutat. Res., 93:339-352 (1982).
15. R. Toper, W. W. Fäth, and M. Brendel, Nucleic acid metabolism in yeast. II. Metabolism of thymidylate during thymidylate excess death, Mol. Gen. Genet., 182:60-64 (1981).

THE EFFECTS OF PYRIMIDINE NUCLEOTIDES ON ALKYLATING AGENT INDUCED CYTOTOXICITY AND SPONTANEOUS AND INDUCED MUTATION TO PURINE ANALOG RESISTANCE IN V79 CELLS

Margaret Fox

Paterson Laboratories
Christie Hospital and Holt Radium Institute
Manchester, M20 9bX
U.K.

SUMMARY

Exposure of three V79 cell lines to dT after treatment with monofunctional alkylating agents resulted in potentiation of alkylation induced cytotoxicity. The degree of potentiation achieved was dependent on the concentration and duration of exposure to dT and was reversed by equimolar concentrations of dCyd. Exposure to dT after UV or X-irradiation or treatment with HN2 or MMC did not affect the cytotoxic response. dT exposure at non-cytotoxic concentrations did not affect DNA synthesis as measured by [^{3}H]-dT incorporation when allowance was made for reductions in specific activity of labelled thymidine. However, dT post treatment reversed the alkylation induced inhibition of DNA synthesis. Toxic concentrations of dT caused an increase in frequency of TG^R colonies but this increase was shown to be due to effects of dT on cell growth rate, and differential sensitivity ot $HGPRT^-$ and $HGPRT^+$ cells. The frequency of spontaneous and alkylation induced AZ^R and to a lesser extent TG^R colonies was also increased by non-toxic dT concentrations. Evidence was obtained which suggests that this increase is more likely to be due to alterations in the selective efficiency of the purine analogs than alterations in coding fidelity due to altered dNTP pools.

1. INTRODUCTION

The potentiating effects of post-treatment with minimally toxic concentrations of thymidine on alkylating agent induced cytotoxicity

TABLE 1. Potentiation of Lethal Damage by dT Exposure and Its Reversal by dCyd

Alkylating Agent	Nucleotide[a] Added	LD 50 Dose. mM[b]: V79A	V79/79	V79S
	dT	4.0	1.0	0.8
	dCyd	4.0		1.0
MNU	-	1.25	0.6	1.0
MNU	dT	0.8	0.25	0.35
	dT + dCyd	1.3	0.6	1.0
ENU	-	4.0	3.0	4.0
ENU	dT	2.5	-	2.0
ENU	dCyd	4.0	-	4.0
ENU	dT + dCyd	4.0		4.0
EMS	-	20.0	16.0	20.0
EMS	dT	12.0	-	12.0

[a] Exposure to dT or dCyd was for 48hrs at a concentration of 2.5mM.

[b] LD 50 doses were determined by measuring inhibition of colony forming ability.

have been observed by several groups of workers (Refs. 4, 12, 15-18). The fact that this effect can generally be reversed by addition of equimolar concentrations of deoxycytidine (Ref. 4) suggests that the enhancement of lethal damage expression is related to the known effects of high dTTP concentrations in producing feed back inhibition of ribonucleotide reductase and deoxycytidylic kinase thus resulting in starvation of cells for deoxycytidine and consequent inhibition of DNA synthesis [8].

Facilitation of alkylation induced mutagenesis by post-treatment with thymidine has also been reported [4, 12, 16-18] and has been interpreted as indicating that the fidelity of DNA replication on an alkylated template can be reduced by alterations in the composition of the deoxyribonucleotide pools in line with the observations in in vitro DNA polymerase assays [1]. An influence of dNTP pool size on the mutagenic response has also been inferred by the studies of Meuth and co-workers in thy$^-$ mutants of CHO cells [13].

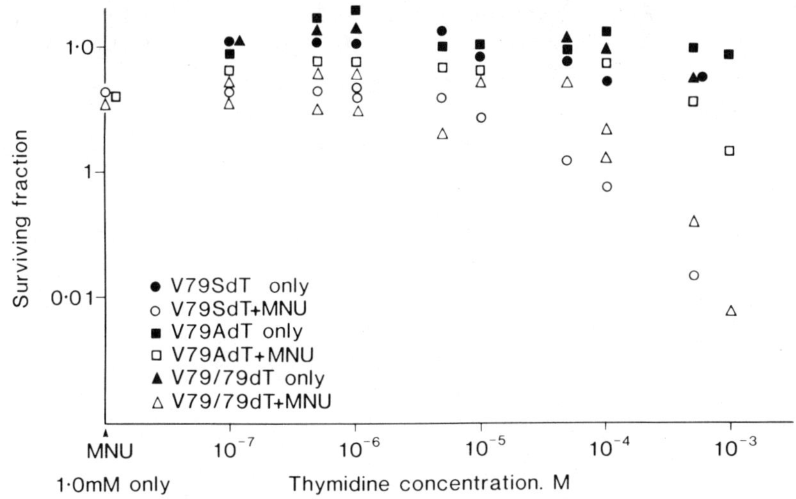

Fig. 1. Dose dependency of dT potentiation of MNU damage in V79 cell lines. The effect of exposure to increasing concentrations of dT for 48 h on the survival of colony forming ability in three different V79 lines after treatment with 1.0 mM MNU. Individual points represents the mean colony count on a minimum of 5 plates at each dose for each cell line.

The data to be presented here however suggest alternative explanations for the observed enhancement of cytotoxic and mutagenic damage.

2. RESULTS AND DISCUSSION

Thymidine post-treatment but not pre-treatment potentiated the lethal effects of a number of monofunctional alkylating agents in V79 cell lines, Table 1, but not UV or X-ray damage [4]. No potentiating effects were observed after exposure to HN2 or mitomcyin C [4]. In all cases the enhancement of lethality observed was reversed by the addition of equimolar concentrations of deoxycytidine (Table 1).

The toxicity of thymidine alone is cell line dependant and lethal potentiation is only observed towards the higher end of the non-toxic dose range (Fig. 1). The time of addition and duration of exposure to dT are also important in determining its potentiating effects (Table 2).

To test whether the synergistic lethal effects of alkylation damage and thymidine post-treatment are mediated via their combined

TABLE 2. Effect of Variation in Duration of Exposure to dT on Survival of Alkylated V79 Cells

Alkylating Agent	Time of addition of dT 2.5mM after treatment	Duration of exposure (hrs)	Surviving Fraction
MNU 1mM			0.47
	0	24	0.12
	0	48	0.07
	24	24	0.38
	48	24	0.50
	72	24	0.55
	0	168	0.07
ENU 3mM	-	-	0.58
	0	24	0.46
	0	48	0.10
	24	24	0.60
	48	24	0.80
	72	24	0.89

inhibitory effects on DNA synthesis, we used a range of MNU, ENU, and dT concentrations. A dose dependent inhibition of [^{3}H]-dT incorporation was evident in two lines of V79 cells after MNU and ENU exposure although the sensitivity of the two lines to such inhibition differed. In V79A cells, exposure to increasing concentrations of unlabelled dT also resulted in a dose dependent decrease in [^{3}H]-dT uptake and a concomitant increase in the incorporation of [^{3}H]-dCyd (Table 3). The decrease in incorporation of [^{3}H]-dT and the increased [^{3}H]-Cyd incorporation are consistent with the known effects of increasing exogenous dT concentrations, i.e., an increase in intracellular dTTP and a decrease in dCTP concentration [8, 13]. These changes occurred on exposure to concentrations of dT and dCyd which had no measurable cytotoxicity in this cell line, which additionally suggest that the changes in incorporation observed reflect changes in dNTP pool size and not inhibition of DNA synthesis per se.

TABLE 3. Effect of Increasing Concentrations of dT on Incorporation of [^{3}H]-dT and [^{3}H]-dCyd

[^{3}H]-Nucleotide Specific Activity	dT added mM	dpm/10^6 cells[a] x 10^{-3}	Final Specific Activity of [^{3}H]-dT
dT	0	12414	
45 Ci/mM	0.004	6500	20Ci/mM
10μCi/ml	0.019	2500	5Ci/mM
	0.099	142.8	100mCi/mM
	0.5	20.0	19.8mCi/mM
	1.0	17.1	9.9mCi/mM
	1.5	17.1	8.6mCi/mM
	2.0	16.4	4.9mCi/mM
	3.0	12.1	3.3mCi/mM
dCyd	0	9.5	
19.3Ci/mM	0.5	9.3	
1.0μCi/mM	1.0	9.8	
	2.0	12.2	
	3.0	16.9	
	4.0	17.2	

[a] Mean incorporation into RNase treated hot acid insoluble material during 30' pulse label given at hourly intervals 0-6 hr after addition of dT.

In V79S cells which are more sensitive to the cytotoxic effects of dT (Table 1) similar results were obtained except that the dose of dT which caused 4-fold elevation of [^{3}H]-dCyd uptake was 10 μM, i.e., similar to that which caused marked changes in dNTP pools in CHO cells [13].

Addition of 2.5 mM dT to V79 cells after MNU and ENU exposure however resulted in an apparent reversal of alkylation induced in-

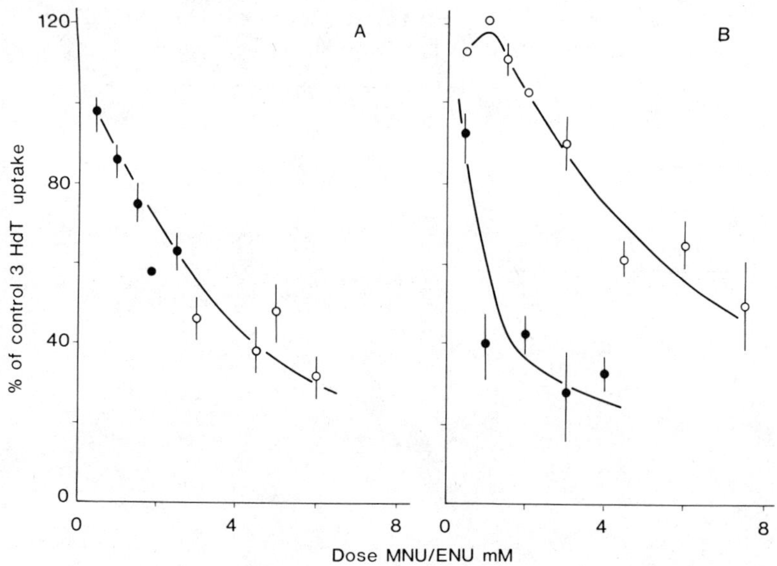

Fig. 2. Inhibition of [^{3}H]-dT incorporation in V79A (A) and V79S (B) cells by MNU and ENU treatment. Cells were treated with appropriate drugs for 1 h then pulse labelled with [^{3}H]-dT 1 μCi/ml 45 Ci/mM for 15 min 1-6 h later. Dose response curves were measured at 2 h the time of maximum inhibition of DNA synthesis. Points plotted represent the mean ± S.E. of at least 3 separate determinations on separate occasions.

hibition of [^{3}H-dT incorporation (Fig. 3). Assuming that the reduced [^{3}H]-dT incorporation in the presence of unlabelled dT is due to expansion of dTTP pools these data would suggest that alkylation damage followed by dT treatment results in a marked decrease in the dTTP pool size, with consequent alteration in the complex feed-back controls normally exerted. Such perturbations and consequent imbalanced growth may be responsible for the observed enhancement of cell lethality. If this hypothesis is correct it is to be expected that dT post-treatment would enhance the lethal effects of any agent which inhibits DNA synthesis. However, dT post-treatment enhanced the cytotoxicity of monofunctional alkylating agents but was without effect after HN2, MMC, X-ray, and UV exposure [4]. dT is also known to be an inhibitor of polyADP ribose polymerase [18] and other inhibitors of this enzyme are known to be markedly cell line and agent specific in their potentiating effects [6].

Caffeine and other methylated xanthines are also known inhibitors of poly(ADP ribose)polymerase [18], and also potentiate cell lethality induced by a wide variety of DNA damaging agents (Refs. 4,

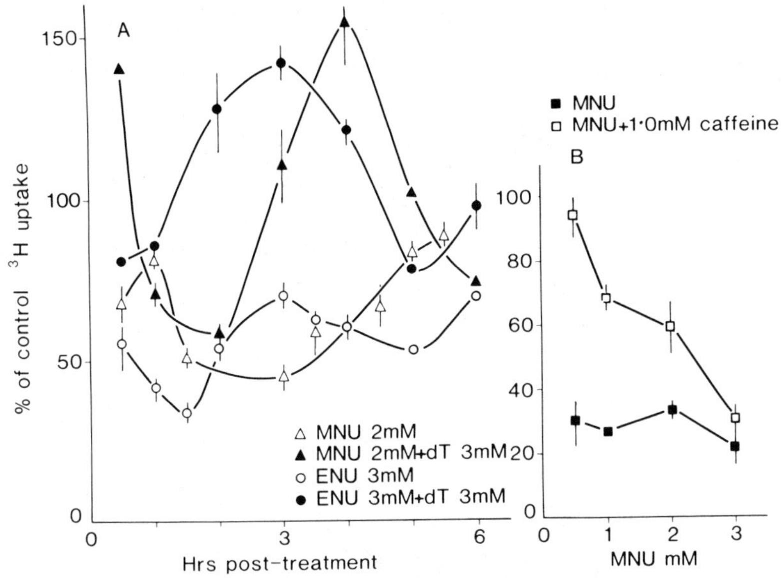

Fig. 3. a) Effect of dT post treatment on the incorporation of [^{3}H]-dT in MNU and ENU treated V79A cells. Cells were pulse labelled with [^{3}H]-dT at indicated times with 10 μCi/ml [^{3}H]-dT. The specific activity of the [^{3}H]-dT was adjusted so that the total dT concentration at the time of the pulse was the same for all treatment conditions. Data are mean ± S.E. of three-four separate determinations. b) Reversal by caffeine 1.0 mM of the inhibitory effect of different doses of MNU on [^{3}H]-dT incorporation. Cells were given a 30' of [^{3}H]-dT pulse 2 h after exposure to MNU.

6, 12, 15-18). Caffeine has been reported to reverse the inhibition of DNA synthesis caused by alkylating agents and UV [15, 19] and the G2 arrest caused by HN2 [9]. Painter has reported that caffeine also reverses the inhibition of replicon initiation caused by X- and UV-irradiation [15]. In one of our V79 cell lines but not the other caffeine reversed the MNU induced inhibition of DNA synthesis as did dT (Fig. 3). These considerable parallels between the effects of thymidine, caffeine and other inhibitors of poly(ADP ribose) polymerase strongly suggest that potentiation of induced cell lethality by dT may be mediated via inhibition of this enzyme which results in failure of the ribosylation reaction and hence failure to activate ligase (Ref. 5) and not via altered pool sizes. In favor of this interpretation is the observation that perturbation of pool sizes via inhibition of de novo DNA synthesis had little effect on UV and EMS induced cell lethality in V79 cells [11].

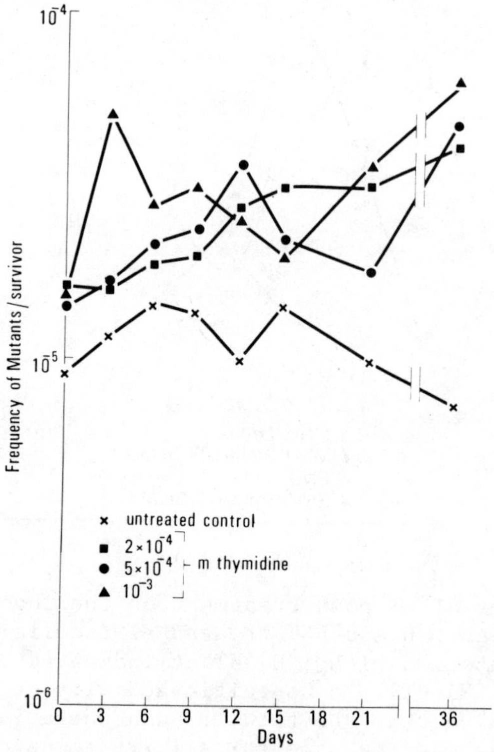

Fig. 4. Expression of thymidine induced TG^R mutants. The effect of 24 h exposure to increasing concentrations of dT on the recovery of TG^R colonies from V79S cells.

The effect of toxic concentrations of thymidine on the frequency of TG^R colonies in V79S cells is shown in Fig. 4. Several features of this response are unusual. The increase in frequency of colonies capable of growing in 6TG is evident immediately after treatment but there is no further increase with time. Although the increase is statistically significant, there was no statistically significant difference between the frequency of TG^R colonies in cell populations exposed to increasing thymidine concentrations. The effects of these concentrations of thymidine on growth rate, colony forming ability and recovery of TG^R mutants was therefore examined. Thymidine treatment increased the recovery of TG^R colonies in reconstruction experiments and inhibited the growth rate of TG^R mutants to a greater extent than Wt cells. Wild type cells were also more sensitive to the cytotoxic effects of thymidine (Table 4).

The effects of exposure to non-toxic dT concentrations for either 24 or 48 h on spontaneous mutant frequencies are shown in Figs. 5 and 6. In both cases the rate of accumulation of resistant

TABLE 4. The Effect of Increasing Concentrations of dT on Growth and Survival of V79 Cells

Thymidine (M)	Frequency of [a] $HGPRT^-$ cells recovered $\times 10^{-2}$	Cell Doubling Time (hrs): $HGPRT^+$	$HGPRT^-$	Colony Forming Ability[c] $HGPRT^+$	$HGPRT^-$
0	1.74 ± 0.17	10.2	10.2		
10^{-4}	1.85 ± 0.15	10.4	10.2	0.98	0.99
2×10^{-4}	1.83 ± 0.11	10.4	10.4	0.60	0.98
5×10^{-4}	2.16 ± 0.14	18.4	11.2	0.25	0.90
10^{-3}	2.62 ± 0.41	21.6	13.6	0.10	0.45

[a] Recovery of TG^R colonies from a mixed population following 24hrs exposure to dT and selection in 5µg/ml TG. Values represent mean of 10 determinations ± one standard deviation.

[b] Relative growth of $HGPRT^-$ and $HGPRT^+$ cells after exposure to dT for 24 hours.

[c] Toxicity in $HGPRT^+$ and $HGPRT^-$ cells of 24h exposure to dT.

colonies was increased. dT post-treatment markedly enhanced the frequency of ENU induced AZ^R colonies as shown in Fig. 7. A similar though less marked enhancement of MNU induced mutation to TG^R but no effect on expression time, was also observed [4]. The effects of dT on the frequency of TG^R and AZ^R colonies were completely reversed by the addition of equimolar concentrations of dCyd as shown in Table 5. Since we had shown [3] that marginally toxic dT concentrations could have effects on cell growth rates without significantly perturbing rates of DNA synthesis, and because of the known effects of high dTTP levels on dNTP pools we measured the effects of dT alone and dT in combination with ENU on the growth and rate of killing of cells by AZ. The data in Figs. 8 and 9 indicate that cells treated with dT alone, ENU alone and ENU + dT grow more slowly than wild-type cells and are killed more slowly by AZ. AZ is known to be degraded by V79 cells to a non-cytotoxic derivative probably due to release of guanine deaminase [Fox, unpublished observations.] Therefore if treated cells are killed more slowly, $HGPRT^+$ cells in the population may not have been eliminated before the AZ has been degraded leading to the apparent enhancement of IMF.

TABLE 5. Effect of dT, dC, and dT + dC Post Treatment on Alkylation Induced Mutagenesis

Treatment	Induced mutant frequency x 10^{-5} at 8 days[a]	
	AZ (40μg/ml)	TG (10μg/ml)
0	1.4	0.6
dT 3mM	10.4	2.9
ENU 3mM	123.5	120.0
ENU 3mM+dT3mM	295.0	230.0
ENU 3mM+dC3mM	133.0	116.0
ENU 3mM+dC+dT	126.0	109.0
0		0.65
dT 10μM		0.80
MNU 0.2mM		65.0
MNU 0.2mM+dT10μM		90.0
MNU 0.2mM+dT10μM+dC10μM		65.0

[a] Cells were treated or not with ENU or MNU and incubated for 48 hours post treatment in 3mM dT, dC or dT + dC (V79A cells) or 10μM dT or dC + dT (V79S cells) then subcultured for 6 days in 10MEM before plating in AZ and TG (V79A cells) or TG only (V79S cells). Results are mean colony frequency from ten plates for each treatment condition.

The cytotoxicity of purine analogs is profoundly dependent on the affinity of the particular analog and its phosphorylated derivatives for HGPRT and subsequent enzymes in the metabolic pathway. Azaguanine has a considerably lower affinity than does thioguanine for HGPRT and its phosphorylation is therefore greatly affected by the presence of hypoxanthine [7]. Similarly, the affinity of the monophosphate for guanylate kinase and nucleoside diphosphate kinase may also be considerably affected by changes in endogenous pools of dGDP and dGTP. Thus the elevation of dTTP pools which occurs in the presence of excess dT with the concomitant reduction in dCTP and elevation of dGTP pools [8] could account for the reduced growth rate in dT treated cells and affect the selective efficiency of the aza-

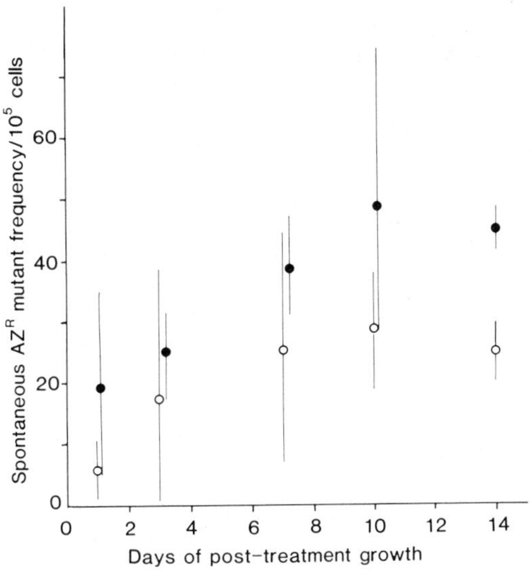

Fig. 5. Effect of 24 h exposure of V79A cells to dT (3 mM) on the frequency of spontaneous AZ^R mutants. 5×10^5 cells were either treated for 24 h with 3 mM dT or left untreated. Immediately after exposure cells were trypsinized and plated at a density of 1×10^5 cells/9 cm plate and 30 μg/ml AZ added. Aliquots of the same suspension were re-seeded at a density of 1×10^6 cells into roller bottles for propagation and subsequent selection at times indicated. Points represent the mean, and error bars the range of values obtained in two separate experiments. ○) Untreated; ●) dT 3 M 24 h.

guanine thus accounting for the reduced rate of kill by AZ observed in such populations.

We therefore suggest that the increase in frequency of AZ^R and to a lesser extent TG^R colonies both spontaneous and induced, may not be due to an alteration in coding fidelity as has been proposed [12, 16-18]. The alternative explanation is that the increase in intracellular deoxyguanine nucleotides which results from inhibition of ribonucleotide reductase reduces the selective stringency of the purine analogs, the result being an artifactual increase in IMF. In support of this hypothesis is our limited analysis of the phenotypes of colonies isolated from azaguanine four of ten of which were not resistant on further testing [4], and the observations of Peterson et al. [17] which also indicated considerable phenotypic heterogeneity amongst clones isolated after azaguanine selection.

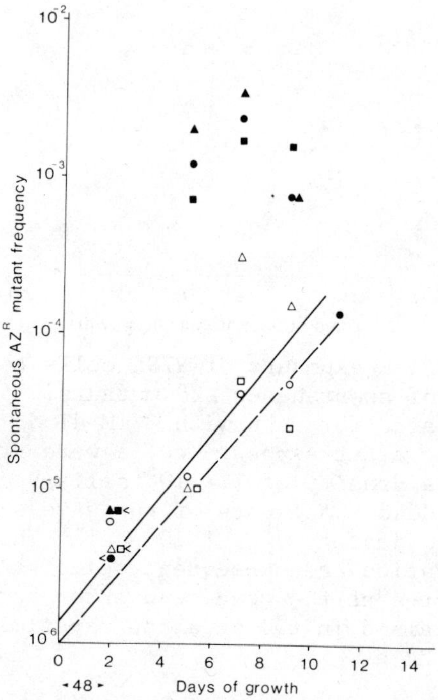

Fig. 6. Effect of 48 h exposure of V79 cells to dT 3 mM on frequency of colonies resistant to different AZ concentrations. Experimental design was as described above except that 2 × 10^6 cells were propagated every 3 days, △, ○, □) selection in 20, 30 or 40 μg/ml AZ (untreated cells) ▲, ●, ■) selection in 20, 30, or 40 μg/ml AZ after exposure to 3 mM dT for 48 h, ----) expected accumulation of spontaneous mutants with successive population doublings calculated assuming 1 mutant/10^6 cells at the start of the experiment and a spontaneous mutation rate of 2.5 × 10^{-7}/cell/generation.

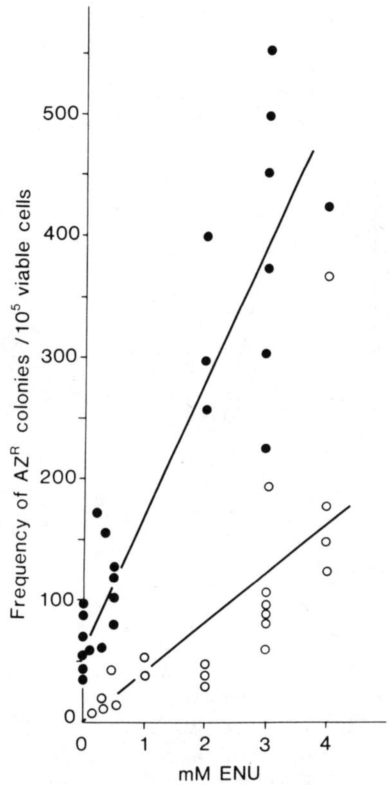

Fig. 7. Effect of 48 h dT (3 mM) exposure on frequency of ENU-induced AZ^R mutants in V79A cells. Individual points represent replicate determinations of frequencies at expression times of 0, 5, 8, or 11 days. ○) ENU only; ●) ENU + 48 h post-treatment exposure to dT (3 mM).

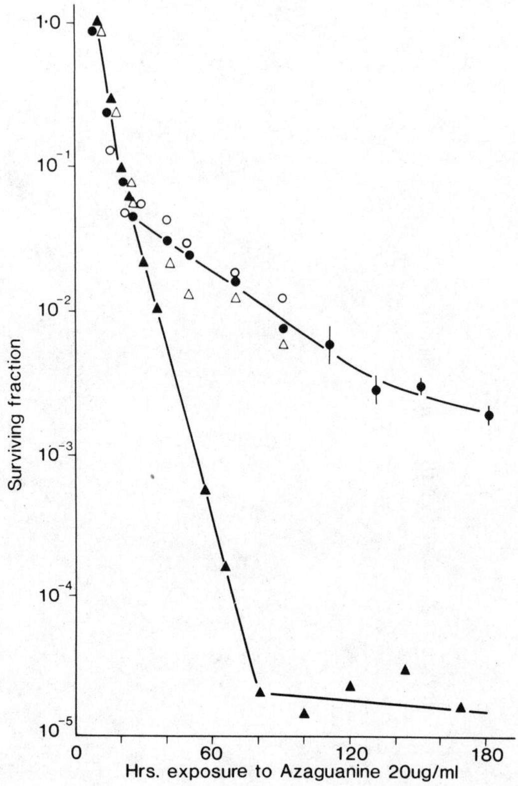

Fig. 8. Effect of dT or ENU + dT exposure on the rate of cell killing by azaguanine. Rate of loss of colony forming ability of V79 cells exposed to 20 μg/ml 8-AZ 5 days after exposure to dT alone, ENU alone or ENU + dT. ▲) Cells exposed to AZ alone 20 μg/ml for increasing periods of time. The AZ-containing medium was then removed and incubation continued in drug free growth medium. ○) Cells exposed to TdR (3 mM 48 h) subcultured for 3 days in drug-free medium then exposed to AZ 20 μg/ml for increasing periods of time, prior to incubation in drug free medium for determination of colony forming ability; ●) cells exposed to 3 mM ENU (1 h) subcultured for 5 days then exposed to 20 μg/ml 8-AZ as above. △) Cells exposed to 3 mM ENU (1 h) post-treated with dT (3 mM 48 h) subcultured for 3 days then exposed to 20 μg/ml AZ as above.

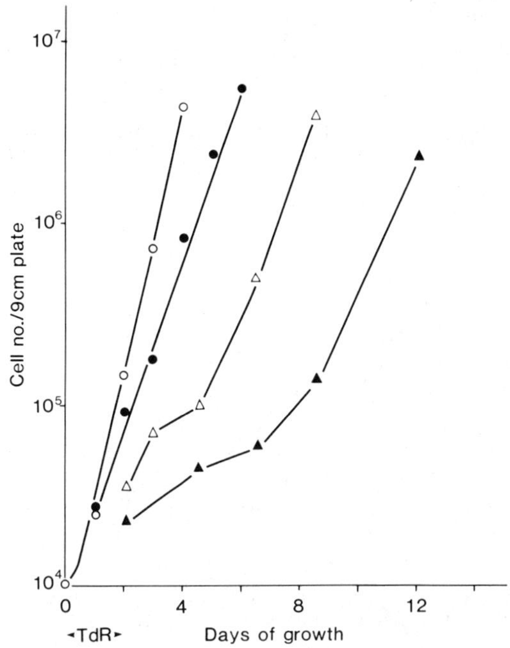

Fig. 9. Effect of 48 h dT exposure on growth rate of untreated and ENU treated V79A cells. ○) Control; ●) 3 mM dT 48 h; △) 4 mM ENU; ▲) 4 mM ENU + 3 mM dT 48 h.

Before any definite conclusions can be drawn with respect to the relative validity of the two alternative hypotheses proposed above it seems important therefore to analyze a large number of mutants both phenotypically and at the molecular level using the cDNA probe for the HGPRT gene [2]. Such studies would enable one to determine whether or not there was a significant proportion of non-mutants surviving after alkylation mutagenesis and dT post-treatment and to possibly analyze any differences between the nature of "mutants" generated in the presence and absence of dT.

REFERENCES

1. P. J. Abbott and R. Saffhill, DNA synthesis with methylated poly)dC-dG) templates: evidence for a competitive nature to miscoding by O^6-methylguanine, Biochem. Biophys. Acta, 562: 51-36 (1979).
2. J. Brennand, A. C. Chinault, D. S. Koneki, D. W. Melton, and C. T. Caskey, Cloned cDNA sequences of the hypoxanthine/ guanine phosphoribosyltransferase gene from a mouse neuroblastoma cell line found to have amplified genomic sequences, Proc. Natl. Acad. Sci., U.S.A., 79:1950-1954 (1982).

3. J. Brennand and M. Fox, Excess thymidine is not mutagenic in Chinese hamster V79 fibroblasts, Cell. Biol. Int. Rep., 4:923-932 (1980).
4. J. Brennand, S. McMillan, and M. Fox, Effects of metabolic inhibitors on cell lethality and mutation induction in Chinese hamster cells. II. The effect of post treatment with non-toxic concentrations of thymidine, Chem. Biol. Int., 36:89-106 (1981).
5. D. Criessen and S. Shall, Regulation of DNA ligase activity by poly(ADP-ribose), Nature, 296:271-272 (1982).
6. L. G. Durrant and J. M. Boyle, Potentiation of cell killing by inhibitors of poly(ADP ribose) polymerase in four rodent cell lines exposed to N-methyl-N-nitrosourea or UV light, Chem. Biol. Int., 38:325-338 (1982).
7. M. Fox and R. J. Hodgkiss, Mechanism of cytotoxic action of azaguanine and thioguanine in wild type V79 cell lines and their relative efficiency in selection of structural gene mutants, Mutation Res., 80:165-185 (1981).
8. B. A. Kunz, Genetic effects of deoxyribonucleotide pool imbalances, Environ. Mut., 4:695-725 (1982).
9. C. C. Lau and A. B. Pardee, Mechanism by which caffeine potentiates lethality of nitrogen mustard, Proc. Natl. Acad. Sci., U.S.A., 79:2942-2946 (1982).
10. V. Levi, E. L. Jacobson, and M. K. Jacobson, Inhibition of poly(ADP ribose) polymerase by methylated xanthines anc cytokinins, FEBS Letters, 88:144 (1978).
11. S. McMillan and M. Fox, Effects of metabolic inhibitors on cell lethality and mutation induction in Chinese hamster cells. 1. Inhibitors of de novo purine synthesis and a comparison of the effects of caffeine, Chem. Biol. Int., 36:71-88 (1981).
12. M. Meuth, Role of deoxynucleoside triphosphate pools in the cytotoxic and mutagenic effects of DNA alkylating agents, Somatic. Cell Genet., 7:89-102 (1981).
13. M. Meuth, Sensitivity of a mutator gene in Chinese hamster ovary cells to deoxynucleoside triphosphate pool alterations, Mol. and Cell Biol., 1:652-660 (1981).
14. J. P. Murnane, J. E. Byfield, J. F. Ward, and Calabro-Jones, Effects of methylated xanthines on mammalian cells treated with bifunctional alkylating agents, Nature, 258:326 (1980).
15. R. B. Painter, Effect of caffeine on DNA synthesis in irradiated and unirradiated mammalian cells, J. Mol. Biol., 143:289-301 (1980).
16. A. R. Peterson, J. R. Landolph, H. Peterson, and C. Heidelberger, Mutagenesis of Chinese hamster cells is facilitated by thymidine and deoxycytidine, Nature, 276:508-510 (1978).
17. A. R. Peterson and H. Peterson, Facilitation by pyrimidine deoxyribonucleotides and hypoxanthine of mutagenic and cytotoxic effects of monofunctional alkylating agents in Chinese hamster cells, Mutation Res., 61:319 (1979).

18. A. R. Peterson and H. Peterson, Differences in temporal aspects of mutagenesis and cytotoxicity in Chinese hamster cells treated with methylating agents and thymidine, Proc. Natl. Acad. Sci., U.S.A., 1643-1647 (1982).
19. J. Priess, R. Schlaeger, and H. Hilg, Specific inhibition of poly(ADP ribose) polymerase by thymidine and nicotinamide in HeLa cells, FEBS Letters, 19:244 (1971).
20. H. W. Van den Berg and J. J. Roberts, Inhibition by caffeine of post replication repair in Chinese hamster cells treated with cis-platinum (JI) diamine dichloride, the extent of platinum binding to template DNA in relation to size of low molecular weight DNA, Chem.-Biol. Int., 12:375 (1976).

ASPECTS OF DNA REPAIR AND NUCLEOTIDE POOL IMBALANCE

Robin Holliday

Genetics Division
National Institute for Medical Research
The Ridgeway
Mill Hill, London NW7 1AA

SUMMARY

Evidence that optimum repair depends on adequate pools of deoxynucleotide triphosphates (dNTPs) comes from the study of pyrimidine auxotrophs of *Ustilago maydis*. These strains are sensitive to UV light and X-rays, and for *pyr1*-1 it has been shown that the intracellular concentration of dTTP is reduced about 7-fold. Evidence has been published that DNA repair synthesis may require higher concentrations of dNTPs, than chromosome replication. The survival curve of *pyr1*-1 after UV-treatment, and split dose experiments with wild-type cells, provide evidence for an inducible repair mechanism, which probably depends on genetic recombination. Although inducible repair saves cellular resources, it has the disadvantage of becoming ineffective at doses which are high enough to inactivate the repressed structural gene(s) for repair enzymes.

To ensure survival it is vital for organisms to preserve the integrity of their DNA, and this is accomplished both by accurate replication and by repair. It is clear that a wide variety of repair mechanisms have evolved to remove lesions which arise either spontaneously (such as the deamination of cytosine to uracil) or as a result of damage from external agents. Nevertheless, it would be incorrect to assume that all species require all possible pathways of repair. It is now well established that the accuracy of

KEY WORDS: DNA repair; *Ustilago maydis*; pyrimidine auxotrophs; uracil glycosylase.

DNA and protein synthesis depends on proof-reading or editing mechanisms (for a review, see Ref. 5). These consume energy and may slow down the rate of synthesis, therefore organisms must invest metabolic resources to achieve any given level of accuracy. Optimum accuracy levels will evolve from the balance between error avoidance in macromolecular synthesis and physiological efficiency in growth and propagation [11, 12, 22]. The same principle would be expected to apply to repair mechanisms, but the efficiency of repair is likely to vary between organisms and relate to the likelihood of DNA damage from the environment. Escherichia coli has a wide range of repair enzymes, but it is far less efficient than Micrococcus radiodurans in removing radiation-induced DNA breaks or lesions. Similarly, the yeast-like basidiomycete, Ustilago maydis, is far more radiation-resistant than Saccharomyces cerevisiae. The reasons for these large differences in repair capacity are not at all obvious, and could probably only be understood by ecological studies of the environments to which these organisms are adapted.

Organisms can save resources by regulating repair pathways. For some kinds of repair it may be essential to have constitutive enzymes (e.g., uracil-DNA glycosylase), but efficiency would be increased if some repair mechanisms are induced only after DNA has been damaged. However, as we shall see, the advantages of inducible repair are lost with saturating damage to DNA.

Pyrimidine Auxotrophs of Ustilago maydis

The control of intracellular nucleotide triphosphate concentrations must be essential to optimize the accuracy of DNA synthesis and the efficiency of DNA repair. An example of this comes from the study of pyrimidine auxotrophs of U. maydis. In the first successful search for radiation-sensitive mutants in any eukaryotic organism [7], it was shown that pyr7 (now designated pyr1-1) was very sensitive to both UV light and X-rays. This strain will grow on uridine, cytidine, uracil, or cytosine, but its radiation sensitivity is scarcely affected by their concentration in the medium (unpublished results). Moore [19] isolated thirteen more pyrimidine mutants, assigned them to four genetic loci and found that all were radiation-sensitive, to a greater or lesser extent. Many of these mutants have very unusual survival curves: at low UV doses killing was exponential, but with increasing dose the viability of cells increased to a given point and then subsequently declined. An example, using pyr1-1, is shown in Fig. 1A. This was interpreted to mean that a constitutive repair pathway was defective in the mutants, but that a proportion of cells were rescued by an inducible repair pathway that was only switched on after a fairly high level of radiation. The interpretation was greatly strengthened when split dose experiments with wild-type cells were subsequently carried out [16]. The kinetics of the response clearly showed that low doses of UV did not protect cells when they were challenged later with a

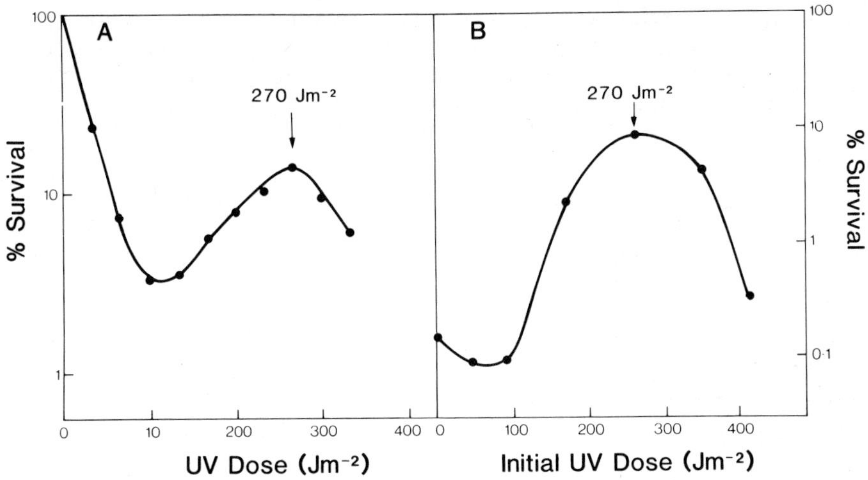

Fig. 1. A) The survival of the pyr1-1 mutant of Ustilago maydis after treatment with UV light. The result is the same as previously published by Moore [19, 20], but the experiment was repeated to check the UV response in relation to the experiment in B. B) Split dose experiments by Lee and Yarranton [16] with wild-type cells of U. maydis. After the initial dose shown the cells were incubated for 3 h in complete medium and then challenged with a dose of 450 J/m^2. Subsequently, the cells were held for 7 h in water containing cycloheximide (to prevent protein synthesis after the challenge dose), before plating for survival.

high dose, but with higher initial doses such protection was observed. The greatest protection occurred after a first dose of 270 J/m^2 which is almost exactly the same which gave the maximum "UV-reaction" in the pyr1-1 survival curve (Fig. 1A and B). Thus the very high level of resistance of U. maydis to UV light is in part due to the existence of an inducible repair pathway. This pathway almost certainly depends on genetic recombination and is absent in rec1 strains, which lack an inducible protein capable of promoting heteroduplex formation [8, 9, 10, 13, 14, 15, 17].

pyr1-1 was shown to be proficient in radiation-induced recombination and also in the excision of pyrimidine dimers [20]. The deoxyribonucleotide pools were measured and it was found that dTTP was reduced 7-fold in comparison to wild-type, whereas the other three deoxynucleotides were not significantly altered. For reasons unknown, it appeared that a block in the endogenous synthesis of pyrimidine led to an imbalance in nucleotide di- or triphosphates.

Since pyr1-1 and the other mutants were defective in a constitutive repair pathway, it seemed a reasonable supposition that

TABLE 1. DNA Polymerase of U. maydis: Kinetic Constants for Limiting Substrate Concentrations in the Presence or Absence of Single-Stranded DNA Binding Protein. Summary of the Results of Yarranton et al., Who Published Lineweaver-Burk Plots [24]

Limiting substrates	DNA substrate	Binding protein	Km (μm)	Vmax*
dTTP+	Denatured	-	1.1	26.7
	(40μM)	+	0.3	41.3
"	Activated	-	1.2	62.5
	(30μM)	+	1.2	62.5
Denatured DNA		-	125	40
		+	31.3	50

+ The other 3 nucleotide triphosphates were in excess. Similar results were obtained with limiting dATP.

* pmoles nucleotide monophosphate incorporated/20 mins.

after the excision of pyrimidine dimers they were unable to fill the gaps by efficient repair synthesis. The cells grow normally in the presence of exogenous pyrimidine, so the favored interpretation is that the reduced level of dTTP has no or little effect on chromosomal replication, but a serious effect on repair replication [20]. This was subsequently tested by in vitro experiments [24].

Interaction of DNA Polymerase and DNA Binding Protein

The major DNA polymerase and a single-stranded DNA binding protein of U. maydis have been purified and characterized, and it was shown that the binding protein significantly increases the activity of the polymerase, when denatured DNA was used as the primer-template [1, 2].

Yarranton, Moore, and Spanos [24] measured the affinity of the DNA polymerase for the deoxyribonucleotide triphosphate substrates (dNTPs), using either denatured or nicked DNA, in the presence or absence of the binding protein. Their results are summarized in Table 1. It can be seen that the binding protein significantly increases the affinity of the DNA polymerase for dNTPs and their rate of incorporation when denatured DNA is the template, but not when nicked DNA is used. The binding protein also increases the affinity of the DNA polymerase for the denatured DNA substrate. They proposed that the denatured DNA substrate (which provides a primer by loop-back of single strands) is comparable to the DNA replication fork, and that double-stranded DNA treated with DNase (nicked or activated) is similar to the gap produced after the excision of pyrimidine dimers. From the results they obtained, it is clear that gap filling is not stimulated by the single-stranded binding protein and requires a higher concentration of dNTPs than does the replication of a primed single-stranded molecule. Thus, the reduced level of dTTP in pyr1-1 may differentially reduce repair synthesis.

Spontaneous DNA Damage

One of the important chemical changes in DNA is the denaturation of cytosine to uracil, which after replication would lead to transition mutations (G-C → A-T). It is therefore not surprising that all organisms so far examined have been shown to have the enzyme, uracil-DNA glycosylase, which can cleave uracil from the deoxyglucose-phosphate backbone of DNA (for a review, see Ref. 18). It is, however, surprising that mutants of *E. coli* lacking uracil-DNA glycosylase (*ung*$^-$) grow well and are relatively weak mutators [3, 4]. It is important to know the physiological consequences of the same defect in eukaryotes. Several *ung*$^-$ mutants of *U. maydis* have been isolated and partially characterized [23]. Colonies grow very slowly and produce fast growing sectors which are usually *ung*$^+$ revertants. The variable morphology and instability of the *ung*$^-$ strains suggest that they are highly mutable, but this has not so far been demonstrated quantitatively. In any event, the phenotype of the *ung*$^-$ strains suggest that the enzyme may play a more important role in maintaining the integrity and function of DNA in eukaryotes than is the case in prokaryotes, but clearly further investigations are necessary.

The Limits to Inducible Repair

It was previously pointed out that inducible repair mechanisms provide a means of saving metabolic resources, until such time as the DNA suffer damage from an extracellular source. However, inducible defence mechanisms run into difficulties with very high doses of DNA-damaging agents, and become less efficient than constitutive repair pathways. The reason is that the repressed structural gene(s) for an inducible repair enzyme is itself a target for DNA damage. An unrepaired lesion which prevents transcription puts the whole pathway out of action, whereas a similar lesion might be repaired by a constitutive pathway.

The kinetics of induced recombination after γ radiation of heteroallelic diploids of *U. maydis* strongly indicated that recombination repair was inducible [8]. When recombination was measured by the production of nitrate reductase (in a diploid heteroallelic for two mutants in the structural gene), it was shown that the increase in recombination was associated with survival, and that cells killed by high doses of γ radiation did not recombine. Furthermore, the survival curve for cells was not markedly different from the survival of a single gene for an inducible metabolic enzyme [8, 21]. Thus, the results suggested that at very high doses (600-750 K rads) the inducible genes for recombination repair were themselves being inactivated. In other words, the target for cell inactivation is approximately the same size as that of a single gene in this organism.

The Efficiency of Repair

In most metabolic reactions, a small decrease in the efficiency of an enzyme will have a relatively small physiological effect. However, the same is not true for the enzymic reactions which remove lesions from DNA. Imagine a dose of radiation which produces 100 potential lethal lesions. With no repair, the cell inevitably dies; with 90% efficiency of repair, 10 lesions are left on average and again the cells will almost always die. Even with 99% efficiency of repair, at least one lethal lesion will often remain in the DNA and viability will be considerably impaired. (For a quantitative treatment, see Ref. 6.) It is common to believe that a radiation-sensitive phenotype is due to the loss of a specific repair enzyme, but for the reasons just outlined it seemed likely that a number of metabolic disturbances could slightly reduce repair efficiency and produce a radiation-sensitive phenotype. The view that repair pathways in yeast are coded by at least 100 *rad* genes is therefore probably fallacious. It seems quite probable that many of these genes have their phenotypic affect by altering the physiological milieu in which repair enzymes act, and one way this can happen is the creation of an imbalance in the intracellular concentration of deoxyribonucleotides.

ACKNOWLEDGMENT

I thank Susan Y. Taylor for carrying out the experiment in Fig. 1A.

REFERENCES

1. G. R. Banks and A. Spanos, The isolation of properties of a DNA-unwinding protein from *Ustilago maydis*, J. Mol. Biol., 93:63-77 (1975).
2. G. R. Banks, W. K. Holloman, M. V. Kairis, A. Spanos, and G. T. Yarranton, A DNA polymerase from *Ustilago maydis*. I. Purification and properties of the polymerase activity, Eur. J. Biochem., 62:131-142 (1976).

3. B. K. Duncan, P. A. Rockstroh, and H. R. Warner, Escherichia coli K12 mutants deficient in uracil-DNA glycosylase, J. Bact., 134:1039-1045 (1978).
4. B. K. Duncan and B. Weiss, Uracil-DNA glycosylase mutants are mutators, in: "DNA Repair Mechanisms" (P. C. Hanawalt, E. C. Friedberg, and C. F. Cox, eds.), pp. 183-186, Academic Press, New York (1978).
5. A. R. Fersht, Enzyme editing mechanisms and the genetic code, Proc. Roy. Soc. B., 212:351-379 (1981).
6. R. H. Haynes, The influence of repair processes on radiobiological survival curves, in: "Cell Survival After Low Doses of Radiation" (T. Alper, ed.), pp. 197-208, J. Wiley, Chichester (1975).
7. R. Holliday, Radiation sensitive mutants of Ustilago maydis, Mut. Res., 2:557-559 (1965).
8. R. Holliday, Biochemical measure of the time and frequency of radiation-induced allelic recombination in Ustilago, Nature New Biol., 232:233-236 (1971).
9. R. Holliday, Further evidence for an inducible recombination repair system in Ustilago maydis, Mut. Res., 29:149-153 (1975).
10. R. Holliday, R. E. Halliwell, M. W. Evans, and V. Rowell, Genetic characterization of rec-1, a mutant of Ustilago maydis defective in repair and recombination, Genet. Res. Camb., 27:413-453 (1976).
11. T. B. L. Kirkwood, Repair and its evolution: survival versus reproduction, in: "Physiological Ecology: An Evolutionary Approach to Resource Use" (C. R. Townsend and P. Calow, eds.), pp. 165-189, Blackwell, Oxford (1981).
12. T. B. L. Kirkwood and R. Holliday, Selection of optimal accuracy and the evolution of ageing, in: "Accuracy in Molecular Biology" (K. B. L. Kirkwood, D. Galas, and R. F. Rosenberger, eds.), in press (1983).
13. E. B. Kmiec, and W. K. Holloman, Homologous pairing of DNA molecules promoted by a protein from Ustilago, Cell, 29:367-374 (1982).
14. E. B. Kmiec and W. K. Holloman, Heteroduplex formation and polarity during strand transfer promoted by Ustilago rec1 protein, Cell, 33:857-864 (1983).
15. E. B. Kmiec, M. Brougham, M. Yarnall, P. Kroeger, and W. K. Holloman, Induction of rec1 protein in irradiated cells of Ustilago, in: "Mechanisms of DNA Replication and Recombination"(N. R. Cozzarelli, ed.), A. R. Liss, New York, pp. 753-755 (1983).
16. M. G. Lee and G. T. Yarranton, Inducible DNA repair in Ustilago maydis, Mol. Gen. Genet., 185:245-250 (1982).
17. M. G. Lee and G. T. Yarranton, in preparation.
18. T. Lindahl, DNA glycosylases, endonucleases for apurinic/apyrimidinic sites and base excision-repair, Prog. Nucl. Acid Res. Mol. Biol., 22:135-192 (1979).

19. P. D. Moore, Radiation-sensitive pyrimidine auxotrophs of Ustilago maydis. I. Isolation and characterization of mutants, Mut. Res., 28:355-366 (1975).
20. P. D. Moore, Radiation-sensitive pyrimidine auxotrophs of Ustilago maydis. II. A study of repair mechanisms and UV recovery in pyr1, Mut. Res., 28:367-380 (1975).
21. M. A. Resnick and R. Holliday, Genetic repair and the synthesis of nitrate reductase in Ustilago maydis after UV irradiation, Mol. Gen. Genet., 111:171-184 (1971).
22. M. A. Savageau and R. R. Freter, On the evolution of accuracy and cost of proofreading tRNA amino acylation, Proc. Natl. Acad. Sci., U.S.A., 76:4507-4510 (1979).
23. Y. Yamamoto and R. Holliday, Uracil-DNA glycosylase defective mutants of Ustilago maydis, Current Genetics, in press (1983).
24. G. T. Yarranton, P. D. Moore, and A. Spanos, The influence of DNA binding protein on the substrate affinities of DNA polymerase from Ustilago maydis: one polymerase implicated in both DNA replication and repair, Mol. Gen. Genet., 145:215-218 (1976).

AGENTS WHICH DECREASE INTRACELLULAR THYMIDINE POOLS CAUSE AN AUGMENTATION IN HUMAN NATURAL KILLER ACTIVITY *IN VITRO**

David S. Matheson, Bridget J. Green,
David I. Hoar, Susan J. Friedman,
and Masafumi Inoue

Departments of Pediatrics and Pharmacology
of the University of Calgary
Calgary, Canada

Nucleotide pool imbalances produce several effects many of which have been referred to in this symposium. Our laboratory has been investigating the effects of thymidylate (TdR) depletion on the function of human natural killer (NK) cells. NK cells are a subpopulation of peripheral blood lymphocytes which lyse several types of target cells without prior sensitization [1]. These NK cells are thought to play a role in the defense against viral infection or malignancies. Some evidence exists to implicate NK cells in the homeostatic control of bone marrow and thymus [2]. One major modulator of NK activity is a set of glycoproteins called interferons (IFN) which are secreted secondary to viral infections or following immune stimulation and which cause an augmentation in NK responsiveness [3, 4].

The modulatory effects of nucleotide pool imbalances on natural killer function were first suspected when it was noted that trimethoprim (TMP), a dihydrofolate reductase (DHFR) inhibitor, caused a significant increase in NK function *in vitro* [5]. Another inhibitor of DHFR, methotrexate (MTX), was demonstrated to have the same enhancing effecton NK activity. Furthermore, 5-fluorodeoxyuridine (FUdR), which more specifically depletes the thymidylate pool by inhibiting thymidylate synthetase directly, also caused an increase in NK activity *in vitro*. The effect of the drugs was shown to be

*Supported in part by the Alberta Heritage Foundation for Medical Research and the Alberta Heritage Savings and Trust Fund Applied Cancer.

TABLE 1

Drug	Lytic Units /10^7 cells Actinomycin-D (1μg/ml) -	+
None	74	77
IFN (200 Units)	161	87
MTX 10^{-7}M	69	87
10^{-8}M	125	74
10^{-9}M	154	74
D-glucosamine 10^{-3}M	62	91
$5X10^{-4}$M	139	87
10^{-4}M	179	80

specifically related to TdR depletion since addition of exogenous thymidine abrogated the drug (MTX, FUdR) effect. This modulation of NK activity was shown to be due to a direct effect on the effector cell population containing the NK cells rather than on the target cell [6].

D-Glucosamine, a naturally occurring amino sugar has a multitude of effects on cellular metabolism, one of which is TdR depletion [7]. D-Glucosamine also stimulated human NK function both in vitro and in vivo [8]. The augmentation of NK by D-glucosamine was also mediated by TdR depletion as exogenous TdR abrogated the effect (unpublished results).

The concentrations of the drugs at which augmentation occurred was as follows: TMP, 10^{-5} and 10^{-6} M; MTX, 10^{-8} and 10^{-9} M; FUdR, 10^{-8} and 10^{-9} M; and D-glucosamine 10^{-4} and 5×10^{-4} M. Inhibition of the drug effects by exogenous thymidine occurred at 10^{-6} to 10^{-8} M [5, 6, 8].

The basic mechanism by which TdR modulates the NK function is unknown. As interferon secretion could be implicated and IFN is known to require transcription for its effect [9], studies were performed using actinomycin-D to determine if MTX and D-glucosamine required transcription to augment the NK response. From Table 1 it can be seen that actinomycin-D impaired the augmentation of NK function induced by MTX or D-glucosamine.

Method: 2×10^6 peripheral blood mononuclear cells (PBMC) were incubated in the drug(s) for 2 h at 37°C and washed. They were added to chromium-51 labelled target cells (K-562, an erythroleukemic cell line) and an 18 h chromium release assay was performed. One lytic unit was defined by the number of cells required to achieve 30% lysis [6].

As interferon may be released from the cells as a response to thymidylate deprivation, the increase in NK activity could be a secondary effect. To examine this possibility, two million PBMC were exposed to the following drugs for 18 h at 37°C: D-glucosamine (10^{-4} and 5×10^{-4}), MTX (10^{-8} and 10^{-9} M) and FUdR (10^{-8} and 10^{-9} M). The supernatants were analyzed for the presence of IFN by an antiviral technique which can measure as low as 1 unit of IFN [10] and in each case we were unable to detect any IFN production.

Thus drugs which lead to TdR depletion can cause augmentation in NK activity in vitro. The changes occurred at concentrations at least one to two logs lower than is necessary to induce significant cell killing or mitochondrial mutagenesis in cultured human cells [11, 12]. These concentrations presumably lead to TdR perturbations as small amounts of exogenous TdR (10^{-6} to 10^{-8} M) could abrogate the effects.

As actinomycin-D inhibited the NK enhancement, transcription and hence, protein synthesis appears to be involved in the process. The fact that no IFN could be detected in the supernatants of the drug treated cells argues against its secretion although these results do not rule out the possibility that IFN is consumed by the effector cells at a greater rate than it is produced during the incubation. It remains to be determined what role if any interferon plays in this phenomenon.

As the levels of TMP achieved during drug therapy are of the same order of magnitude as those used here [8], these observations may have in vivo relevence. In vitro, our data demonstate that "trivial" changes in the intracellular TdR pool can cause significant perturbations in important biological functions (i.e., NK activity).

REFERENCES

1. J. C. Roder, K. Karre, and R. Riessling, Natural killer cells, Prog. Allergy, 28:66 (1981).
2. M. Hanson, R. Kiessling, and B. Anderson, Human fetal thymus and bone marrow contain target cells for natural killer cells, Eur. J. Immunol., 11:8 (1981).
3. G. Trinchieri and D. Santoli, Enhancement of human natural iller cell activity by interferon, J. Immunol., 120:1845 (1978).
4. D. S. Matheson, B. Green, and Y. H. Tan, Human interferons α and b inhibit T-cell-dependent and stimulate T-cell-independent mitogenesis and natural cytotoxicity: Relationship to chromosome 21, Cell Immunol., 65:366 (1981).
5. D. S. Matheson, R. W. Wilson, and B. Green, Trimethoprim-induced augmentation of human natural killer activity in vitro, Immunopharm., 4:73 (1982).
6. D. S. Matheson, B. Green, and D. I. Hoar, The influence of methotrexate and thymidine on the human natural killer cell function in vitro, J. Immunol., 131:1619 (1983).

7. S. J. Friedman, C. D. Trotter, T. Kimball, and P. Skehan, The inhibition of thymidine metabolism in tumor cells treated with D-glucosamine, Cancer Res., 37:1141 (1977).
8. D. S. Matheson, E. Wong, B. J. Green, M. Van Olm, and S. J. Friedman, Effect of D-glucosamine on human natural killer cell activity, Proceedings of the 13th International Congress of Chemotherapy, Vienna, Sept.,1983, 265:32 (1983).
9. J. R. Ortaldo, W. Phillips, K. Wasserman, and R. B. Herberman, Effects of metabolic inhibitors on spontaneous and interferon boosted human natural killer cell activity, J. Immunol., 125: 1839 (1980).
10. Y. H. Tan and W. Berthold, A mechanism for the induction and regulation of human fibroblastoid interferon genetic expression, J. Gen. Virol., 34:401 (1977).
11. L. S. Dimnik, R. B. Church, and D. I. Hoar, Induction of mitochondrial mutations by methotrexate, Cancer Research, in press.
12. D. I. Hoar and L. S. Dimnik, Induction of mitochondrial mutations in human cells by methotrexate (this volume).

ONCOGENIC TRANSFORMATION OF C3H/10T1/2 C1 8 MOUSE EMBRYO FIBROBLASTS BY INHIBITORS OF NUCLEOTIDE METABOLISM

A. R. Peterson,* C. Heidelberger,** and W. F. Benedict†

*The Cancer Research Laboratory
University of Southern California Cancer Center and
The Institute for Toxicology
School of Pharmacy
1303 N Mission Road
Los Angeles, California 90033

†Division of Hematology-Oncology
Department of Medicine
Childrens Hospital of Los Angeles, and
Department of Pediatrics
University of Southern California School of Medicine
Los Angeles, California 90027

INTRODUCTION

Since 1956, when Coughlin and Adelberg demonstrated that thymine starvation is mutagenic in bacteria, extensive evidence has accumulated to show that imbalances in deoxyribonucleotide (dNTP)‡ pools can induce mutations in bacteriophages, bacteria, fungi, higher plants and in cultured mammalian cells (reviewed in Ref. 36). Mutations are here defined in the broadest sense: as conservable alteractions in the sequence of bases in the genome, and in this sense

---

**Deceased, January 18, 1983.

‡Abbreviations: ara-C, 1-β-D-arabinofuranosyl-cytosine; 5-aza-CR, 5-azacytidine; BrdUrd, 5-bromo-deoxyuridine; BrU, 5-bromouracil; DMBA, dimethyl benzanthracene; dCyd, deoxycytidine; dCMP, deoxycytidine 5'-monophosphate; dCTP, deoxycytidine-5'-triphosphate; dNTP, deoxyribonucleoside 5'-triphosphate, dTTP, deoxythymidine 5'-triphosphate, FdUrd, 5-fluorodeoxyuridine; FU, 5-fluorouracil; IdUrd, 5-iododeoxyuridine; MTX, ethotrexate; MCA, 3-methylcholanthrene.

there is now impressive evidence indicating that one or more mutations are necessary events in some forms of carcinogenesis (reviewed in Ref. 61). These considerations suggest that agents that cause dNTP pool imbalances, which, in turn, induce mutations, could also cause cancer. The fact that some of these agents are antimetabolites used in cancer chemotherapy raises concern that cancer, already a recognized side-effect of treatment with alkylating agents [29], may also be a consequence of therapy with antimetabolites. Therefore, with these antimetabolites, objectives of practical necessity are: to define their carcinogenic potential, and to explore the possibility of selectively modulating their chemotherapeutic and potentially carcinogenic effects.

An experimental approach to these objectives is to study oncogenic transformation induced by the antimetabolites in the C3H/10T1/2 Cl 8 line of mouse embryo fibroblasts. These cells, which were developed in the laboratory of one of the authors (CH) [23, 26, 52, 53], can be morphologically transformed by a variety of chemical carcinogens and by ultraviolet and ionizing radiation but rarely undergo spontaneous transformation. The transformed cells produce foci that can be readily recognized against the background monolayer of contact inhibited cells, and ∿75% of the foci with type III morphology produce malignant fibrosarcomas in irradiated, syngeneic mice [46, 52]. Mutagenesis to ouabain resistance can also be studied in the C3H/10T1/2 system [26, 37]. Furthermore, because the system is a cell line, the molecular genetics of oncogenesis can be studied in homogeneous populations of cells under conditions that can be more precisely controlled in tissue culture than is possible in whole animals [23].

In this article, we have collected together the results of our studies of the oncogenic transformation and cytotoxicity produced by antimetabolites of pyrimidine metabolism in the C3H/10T1/2 cell line. We also discuss our results in the light of published data on mutagenesis, clastogenesis and perturbations of nucleotide metabolism in mammalian cells treated with these antimetabolites. We have not yet studied nucleotide metabolism and its perturbations in C3H/10T1/2 cells, but this field in relation to other mammalian cells and prokaryotic systems has been reviewed in this volume (see articles by G. O'Donovan, P. Reichard, and C. K. Mathews) and repeatedly in the literature [8-14, 21, 22, 24, 25, 36, 42, 49, 51, 55], so, to avoid undue repetition, we have drawn extensively on these articles. Some of our studies with C3H/10T1/2 cells have been reviewed by the International Agency for Research on Cancer in the context of evaluating the carcinogenic risks of antineoplastic agents to humans [29].

METHODS FOR TRANSFORMATION AND CYTOTOXICITY ASSAYS

The C3H/10T1/2 Cl 8 cells, between the 12th and 18th passage, were grown in Eagle's basal medium containing 10% heat inactivated fetal calf serum in 60 mm plastic petri dishes [6, 31, 32, 46]. Twenty-four hours after seeding 1000 cells/dish, 12 dishes were treated with antimetabolites for an additional 24 h for transformation assays. The medium was then removed, the dishes were washed once with phosphate buffered saline, and fresh medium was added. Medium was changed weekly for 4-6 weeks, after which the cultures were fixed with methanol and stained with Giemsa, and the stained dishes were examined for dense colonies of crisscrossed cells (type III foci). The cell cycle dependence of oncogenic transformation was determined by treating cultures that had been synchronized either by release from postconfluence inhibition of division or by release from isoleucine deprivation [31, 32]. Tumorigenicity of morphologically transformed cells was determined by measuring the induction of fibrosarcomas in 4-6 day old or 3-6 week old C3H mice. The mice were immunosuppressed by whole body irradiation with 550 rads of X-rays [46] or by rabbit anti-mouse thymocyte serum [31, 32] prior to s.c. injection of $1\text{-}2.5 \times 10^6$ transformed cells. Under these conditions $1\text{-}10 \times 10^6$ untreated C3H/10T1/2 Cl 8 cells were not tumorigenic [31, 32, 46].

Cytotoxicity was determined by the reduction in plating efficiency in dishes seeded with 200-1000 cells [6, 31, 32, 46]. The cytotoxicity of thymidine (dThd) and deoxycytidine (dCyd) in C3H/10T1/2 Cl 8 cells and in MCA- and DMBA-transformed C3H/10T1/2 cells [46] was determined by treating 200-1000 cells/dish with the nucleosides for 7 days.

DEOXYTHYMIDINE AND DEOXYCYTIDINE

Treatment of mammalian cells with 10^{-4}-10^{-3} M dThd has been found to result in elevated levels of dTTP, and the dNTP pools of the cells are further imbalanced by the inhibition by dTTP of enzymes of nucleotide metabolism (Reichard, in this volume) [36, 42, 51]. These pool imbalances have been considered associated with the cytotoxicity, clastogenesis and mutagenesis induced by dThd [36]. However, although 24 h treatments of C3H/10T1/2 Cl 8 cells with 10^{-4}-10^{-3} M dThd were cytotoxic, suggesting that dNTP pool imbalances did occur, no transformation was detected [32]. Furthermore, dThd inhibited oncogenic-transformation induction by FdUrd and possibly by FU (Table 1). On the other hand, dThd is selectively lethal to cells derived from human and mouse tumors [38, 42], and we found that clones of morphologically transformed C3H/10T1/2 cells that were tumorigenic were 2-8 fold more sensitive to dThd than were C3H/10T1/2 Cl 8 cells or clones of non-tumorigenic, morphologically transformed cells (Table 2) [46]. The plating efficiencies of the transformed clones in medium lacking dThd did not correlate with their dThd sensitivity nor with their tumorigenicity, and the doubling times of the highly tumorigenic cells and nontumorigenic cells were not significantly different (16.7 ± 0.24 h, MCA 16; 18.3 ± 2.0

TABLE 1. Oncogenic Transformation of C3H/10T1/2 Cells by Pyrimidine Analogs and Antimetabolites

Compound	Concentration (M)	Surviving Fraction	Transformation[d] Frequency (%)
dThd[a]	10^{-3}	0.42	0
dCyd[c]	10^{-3}	1.00	0
BrdUrd[a]	10^{-4}	0.62	0
IdUrd[a]	10^{-4}	0.15	0
F_3dThd[a]	10^{-5}	0.40	0
FdUrd[a]	10^{-6}	0.63	0.55
FdUrd[a] + dThd	10^{-5} } 10^{-4} }	0.70	0
FU[a]	10^{-5}	0.63	0.07
FU[a] + dThd	10^{-5} } 10^{-4} }	0.80	0
MTX[b]	$2.2.10^{-4}$	0.85	0.16
5-aza-CR[b]	$3.3.10^{-7}$	0.70	0
5-aza-CR[b]	$1.1.10^{-6}$	0.40	1.75
ara-C[c]	10^{-7}	0.70	0
ara-C[c]	10^{-3}	0.01	1.30

a: From ref. 32; b: from ref. 6; c: from ref. 31; d: transformation frequency = % of surviving cells that form type III foci.

h, C3H/10T1/2 Cl 8) [46]. These observations suggest that dThd selectively inhibits the growth of oncogenically transformed C3H/10T1/2 cells that are highly tumorigenic.

Treatment of mammalian cells with $\sim 10^{-5}$ M dCyd results in elevated levels of dCyd phosphates and, via deamination by dCMP deaminase and methylation of the resultant dUMP by thymidylate synthetase, dCyd nucleotides are converted into dTTP [21, 36, 51]. Constitutively elevated dCTP pools in mutant Chinese hamster ovary cells are associated with increased rates of spontaneous mutation in these cells [43]. Moreover, dCyd is clastogenic and facilitates mutagenesis by alkylating agents and FdUrd in Chinese hamster cells [36, 47], and the facilitation of mutagenesis cannot entirely be attributed to conversion of dCyd into thymine nucleotides (Peterson, Danenberg, et al., this volume). However, 10^{-3} M dCyd did not induce transformation in C3H/10T1/2 cells (Table 1), suggesting that

TABLE 2. Tumorigenicity and Sensitivity to the Cytotoxicity of Thymidine of C3H/10T1/2 Cl 8 Cells and of Morphologically Transformed Clones Derived from Them. From Ref. 46

Clone	D_o[a] $M \times 10^{-5}$	Focus Morphology Type[c]	Tumorigenicity Assays[e] 1973[c]	1980[d]
C3H/10T1/2 Cl 8	32 ± 6.9	-	0/5	0/10
DMBA 2	38 ± 6.8	I	0/2	0/10
DMBA 5	36 ± 3.0	II	0/2	0/10
DMBA 8	30 ± 2.8	III	0/2	0/10
DMBA 4	12 ± 2.8	II	1/2	0/10
DMBA 3	8.9 ± 0.59	III	2/2	0/10
MCA 15	4.7 ± 0.68	III	2/2	10/10
MCA 16	3.6 ± 0.18	III	2/2	10/10
MCA 16	20[b]	III	2/2	10/10

a : Thymidine concentration that reduces the surviving fraction by 1/e in the linear portion of the survival curve; b : 20×10^{-5}M dThd + 9×10^{-5}M dCyd; c : data from ref. 52; d : data from ref. 46; e : Mice with fibrosarcomas/mice injected.

dNTP pool imbalances produced by treatment with dCyd alone are not sufficient to induce oncogenic transformation of the C3H/10T1/2 cells. On the other hand, dCyd inhibited the toxicity of dThd to tumorigenic, transformed C3H/10T1/2 cells (Table 2).

HALOGENATED PYRIMIDINES

5-Bromodeoxyuridine (BrdUrd) and 5-iododeoxyuridine (IdUrd) are mutagenic in a variety of organisms [49]. The mutagenicity of BrdUrd in mammalian cells has been ascribed to two molecular effects of the compound: a) incorporation of bromouracil (BrU) into DNA as a mispairing analog of thymine; b) inhibition, by BrU nucleotides, of the enzymatic reduction of CDP resulting in depletion of dCTP pools (Ashman et al., and Kaufman et al., this volume) [36]. IdUrd induced murine leukemia virus production but did not induce oncogenic transformation in C3H/10T1/2 cells (Table 1), adding to the weight of evidence that shows that chemically induced transformation is independent of virus production in these cells [50]. Recently, BrdUrd was shown to be mutagenic in C3H/10T1/2 cells [37] but did not induce detectable oncogenic transformation (Table 1), suggesting that neither incorporation of BrU into DNA nor depletion of dCTP pools could induce the transformation. However, ultraviolet-light-induced photolysis of BrU that was incorporated into the DNA of Syrian hamster cells induced mutation and oncogenic transformation in those cells [3].

In C3H/10T1/2 cells, the oncogenic transformation assay has been performed with three fluorinated pyrimidine derivatives: 5-fluorouracil (FU), 5-fluorodeoxyuridine (FdUrd) and trifluorothymidine (F_3dThd). In mouse cells, a fraction of the FdUrd (about 10% in C3H/10T1/2 cells) [32] is enzymatically converted into FU, which becomes incorporated instead of uracil into RNA [9, 22, 24]. By contrast, F_3dThd is not incorporated into RNA, but, because of the similarity in size between the methyl and trifluoromethyl groups, F_3dThd is an analog of dThd and is incorporated at a low frequency into DNA [25]. Recently, FdUrd also has been found to be incorporated into the DNA of mouse cells, but there is disagreement about the extent and biological significance of the incorporation [15, 28, 35]. An effect that these three fluorinated pyrimidines produce in common is that their administration to mammalian cells leads to a powerful inhibition of thymidylate synthetase [9, 14, 22, 24, 25]. Since thymidylate synthetase is the only de novo source of dTTP, inhibition of this enzyme decreases the availability of dTTP for DNA synthesis and thereby effectively perturbs intracellular dNTP pools in mouse cells [9, 15, 18, 24].

By co-treatment of the cells with dCyd to exacerbate the expected pool imbalance, FdUrd was shown to be strongly mutagenic in Chinese hamster V79 cells (Peterson, Danenberg, et al., in this volume) [47]. By contrast, FdUrd and F_3dThd are only weakly mutagenic in synchronized V79 cells (1), and in asynchronous V79 cells without added dCyd less than 10^{-6} M FdUrd is not mutagenic (Grosovsky and Little, in this volume) [1, 27, 37, 47] and F_3dThd did not induce mutations [27]. However, F_3dThd did mutate bacteriophage T_4 [20], FU and FdUrd are clastogenic in cultured mammalian cells [40, 59], and FdUrd is recombinogenic in yeast [57].

At equitoxic concentrations, F_3dThd did not induce oncogenic transformation in C3H/10T1/2 cells, whereas FU and FdUrd induced relatively low and relatively high levels of transformation, respectively (Table 1). The transformation was dose and time dependent, and that by FdUrd was cell cycle dependent with maximum transformation occurring when cells were treated in S-phase [32]. Cells derived from 60% of the morphologically transformed foci that were tested for tumorigenicity produced tumors, but these cells were not less sensitive to the cytotoxicity of FdUrd than were the C3H/10T1/2 Cl 8 cells, indicating that the transformation was in most cases tumorigenic and directly induced and not a selection of pre-existing transformed cells [32]. These findings suggest that FdUrd and possibly FU could be carcinogenic under the proper conditions. However, there is no evidence for carcinogenicity induced by FU [29].

METHOTREXATE

Methotrexate (MTX) is an analog of folic acid that inhibits the enzyme dihydrolate reductase. Thus, MTX inhibits the de novo synthesis of thymidylate and also of purines and thereby inhibits RNA and DNA synthesis [8, 10]. The inhibition of thymidylate synthesis results from a depletion of 5-10 methylene tetrahydrofolate, which is the obligatory cofactor for thymidylate synthetase, so MTX treatment has been found to produce dNTP pool imbalances similar to those produced by FdUrd [8, 10, 24, 30, 58]. In cultured mouse and hamster cells, MTX was clastogenic and induced sister chromatid exchanges [2, 6, 36, 54]. It induces dominant lethal mutations in mice [17, 48] is recombinogenic in yeast [57] but is not mutagenic in conventional bacterial or mammalian cell assays of mutagenicity [5, 29]. However, MTX induces thymine starvation, and in folic-acid-requiring Streptococcus fecalis MTX induced mutations [19].

In asynchronous C3H/10T1/2 Cl 8 cells, MTX produced as much morphological transformation as did similarly toxic treatments with FdUrd (Table 1). This observation together with those made with FdUrd and FU suggest that the dNTP pool imbalances created by inhibition of thymidylate synthetase can facilitate oncogenic transformation in cultured C3H/10T1/2 Cl 8 cells. However, the data currently available on the carcinogenic effects of MTX in laboratory animals or in human beings has been judged to be inadequate to evaluate the carcinogenic potential of MTX [29].

ANALOGS OF CYTOSINE NUCLEOSIDES

In mammalian cells the cytidine analog 5-azacytidine (5-aza-CR) is incorporated into RNA to a greater extent than into DNA [9, 12, 55, 60]. By contrast the deoxycytidine analog 1-β-D-arabinofuranosylcytosine (ara-C) is incorporated into DNA to a greater extent than into RNA [7, 11, 13]. The cytotoxic and antitumor activities of both analogs are associated with inhibition of DNA synthesis resulting from their incorporation into nucleic acids [9, 11, 13]. The 5-azacytosine moiety of the incorporated 5-aza-CR is thought to be spontaneously cleaved at neutral pH [9, 12, 55], but although 5-aza-CR is unstable in solution [9, 12, 55] more than 85% of the s-triazine moiety was recovered intact from C3H/10T1/2-cell DNA that contained incorporated [^{14}C]-5-aza-CR [34]. The incorporated ara-C blocks ligation and synthesis of Okazaki fragments of DNA [11, 13]. Both compounds are clastogenic [6, 7] but only weakly and non-reproducibly mutagenic in mammalian cells [37].

Both compounds cause perturbations in nucleotide metabolism in mammalian cells [9, 11, 12]. In mouse embryo cells, the perturbations induced by ara-C include increased dATP, dGTP, and dTTP levels, which result from the inhibition of DNA synthesis, and a small initial decrease in the dCTP level, which recovers within 4 h [56]. However, in the latter study [56], ara-CTP in the pools increased progressively during the incubation with ara-C and was not significantly degraded to arabinosyluracil phosphates, so after 2 h of incubation, the sum of the concentrations of dCTP and of the dCTP analog, ara-CTP, were greater than the concentration of dCTP in untreated cells. These findings in mouse embryo cells treated with 10 μM ara-C suggest that treatment of C3H/10T1/2 cells with the 1-3 orders of magnitude higher concentrations of ara-C necessary for oncogenic transformation (Table 1) could produce levels of ara-CTP and dCTP that in total are substantially greater than the dCTP levels in untreated cells. The effects of 5-aza-CR on nucleotide metabolism in mouse embryo cells have not been as extensively studied as have the effects of ara-C, but the fact that 5-aza-CR is converted by mammalian cells into 5-aza-2'-deoxy CTP [39, 60] suggests that the sum of 5-aza-2'-deoxy CTP levels and dCTP levels in C3H/10T1/2 cells treated with 5-aza-CR may be higher than the level of dCTP that occurs in untreated cells.

In C3H/10T1/2 cells, 5-aza-CR inhibited DNA methylation and affected the stability of the differentiated state of the cells, whereas ara-C did not produce these effects [33]. However, both compounds induced morphological transformation, although the concentrations required to do this were more cytotoxic than the concentrations of MTX, FU, or FdUrd that produced oncogenic transformation (Table 1). Transformation by ara-C was maximal when cells were treated in S-phase, but some transformation was seen in cells in the G_1-phase of the cell-cycle, and the transformed cells were tumorigenic and grew in soft agarose. Furthermore, cells from transformed foci induced by 5-aza-CR were tumorigenic in nude mice (Benedict and Jones, unpublished). There is also limited evidence indicating that 5-aza-CR is carcinogenic in laboratory animals [29], and ara-C has been shown to induce oncogenic transformation in cultured hamster embryo cells and rat cells [7, 41].

IMPLICATIONS FOR SELECTIVE MODULATION OF ONCOGENIC TRANSFORMATION AND CYTOTOXICITY

Before discussing the implications of our studies on C3H/10T1/2 Cl 8 cells, attention should be given to the unusually low sensitivity of these cells to the fluorinated pyrimidines and methotrexate. The LD_{50} of a 24 h treatment with FdUrd in these cells is $\sim 10^{-6}$ M, and that of a 24 h treatment with FU is $\sim 10^{-5}$ M [32]. By contrast, the LD_{50} of a 2 h treatment with FdUrd in V79 cells is $\sim 10^{-8}$ M [47], and the IC_{50} of 48 h treatments with FdUrd and FU in murine leukemia cell lines are $\sim 5 \times 10^{-10}$ M and $\sim 4 \times 10^{-7}$ M, respectively [44]. In V79 Chinese hamster cells, the LD_{50} of MTX is four orders of magnitude lower than in C3H/10T1/2 cells (Peterson, manuscript in preparation). The differences of sensitivity of the C3H/10T1/2 cells from those of the other cells may be due to: a) the use of undialyzed serum in the culture medium of the C3H/10T1/2 Cl 8 cells, whereas the media for the other cells contained dialyzed serum, or b) intrinsic differences of biochemistry of the C3H/10T1/2 cells from that of the other cells. In view of these differences in sensitivity, we will restrict our discussion of oncogenesis to oncogenic transformation of the C3H/10T1/2 cells.

All the compounds discussed in the present article are clastogenic or mutagenic, and we believe that specific chromosomal changes may be particularly important in chemically induced oncogenic transformation [4]. All the base and nucleoside analogs that we studied and dThd are incorporated into DNA and inhibit DNA synthesis in cultured mammalian cells. However, only FdUrd, FU, 5-aza-CR, ara-C, and MTX, which is neither a base nor nucleoside analog and is not incorporated into DNA, induce transformation, whereas dThd inhibits transformation (Table 1) and is selectively toxic to transformed cells (Table 2). Thus, the incorporation of these compounds into DNA and their inhibition of DNA synthesis are not uniformly associated with oncogenic transformation. Furthermore, FU is incorporated into RNA to a greater extent than is FdUrd from an equitoxic dose in mammalian cells [9, 22, 24], yet at equitoxic concentrations FdUrd induced more oncogenic transformation than FU (Table 1), indicating that the incorporation of FU into RNA is not a major factor necessary for the transformation induced by FdUrd and FU.

On the other hand, the published results of experiments with mammalian cells indicate that all the compounds discussed in the present article perturb dNTP pools [8-14, 24, 39, 56, 60]. These results suggest that compounds, such as ara-C and 5-aza-CR, that introduce dCTP analogs into the pools [39, 56, 60], effectively increase the dCTP: dTTP ratio. Furthermore, FU and FdUrd are converted to analogs of dUMP and their effect and that of MTX are to inhibit thymidylate synthetase and thereby to increase the dCTP: dTTP ratio [30, 58]. By contrast, compounds such as BrdUrd and IdUrd that are converted by cells to analogs of dTTP [25, 49] effectively decrease the ratio. Moreover, TTP and BrdUTP inhibit de novo synthesis of dCTP and this inhibition also decreases the ratio (Reichard, Ashman et al., and Kaufman et al, this volume). The compound F_3dThd is a thymidine analog, but in some mouse cell lines it is toxic at concentrations too low ($\sim 10^{-7}$ M) to produce sufficient F_3dTTP to have the effect of decreasing the dCTP:dTTP ratio [45]. However, in C3H/10T1/2 cells, the much higher concentrations ($\sim 10^{-5}$ M) required to produce cytotoxicity (Table 1), may produce sufficient F_3dTTP to effectively decrease the dCTP:dTTP ratio even in the face of the inhibition of thymidylate synthetase.

It is interesting to note that the pyrimidine analogs and antimetabolites that are associated with decreases in the dCTP:dTTP ratio do not transform C3H/10T1/2 cells, whereas those compounds that are associated with increases in the ratio do transform the cells (Table 1). One exception to these associations is dCyd, which increases the dCTP:dTTP ratio in mammalian cells [51] but does not transform C3H/10T1/2 cells (Table 1). The results with dCyd suggest that increasing the dCTP:dTTP ratio is not itself sufficient to cause transformation. However, our results on transformation of C3H/10T1/2 cells (Table 1) together with published studies of perturbations of nucleotide metabolism in other mammalian cells suggest a hypothesis that oncogenic transformation induced by MTX, FU, FdUrd, 5-aza-C, and ara-C, all of which are used or have been used in cancer chemotherapy, can be selectively inhibited by treatments, such as with dThd, that inhibit dCTP synthesis and reduce either the dCTP:dTTP ratio or the dCTP-analog:dTTP ratio in C3H/10T1/2 cells.

An alternative hypothesis suggested by our results is that dThd is selectively toxic only to established, highly tumorigenic cell lines (Table 2), and the inhibition by dThd of FdUrd-induced transformation (Table 1) results only from a reversal of the inhibition of thymidylate synthetase. Moreover, by inhibiting de novo dCTP synthesis dThd potentiates the incorporation of 5-aza-CR and ara-C into nucleic acids [9, 11-13, 16]. Therefore, according to the latter hypothesis, dThd should actually potentiate the induction of morphological transformation in experiments in which C3H/10T1/2 cells are treated with dThd and the cytosine-nucleoside analogs. These critical experiments and careful measurements of dNTP pool perturbations in C3H/10T1/2 cells will enable us to test the above hypotheses. This in turn will provide further insights into the possibility of selectively modulating cytotoxicity and oncogenic transformation induced by cancer chemotherapeutic agents in C3H/10T1/2 cells.

SUMMARY

Morphological or oncogenic transformation of mouse embryo, C3H/10T1/2 Cl 8 fibroblasts was induced by methotrexate, 5-fluorouracil and 5-fluorodeoxyuridine. It is known that these compounds cause inhibition of thymidylate synthetase and, hence, depletion of deoxythymidine triphosphate (dTTP) and an increased ratio of deoxycytidine triphosphate (dCTP) to dTTP in the deoxyribonucleotide pools that are used for DNA synthesis in mammalian cells. This ratio is, in effect, increased by treating mammalian cells with arabinosyl cytosine and 5-azacytidine, which are converted into analogs of dCTP in mammalian cells and also induce oncogenic transformation of C3H/10T1/2 cells. By contrast, trifluorothymidine, 5-bromodeoxyuridine and 5-iododeoxyuridine, which are analogs of thymidine that in effect reduce the dCTP:dTTP ratio, did not induce oncogenic transformation. Moreover, thymidine was selectively lethal to tumorigenic C3H/10T1/2 cells and inhibited oncogenic transformation in cells treated with 5-fluorodeoxyuridine. These observations suggest that treatments that effectively increase the dCTP:dTTP ratio in mammalian cells facilitate oncogenic transformation of C3H/10T1/2 cells, whereas treatments that have the effect of decreasing this ratio inhibit transformation. However, dCyd did not induce oncogenic transformation of C3H/10T1/2 cells, although it has been shown to increase the dCTP:dTTP ratio in mammalian cells. Thus, increasing this ratio may not be sufficient to cause the transformation.

ACKNOWLEDGEMENTS

We gratefully acknowledge the contributions of B. N. Ames, M. S. Baker, A. Banerjee, J. S. Bertram, E. Choi, M. Fisher, P. A. Jones, S. Mondal, and U. Rapp, who collaborated with us on various aspects of the published work reviewed in this article. Supported in part by grants RD-174 and BC-441 from the American Cancer Society to A.R.P.

REFERENCES

1. P. M. Aebersold, Mutation Induction by 5-Fluorodeoxyuridine in Synchronous Chinese Hamster Cells, Cancer Res., 39:808-810 (1979).
2. A. Banerjee and W. F. Benedict, Production of Sister Chromatid Exchanges by Various Cancer Chemotherapeutic Agents, Cancer Res., 39:797-799 (1979).
3. J. C. Barrett, T. Tsutsui, and P. O. P. Ts'o, Neoplastic Transformation Induced by a Direct Perturbation of DNA, Nature, 274: 229-232 (1978).
4. W. F. Benedict, The Importance of Chromosomal Changes in the Expression of Malignancy, in: "The Molecular Biology of the Mammalian Genetic Apparatus, Part III: Relationship to Somatic Genetics, Cancer and Aging" (P. O. P. Ts'o, ed.), pp. 215-226, Elsevier, Amsterdam (1977).
5. W. F. Benedict, M. S. Baker, L. Haroun, E. Choi, and B. N. Ames, Mutagenicity of Cancer Chemotheraphy Agents in the Salmonella/Microsome Test, Cancer Res., 37:2209-2213 (1977).
6. W. F. Benedict, A. Banerjee, A. Gardner, and P. A. Jones, Induction of Morphological Transformation in Mouse C3H/10T1/2 Clone 8 Cells and and Chromosomal Damage in Hamster A (T_1) Cl-3 Cells by Cancer Chemotherapeutic Agents, Cancer Res., 37:2202-2208 (1977).
7. W. F. Benedict and P. A. Jones, Mutagenic, Clastogenic, and Oncogenic Effects of 1-β-D-Arabinofuranosylcytosine, Mutation Res., 65:1-20 (1979).
8. J. R. Bertino, Methotrexate: Molecular Pharmacology, in: "Cancer and Chemotherapy," Vol. 3 (S. T. Crooke, A. W. Prestayko, and N. Alder, eds.), pp. 311-322, Academic Press, New York (1981).

9. B. Chabner, Pyrimidine Antagonists, in: "Pharmacologic Principles of Cancer Treatment" (B. Chabner, ed.), pp. 183-212, Saunders, Philadelphia (1982).
10. B. Chabner, Methotrexate, in: "Pharmacologic Principles of Cancer Treatment" (B. Chabner, ed.), pp. 229-255, Saunders, Philadelphia (1982).
11. B. Chabner, Cytosine Arabinoside, in: "Pharmacologic Principles of Cancer Treatment" (B. Chabner, ed.), pp. 387-401, Saunders, Philadelphia (1982).
12. A. Cihak, Biological Effects of 5-Azacytidine in Eukaryotes. A Review, Oncology, 30:405-422 (1975).
13. W. A. Creasey, Arabinosylcytosine, in: "Antineoplastic and Immunosuppressive Agents," Vol. 2 (A. C. Sartorelli and D. G. Johns, eds.), pp. 232-256, Springer-Verlag, New York (1975).
14. P. V. Danenberg, Thymidylate Synthetase: A Target Enzyme for Cancer Chemotherapy, Biochem. Biophys. Acta, 473:73-85 (1977).
15. P. V. Danenberg, C. Heidelberger, M. Mulkins, and A. R. Peterson, The Incorporation of 5-Fluoro-2'-deoxyuridine into DNA of Mammalian Tumor Cells, Biochem. Biophys. Res. Commun., 102: 617-623 (1981).
16. L. L. Danhauser and Y. M. Rustum, Effects of Thymidine on the Toxicity, Antitumor Activity and Metabolism of 1-β-D-Arabinofuranosylcytosine in Rats Bearing a Chemically Induced Colon Carcinoma, Cancer Res., 40:1274-1280 (1980).
17. S. S. Epstein, E. Arnold, J. Andrea, W. Bass, and Y. Bishop, Detection of Chemical Mutagens by the Dominant Lethal Assay in the Mouse, Toxicol. and Appl. Pharmacol., 23:2139-2141 (1977).
18. R. M. Evans, J. D. Laskin, and M. T. Hakala, Assessment of Growth-Limiting Events Caused by 5-Fluorouracil in Mouse Cells and Human Cells, Cancer Res., 40:4113-4122 (1980).
19. C. S. Genther, R. S. Schoeny, J. C. Loper, and C. C. Smith, Mutagenic Studies of Folic Acid Antagonists, Antimicrob. Agents. Chemother., 12:84-92 (1977).
20. H. Gottschling and C. Heidelberger, Fluorinated Pyrimidines. XIX. Some Biological Effects of 5-Trifluoromethyl-2'-deoxyuridine on Escherichia coli and Bacteriophage T4B, J. Mol. Biol., 7:541-560 (1963).
21. P. V. Hauschka, Analysis of Nucleotide Pools in Animal Cells, Methods in Cell Biol., 7:361-462 (1973).
22. C. Heidelberger, Fluorinated Pyrimdines and Their Nucleosides, in: "Antineoplastic and Immunosuppressive Agents II" (A. C. Sartorelli and D. G. Johns, eds.), pp. 193-231, Springer-Verlag, New York (1975).
23. C. Heidelberger, Relationship between Carcinogenesis and Transformation of Cell Cultures, in: "Mechanisms of Chemical Carcinogenesis" (C. C. Harris and P. A. Cerutti, eds.), pp. 563-573, Liss, New York (1982).

24. C. Heidelberger, P. V. Danenberg, and R. G. Moran, Fluorinated Pyrimidines and Their Nucleosides, in: "Advances in Enzymology and Related Areas in Molecular Biology" (A. Meister, ed.), pp. 57-118, Wiley, New York (1983).
25. C. Heidelberger and D. H. King. Trifluorothymidine, Pharmac. Ther., 6:427-442 (1979).
26. C. Heidelberger, J. R. Landolph, R. E. K. Fournier, A. Fernandez, and A. R. Peterson, Genetic and Probability Aspects of Cell Transformation by Chemical Carcinogens, Prog. Nucleic Acid Res. and Mol. Biol., 29:87-98 (1983).
27. E. Huberman and C. Heidelberger, The Mutagenicity to Mammalian Cells of Pyrimidine Nucleoside Analogs, Mutation Res., 14: 130-132 (1972).
28. H. A. Ingraham, B. Y. Tseng, and M. Goulian, Nucleotide Levels and Incorporation of 5-Fluorouracil and Uracil into DNA of Cells Treated with 5-Fluorodeoxyuridine, Mol. Pharmacol., 21: 211-216 (1981).
29. IARC Monographs on the Evaluation of the Carcinogenic Risk of Chemicals to Humans: Some Antineoplastic and Immunosuppressive Agents, Vol. 26, pp. 115-136 and 165-202, IARC, Lyon (1981).
30. R. C. Jackson, The Regulation of Thymidylate Biosynthesis in Novikoff Hepatoma Cells and the Effects of Amethopterin, 5-Fluorodeoxyuridine and 3-Deazauridine, J. Biol. Chem., 253: 7440-7446 (1978).
31. P. A. Jones, M. S. Baker, J. S. Bertram, and W. F. Benedict, Cell Cycle-Specific Oncogenic Transformation of C3H/10T1/2 Clone 8 Mouse Embryo Cells by 1-β-D-Arabinofuranosylcytosine, Cancer Res., 37:2214-2217 (1977).
32. P. A. Jones, W. F. Benedict, M. S. Baker, S. Mondal, U. Rapp, and C. Heidelberger, Oncogenic Transformation of C3H/10T1/2 Cl 8 Mouse Embryo Cells by Halogenated Pyrimidine Nucleosides, Cancer Res., 36:101-107 (1976).
33. P. A. Jones and S. M. Taylor, Cellular Differentiation, Cytidine Analogs and DNA Methylation, Cell, 20:85-93 (1980).
34. P. A. Jones and S. M. Taylor, Hemimethylated Duplex DNAs Prepared from 5-Azacytidine Treated Cells, Nucleic Acids Res., 9: 2833-2987 (1981).
35. D. Kufe, P. Major, E. Egan, and E. Loh, 5-Fluorodeoxyuridine Incorporation in L1210 DNA, J. Biol. Chem., 256:8885-8888 (1981).
36. B. A. Kunz, Genetic Effects of Deoxyribonucleotide Pool Imbalances, Environ. Mutagen., 4:695-725 (1982).
37. J. R. Landolph and P. A. Jones, Mutagenicity of 5-Azacytidine and Related Nucleosides in C3H/10T1/2 Clone 8 and V79 Cells, Cancer Res., 42:817-823 (1982).
38. S. S. Lee, B. C. Giovanella, and J. J. Stehlin, Jr., Selective Lethal Effect of Thymidine on Human and Mouse Tumor Cells, J. Cell. Physiol., 92:401-405 (1977).

39. T. Lee, M. Karon, and R. L. Momparler, Cellular Phosphorylation of 1-β-D-Arabinofuranosyl cytosine and 5-Azacytidine with Intact Fibrosarcoma and Leukemic Cells, Cancer Res., 35:2506-2510 (1975).
40. P. Maier and W. Schmid, Ten Model Mutagens Evaluated by the Micronucleus Test, Mutation Res., 40:325-328 (1976).
41. H. Marquardt and H. Marquardt, Induction of Malignant Transformation and Mutagenesis in Cell Cultures by Cancer Chemotherapeutic Agents, Cancer, 40:1930-1934 (1977).
42. D. S. Martin, R. L. Stolfi, R. C. Sawyer, R. Nayak, S. Spiegelman, C. W. Young, and T. Woodcock, An Overview of Thymidine, Cancer, 45:1117-1128 (1980).
43. M. Meuth, N. L'Hereux-Huard, and M. Trudel, Characterization of a Mutator Gene in Chinese Hamster Ovary Cells, Proc. Natl. Acad. Sci., U.S.A., 76:650-659 (1979).
44. M. A. Mulkins and C. Heidelberger, Isolation of Fluoropyrimidine-Resistant Murine Leukemic Cell Lines by One-Step Mutation and Selection, Cancer Res., 42:956-964 (1982).
45. J. R. Parkhurst, P. V. Danenberg, and C. Heidelberger, Growth Inhibition of Cells and of Vaccinia Virus Infected HeLa Cells by Derivatives of Trifluorothymidine, Chemotherapy, 22:221-231 (1976).
46. A. R. Peterson, M. Fisher, and C. Heidelberger, Association between the Cytotoxicity of Thymidine and Tumorigenicity of Clones Derived from C3H/10T1/2 Mouse Embryo Fibroblasts, Biochem. Biophys. Res. Commun., 95:182-186 (1980).
47. A. R. Peterson, H. Peterson, and P. V. Dananberg, Induction of Mutations by 5-Fluorodeoxyuridine: A Mechanism of Self-Potentiated Drug Resistance? Biochem. Biophys. Res. Commun., 110:573-577 (1983).
48. P. Propping, G. Rohrbon, and W. Buselmaier, Comparative Investigations on the Chemical Induction of Point Mutations and Dominant Lethal Mutations in Mice, Mol. Gen. Genet., 117:197-209 (1972).
49. W. H. Prusoff and B. Goz, Halogenated Pyrimidine Deoxyribonucleosides, in: "Antineoplastic and Immunosuppressive Agents II" (A. C. Sartorelli and D. G. Johns, eds.), pp. 272-347, Springer-Verlag, New York (1975).
50. U. R. Rapp, R. C. Nowinski, C. A. Reznikoff, and C. Heidelberger, Endogenous Oncornaviruses in Chemically Induced Transformation. 1. Transformation Independent of Virus Production, Virology, 65:392-409 (1975).
51. P. Reichard, From Deoxynucleotides to DNA Synthesis, Fed. Proc., 37:9-15 (1978).
52. C. A. Reznikoff, J. S. Bertram, D. W. Brankow, and C. Heidelberger, Quantitative and Qualitative Studies of Chemical Transformation of Cloned C3H Mouse Embryo Cells Sensitive to Post-Confluence Inhibition of Cell Division, Cancer Res., 33:3239-3249 (1973).

53. C. A. Reznikoff, D. W. Brankow, and C. Heidelberger, Establishment and Characterization of a Cloned Line of C3H Mouse Embryo Cells Sensitive to Post-Confluence Inhibition of Cell Division, Cancer Res., 33:3231-3238 (1973).
54. T. J. Ryan, M. M. Boddington, and A. I. Spriggs, Chromosomal Abnormalities Produced by Folic Acid Antagonists, Br. J. Dermatol., 77:541-555 (1965).
55. J. Skoda, Azapyrimidine Nucleosides, in: "Antineoplastic and Immunosuppressive Agents II" (A. C. Sartorelli and D. G. Johns, eds.), pp. 348-372, Springer-Verlag, New York (1975).
56. L. Skoog and B. Nordenskjöld, Effects of Hydroxyurea and 1-β-D-Arabinofuranosyl cytosine on Deoxyribonucleotide Pools in Mouse Embryo Cells, Eur. J. Biochem., 19:81-89 (1979).
57. V. F. Simmon, In vitro Assays for Recombinogenic Activity of Chemical Carcinogens and Related Compounds with Saccharomyces cerevisiae D3, J. Natl. Cancer Inst., 62:901-909 (1979).
58. M. H. Tattersall, R. C. Jackson, S. T. M. Jackson, and K. R. Harrap, Factors Determining Cell Sensitivity to Methotrexate: Studies of Folate and Deoxyribonucleoside Triphosphate Pools in Five Mammalian Cell Lines, Eur. J. Cancer, 10:819-826 (1974).
59. J. H. Taylor, W. F. Haut, and J. Tung, Effects of Fluorodeoxyuridine on DNA Replication, and Chromosome Breakage and Reunion, Proc. Natl. Acad. Sci., U.S.A., 48:190-198 (1962).
60. J. Vesely and A. Cihak, Incorporation of a Potent Antileukemic Agent, 5-Aza-2'-deoxycytidine, into DNA of Cells from Leukemic Mice, Cancer Res., 37:3684-3689 (1977).
61. I. B. Weinstein, Current Concepts and Controversies in Chemical Carcinogenesis, J. Supramol. Struct. Cell. Biochem., 17:99-120 (1981).

REGULATION OF EXTRACELLULAR THYMIDINE POOLS IN HUMAN BLOOD SAMPLES

Ronald W. Pero,* Anders Olsson,*
and Desmond Johnson†

*Wallenberg Laboratory
Biochemical and Genetic Ecotoxoicology
University of Lund
Box 7031, S-220 07 Lund, Sweden, and

†Biochemistry Department
University College
Galway, Ireland

Human blood samples are the basis for nearly all the techniques available that assess genotoxic risk in man [1-3]. Many of these procedures are influenced by exogenously supplied thymidine because of their dependence on either scheduled or unscheduled DNA synthesis. It is well known that there are significant levels of thymidine in human sera [4]. In addition, infusions of large doses of thymidine into the circulation are used chemotherapeutically as a "rescue agent" to protect normal tissues from the toxicity of methotrexate, and as a potentiator of the anti-cancer agent, cyclophosphamide [5]. Hence, we have viewed it as important to evaluate if there is any catabolic activity of thymidine in whole blood that may, in turn, affect either genotoxic-test procedures or chemotherapeutic protocols in man.

Assay systems involving _in vitro_ culturing of blood cells where ^{3}H-thymidine pulses are used to quantify DNA synthesis, are particularly sensitive to exogenous thymidine pools. These include estimations of mitogenic responsiveness and unscheduled DNA synthesis (UDS) (i.e., DNA repair synthesis). Moreover, sister chromatid exchanges [6], chromosome aberrations [7, 8] and mutational frequencies [7] have also been shown to increase in proportion to increases in extracellular thymidine.

Previously we have shown that the platelets present in peripheral blood samples inhibit UDS [9]. Since platelets are known

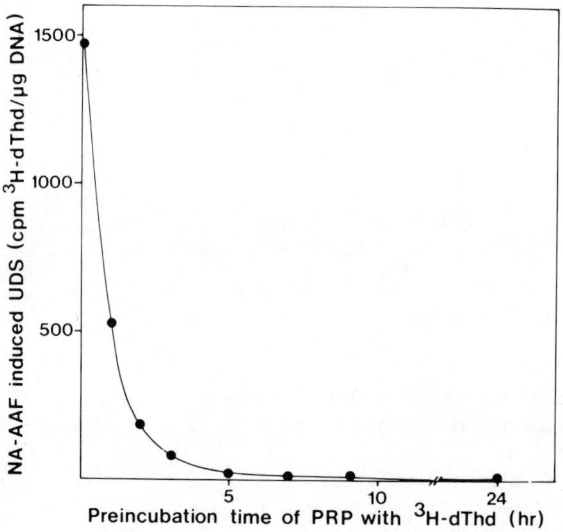

Fig. 1. The effect of preincubating ^{3}H-thymidine with platelet rich plasma (PRP) on the unscheduled incorporation of thymidine into DNA after a standardized dose of 10 μM N-acetoxy-2-actylaminofluorene (NA-AAF).

to have thymidine phosphorylase [10], we have examined if there were any effects of thymidine catabolism on the determination of UDS in mononuclear leukocytes.

For this purpose, we have isolated platelet rich plasma (PRP) as already described [9] and then 1.5 ml aliquots of PRP were incubated at 37°C, with 3.6 μM ^{3}H-thymidine for the various time periods indicated in Fig. 1. After preincubation the ^{3}H-thymidine-treated PRPs were then used to estimate UDS induced by a 1 h exposure to a standardized dose of N-acetoxy-2-acetylaminofluorene (NA-AAF). Figure 1 clearly demonstrates that ^{3}H-thymidine treated PRP preincubated for 5 h or more does not result in any unscheduled incorporation of ^{3}H-thymidine into DNA.

We next compared the 1 h preincubation of ^{3}H-thymidine-treated PRP with the 5 h preincubation point by ion exchange column chromatography on Dowex AG 50 W × 4. Eluation of this resin with 0.1 M ammonium formate buffer at pH = 3.2 has been shown to effectively separate thymidine from thymine [11, 12]. The data are presented in Fig. 2. After 1 h of preincubation of ^{3}H-thymidine with PRP, there was a detectable degradation of the ^{3}H-thymidine to ^{3}H-thymine but this was only sufficient to reduce the estimation of UDS by about 50% (see Fig. 1). However, after a 5 h preincubation of PRP

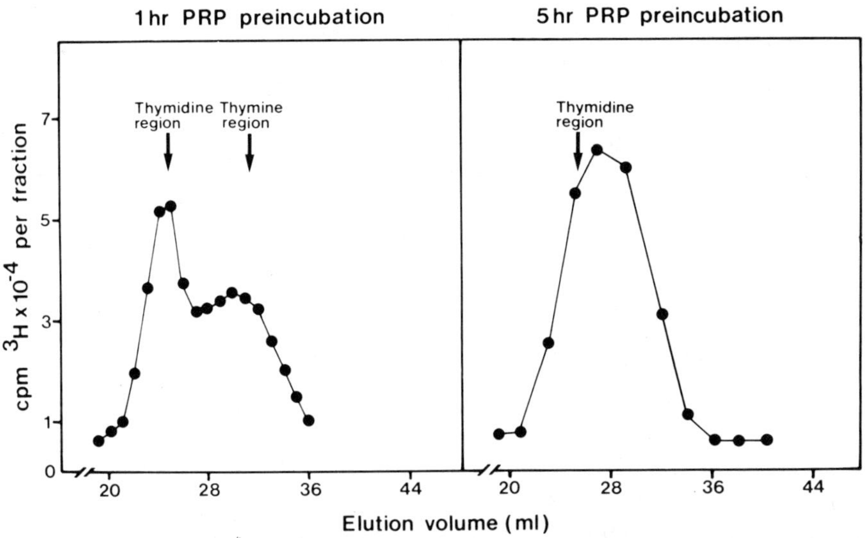

Fig. 2. Ion exchange column chromatography of the 1 h and 5 h ^{3}H-thymidine preincubations with platelet rich plasma (PRP) that were used in Fig. 1. Dowex AG 50 W × 4, 400 mesh, 1 × 24 cm column, flowrate at 53 ml/h, eluation in 0.1 M ammonium formate buffer pH = 3.2.

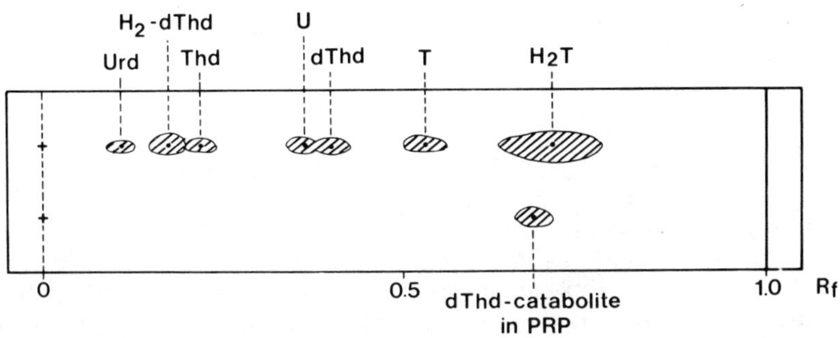

Fig. 3. Identification of the thymidine catabolite produced in platelet rich plasma (PRP) by thin layer chromatography on DEAE MN-300 plates as dihydrothymine. The solvent system was the upper phase of ethylacetate:water:formic acid, 60:35:5. Uridine (Urd), dihydrothymidine (H_2-dThd), ribothymine (Thd), uracil (U), thymidine (dThd), thymine (T), and dihydrothymine (H_2T).

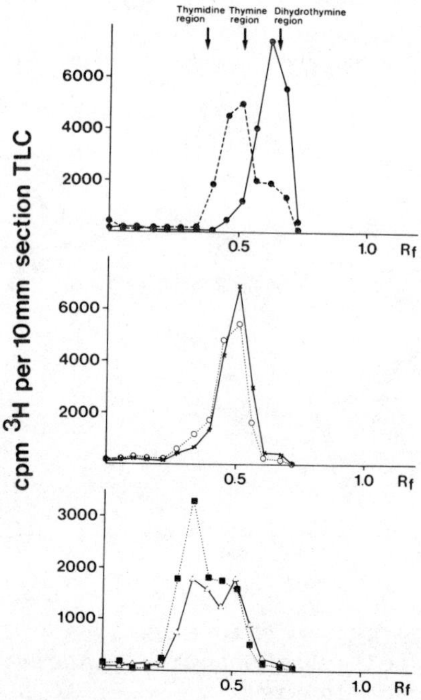

Fig. 4. Thymidine catabolic capacity of platelet rich plasma (PRP). Two ml aliquots of PRP were incubated for 5 h at 37°C with 4 μM (●—●), 22 μM (●---●), 107 μM (×—×), 202 μM (○···○), 500 μM (Δ—Δ), and 1000 μM (■···■) ^{3}H-thymidine. Thymidine breakdown was assessed by TLC on DEAE MN-300 plates as in Fig. 3. Reprinted with permission, Pero et al., 1984.

with ^{3}H-thymidine practically no UDS could be measured even though there was only one peak on the ion exchange column that co-chromatographed with undegraded ^{3}H-thymidine. We have concluded from these data that platelets degrade thymidine to thymine, and then further breakdown occurs to a product that co-chromatographs with thymidine on the ion exchange column but does not incorporate into DNA.

Final identification of the thymidine catabolite produced by platelets has been accomplished by using thin layer chromatography (TLC) on cellulose DEAE MN-300 in a solvent system consisting of the upper phase of ethylacetate:water:formic acid, 60:35:5 [13]. The chromatograms were divided into 1 cm sections and the radioactivity in each section was determined. The tracing of a typical TLC is shown in Fig. 3. The thymidine catabolite has been identified as dihydrothymine by virtue of its co-chromatography (R_f = 0.701). Moreover, we have biochemically verified the degradative pathway of thymidine in platelets as follows (data to be published):

$$\text{Thymidine} \underset{\text{Pi}}{\overset{\text{Thymidine Phosphorylase}}{\rightleftharpoons}} \text{Thymine} \underset{\text{NADPH}}{\overset{\text{Dihydrothymine Dehydrogenase}}{\rightleftharpoons}} \text{Dihydrothymine}$$

Our next concern was to evaluate the influence thymidine degradation by platelets might have on the regulation of extracellular thymidine pools in whole blood or in platelet-contaminated blood cell cultures. Hence, we have studied the thymidine catabolic capacity of PRP. Two milliliter aliquots of PRP were incubated for 5 h at 37°C with concentrations of ^{3}H-thymidine ranging from 4 μM to 1000 μM. After incubation about 1-5 μl portions of the various ^{3}H-thymidine-PRP incubations were subjected to TLC on DEAE MN-300 plates as already described above. The data are shown in Fig. 4. PRP completely converted 4 μM ^{3}H-thymidine to dihydrothymine, and up to 202 μM ^{3}H-thymidine treated PRP degraded all the thymidine to thymine. Even when PRP was incubated with 1000 μM ^{3}H-thymidine a substantial fraction was catabolized to thymine. It was concluded that platelets have a tremendous capacity to regulate extracellular thymidine pools, and thus, they have a potential to influence the genetic consequences of assay procedures dependent on blood samples.

Our data deserve special consideration when it is realized that nearly all cytogenetic, UDS and mitogenic procedures do not take precautions to remove platelets from the final blood cell preparations [1-3].

ACKNOWLEDGMENTS

This study was supported by the Swedish Council for Planning and Coordination of Research in "Chemical Health Risks in our Environment," by the Swedish Workers' Protection Fund and by the Swedish National Association Against Heart and Lung Diseases, and by a special grant from the Health Sciences Centre in Dalby, Sweden.

REFERENCES

1. A. D. Bloom. ed., Guides for studies of human populations exposed to mutagenic and reproductive hazards. March of Dimes Birth Defects Foundation, White Plains, New York, p. 163 (1981).
2. B. A. Bridges, B, E. Butterworth, and I. B. Weinstein, eds., Indicators of genotoxic risk, Banbury Report 13, Cold Spring Harbor Laboratory, p. 580 (1982).
3. T. C. Hsu, ed., Cytogenetic assays of environmental mutagens, Allanheld, Osmun, and Co., Totowa, New Jersey, p. 430 (1982).
4. L. Holden, A. V. Hoffbrand, and M. H. N. Tattersal, Thymidine concentrations in human sera: Variations in patients with leukemia and megaloblastic anemia, Europ. J. Cancer, 16:115-21 (1980).

5. D. S. Martin, R. L. Stolfi, R. C. Sawyer, R. Nayak, S. Spiegelman, C. W. Young, and T. Woodcock, Cancer, 45:1117-1128 L980).
6. P. E. Perry, Induction of sister-chromatid exchanges (SCEs) by thymidine and the potentiation of mutagen-induced SCEs in Chinese hamster ovary cells, Mutation Res., 109:219-229 (1983).
7. D. Anderson, C. R. Richardson, and P. J. Davies, The genotoxic potential of bases and nucleosides, Mutation Res., 91:265-272 (1981).
8. S. J. Yang, G. M. Hahn, and M. A. Bagshow, Chromosome aberrations induced by thymidine, Exp. Cell Res., 42:130-135 (1966).
9. R. W. Pero and C. Vopat, A human platelet-derived inhibitor of unscheduled DNA synthesis in resting lymphocytes, Carcinogenesis 2:1103-1110 (1981).
10. C. Desgranges, G. Razoka, M. Rabaud, and H. Bricaud, Catabolism of thymidine in human blood platelets. Purification and properties of thymidine phosphorylase, Biochimica Biophysica Acta, 654:211-218 (1981).
11. G. Nilsson, Thymidine and uridine metabolism oat cell growth inhibition of HeLa cells by human liver extract, Expt. Cell Res., 59:207-216 (1970).
12. C. H. Heldin, Å. Wasteson, and B. Westermark, Partial purification and characterization of platelet factors stimulating the multiplication of normal human glial cells, Expt. Cell Res., 109:429-437 (1977).
13. N. C. Bals, B. W. Bawen, K. G. Khar, and S. A. Baliska, Thin layer chromatography of thymidine and thymidine metabolites in enzyme assays and cell extracts, Anal. Biochem., 106:230-237 (1980).

PURINE NUCLEOTIDE IMBALANCE IN IMMUNODEFICIENCY DISORDERS

Arthur J. Ammann
Dir. Pediatric Immunology/Rheumatology and
Pediatric Clinical Research Center
University of California
San Franscisco, California

1. HISTORICAL PERSPECTIVE

In 1972, Dr Hilaire Meuwissen of Albany, New York, sent some blood samples to Eloise Giblett of the Puget Sound Blood Center, Seattle, Washington, for analysis of genetic markers. One of the blood samples was from a patient with severe combined immunodeficiency disease and the other samples were from the patient's parents. Dr. Meuwissen was prepared to perform a bone marrow transplant in the patient and in preparation for the eventual necessity of documenting cell chimerism, was attempting to obtain as many genetic markers as possible to prove establishment of a bone marrow graft. Much to Dr. Giblett's surprise, one of the enzymes, adenosine deaminase (ADA) (EC 3.5.4.4) was absent while the parents had approximately one-half normal activity. Several weeks later, Dr. Flossie Cohen of Detroit sent a blood specimen to Dr. Giblett from an additional patient with severe combined immunodeficiency disease. This patient also lacked ADA activity, while the parents had levels lower than normal. In 1973, a symposium on ADA deficiency and severe combined immunodeficiency disease was held in Albany, New York. After a careful review of laboratory and biochemical data, it was determined that 12 of 22 patients with SCID who had been tested for ADA activity in North American and Western Europe were ADA-deficient. Also in 1973, Jenkins reported finding a !Kung aborigine boy with absent red cell ADA activity but with normal function. Subsequently, it was determined that the patient's lymphocytes and fibroblasts had 10-30% of normal ADA activity. Recent reviews summarize these and additional cases [24].

Early in 1975, Dr. Louis Diamond evaluated a patient who had been referred to him with a diagnosis of Diamond-Blackfan syndrome.

Based on the clinical history and the initial laboratory evaluation, Dr. Diamond felt that the patient had an immunodeficiency disorder rather than the Diamond-Blackfan syndrome. Accordingly, he referred the patient to us for more extensive immunologic evaluation. There were several unusual aspects of the patient's clinical course and subsequent immunologic evaluation. The patient, who was five years old at the time of our first immunologic evaluation, had been free of significant infection during the first four years of life. Laboratory studies showed that the patient had severe lymphopenia, elevated immunoglobulins, a normal antibody response following immunization, and severely deficient numbers of T cells and lymphocyte response and alloantigen. We had not previously encountered a patient with such a degree of T cell immunodeficiency who was capable of forming normal amounts of antibody. The pattern of immunodeficiency was certainly not like that observed in SCID with ADA deficiency. Once again, a sample of blood was sent to Dr. Giblett, who on this occasion found a complete absence of the enzyme purine nucleoside phosphorylase (PNP) (EC 2.4.2.1). Analysis of blood samples on the parents demonstrated approximately 50% of normal activity [2, 24].

Soon after the first report of PNP deficiency, a family with three PNP-deficient siblings was reported by Stoop [39]. An additional two patients in a single family were described by Drs. Douglas Biggar and Erwin Gelfand of Toronto. Unlike the previous patients, the two brothers had some evidence of PNP activity (<1%) and a more moderate immunodeficiency [4].

The excitement of the discovery of deficiencies of two enzymes in the purine metabolic pathway, both associated with immunodeficiency, brought together researchers in many different specialties. A close cooperative effort between investigators in immunology, biochemistry, genetics and hematology resulted in significant contributions to the understanding of the pathogenesis of immundeficiency disorders and the importance of the purine metabolic pathway in the regulation of normal T and B cell immunity. The only discouraging note to be sounded was the observation that fewer than 20% of patients with combined B and T cell immunodeficiency had an associated enzyme deficiency. Thus, the etiology of immunodeficiency in the remaining 80% is yet to be discovered.

2. ADENOSINE DEAMINASE DEFICIENCY

2.1. Clinical Features

ADA deficiency is inherited in an autosomal recessive manner. Eighty to ninety percent of patients have the clinical and laboratory features of SCID and as such, become ill within the first several months of life [24]. Initial abnormalities may include multiple, severe, recurrent, upper respiratory tract infections, candida infection of the skin and mucus membranes, acute and chronic diarrhea, and general failure to thrive. Infectious agents include fungi, protozoa, viruses, and bacteria. Without treatment, most patients die within the first year of life. Organ systems other than the immune system may be abnormal and include the skeletal system, the kidneys and the central nervous system. The skeletal abnormalities include concavity and flaring of the anterior ribs, abnormal bony pelvis, abnormal contour and articulation of posterior ribs in the transverse processes, platyspondyly and growth arrest lines. The radiologic abnormalities are felt not to be specific for ADA deficiency, as they have been observed in other chronic illnesses and metabolic disorders. The cause of the neurologic abnormalities has not been ascertained.

Approximately 10-15% of patients with ADA deficiency have a delayed presentation, sometimes beyond one year of age [24]. As the patients become older, however, there is an increasing susceptibility to infections, and a clinical picture evolves which closely resembles the patients with early presentation. Recently, we evaluated a five-year old child with an insignificant history of susceptibility to infection but with chronic lung disease which was attributed initially to asthma. The presence of severe lymphopenia suggested the possibility of immunodeficiency. Subsequent evaluation demonstrated absent ADA activity in red blood cells with approximately 2% activity in lymphocytes [3]. The !Kung child described by Jenkins and an additional infant described by Hirschhorn with absent red cell ADA activity and 15-30% ADA activity in lymphocytes have remained clinically well for many years following their initial description [27].

2.2. Immunologic Abnormalities

The degree of immunodeficiency in ADA deficiency closely parallels the severity of the clinical presentation, subsequent course, ADA levels in lymphocytes and toxic metabolites in red blood cells and urine (discussed under biochemical abnormalities). Patients with early onset symptomatology have severe degrees of T cell immunodeficiency associated with lymphopenia, absent delayed hypersensitivity skin tests, and depressed to absent lymphocyte responeses to antigen, mitogens, and alloantigen. Infants less than four months of age may have levels of IgG which are normal for age but which reflect maternal transfer of immunoglobulin. IgM and IgA are usually low to absent in spite of a history significant infection. In older infants hypogammaglobulinema is usually severe. Infants who present beyond one year of age may have less severe degrees of T and B cell immunodeficiency with varying levels of immunoglobulins, occasional positive antibody responses following immunization and varying degrees of lymphocyte responsiveness to antigen, mitogen, and alloantigens. Patients nevertheless succumb at an early age to over-

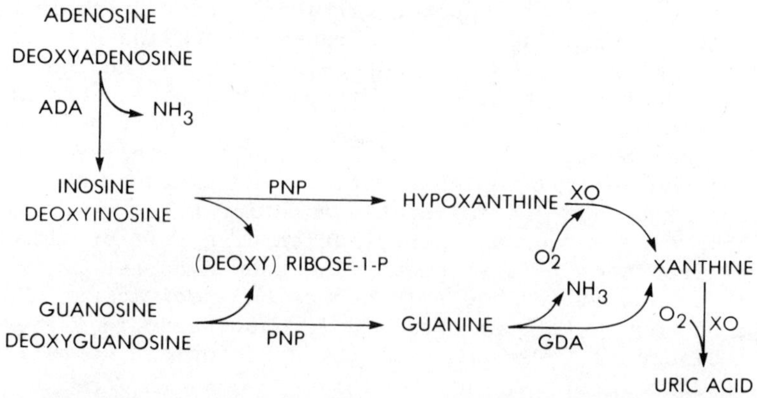

Fig. 1. Purine catabolism. The substrates for ADA are adenosine and deoxyadenosine while the substrates for PNP are inosine, guanosine, and deoxyguanosine.

whleming infection unless adequate treatment is provided. Our recently evaluated patient, with 2% lymphocyte ADA activity, had normal immunoglobulin levels with depressed antibody responses to protein and polysaccharide antigens but severely depressed T cell numbers and lymphocyte response to mitogens, antigens, and alloantigens [3]. In contrast, patients with 15-30% lymphocyte ADA activity have entirely normal immunologic function when followed for several years after initial diagnosis [5].

2.3. Biochemical Abnormalities

Adenosine deaminase catalyzes the deamination of adenosine and deoxyadenosine to inosine and deoxyinosine in the purine metabolic pathway (Fig. 1). The gene for ADA is located on chromosome 20 in the human with a single locus and two alleles [13]. Family studies of patients with ADA deficiency have shown an anomalous inheritence of ADA allelic types, suggesting the presence of a null or silent gene in addition to the normal alleles. Some patients have been found to have residual enzyme activity in nonlymphoid tissue [25]. In some patients this has been due to a structurally abnormal enzyme, while in others this is a result of small amounts of normal enzyme. A high and low molecular weight form of ADA exists. Erythrocytes and lymphocytes contain the low molecular weight ADA (32,000 daltons) while other tissue contains high molecular weight forms (100,000-300,000 daltons) [24].

In screening for ADA deficiency, most patients with absent red blood cell ADA activity have had SCID. Several patients have been described, however, who have absent red cell ADA activity, but normal immunologic function. Detailed biochemical analysis of cells

TABLE 1

	Red Cell ADA Activity	White Cell ADA Activity	Red Cell dATP Level*	Immunologic Status
Normals	100%	100%	<4	Normal
Borkowsky (5)	2-3%	15-30%	<4	Normal
Ammann (3)	0%	2%	150	Absent T cell immunity; B cell immunity mildly impaired
Classic (26)	0%	<2%	>200	Absent T cell immunity; absent B cell immunity.

* nmol/ml packed reb blood cells

and tissue from these patients indicate that while red blood cell ADA activity is absent, white blood cell and fibroblast activity ranges from 10-30% of normal. Detailed biochemical analysis of nucleosides and nucleotides in the various ADA-deficient patients have been interesting and suggest a direct relationship between the degree of enzyme deficiency, the elevation of deoxynucleotide in red blood cells, and the excretion of nucleosides in the urine (Table 1).

Patients with normal red blood cell ADA activity have <0.1 dATP pmol/10^6 cells in red cells. They excrete <0.1 mg/24 h deoxyadenosine in the urine. Two patients with 0-3% red blood cell ADA activity and 10-30% white blood cell ADA activity have been described. No increased dATP is detected in red blood cells and the excretion of deoxyadenosine in the urine is within normal range. Our recently diagnosed patient with absent red blood cell ADA activity and 2% activity in while blood cells has a dATP concentration of 150 nmol/ml packed RBC and a 24 h urine excretion of deoxyadenosine of 54 nmol/mg Cr. Patients with complete ADA deficiency in both red cells and white cells have dATP levels of >200 nmol/ml packed RBC with 24 h urine excretion of deoxyadenosine of >200 nmol/mg Cr. The severity of immunodeficiency correlates with the degree of dATP elevation and amount of deoxyadenosine excretion.

Elevations of plasma adenosine and deoxyadenosine have been found in a number of ADA-deficient patients. However, many patients do not have detectable levels of nucleosides in plasma and it would therefore appear that dATP levels in red blood cells and (d)nucleoside excretion in urine are more reliable indicators of significant ADA deficiency.

A diagnosis of ADA deficiency can be made in utero. Fibroblast cultures were obtained from the amniotic fluid of a fetus who was subsequently shown to have ADA deficiency at birth [25]. Biochemical studies demonstrated elevated plasma deoxyadenosine and adenosine levels and elevated red blood cell dATP at birth, although the levels were somewhat lower than those observed in older ADA-deficient patients. The patient was also found to have significant immunodeficiency at birth.

3. PURINE NUCLEOSIDE PHOSPHORYLASE DEFICIENCY

3.1. Clinical Features

In 1975, we described a patient with immunodeficiency and purine nucleoside phosphorylase (PNP) (E.C. 2.4.2.1) deficiency [18]. The patient was first diagnosed at five years of age and did not have a history of significant infection prior to three years of age. Initially, she had difficulty with anemia which was diagnosed as aplastic anemia and treated with multiple blood transfusions. The fact that the patient survived normal childhood immunizations with live viruses and multiple unirradiated blood transfusions indicated that the immune system must have been relatively intact until three to four years of age. We became suspicious that we were dealing with a new form of immunodeficiency when the results of initial immunologic evaluation were obtained. The patient had elevated immunoglobulin levels with a normal antibody-forming capability following immunization with nonreplicating antigens. However, T cell immunity was severely depressed with lymphopenia, reduced numbers of T cells and absent lymphocyte response to mitogen and alloantigen. The delayed onset of immunodeficiency and the presence of normal B cell immunity with severely depressed T cell immunity was felt to be unique. Screening of the patient for abnormalities of enzymes in the purine metabolic pathway resulted in the discovery of PNP deficiency. The parents had approximately 50% of normal activity. Since the original description of PNP deficiency, over 12 additional cases have been described [46, 20, 32, 36, 39, 43].

The clinical phenotype of PNP-deficient patients is more consistent than that of ADA-deficient patients [1]. The earliest onset of clinical symptomatology has been one year of age, with most patients developing clinical difficulties by five to six years of age. Recurrent infections with a variety of viral, bacterial, and fungal agents occurs. Progressive vaccinia, chronic cytomegalovirus and overwhelming varicella infection have been described. Patients who receive unirradiated blood products at the time of immunodeficiency have succumbed to fatal graft versus host reactions. Central nervous system abnormalities have been described in several patients consisting of ataxia, tremor, and spasticity. The etiology of these abnormalities has not been uncovered. Patients also appear susceptible to the development of lymphosarcoma.

A unique clinical feature of PNP-deficient patients is an apparent susceptibility to autoimmune disease. At least three patients have been described with severe hemolytic anemia and one patient has a systemic lupus erythematosus-like disorder [4]. Two long-term survivors, ages 10 and 12 years, have been described. They continue to have recurrent infection and autoimmune disease.

3.2. Immunologic Abnormalities

As with clinical features, the immunologic abnormalities found in PNP-deficient patients exhibit a more consistent pattern. All patients have had severe lymphopenia, significant reduction in T cell numbers, and absent lymphocyte response to antigen, mitogen, and alloantigen [1]. Immunoglobulin levels have been normal to elevated and patients have formed antibodies following immunization. Evidence of abnormal antibody production has also been found. Patients may ahve a Coombs' positive hemolytic anemia and positive antinuclear antibody. Autobodies are considered to indicate a form of immunologic dysregulation. One patient developed typical allergic reactions associated with overproduction of IgE following attempted treatment with bovine thymosin fraction V [2]. Another patient developed a monoclonal gammopathy.

A particularly interesting patient with PNP deficiency was described by Stoop et al. [39]. Two previous siblings had died with immunodeficiency and PNP deficiency. The patient was diagnosed as having PNP deficiency at birth. Immunologic studies performed at this time were entirely normal as was the total lymphocyte count. By six months of age, the lymphocyte count had decreased significantly and studies of T cell immunity were only 50% of normal. By one year of age, severe lymphopenia was present and functional studies were 10% of normal.

3.3. Biochemical Abnormalities

PNP deficiency is inherited in an autosomal recessive manner. Obligate carriers have 50% of normal PNP activity. Patients have severely reduced (<.01%) activity. Two patients with delayed onset and prolonged survival have 0.5% PNP activity. Unlike ADA deficiency, no patients have been described with absent red blood cell PNP activity and partially reduced white cell PNP activity. There is genetic heterogeneity of the mutant PNP with at least three different electrophoretic types described [32].

In the absence of PNP, (d)nucleosides accumulate in the plasma and urine of PNP-deficient patients [10]. Deoxyguanosine is the major deoxynucleoside phosphorylated and accumulates as dGTP in red blood cells. Inosine, guanosine, and guanine accumulate in the plasma and urine. Nucleosides in the plasma range in concentration from 3-70 μM. Nucleosides are excreted in the urine in concentrations ranging from 100-700 mg/24 h. Because patients cannot metabolize accumulated nucleosides to hypoxanthine and guanine, serum uric acid and 24 h urine uric acid excretion is low.

4. NUCLEOTIDASE DEFICIENCY

Several investigators have reported decreased 5'-nucleotidase (5'-NT) (EC 3.1.3.5) activity in patients with hypogammaglobulinemia [15, 44]. 5'-NT is responsible for the dephosphorylation of 5'-mononucleotides (AMP, IMP, GMP) to nucleosides and has been implicated as an important determinant in the toxicity of nucleotides to individual cells. Activity of 5'-NT has been localized to the plasma membrane although recent data indicate that a soluble form exists [9]. Enzymatic activity of 5'-NT varies considerably in subpopulations of immunologically competent cells. B cells have been shown to have the highest activity, mature T cells intermediate, and immature T cells the least activity [8]. Because of the significant differences in 5'-NT in T versus B cells, it has been posulated that the decreased enzyme activity in hypogammaglobulinemia is a result of a diminished number of B cells. Two observations argue against this possibility. First, additional patients have been described with 5'-NT deficiency who had normal percentages of T cells and B cells [17]. These include individuals with acquired hypogammaglobulinemia and selective IgA deficiency. Second, when T and B cells were isolated from patients with decreased 5'-NT activity, it was demonstrated that purified T cells had diminished 5'-NT activity [16].

A number of arguments can be presented against a primary role for 5'-NT in immunodeficiency. Female carriers of X-linked hypogammaglobulinemia have normal 5'-NT. Not all patients with similar phenotypes have decreased 5'-NT activity. Although levels of 5'-NT are reduced in some immunodeficiency disorders, the activity remains at a level which should not significantly interfere with intracellular dephosphorylation and, in fact, no abnormalities of purine metabolism have been detected in patients [17]. Finally, genetically dissimilar diseases (autosomal recessive and X-linked) have been shown to be associated with decreased 5'-NT and therefore a primary role for this enzyme as a cause of immunodeficiency is unlikely.

5'-NT appears to be a marker for cell differentiation. Mature T cells have greater activity than immature T cells, while in the peripheral blood B cells have even greater activity. These observations may offer an explanation for the diminished 5'-NT observed in certain immunodeficiency diseases. Some of these disease are associated with a failure of maturation of immunocompetent cells and thus immature T cells and B cells; hence, cells with decreased 5'-NT may be found in the peripheral blood.

A decrease in 5'-NT in cultured B cells of carriers of X-linked hypogammaglobulinemia has been reported [40]. This abnormality may again reflect the presence of an increased number of immature B cells rather than a primary abnormality. Decreased 5'-NT activity has also been described in acute and chronic lymphatic leukemia and acute infectious mononucleosis [34]. The role of 5'-NT in determining the cellular toxicity of deoxynucleotides is discussed in Section 5.

5. ETIOLOGY OF THE IMMUNODEFICIENCY ASSOCIATED WITH ADENOSINE DEAMINASE AND PURINE NUCLEOSIDE PHOSPHORYLASE DEFICIENCY

Several theories have been proposed to explain the selective toxicity of ADA and PNP deficiency to cells of the immune system. Utilizing cell cultures, a toxic effect of adenosine was described. This toxic effect could be reversed by the addition of uridine, suggesting that ADA deficiency, with a subsequent accumulation of adenosine, might result in pyrimidine starvation [19, 21]. Although other studies subsequently confirmed this early observation, uridine was not capable of preventing all of the toxicity of adenosine [22]. Additional studies indicated that deoxyadenosine was a significantly more toxic compound to cultured cell lines and that uridine was incapable of reversing these effects [22].

An additional theory proposed that increased adenosine resulted in increased ATP with subsequent inhibition of PRPP synthetase resulting in inhibition of de novo purine biosynthesis. However, studies have been performed in fibroblasts from ADA-deficient patients demonstrating normal PRPP levels with normal synthetic activity. In addition, significant inhibition of PRPP synthetase would be expected to result orotic aciduria in patients. Only trace amounts of orotic acid have been described in a few patients with ADA and PNP deficiency [11, 14].

An increase in cyclic AMP (cAMP) has been demonstrated in the red cells of patients with ADA deficiency [25]. Cells are known to have adenosine receptors on their surfaces which modulate intracellular cAMP levels [31, 37]. Elevated levels of cAMP have been associated with suppression of a number of different immune responses. Several arguments can be presented against a primary role for increased cAMP in the severe immunodeficiency of ADA-deficient patients. For example, increased cAMP, under certain circumstances, may stimulate the immune system [37]. Utilizing S49 cells lines, investigators have demonstrated that the toxicity of deoxyadenosine can be effected in mutant cell lines incapable of increasing intracellular cAMP.

The theory which has received the widest attention has been that of inhibition of ribonucleotide reductase (RR) by increased amounts of deoxynucleotides [12, 41]. Both dATP and dGTP, in increased amounts, inihibit RR activity. In the absence of adenosine deaminase, deoxyadenosine is phosphorylated to dATP while deoxyguanosine is phosphorylated to dGTP. Thus, in both ADA and PNP deficiency, increased deoxynucleotides would result in inhibition of RR with subsequent decreased formation of dCTP. Support for this theory is obtained in cell culture models where increased deoxyadenosine or deoxyguanosine results in increased dATP and dGTP respectively, with RR inhibition and cell cytotoxicity [41, 42]. The adverse effects of deoxyadenosine and deoxyguanosine can be abrogated with the use of deoxycytidine. In spite of the attractiveness of this theory, successful in vivo treatment of patients with ADA or PNP deficiency utilizing deoxycytidine has not been accomplished. This may be due to the fact that deoxycytidine is capable of preventing deoxyadenosine or deoxyguanosine toxicity, but is incapable of reversing established toxicity. Most of the patients who have been treated have had severe immunodeficiency. Early treatment of patients, or treatment of patients with less sever forms of immunodeficiency may be more successful. That this is possible is suggested by the results of a recent study using intravenous deoxycytidine in a patient with incomplete immunodeficiency where the immunodeficiency was partially reversed [3].

A final theory for the mechanism of immunodeficiency in ADA and PNP deficiency relates the toxicity of both deoxyadenosine and deoxyguanosine to an inhibition of methylation. S-adenosylhomocysteine hydrolase (SAHase). S-adenosine-L-methoionine (SAM) serves as a methyl donor for the methylation of a large variety of substrates. SAM-dependent methylations are a source of adenosine, since the only metabolic fate of S-adenosyl-L-homocysteine (SAH) is its hydrolysis to adenosine and L-homocysteine by SAHase [23, 29, 30]. The action of SAHase is readily reversible. It has been shown in both cell culture models and cultures of cells from ADA-deficient patients that adenosine, deoxyadenosine and deoxyguanosine inhibit SAHase activity [23, 28]. A postulated mechanism of toxicity in ADA and PNP deficiency is therefore related to the inhibition of critical methylation reactions. It is possible that more than one mechanism of toxicity exists. For example, both RR and SAHase inhibition could result in significant cell toxicity and subsequent immunodeficiency.

The selective toxicity of ADA deficiency for T and B cells and PNP deficiency for T cells is not easily explained. A "trapping" mechanism has been proposed [7]. T cells have greater kinase activity for both deoxyadenosine and deoxyguanosine than B cells or other cells from a variety of organs. Increased amounts of dexyadenosine and deoxyguanosine would therefore be trapped within T cells, phosphorylated to deoxynucleotides and result in toxicity to those cells. An alternative theory proposes that selective toxicity is related to differences in the activity of 5'-NT. Mature T cells have increased amounts of 5'-NT activity; B cells have even greater 5'-NT.

As deoxynucleotides accumulate in cells, the larger amounts of 5'-NT in mature T and B cells might be capable of preventing deoxynucleotide toxicity [45].

6. TREATMENT OF ADENOSINE DEAMINASE AND PURINE NUCLEOSIDE PHOSPHORYLASE DEFICIENCY

Several approaches to treating enzyme-deficient patients have been utilized in an attempt to restore immunologic function and prevent the eventual death of the patient from overwhelming infection. The most successful treatment to date has been the use of histocompatible bone marrow transplanation using a sibling donor [24]. Under these ideal circumstances there is little risk of fatal graft versus host reaction. Patients have had reconstitution of enzyme activity associated with normal T and B cell immunity. Transplantation resulted in a reduction of accumulated dATP in the red cells and purine metabolites in the plasma and urine.

Histocompatible bone marrow transplant donors cannot be found for the majority of enzyme-deficient patients. Accordingly, alternative forms of therapy have been investigated. In an attempt to provide a source of enzyme replacement, the regular infusion of normal red blood cells has been utilized. Both ADA and PNP activity are high in normal red blood cells. In both ADA and PNP deficiency, monthly infusion of red blood cells has resulted in decreased dATP and dGTP levels, respectively, and improved immunologic function [33, 35, 38]. Unfortunately, the number of patients who have responded to this form of treatment has been low. An explanation for the failure of certain patients to respond has not been found. A disadvantage of this form of treatment is the high risk of subsequent transfusion reactions and/or the transmission of a fatal infectious agent.

Several biochemical approaches have been used. Uridine, given orally, did not have a significant effect on reconstitution of immunity in a patient with PNP deficiency [2]. Deoxycytidine, given intravenously, did not reverse the immunodeficiency in a patient with PNP deficiency. However, in a patient with partial enzyme defiency and severe T cell deficiency, intravenous deoxytidine caused significant enhancement of T cell function, although immunity did not approach normal levels [3]. In this same patient an alternative form of treatment was evaluated. Dipyridamole is an inhibitor of purine and pyrimidine transport across cell membranes. This compound is capable of inhibiting the uptake of deoxyadenosine into cells. Use of this compound in an ADA-deficient patient resulted in a decrease in the dATP/ATP ratio and an increase in T cell function. However, after three months of treatment, T cell studies did not return to normal values. Further treatment of this patient with more potent inhibitors of purine transport are under investigation.

7. CONCLUSIONS

Biochemical and immunologic studies in ADA and PNP deficient patients demonstrate a clear relationship between the degree of accumulation of purines and deoxynucleotides and the severity of immunodeficiency. The toxic effects of accumulated metabolites are probably mediated through the combined effects of several different mechanisms including the inhibition of ribonucleotide reductase and SAHase. Although many studies have been performed using cell models and cell cultures from affected patients, no completely satisfying theory has evolved to explain the selective toxicity of ADA deficiency for T and B cells and PNP deficiency for T cells. In vitro studies have provided an important basis for biochemical approaches to the treatment of enzyme-deficient patients. Additional studies will be required to determine if a biochemical approach can be utilized to reverse the adverse effects of ADA and PNP deficiency.

Studies in ADA and PNP deficient patients in vitro and in vivo have provided important information regarding the mechanisms of toxicity of nucleosides and nucleotides as well as providing a basis for the investigation of immunopharmacologic agents designed to selectively modulate cells of the immune systems.

REFERENCES

1. A. J. Ammann, in: "Enzyme Defects and Immune Dysfunction," Ciba Foundation Symposium 68, p. 55, Excerpta Medica, Amsterdam (1979).
2. A. J. Ammann, D. W. Wara, and T. Allen, Immunotherapy and immunopathologic studies in a patient with nucleoside phosphorylase deficiency, Clin. Immunol. and Immunopathol., 10:262-269 (1978).
3. A. J. Ammann, M. J. Cowan, D. W. Martin, and D. W. Wara, Dipyridamole and intravenous deoxycytidine therapy in a patient with adenosine deaminase deficiency, March of Dimes Symposium, in press (1983).
4. W. D. Biggar, E. R. Giblett, R. L. Ozere, and B. D. Grove, A new form of nucleoside phosphorylase deficiency in two brothers with defective T-cell function, J. Pediatr., 92:354-357 (1978).
5. W. Borkowsky, A. A. Gerson, L. Shenkman, and R. Hirschhorn, Adenosine deaminase deficiency without immunodeficiency: Clinical and metabolic studies, Pediatr. Res., 14:885-889 (1980).
6. E. Carapella-de Luca, F. Aiuti, P. Lucarelli, L. Bruni, and C. D. Baroni, C. Imperato, D. Roos, and A. Astaldi, A patient with nucleoside phosphorylase deficiency, selective T-cell deficiency, and autoimmune hemolytic anemia, J. Pediatr., 93: 1000-1003 (1978).

7. D. A. Carson, J. Kaye, and J. E. Seegmiller, Lymphospecific toxicity in adenosine deaminase deficiency and purine nucleoside phosphorylase deficiency: Possible role of nucleoside kinase(a), Proc. Natl. Acad. Sci., U.S.A., 74:5677-5681 (1977).
8. D. A. Carson, J. Kaye, and D. B. Wasson, Differences in deoxyadenosine metabolism in human and mouse lymphocytes, J. Immunol., 124:8-12 (1980).
9. D. A. Carson, J. Kaye, and D. B. Wasson, The potential importance of soluble deoxynucleotidase activity in mediating deoxyadenosine toxicity in human lymphoblasts, J. Immunol., 126: 348-352 (1981).
10. A. Cohen, D. Doyle, D. W. Martin, Jr., and A. J. Ammann, Abnormal purine metabolism and purine overproduction in a patient deficient in purine nucleoside phosphorylase, N. Engl. J. Med., 295:1449-1454 (1976).
11. A. Cohen, G. E. J. Staal, A. J. Ammann, and D. W. Martin, Jr., Orotic aciduria in two unrelated patients with inherited deficiencies of purine nucleoside phosphorylase, J. Clin. Invest., 60:491-494 (1977).
12. A. Cohen, R. Hirschhorn, S. D. Horowitz, A. Rubinstein, S. H. Polmar, R. Hong, and D. W. Martin, Jr., Deoxyadenosine triphosphate as a potentially toxic metabolite in adenosine deaminase deficiency, Proc. Natl. Acad. Sci., U.S.A., 75:472-476 (1978).
13. R. P. Creagan, J. A. Tischfield, E. A. Nichols, and F. H. Ruddle, Autosomal assignment of the gene for the form of adenosine deaminase which is deficient in patients with combined immunodeficiency syndrome, Lancet II, 1449 (1973).
14. N. L. Edwards, E. W. Gelfand, D. Biggar, and I. H. Fox, Partial deficiency of purine nucleoside phosphorylase: studies of purine and pyrimidine metabolism, J. Lab. Clin. Med., 91:736-749 (1978).
15. N. L. Edwards, D. B. Magilavy, J. T. Cassidy, and I. H. Fox, Lymphocyte Ecto-5'-nucleotidase deficiency in agammaglobulinemia, Science, 201:628-630 (1978).
16. N. L. Edwards, E. W. Gelfand, L. Burk, H.-M. Dosch, and I. H. Fox, Distribution of 5'-nucleotidase in human lymphoid tissues, Proc. Natl. Acad. Sci., U.S.A., 76:3474-3476 (1979).
17. N. L. Edwards, J. T. Cassidy, and I. H. Fox, Lymphocyte 5'-nucleotidase deficiency in hypogammaglobulinemia: Clinical characteristics, Clin. Immunol. and Immunopathol., 17:76-88 (1980).
18. E. R. Giblett, A. J. Ammann, D. W. Wara, R. Sandamn, L. K. Diamond, Nucleoside phosphorylase deficiency in a child with severely defective T-cell immunity and normal B-cell immunity, Lancet I, 1010-1013 (1975).
19. H. Green and T. S. Chan, Pyrimidine starvation induced by adenosine in fibroblasts and lymphoid cells: Role of adenosine deaminase, Science, 182:836 (1973).

20. M. Hamet, C. Griscelli, P. Cartier, J. Ballay, C. de Bruyn, and P. Hosli, A second case of inosine phosphorylase deficiency with severe T-cell abnormalities, Adv. Exp. Med. Biol., 76A: 477-480 (1977).
21. L. J. Gudas, A. Cohen, B. Ullman, and D. W. Martin, Jr., Analysis of adenosine-mediated pyrimidine starvation using cultured wild-type and mutant mouse T-lymphoma cells, Somat. Cell Genet., 4:201-219 (1978).
22. L. J. Gudas, B. Ullman, A. Cohen, and D. W. Martin, Jr., Deoxyguanosine toxicity in a mouse T lymphoma: Relationship to purine nucleoside phosphorylase-associated immune dysfunction, Cell, 14:531-538 (1978).
23. M. S. Hershfield, Proposed explanation for S-adenosylhomocysteine hydrolase deficiency in purine nucleoside phosphorylase and hypoxanthineguanine phosphoribosyltransferase-deficient patients, J. Clin. Invest., 67:696-701 (1981).
24. R. Hirschhorn and D. W. Martin, Jr., Enzyme defects in immunodeficiency diseases, Springer Sem. in Immunopathol., 1: 299-321 (1978).
25. R. Hirschhorn, F. Martiniu, and F. S. Rosen, Adenosine deaminase activity in normal tissues and tissues from a child with severe combined immunodeficiency and adenosine deaminase deficiency, Clin. Immunol. and Immunopathol., 9:287-292 (1978).
26. R. Hirschhorn, V. Roegner, A. Rubinstein, and P. Papageorgiou, Plasma deoxyadenosine, adenosine, and erythrocyte deoxyATP are elevated at birth in an adenosine deaminase-deficient child, J. Clin. Invest., 65:768-771 (1980).
27. T. Jenkins, A. R. Rabson, G. T. Nurse, A. B. Lane, and D. A. Hopkinson, Deficiency of adenosine deaminase not associated with severe combined immunodeficiency, J. Pediatr., 89:732-736 (1976).
28. J. M. Johnston and N. M. Kredich, Inhibition of methylation by adenosine in adenosine deaminase-inhibited, phytohemmagglutinin-stimulated human lymphocytes, J. Immunol., 123:97-103 (1979).
29. N. M. Kredich and M. S. Hershfield, S-adenosylhomocysteine toxicity in normal and adenosine kinase-deficient lymphoblasts of human origin, Proc. Natl. Acad. Sci., U.S.A., 76:2450-2454 (1979).
30. N. M. Kredich and D. W. Martin, Jr., Role of S-adenosylhomocysteine in adenosine mediated toxicity in cultured mouse T lymphoma cells, Cell, 123:931-938 (1977).
31. G. Marone, M. Plaut, and L. M. Lichtenstein, Characterization of a specific adenosine receptor on human lymphocytes, J. Immunol., 11:2153-2159 (1978).
32. W. R. A. Osborne, S.-H. Chen, E. R. Giblett, W. D. Biggar, A. J. Ammann, and C. Ronald Scott, Purine nucleoside phosphorylase deficiency, J. Clin. Invest., 60:741-746 (1977).
33. S. H. Polmar, R. C. Stern, A. L. Schwartz, E. M. Wetzler, P. A. Chase, and R. Hirschhorn, Enzyme replacement therapy for adeno-

sine deaminase deficiency and severe combined immunodeficiency, N. Engl. J. Med., 295:1337 (1976).

34. F. Quagliata, D. Faig, M. Conklyn, and R. Silbert, Studies on the lymphocyte 5'-nucleotidase in chronic lymphocyte leukemia, infectious mononucleosis, normal subpopulations, and phytohemagglutinin-stimulated cells, Cancer Res., 34:3197-3203 (1974).
35. K. C. Rich, E. Mejias, and I. H. Fox, Purine nucleoside phosphorylase deficiency: Improved metabolic and immunologic function with erythrocyte transfusions, Med. Intell., 303:973-976 (1980).
36. K. C. Rich, W. J. Arnold, T. Palella, and I. H. Fox, Cellular immune deficiency with autoimmune hemolytic anemia in purine nucleoside phosphorylase deficiency, Am. J. Med., 67:172-176 (1979).
37. A. L. Schwartz, R. C. Stern, and S. H. Polmar, Demonstration of an adenosine receptor on human lymphocytes in vitro and its possible role in the adenosine deaminase-deficient form of severe combined immunodeficiency, Clin. Immunol. and Immunopathol., 9:499-505 (1978).
38. G. E. J. Staal, J. W. Stoop, B. J. M. Zegers, L. H. Siegenbeek von Heukelom, M. J. M. van der Vlist, S. K. Wadman, and D. W. Martin, Erythrocyte metabolism in purine nucleoside phosphorylase deficiency after enzyme replacement therapy by infusion of erythrocytes, J. Clin. Invest., 65:103-108 (1980).
39. J. W. Stoop, B. J. M. Zegers, G. F. M. Hendrickx, L. H. Siegenbeek van Heukelom, G. E. J. Staal, P. K. de Bree, S. K. Wadman, and R. E. Ballieux, Purine nucleoside phosphorylase deficiency associated with selective cellular immunodeficiency, N. Engl. J. Med., 296:651-655 (1977).
40. L. F. Thompson, G. R. Boss, H. L. Spiegelberg, A. Bianchino, and J. E. Seegmiller, Ecto-5'-nucleotidase activity in lymphoblastoid cell lines derived from heterozygotes for congenital X-linked agammaglobulinemia, J. Immunol., 125:190-193 (1980).
41. B. Ullman, A. Cohen, and D. W. Martin, Jr., Characterization of a cell culture model for the study of adenosine deaminase- and purine nucleoside phosphorylase-deficient immunologic disease, Cell, 9:205-211 (1976).
42. B. Ullman, L. J. Gudas, A. Cohen, and D. W. Martin, Jr., Deoxyadenosine metabolism and cytotoxicity in cultured mouse T lymphoma cells: a model for immunodeficiency disease, Cell, 14:365-375 (1978).
43. A. R. Watson, D. I. K. Evans, H. B. Marsden, V. Miller, and P. A. Rogers, Purine nucleoside phosphorylase deficiency associated with a fatal lymphoproliferative disorder, Arch. Dis. Chil., 563-565 (1981).
44. A. D. B. Webster, M. North, J. Allsop, G. L. Asherson, and R. W. E. Watts, Purine metabolism in lymphocytes from patients with primary hypogammaglobulinaemia, Clin. Exp. Immunol., 31:456-463 (1978).

45. R. L. Wortmann, B. S. Mitchell, N. L. Edwards, and I. H. Fox, Biochemical basis for differential deoxyadenosine toxicity to T and B lymphoblasts: Role for 5'-nucleotidase, Proc. Natl. Acad. Sci., U.S.A., 76:2434-2437 (1979).

CONTRIBUTORS

Arthur J. Ammann, Rheumatology and Pediatric Clinical Research Center, University of California at San Francisco, San Francisco, CA. 94143.

Diana Anderson, The British Industrial Biological Research Association, Woodmasterne Road, Carshalton, Surrey, SM5 4DS, United Kingdom.

Charles R. Ashman, Center for Genetics, University of Illinois College of Medicine, 808 South Wood Street, Chicago, IL. 60612.

Dai Ayusawa, Department of Immunology and Virology, Saitma Cancer Center Research Institute, Ina-machi, Saitma-ken 362, Japan.

William F. Benedict, Division of Hematology-Oncology, Department of Medicine, Childrens Hospital of Los Angeles, and Department of Pediatrics, University of Southern California School of Medicine, Los Angles, CA. 90027.

Martin Brendel, Institut für Mikrobiologie, J. W. Goethe-Universität, Theodor-Stern-Kai 7, Haus 75, D-6000 Frankfurt/Main, Federal Republic of Germany.

Herman E. Brockman, Department of Biological Sciences, Illinois State University, Normal, IL. 61761.

Oliver E. Brown, Departments of Medicine and Microbiology-Immunology, Duke University Medical Center, Durham, NC. 27710.

Voon-Loong Chan, Department of Microbiology, Fitzgerald Building, University of Toronto, Toronto, Ontario, Canada.

James E. Cleaver, Laboratory of Radiobiology and Environmental Health, University of California at San Francisco, San Francisco, CA. 94143.

P. V. Danenberg, Department of Biochemistry, School of Medicine and the Cancer Center, University of Southern California, Los Angeles, CA. 90027.

Shishir K. Das, Department of Pathology, University of Washington School of Medicine, Seattle, Washington 98195.

Richard L. Davidson, Center for Genetics, University of Illinois, College of Medicine, 808 South Wood Street, Chicago, Il. 60612.

Frederick J. de Serres, Office of the Director, National Institute of Environmental Health Sciences, PO Box 12233, Research Triangle Park, NC. 27709.

Leo S. Dimnik, Departments of Medical Biochemistry and Pediatrics, University of Calgary, Calgary, Alberta, T2N 4N1, Canada.

John W. Drake, Laboratory of Genetics, National Institute of Environmental Health Sciences, PO Box 12233, Research Triangle Park, NC. 27709.

Margaret Fox, Paterson Laboratories, Christie Hospital and Holt Radium Institute, Manchester M20 9BX, United Kingdom.

Susan J. Friedman, Departments of Pediatrics and Pharmacology, University of Calgary, Calgary, Alberta, T2N 4N1, Canada.

Barry W. Glickman, Laboratory of Genetics, National Institute of Envirnonmental Health Sciences, PO Box 12233, Research Triangle Park, NC. 27709.

Otelinda Goncalves, Clinical Research Institute of Montreal, 110 Pine Avenue West, Montreal, Quebec, H2W 1R7, Canada.

Myron F. Goodman, Department of Biological Sciences, University of Southern California, Los Angeles, CA. 90089-1481.

Bridget J. Green, Departments of Pediatrics and Pharmacology, University of Calgary, Calgary, Alberta, T2N 4N1, Canada.

Robert H. Haynes, Department of Biology, York University, Toronto, Ontario, M3J 1P3, Canada.

Charles Heidelberger (Deceased), Kenneth Norris Jr. Cancer Research Institute, University of Southern California, Los Angeles, CA. 90031.

H. Jeok Ho, Department of Microbiology, Fitzgerald Building, University of Toronto, Toronto, Ontario, Canada.

David I. Hoar, Departments of Medical Biochemistry and Pediatrics, University of Calgary, Calgary, Alberta, T2N 4N1, Canada.

Robin Holliday, Genetics Division, National Institute for Medical Research, The Ridgeway, Mill Hill, London NW7 1AA, United Kingdom.

Randi L. Hopkins, Department of Biological Sciences, University of Southern California, Los Angeles, CA. 90089-1481.

Tada-aki Hori, Division of Genetics, National Institute of Radiological Sciences, Chiba 280, Japan.

Ching-Yuan Hung, Department of Biological Sciences, Illinois State University, Normal, IL. 61761.

L. L. V. Ibric, Department of Biochemistry, School of Medicine and the Cancer Center, University of Southern California, Los Angleles, CA. 90027.

Masafumi Inoue, Departments of Pediatrics and Pharmacology, University of Calgary, Calgary, Alberta, T2N 4N1, Canada.

Desmond Johnson, Biochemistry Department, University College, Galway, Ireland.

Elliot R. Kaufman, Center for Genetics, University of Illinois College of Medicine, 808 South Wood Street, Chicago, Il. 60612.

Hideki Koyama, Department of Biochemistry, Cancer Institute, Toshima-ku, Tokyo 170, Japan.

Thomas A. Kunkel, Laboratory of Genetics, National Institute of Environmental Health Sciences, PO Box 12233, Research Triangle Park, NC. 27709.

Bernard A. Kunz, Biology Department, York University, 4700 Keele Street, Toronto, Ontario M3J 1P3, Canada.

Roger Lasken, Department of Biological Sciences, University of Southern California, Los Angeles, CA. 90089-1481.

J. G. Little, Department of Biology, York University, 4700 Keele Street, Toronto, Ontario M3J 1P3, Canada.

Lawrence A. Loeb, Department of Pathology, University of Washington School of Medicine, Seattle, Washington 98195.

David S. Matheson, Departments of Pediatrics and Pharmacology, University of Calgary, Calgary, Alberta, T2N 4N1, Canada.

Christopher K. Mathews, Department of Biochemistry and Biophysics, Oregon State University, Corvallis, Oregon 97331.

Robert J. Melamede, Department of Microbiology, New York Medical College, Valhalla, NY. 10595.

Dhan N. Mhaskar, Department of Biological Sciences, University of Southern California, Los Angeles, CA. 90089-1481.

Kathyrn M. Milam, Laboratory of Radiobiology and Environmental Health, University of California at San Francisco, San Francisco, CA. 94143.

J. H. Miller, Battelle, Pacific Northwest Laboratory, Richland, WA. 99352.

Mark Meuth, Clinical Research Institute of Montreal, 110 Pine Avenue West, Montreal, Quebec H2W 1R7, Canada.

Chester N. Newman, Battelle, Pacific Northwest Laboratory, Richland, WA. 99352.

Binh Trong Nguyen, School of Pharmacy, University of California at San Francisco, San Francisco, CA. 94143.

Anders Olsson, Wallenberg Laboratory, Biochemical and Genetic Eco-toxicology, University of Lund, S-220 07, Lund, Sweden.

Laurie K. Overton, Department of Biological Sciences, Illinois State University, Normal, Il. 61761.

Ronald W. Pero, Wallenberg Laboratory, Biochemical and Genetic Eco-toxicology, University of Lund, S-220 07, Lund, Sweden.

Andrew R. Peterson, Kenneth Norris Jr. Cancer Research Institute, University of Southern California, Los Angeles, CA. 90033.

Hazel Peterson, Kenneth Norris Jr. Cancer Research Institute, University of Southern California, Los Angeles, CA. 90033.

Tashmeem Razzaki, Department of Genetics, University of Alberta, Edmonton, Alberta, Canada.

Peter Reichard, Medical Nobel Institute, Department of Biochemistry, Karolinska Institutet, S-104 01, Stockholm, Sweden.

Randall G. Richards, Departments of Medicine and Microbiology-Immunology, Duke University Medical Center, Durham, NC. 27710.

Wolfgang Sadée, School of Pharmacy, University of California at San Francisco, San Francisco, CA. 94143.

W. David Sedwick, Departments of Medicine and Microbiology-Immunology, Duke University Medical Center, Durham, NC. 27710.

Takeshi Seno, Department of Immunology and Virology, Saitma Cancer Center Research Institute, Ina-machi, Saitma-ken 362, Japan.

Kimiko Shimizu, Department of Immunology and Virology, Saitma Cancer Center Research Institute, Ina-machi, Saitma-ken 362, Japan.

Navin K. Sinha, Waksman Institute of Microbiology, Rutgers University, Piscataway, NJ. 08854.

Ronald D. Snyder, Stauffer Chemical Company, 400 Farmington Avenue, Farmington, CT. 06032.

Keiichi Takeishi, Department of Immunology and Virology, Saitma Cancer Center Research Institute, Ina-machi, Saitma-ken 362, Japan.

Michael D. Topal, Department of Pathology and the Cancer Research Center, University of North Carolina Medical School, Chapel Hill, NC. 27514.

Marie Trudel, Clinical Research Institute of Montreal, 110 Pine Avenue West, Montreal, Quebec H2W 1R7, Canada.

Susan S. Wallace, Department of Microbiology, New York Medical College, Valhalla, NY. 10595.

NIEHS CONFERENCE ON
GENETIC CONSEQUENCES OF
NUCLEOTIDE POOL IMBALANCE

INDEX